XV International Scientific and Engineering Conference Hermetic Sealing, Vibration Reliability and Ecological Safety of Pump and Compressor Machinery (HERVICON+PUMPS-2017)

IOP Conference Series: Materials Science and Engineering
Volume 233

Sumy, Ukraine
5 – 8 September 2017

Editors:

Volodymyr Martsynkovskyy
Igor Tverdokhleb
Andriy Zahorulko
Serhii Hudkov

ISBN: 978-1-5108-4776-7
ISSN: 1757-8981

Printed from e-media with permission by:

Curran Associates, Inc.
57 Morehouse Lane
Red Hook, NY 12571

CURRAN ASSOCIATES INC.
proceedings
.com

Some format issues inherent in the e-media version may also appear in this print version.

Printed by Curran Associates, Inc. (2017)

For permission requests, please contact the Institute of Physics
at the address below.

Institute of Physics
Dirac House, Temple Back
Bristol BS1 6BE UK

Phone: 44 1 17 929 7481
Fax: 44 1 17 920 0979

techtracking@iop.org

Additional copies of this publication are available from:

Curran Associates, Inc.
57 Morehouse Lane
Red Hook, NY 12571 USA
Phone: 845-758-0400
Fax: 845-758-2633
Email: curran@proceedings.com
Web: www.proceedings.com

Table of contents

Volume 233

XV International Scientific and Engineering Conference Hermetic Sealing, Vibration Reliability and Ecological Safety of Pump and Compressor Machinery HERVICON+PUMPS-2017

5–8 September 2017, Ukraine

Accepted papers received: 18 August 2017
Published online: 30 August 2017

Preface

Issues Of Pumps Flow Parts Hydrodynamics

Research And Modeling Of Compressor Equipment

Rotor Seals

Bearings

Some Problems of Rotordynamics

IOP Conf. Series: Materials Science and Engineering **233** (2017) 011001 doi:10.1088/1757-899X/233/1/011001

HERVICON+PUMPS-2017

The XV International Scientific and Engineering Conference
Hermetic Sealing, Vibration Reliability and Ecological Safety of Pump and Compressor Machinery
September 5-8, 2017, Sumy State University, Sumy, Ukraine

PROCEEDINGS EDITED BY

Volodymyr MARTSYNKOVSKYY, Prof., Dr., Dr.h.c., Sumy State University

Igor TVERDOKHLEB, Ass. prof., Dr.-Ing., HMS GROUP

Andriy ZAHORULKO, Ass. prof., Dr.-Ing., Sumy State University

Serhii HUDKOV, Dr.-Ing., Sumy State University

Selected, peer reviewed papers from the XV International Scientific and Engineering Conference HERVICON+PUMPS-2017

FOREWORD

The Conference Proceedings devoted to problems of reliability and economy of pump and compressor equipment are offered to the attention of specialists. The book presents 58 papers of various authors, included in the Program of the XV International Scientific and Engineering Conference HERVICON+PUMPS-2017, which, since 1976, with an enviable constancy passes every three years in Sumy on the basis of the General Mechanics and Machine Dynamics Department of the Sumy State University. The main sponsor that provided holding of the current conference, as well as the previous one, was the HMS Group, demonstrating an effective interest in the technical progress development.

Published papers are grouped in five sections, which cover a wide range of issues of modern theory and practice of centrifugal machines. At the present stage of its development, the theory, thanks to the successes of computational mathematics and computer technology, is experiencing a period of prosperity. The possibility of numerical analysis of more perfect mathematical models of unsteady flows of real liquids and gases in flow channels has appeared. The existing solutions are specifying, the new physical processes laws accompanying the operation of pumps and compressors are finding. In the published papers, at least to a small extent, the present state of the theory of these machines is reflected. Part of the papers is devoted to the implementation of new design solutions. The results of experimental investigations and operational experience of the proposed designs are given (Sections 1 and 2).

Centrifugal pumps and compressors - rotary machines. The most critical units of these machines are bearings and seals, which are complex tribomechanical systems. The papers given in Section 3 and 4 are devoted to the problems of calculating and designing of bearing and seals, increasing of the wear resistance of friction pairs.

The unbalanced rotor of a centrifugal machine is the source of vibration disturbances, which determine the vibration state of the machine as a whole. At the same time, the level of vibration is one of the main indicators of reliability, therefore reliability problems are directly related to the rotor dynamics in the bearings and annular seals. The rotor dynamics are discussed in Section 5.

On behalf of the authors' team, I hope that the papers published in the book will awaken the interest of young research engineers and will be useful to experienced designers in creating new reliable and economical pumps and compressors. In addition, the technical personnel involved in the maintenance and repair of units, can to get useful information from the book.

Professor Volodymyr A. Martsynkovskyy

IOP Conf. Series: Materials Science and Engineering **233** (2017) 011001 doi:10.1088/1757-899X/233/1/011001

SCIENTIFIC COMMITTEES AND CHAIRS:

Chairmen:

Prof. Anatoliy V. VASILIEV
Sumy State University
Prof., Dr. Volodymyr A. MARTSYNKOVSKYY, dr.h.c.
Sumy State University
Ass. prof., Dr.-Ing. Igor B. TVERDOKHLEB
Hydraulic Machines and Systems Group
Dr.-Ec. Vladimir N. YAMBURENKO
Hydraulic Machines and Systems Group

Organizational Committee Members:

Prof., Dr. Stanislaw ADAMCZAK, dr.h.c.multi
Kielce University of Technology
Prof., Dr. Bogdan ANTOSZEWSKI, dr.h.c.
Kielce University of Technology
Dr.-Ing. Andriy K. DAVIDENKO
PJSC VNIIAEN
Dr. Grigoriy V. KIRIK
Concern NICMAS
Prof., Dr. Andrzej KORCZAK
Silesian University of Technology
Prof., Dr. Milada KOZUBKOVÁ
Technical University of Ostrava
Prof., Dr. Czeslaw KUNDERA, dr.h.c.
Kielce University of Technology
Ass. prof., Dr.-Ing. Vasyl S. MARTSYNKOVSKYY
TRIZ Ltd.
Prof., Dr. Viacheslav B. TARELNIK
Sumy National Agrarian University
Prof., Dr. Anatoliy N. CHORNOUS
Sumy State University

Executive secretaries:

Ass. prof., Dr.-Ing. Andriy V. ZAHORULKO
Sumy State University
Sergey B. SHRAMKO
JSC Nasosenergomash Sumy

Work group of the Conference:

Dr.-Ing. Serhii M. HUDKOV
Svitlana O. MISCHENKO
Sumy State University

PROGRAM COMMITTEE

Ass. prof., Dr.-Ing. Olexander G. GUSAK
Sumy State University
Prof., Dr. Anatoljj P. DZUBA
Dnepropetrovsk National University of Oles Gonchar

IOP Conf. Series: Materials Science and Engineering **233** (2017) 011001 doi:10.1088/1757-899X/233/1/011001

Ass. prof., Dr. Mykhailo Ya. DOVZHIK
Sumy National Agricultural University
Prof., Dr.-Ing. Ivan B. KARINCEV
Sumy State University
Prof., Dr. Volodymyr P. PARAFEJNIK
Sumy NPO
Ass. prof., Dr.-Ing. Grzegorz PECZKIS
Silesian University of Technology
Ass. prof., Dr.-Ing. Olexander V. RADIONOV
SIE Ferrogidrodinamika Ltd.
Ass. prof., Dr.-Ing. Andriy V. SMIRNOV
Sumy NPO

CO-ORGANIZERS AND SPONSORS:

Sumy State University
http://sumdu.edu.ua/int/en/

Hydraulic Machines and Systems Group
http://grouphms.com/

TRIZ Ltd.
http://www.triz.sumy.ua/en.html

Kielce University of Technology
http://tu.kielce.pl/

Silesian University of Technology
http://www.polsl.pl/

Sumy National Agrarian University
http://www.sau.sumy.ua/index.php?lang=en

Sumy NPO
http://frunze.com.ua/en/main/

International Institute of Compressor and Power Engineering
http://mikem.com.ua/

PJSC VNIIAEN
http://vniiaen.sumy.ua/en/index.html

JSC Nasosenergomash Sumy
http://nempump.com/en/

Informational Support:

Journal of Pumps and Equipment
http://www.allpumps.kiev.ua/index.php?sid=2
Sealing Technology
http://www.sciencedirect.com/journal/sealing-technology
Journal of Compressor and Power Engineering
http://mikem.com.ua

Peer review statement

All papers published in this volume of *IOP Conference Series: Materials Science and Engineering* have been peer reviewed through processes administered by the proceedings Editors. Reviews were conducted by expert referees to the professional and scientific standards expected of a proceedings journal published by IOP Publishing.

IOP Conf. Series: Materials Science and Engineering **233** (2017) 012001 doi:10.1088/1757-899X/233/1/012001

Development (design and systematization) of HMS Group pump ranges

I Tverdokhleb[1], V Yamburenko[1]

[1] HMS Group Management Company LLC, 7, Chayanova st., Moscow, 125047, Russia

E-mail: tib@hms.ru

Abstract. The article reveals the need for pump range charts development for different applications and describes main principles used by HMS Group. Some modern approaches to pump selection are reviewed and highlighted the need for pump compliance with international standards and modern customer requirements. Even though pump design types are similar for different applications they need adjustment to specific requirements, which gets manufacturers develop their particular design for each pump range. Having wide pump ranges for different applications enables to create pump selection software, facilitating manufacturers to prepare high quality quotations in shortest time.

It is obvious that perfect choice of pump equipment complying with the customer's hydraulic technological system requirements is the key for its most efficient application. The customer will choose those pump producers who may offer the most efficient pumping equipment capable to work in required operation modes and comply with modern technical standards and requirements. As a result nowadays the availability of a proper pump which meets the customer requirements is a must for any pump manufacturer.

In former USSR the pumping equipment used to be specifically designed and developed for particular needs and technological systems. Unfortunately such approach has led to scarce range of the pumping equipment. That is why Customers often had to adopt and develop their own hydraulic system in accordance with available pump range which certainly affected the whole system's efficiency and energy consumption. So that huge energy saving potential can be utilized after Customer hydraulic system energy audits and proper selection of modern pumps and their replacement as a result.

Today the situation has ultimately changed. Actually the market is over saturated with the pumping equipment produced by numerous overseas companies, which may offer a wide range of all types and models of the pumping equipment. Such assortment has been developed for decades, counting for the world market pumping equipment needs. Naturally a number of national pump manufacturers couldn't meet the competition and either closed down or dramatically decreased their production facilities and sales. Now they are faced up with a challenge to catch up with competitors and keep to the pace of the world technological progress. This challenge first of all imposes the urgent need for proper pump designing.

How has HMS Group taken up that challenge and defined strategy for working out ultimately new approach to pump development? How the process shall be optimized and how to become competitive

IOP Conf. Series: Materials Science and Engineering **233** (2017) 012001 doi:10.1088/1757-899X/233/1/012001

in the world market of the pumping equipment? These are the rhetorical questions. But the implementation of such ambitious plans involves considerable resources, primarily intellectual ones.

In the highly competitive market it is essential to understand what customers are looking for and what are the criteria of their choice? Obviously the pumps must comply with international standards for certain industry as well as availability R&D potential and a wide range of pump types. Naturally the pump industry is steadily developing, existing standards are being regularly updated and new editions are released. This pushes the pump manufacturers to rise their awareness of innovations as well as to join the new standards development process. It enables them to upgrade timely their pumping equipment. That is why most companies strive to become members of national and international professional associations such as RPMA in Russia, HI in USA and Europump in Europe. Analyzing the production ranges of the leading pump companies it is highlighted that despite the similar type of design, companies are obliged to develop in each pump range different real design and use a wide range of materials in compliance with a particular standard. The costs vary accordingly. For example, the cost of overhung pump for pumping water and API 610 overhung pump for the same parameters differs considerably. Hence currently it is impractical to develop universal pump range for different applications and pump manufacturers have to have specific pump ranges to comply with particular standards (API, DIN, EN, GOST), as well special technical requirements of Customers, such as nuclear power stations, oil transportation, etc. For example, HMS Group pump types for different applications are shown in Figure 1.

However, having the most modern pump which corresponds to the latest standards, but if it is not optimum with the parameters Q, H and efficiency for the required operating point, it will not be accepted by the customer or lose in competition: Figure 2 [1].

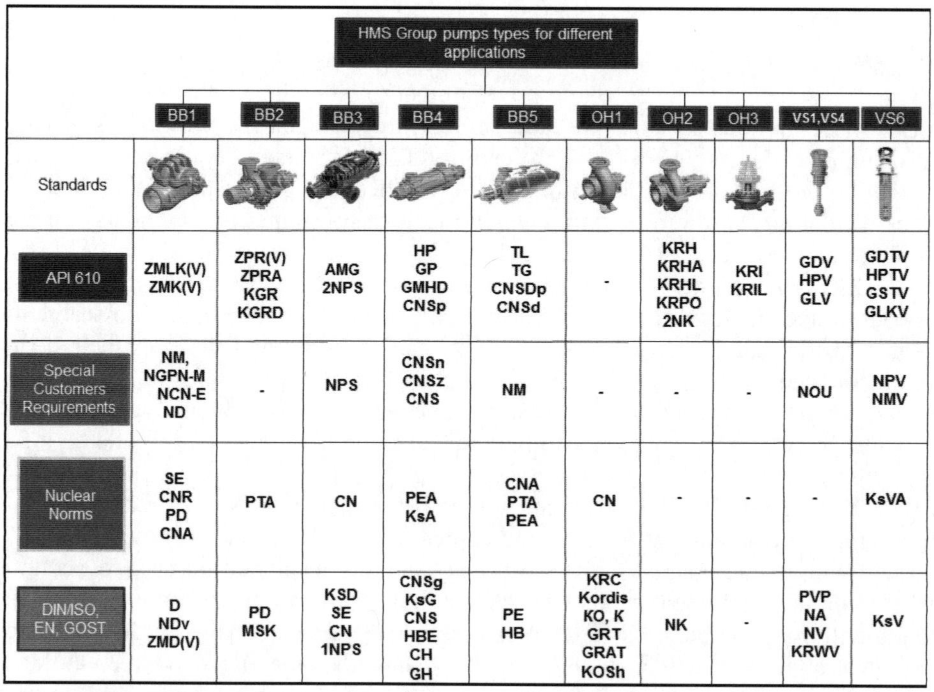

Figure 1. HMS Group pump types for different applications.

IOP Conf. Series: Materials Science and Engineering **233** (2017) 012001 doi:10.1088/1757-899X/233/1/012001

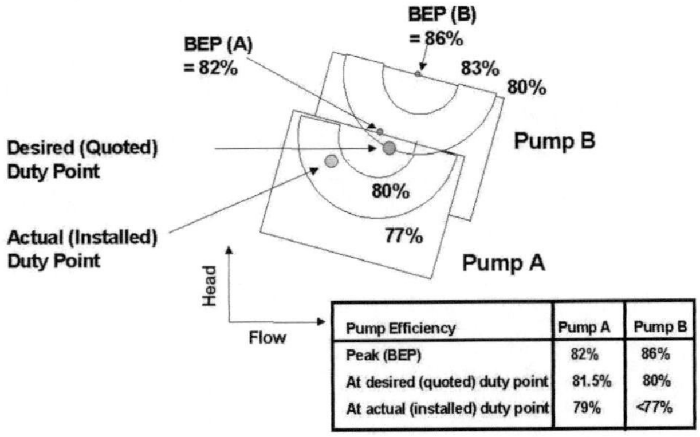

Pump Efficiency	Pump A	Pump B
Peak (BEP)	82%	86%
At desired (quoted) duty point	81.5%	80%
At actual (installed) duty point	79%	<77%

Figure 2. Selection of pumps.

Despite the fact that the pump B has a higher Efficiency at BEP, pump A actually has a higher Efficiency than pump B for the required flow rate. This raises the task of the "right" designing of the whole pump range. In the professional literature there are various tips how to design pump ranges. For example, in the reference [2] authors determine the basic pump range with flaw rate ratio 1.75, the head with coefficient 1.45 (Figure 3), and develop a whole pump range, using the laws of similarity, that only enables to design a few basic hydraulics, and to apply the results of their tests to other pump sizes in this pump range.

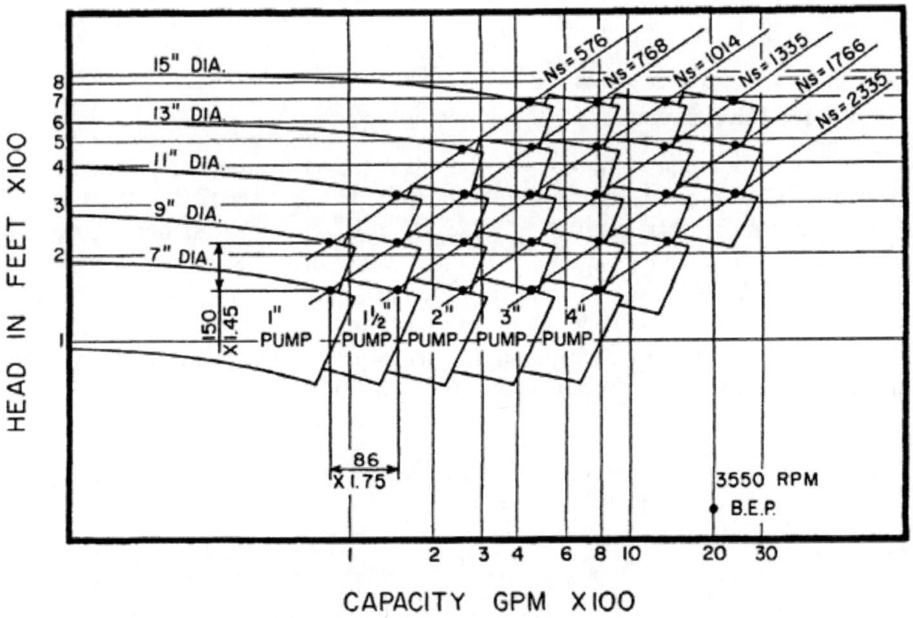

Figure 3. Example of traditional Performance Chart.

Reference [3] specifies the coefficients 1.5…2 for flow rate and 1.26...1.75 for head and reference [4] proposes to take variable coefficients. Author [5] makes accent on modern capabilities of CFD and fast wetting parts design using CFD analysis. The diversity of these recommendations is apparent and caused by the desire of different pump companies to find some optimized quantity of pumps for a certain pump range, which on the one hand would meet the customer needs while on the other hand does not require the development of additional pump sizes, because the amount of investing costs is

IOP Conf. Series: Materials Science and Engineering **233** (2017) 012001 doi:10.1088/1757-899X/233/1/012001

proportional to the number of designed pumps. Systematization and standardization of pumps in the form of pump range facilities companies to improve planning R&D work, working out better hydraulics, mechanical parts of different pump types, various calculating methods and regularly develop the necessary test facilities, etc.

Currently, the technical requirements of Customers dictate the desired location of the pump duty point. This is usually 0.8…1.1Qbep, as seen on Figure 4.

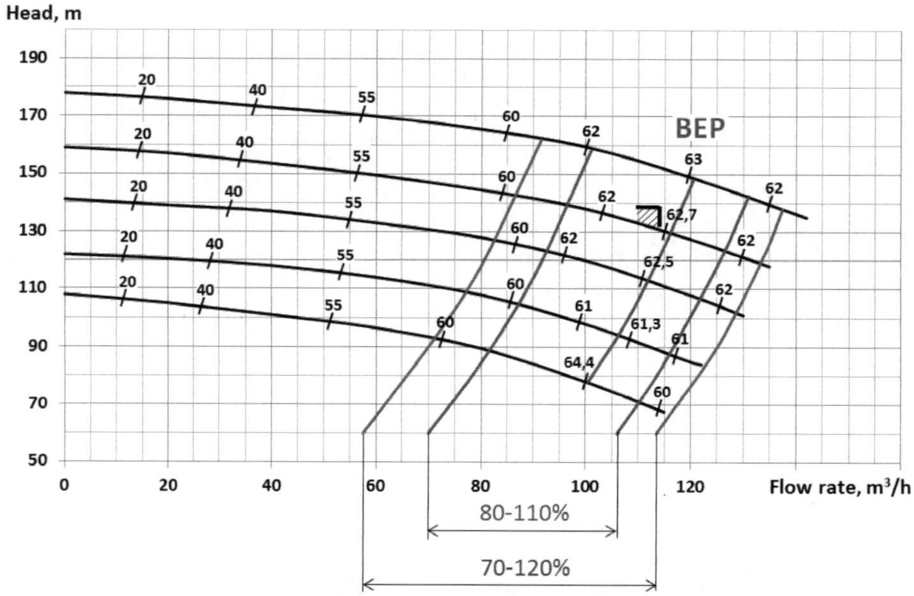

Figure 4. Requirements for the duty point location.

If one accepts this requirement as a base, the boundary of each pump range should be chosen so that the point with 1.1Q of previous pump size bordered with 0.8Q for the next size. I.e. the step of flow rate shall be not more than 1.1/0.8 = 1.375. In this way the number of pumps increases significantly and many manufacturers are going to create additional rotors or special wetting parts [6] within one pump. As a result, the distance between the basic pumps' flow rates can be increased, but the costs of neighboring sizes it shall be taken into account. Most of the HMS Group pump range charts have 2 or 3 flow part within each pump size as Figure 5 shows. In addition to reducing the number of pumps such approach helps to obtain more flexibility to select pumps by the Customer requirements.

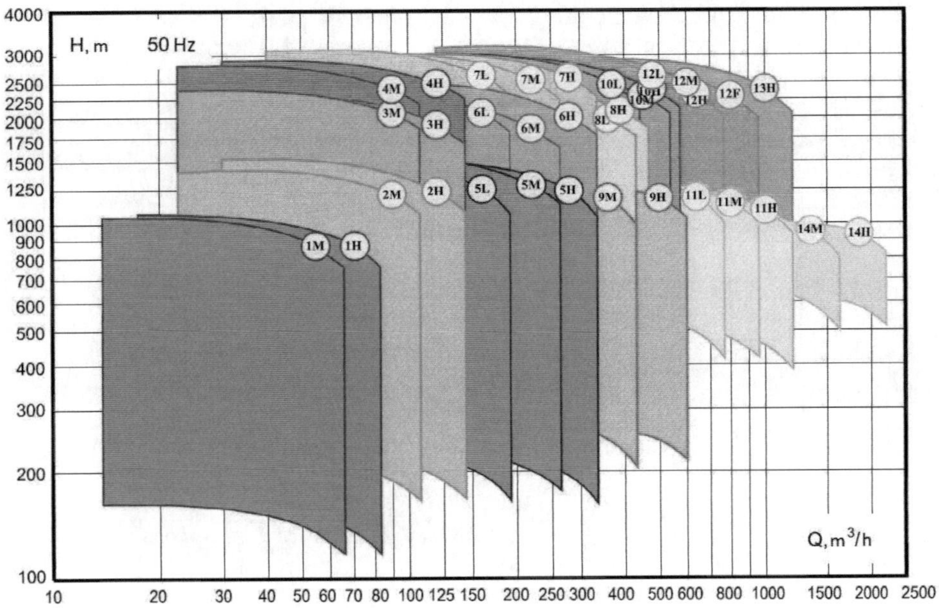

Figure 5. A range chart of HMS Group multi-stage pumps.

In our view, the following approach especially for engineering the pumps can be quite perspective when a pump's wetting parts within its range are "redesigned" according to customer requirements, and the impeller and casing or diffuser are manufactured using modern additive technology or other methods of quick 3D manufacturing. Thus, one can get an endless quantity of flowing parts within the same pump size. Such approach will involve consideration the other principles of working out the range chart when characteristics and BEP flow rates of neighbor pumps are not at first place, but the ability of each pumps to provide a wide range of Q and H parameters as well as the cost of the basic pump, depending on the pump size.

In some cases, for pumps of special applications, for example, slurry pumps, each pump range chart shall include information on the narrower operating range and enables pumps to work at lower speed in extreme conditions with varying concentrations of additives in pumped liquid, etc.

Having a developed pump range, each pump characteristics, design and estimated pump costs the selection process is easy to automate. Today there are many commercial software products using manufacturers' databases, which have become a handy tool. For example, HMS Group has been using for many years the SPAIX program for industrial pumps selection (Figure 6) in addition to some own software for the selection of specially engineered pumps and the ones complying to the API standards.

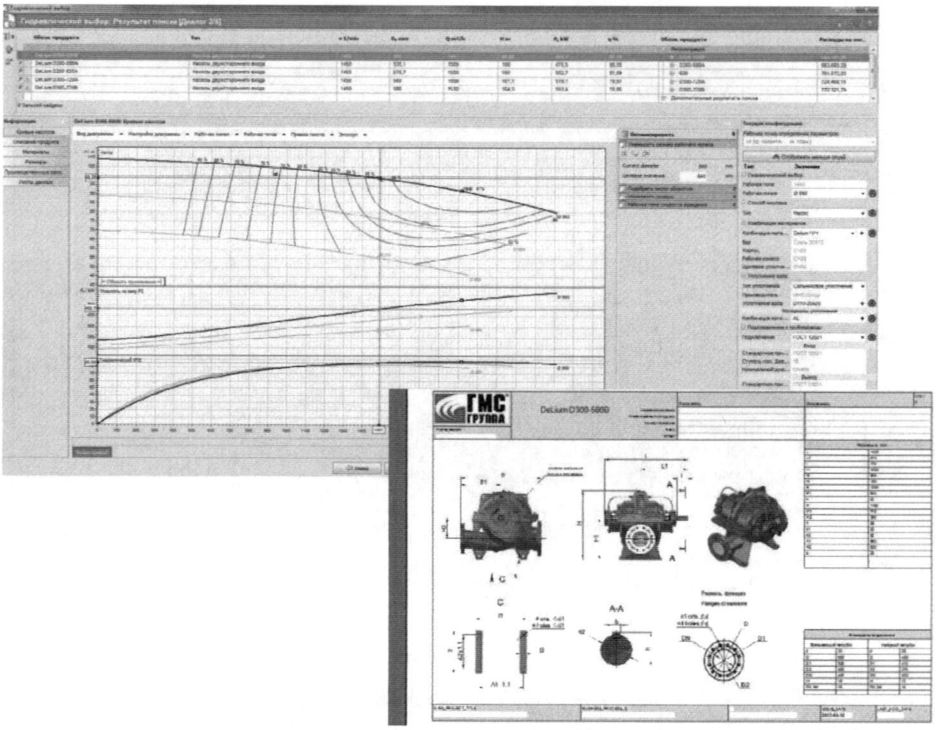

Figure 6. Pump selection using SPAIX software.

This software helps pump manufacturers to be competitive and prepared with high quality quotation in the shortest time. At the same time the designers, whose working time is very expensive, are not involved in the selection process of the pumping equipment, so the sales persons using this software can quickly respond to customer requirements. In fact, the pump selection software is a logical embodiment of knowledge and experience possessed by the manufacturer of pumping equipment.

Conclusions

1. The availability of the modern optimized pump ranges is currently a must for successful work in the market.

2. Pumps with certain design type shall not only pursue obtaining the necessary parameters at any duty point within the chart range, but also comply with international standards and the specific technical requirements of different customers.

3. Pump ranges shall not be considered as something immutable. They can and shall be updated in line with advances in technology and changes according to Customer requirements.

4. The wide pump ranges enables pump manufacturers to develop the software on their base and provide the pumping equipment selection with a wide functionality and maximum customization.

References

[1] System Efficiency. A Guide for Energy Efficient Rotodynamic Pumping Systems. (Europump, Brussels, Belgium)

[2] Lobanoff Val S, RossRobert R. Centrifugal Pumps *Design & Application. Secnd Edition* (Butterworth-Heinemann. An Imprint of Elsevier)

[3] Guilich J F 2010 *Centrifugal pumps* (Springer) p 963

[4] Tverdokhleb I, Semenov A, Ivanyushin A, Niemtsev O, Rudenko A 2014 Creating a Standard Size Range as One of the Factors Reducing Production Time of Modern Pumping Equipment *Applied*

IOP Conf. Series: Materials Science and Engineering **233** (2017) 012001 doi:10.1088/1757-899X/233/1/012001

Mechanics and Materials **630** pp 137-142

[5] Philippe Dupond 2012 A Novel Standardization Approach for the Development of a Single Stage Double Flow Axially Split Pump *Pump Users International Forum* (Winterthur, Switzerland) pp 247-256

[6] Birukov A, Knyazeva E, Rudenko A, Tverdokhleb I 2012 Increasing Pump Performance in the Low Flow-Rate Mode by Applying Special Volute Inserts. Research Study and Design Optimization *Pump Engineer Magazine* pp 22-25

IOP Conf. Series: Materials Science and Engineering **233** (2017) 012002 doi:10.1088/1757-899X/233/1/012002

Problems of energy efficiency of pumps and pumping systems

I Tverdokhleb[1], A. Kostyuk[1], S. Sokolov[2]

[1] HMS Group Management Company LLC, 7, Chayanova st., Moscow, 125047, Russia
[2] Subsidiary of JSC «HMS Livgidromash» in Moscow, 7, Chayanova st., Moscow, 125047, Russia

E-mail: tib@hms.ru

Abstract. The total energy consumption of various types of pumps is significant and the energy saving potential in this area is also huge. The transition from consideration of the characteristics of a separate pump to the features of its operation with a drive, taking into account the operation in different modes in conditions of a specific hydraulic network, allows obtaining a much greater potential for reducing energy consumption, significantly increasing the efficiency and reliability of pumping equipment.

Efficiency of pumps and pumping systems is a key factor of for presence on the world market. Application of efficient pumps allows cut cost of operation. In addition, reduction of energy consumption helps to solve ecological problems.

Until recent days the main approach for increasing efficiency of pumps was the upgrading of hydraulics of pumps and increasing of efficiency of drives. This traditional approach among the experts was named as a Product approach because considers an improving of properties only a product.

Today the design of most widely used type of pumps manufactured by leading companies have efficiency very close to the maximum achievable efficiency.

The estimated potential of energy saving for these types is approximately 3 ÷ 5 %. The situation with electric motors and drives is the same. The Product approach can give up to 3% of energy reduction.

European experts estimate the energy consumption of pumps for water in Europe Union as a 226.2 TW*h per year [1] what at price for kWh costs 226.2 TWh year.

The potential of energy saving is huge. That is why since the 2000 the work of pumps efficiency has been going on.

In Europe since 2009 with the methodological and organizational support of Europump, the Directive 2009/125 / EC, Commission Regulation No. 622/2012 and the European Standard EN 16297: 2012, have been implemented. These standards define the energy efficiency requirements for low-capacity circulating pumps used in the EU.

The draft of European standard EN 16480 "Pumps - Minimum required efficiency. Centrifugal pumps for water "[2] has been developed. The standard defines the efficiency requirements of pumps for clean water by introducing the required minimum efficiency index MEI.

The scope of the standard includes end suction pumps, monoblock pumps, vertical multistage and borehole multistage pumps. The introduction of national standards based on this standard is being

IOP Conf. Series: Materials Science and Engineering **233** (2017) 012002 doi:10.1088/1757-899X/233/1/012002

planned in Europe countries. The actual objective of the standard is to increase the requirements for pump efficiency, which are presented on the market by step-by-step banning of sales of low-efficiency pumps.

In the standard as low efficiency pumps considered pumps with efficiency in three points with flows Q=QBEP (best efficiency point), QPL=0,75QBEP (part load) and QOL=1,1QBEP (overload) lower than defined values. Standard EN 16480 defines values of efficiency in these points (house of efficiency). Lower values of efficiencies for these points are not allowed for sale (Figure 1). Formulas (1). (3) for calculation of efficiency levels are included in standard.

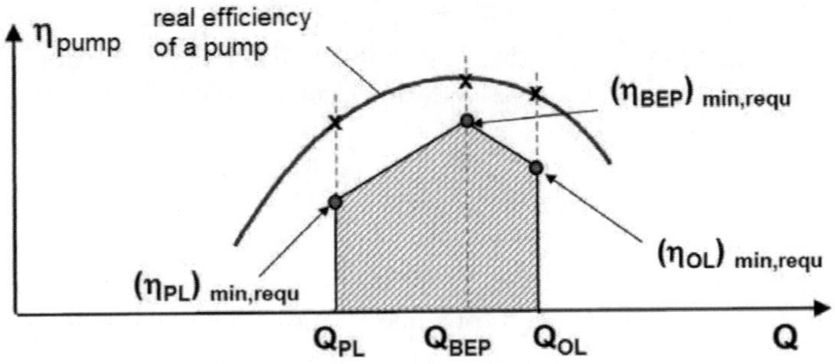

Figure 1. «House» of efficiency.

Minimum allowable values of efficiency should be calculated by the formula depend on flow at a BEP and specific speed n_Q.

$$\eta_{BEP} = -11,48x^2 - 0,85y^2 - 0,38xy + 88,59x + 13,46y - C \tag{1}$$

$$(\eta_{PL})_{min,\_requ} = 0,947 \cdot (\eta_{BEP})_{min,\_requ} \tag{2}$$

$$(\eta_{OL})_{min,\_requ} = 0,987 \cdot (\eta_{BEP})_{min,\_requ} \tag{3}$$

where $x = \ln(n_Q)$, $y = \ln(Q_{BEP})$.

In order to take into account the current technical level of the pumps produced by European manufacturers, in formula the correction factor C is used (1), by the amount of which the required minimum value of the efficiency is reduced.

The coefficient is given in the standard in tabular form and depends on the pump type, speed and MEI index (minimum efficiency index).

The MEI index reflects the quantitative ratio of products of different technical level presented on the market, and in fact determines the levels of efficiency of these types of pumps for the next few years. Based on the results of the review of all pumps of these types on the market since the introduction of the above-mentioned EU regulations in 2013, 10% of pumps with low efficiency have been excluded from the market. (MEI = 0.1), then from 2015. - 40% of the pumps (MEI = 0.4).

Practice shows that the switch over from the characteristics of only pump to the evaluation of its operation, taking into account specific conditions in a particular system, allows to obtain a much greater potential for reducing of energy consumption.

It can be achieved by the correct selection of a pump in accordance with system parameters, the application of efficient drives, control methods and control algorithms. This glance on efficiency of pump systems has been named as Extended Product Approach. In this case as the object for energy saving considered the pump unit with the drive and the control system. The potential of energy saving at EPA is 15 ÷ 25%.

System approach considers a complete hydraulic system and takes into account changes in the characteristics of pipelines, requirements of technological processes, as well as application of effective valves. System approach can additionally give 10-15% of energy savings.

But it is necessary to notice that the System Approach usually is out of scope of responsibility of pump manufacturers and usually is the responsibility of end users of pump systems. This approach is more complicated for implementation.

To evaluate the efficiency of pump equipment by using the extended product approach for different types of pumps, it is necessary to calculate the index of extended product approach EEI_{EP} in accordance with formula (4)

$$EEI_{EP} = \frac{P_{1,avg}}{P_{1,ref}} \quad (4)$$

The index is calculated taking into account the load profile and control method.

The power consumption of the pump $P_{1,avg}$, is defined during its operation for a given load profile for a specified period of time.

The referenced (or calculated) power $P_{1,ref}$ of the pump is fixed legislatively and is determined on the basis of the analysis of high-efficiency pumps and electric motors taking into specific speed of the pump and flow.

In accordance with extended product approach a pump unit is considered as a high efficient, if the value of EEI_{EP} is higher than the value defined for same system.

The value of EEI_{EP} is used as a criteria of pump efficiency. The pump unit considered as high efficient in accordance with the extended approach if it has EEI_{EP} higher then certain by the standard for a system of this type. Otherwise, restrictive sanctions determined by law can be applied.

The experience of the USA in increasing the efficiency of pumps is very interesting. HI with support of DOE has developed the standards on energy savings.

In 2016 Standard [3] of energy saving of pumping equipment was adopted. The scope of the standard includes pump types included in the standard EN 16480. The performance of the pumping unit is characterized by the dimensionless PEI coefficient calculated for the operating modes with constant and variable flow according to formulas (5.1) and (5.2), respectively.

$$PEI_{CL} = \frac{PER_{CL}}{PER_{STD}} \quad (5.1);$$

$$PEI_{VL} = \frac{PER_{VL}}{PER_{STD}} \quad (5.2)$$

In accordance with HI/ANSI the average value of pump power is calculated. The calculation is made on base of test results at either constant or changing load.

For every pump, a certain standard average power consumption of the motor is calculated PER_{STD} at operation on different modes at constant load (CL). At operation in different modes at constant load or with VSD with variable load (VL).

For the constant load the formula has

$$PER_{STD} = \sum_{i=75\%,100\%,110\%} \omega_i \cdot \left(\frac{P_{u,i}}{a_i \cdot \eta_{pump,STD}} + L_i \right) \quad (6)$$

where

ω_i – weight load factor at each point (for this case it is assumed equal to 0,3333),

$P_{u,i}$ – measured hydraulic output power at ith mode of the pump in question (hp)

a_i – 0,947 for operating point with flow 75% Q_{BEP}, 1,000 for operating point 100% Q_{BEP} and 0,985 for operating point 110% Q_{BEP}

L_i – losses in electric motor at ith mode calculated in accordance with DOE approach.

i – points of loads, corresponding to modes with 75,100 и 110% Q_{BEP}.

Minimum allowable efficiency of a pump $\eta_{pump,STD}$ defined as function of specific speed nq and value $Q_{100\%}$ at BEP.

$$\eta_{pump,STD} = -0{,}85 \cdot \ln(Q_{100\%})^2 - 0{,}38 \cdot \ln(Ns) \cdot \ln(Q_{100\%}) - 11{,}48 \cdot \ln(Ns)^2 + \\ + 17{,}8 \cdot \ln(Q_{100\%}) + 179{,}8 \cdot \ln(Ns) - (C + 555{,}6) \quad (7)$$

Coefficient C is defined by the legislation [3]. Its physical meaning and order of application is similar to the coefficient of formula (1).

A mandatory condition for the sale in the United States is $PEI < 1{,}0$

The next step on the way of increasing the efficiency of pumps is the introduction in the US of the so-called high energy efficiency labeling (hereinafter referred to as MVE) with an energy rating of ER.

Any company that manufactures and sells pumps, follows requirements and procedures specified in the HI Guidelines for the High Energy Efficiency Labeling Program [4], may use this marking to inform the buyer of the high energy-saving properties of the products supplied.

The key factor in the application of MVE is the comparison of the PEI index of the pump to be marked and the base index of the pump that meets the minimum energy efficiency requirements. In accordance with formula (5.1), the base index is calculated as:

$$PEI_{Baseline} = \frac{PER_{Baseline}}{PER_{STD}} \quad (8)$$

$PER_{Baseline}$ is the weighted average power consumption of the unit (with the electric motor) with efficiency of basic level, determined uniquely for each pump by formulas (6) and (7) at nominal speed and flow Q_{BEP}, hydraulic power at 75.100 and 110% Q_{BEP}, and given in [4] in a table for the base level of energy efficiency coefficient "C".

The energy rating of ER is calculated by the formula:

$$ER = (PEI_{Baseline} - Rated\_PEI_{CL/VL}) \times 100 \quad (9).$$

Thus, ER is a tool for calculating energy savings when operating a pump in comparison to a "basic" pump that meets the minimum energy efficiency requirements. The main idea of the introduction of ER is the use of a user-friendly marking (Figure 2), which makes it possible to estimate the energy savings of the acquired equipment.

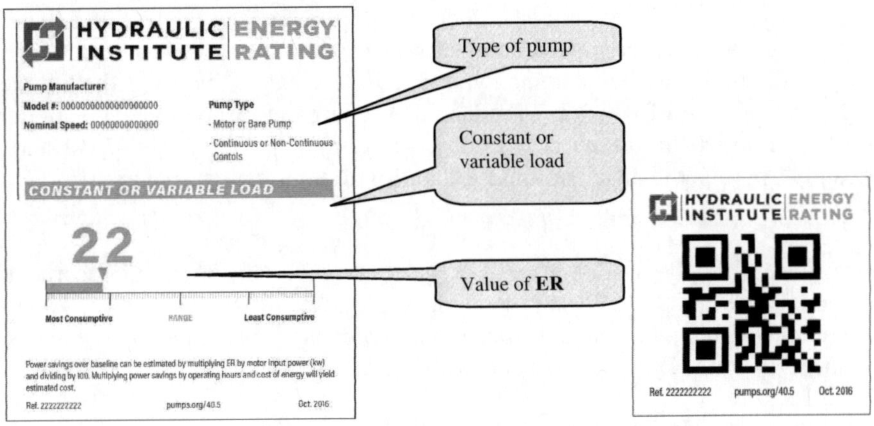

Figure 2. Labeling of ER and QR code.

It is possible to calculate the energy savings of a pump relatively to pump with low (basic) to level of energy efficiency by using the ER labeling according to formula (10).

$$Saving(kW \cdot h) = \frac{ER}{100} \times P_M \times h \tag{10}$$

where P_M –power of a motor (kW)

h – time of operation (h)

For example, for a pump with the label shown in Figure 2 with an electric motor 37 kW, the forecasted savings for 1 year (8760 hours) of continuous operation will be

$$\frac{22}{100} \times 37 \times 8760 = 71306{,}4 kW \cdot h$$

In Russia, the legislative approach to energy efficiency differs from the US and European countries. Documents which limit the energy consumption of pumping equipment, includes only recommendations.

For example, the Decree of the Government of the Russian Federation of June 17, [5] provides tax benefits to enterprises and organizations operating energy-efficient equipment and systems. The standard defines requirements of efficiency to centrifugal multi-stage sectional, oil trunk and retaining, as well as piston and rotary pumps.

The efficiency of pump was selected as an indicator of energy efficiency. The value of pump efficiency should be not less than the values specified for different types of pumps in the Decree. This decision caused an ambiguous reaction among pump industry specialists, as the same requirements are imposed on pumps for various purposes and design schemes, the specific speed of the pumps was not taken into account. Often these requirements are either overstated or vice versa, underestimated.

The Russian association of pump manufacturers applies efforts to perform work in of increasing the pumps efficiency within the framework of world legislative initiatives.

National standard 16297 parts 1, 2, 3 Pumps. Rotodynamic pumps. Glandless circulators. . General requirements and procedures for testing and calculation of energy efficiency index (EEI)

At the stage of acceptance in the Comission of standardization there is a modified GOST EN 16480 "Minimum required index of energy efficiency (MEI) for water pumps", developed on the basis of the above-mentioned European standard.

On the basis of the international standard, ISO/ASME 14414 "Evaluation of the energy efficiency of the pumping system" is prepared, which integrates and makes available for a wide range of specialists the basic elements for assessing the energy efficiency of pumping equipment. The standard is useful from a practical point of view, since it is focused on pumping equipment and is developed taking into account the world experience in conducting assessments of pumping systems.

Throughout the world, work has been going on for a long time to improve the efficiency of pumping systems, the key element of which is to reduce their energy consumption. The application of the extended approach, in which the pump is considered in conjunction with the drive and control system, can significantly improve the efficiency and reliability of pumping equipment.

References

[1] Ecodesign Pump Review. Study of Commission Regulation (EU) No. 547/2012 incorporating the preparatory studies on 'Lot 28' and 'Lot 29' (Pumps). Final Progress report. December 2015. Annex 4

[2] prEN 16480:2014. Pumps - Minimum required efficiency of rotodynamic water pumps

[3] Energy Conservation Program: Energy Conservation Standards for Pumps [Docket Number EERE–2011–BT–STD–0031]

[4] Hydraulic Institute Program Guide for HI Energy Rating Program (HI 40.5-2016)

[5] Decree of the Government of the Russian Federation of June 17, 2015. No. 600 "On approval of the list of objects and technologies that relate to objects and technologies of high energy efficiency"

Effect of vanes diffuser geometric components on shape of the Q-H-curve

A S Kostornoi[1], S O Lugova[1], P L Olshtynskyi[1]

[1] JSC "VNIIAEN", 2, 2nd Zheleznodorozhnaya str., Sumy, 40003, Ukraine

E-mail: admin@vniiaen.sumy.ua

Abstract. The article deals with the effect of the vanes diffuser geometric components at the inlet (downstream of an impeller) and at the outlet (upstream of the impeller) on the shape and steepness of the Q-H-curve. Intermediate stage comparative curves based on centrifugal pump tests are given with corrections to the vanes diffuser components were made. This allows changing steepness of the Q-H-curve. Comparative tests of the intermediate stage with different types of vanes diffusers (with continuous crossover channels and discontinuous crossover channels) are shown.

1. Introduction

Requirements on the steepness and the shape of the Q-H-curve are imposed upon for the majority of pumps being operated at thermal and nuclear power plants (TPP and NPP). First of all, this refers to centrifugal boiler feed pumps being operated in parallel, which are supplying in common network according to the power plant scheme. In this case, the saddle Q-H-curve can result in surging [1] leading to unstable operation of pumps. Therefore, the stable pump operation needs the flat Q-H-curve.

Another important requirement imposed on the Q-H-curve is to ensure a certain slope. So, according to GOST 22337-77 [2] and GOST 24464-80 [3], when designing the centrifugal boiler feed pumps, it is necessary to obtain the slope of Q-H-curve that doesn't exceed 30%, and in some cases – not above 18%.

This is mainly due to the fact that a power plant pipeline system is designed for a certain pressure.

Complete replacement of the whole pipeline system at the power plant is economically irrational decision. Further, entering the corrections to the geometry of hydraulic part of the pump allows changing the slope of the Q-H-curve without significant costs.

There is a fair amount of recommendations on hydraulic part design or further development of geometry after the pump tests to obtain a stable Q-H-curve with required steepness [4-6]. In this article, the effect of the vanes diffuser geometric elements on the shape and steepness of the Q-H-curve is considered.

2. Theoretical background

The physical origin of the saddle Q-H-curve is described in the literature sources [1,7]. The study of the impeller inlet shows that backflow is observed at pump partload. Backflow moves from the impeller outlet towards the diffuser return channels of the vanes diffuser and forms vortices that cause the head decrease at partload. The vortex at the impeller outlet also promotes an increase of hydraulic losses at partload (Figure 1).

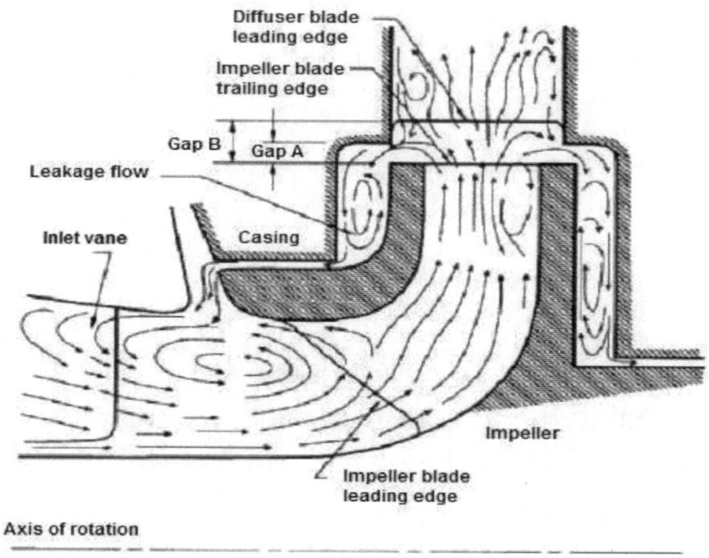

Figure 1. Schematic of fluid flow field within the hydraulic part of multistage centrifugal pump at partload.

In order to reduce vortex activity that can lead to the occurrence of the saddle Q-H-curve, a stationary vane cascade is usually installed upstream the impeller. Its function in the intermediate stage is performed by return channels vanes of the vanes diffuser.

3. Study of the vanes diffuser outlet components

The effect of the return channels vanes on the shape and steepness of the Q-H-curve was studied. Therefore, it is assumed that the vortex activity at low flows will increase with growing distance between rotating and stationary vane cascade (between the impeller blades at the inlet and the return channels vanes at the outlet); this will reduce the slope of the Q-H-curve due to the increase in hydraulic losses. However, as previously noted, increase of the vortex activity can lead to the saddle Q-H-curve. To obtain the flat Q-H-curve with the minimum steepness, the intermediate stage was designed in assembly with the impeller and vanes diffuser and tested at the testing facility as per single-stage scheme. This scheme means the vanes diffuser is installed upstream the stage to ensure flow parameters at the impeller input. In order to reduce the slope of the Q-H-curve and prevent the occurrence of the saddle in this curve the diameter d6 of the eight vanes out of the twelve has been trimmed at the outlet - for two vanes per one vane (Figure 2).

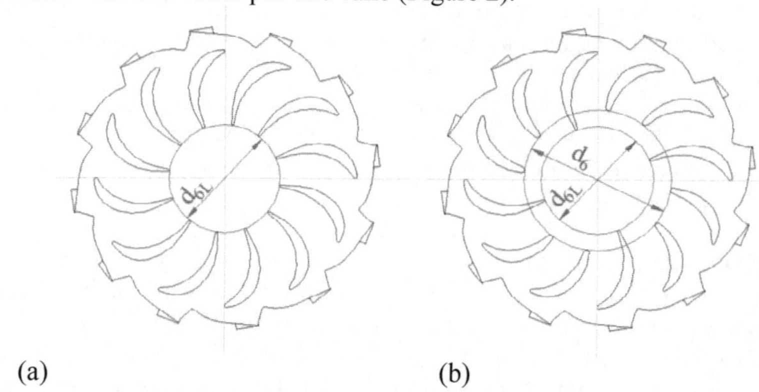

Figure 2. Vanes diffuser before trimming of the return channel vanes at the outlet (a) and after trimming of them (b).

In this case, ratio of trimmed return channel vanes at the outlet to the impeller eye was $d_6/D_0 = 1.08$, and ratio of the untrimmed ones was $d_{6L}/D_0 = 0.76$.

Trimming of the return channel vanes was performed according to this scheme with understanding that four untrimmed vanes would sufficiently reduce the vortex activity for obtaining the flat Q-H-curve, and eight trimmed vanes would form a vortex zone that increases hydraulic losses at partload. Consequently, the head as well as the steepness of the Q-H-curve would decrease.

Comparative analysis of the test results (Figure 3) showed that the Q-H-curve after trimming of the diffuser vanes remained the flat one at partload, and Q-H-curve steepness decreased by 6%. The efficiency value and best efficiency point (BEP) were not changed.

Based on obtained results it could be make a conclusion that a judicious mix of blade diameters of return channels at the outlet and their quantity (trimmed and untrimmed return channels) allows to decrease the steepness of the Q-H-curve and prevent the occurrence its saddle shape simultaneously.

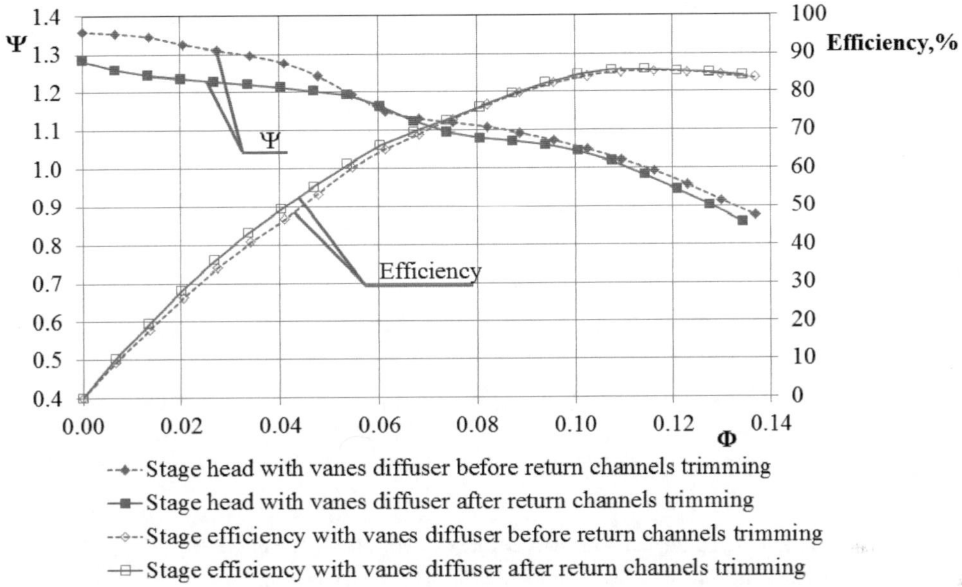

Figure 3. Comparative curves of pump stage being tested with vanes diffuser before trimming of return channels and after trimming of them.

4. Studies of vanes diffuser components at the inlet

As previously noted, vortices form at partload not only at inlet of the impeller, but at outlet of it.

Considering diffuser channels of the vanes diffuser are a stationary vane cascade placed downstream of the impeller it could be supposed that they also have an effect on the vortex activity. It is possible to extend the inner space between grids due to trimming of leading edges of the diffuser channels (Figure 4). Herein, it could be supposed the the inner space between a rotating vane cascade (the impeller outlet) and stationary vane cascade (the diffuser inlet) extends and results in increasing of the vortex activity at partload and hydraulic losses. Due to this the head decreases at partload. At the same time the selection of the judicious mix d_3/d_2 (the gap between the diameter of the vanes diffuser channels - d_3 and diameter of impeller outlet – d_2) [8] allows for vortex intensity to be not so significant, and it allow to avoid the occurrence of saddle Q-H-curve.

IOP Conf. Series: Materials Science and Engineering **233** (2017) 012003 doi:10.1088/1757-899X/233/1/012003

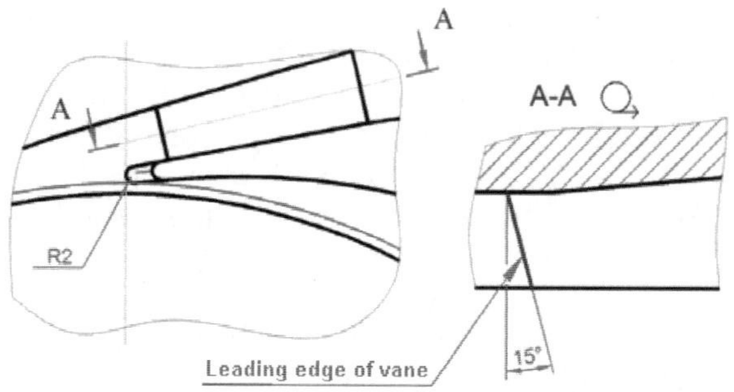

Figure 4. Sketch of trimming of vanes diffuser leading edges.

Figure 5 shows comparative curves of the stage with trimmed leading edges of the diffuser channels in the vanes diffuser.

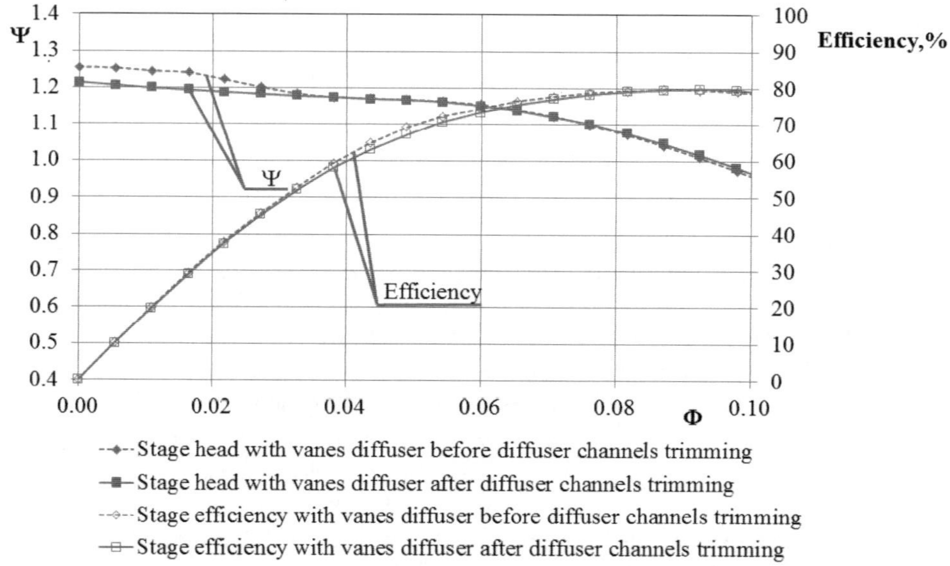

Figure 5. Comparative curves of pump stage tested with vanes diffuser before trimming of vanes diffuser leading edges and after trimming.

Comparative analysis showed that trimming of the vanes diffuser leading edges resulted in Q-H-curve decrease of 5%, and efficiency wasn't changed.

5. Study of effect of vanes diffuser design type

Two types of the vanes diffusers can be applied in a hydraulic part of a multistage pump (Figure 6): a vanes diffuser with continuous crossover channels (a – the continuous vanes diffuser) and a vanes diffuser with discontinuous crossover channels (b – the discontinuous vanes diffuser).

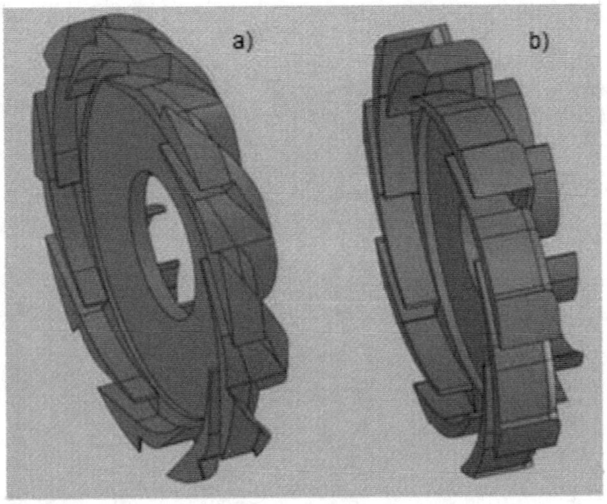

Figure 6. Design variants of vanes diffusers: (a) continuous vanes diffuser; (b) discontinuous vanes diffuser.

When studying a working process a numerical calculation was carried out for the vanes diffuser [9] by simulation of fluid flow. It was indicated that the flow pattern within discontinuous and continuous vanes diffuser has the main difference within the crossover channel at the BEP. The flow within the channel is observed in the continuous vanes diffuser, and mixing of flows effluent the diffuser channels in combination with the flow rotating in annular space is observed in the discontinuous vanes diffuser. We can suppose that a separation flow forming the vortex zone at partload appears in the turning area of the crossover channel of the continuous vanes diffuser that promotes to increase velocity and hydraulic losses and, consequently, decrease in the head. Therefore, absence of the continuous channel and presence of the annular space in the discontinuous vanes diffuser promote to reduce the vortex zone at partload. Moreover, fundamental hydraulic losses in the crossover channel of the discontinuous vanes diffuser are associated with matching of flow angle and inlet blade angles. According to the studies [10], the blade angle of the discontinuous vanes diffusers is essentially independent of the operation conditions. Therefore, hydraulic shock losses at partload that depend on the flow angle are expected to be virtually the same as the hydraulic losses at the BEP. It means that total hydraulic losses occurring at partload are less in the discontinuous vanes diffusers than in discontinuous ones. Consequently, the stages with the discontinuous vanes diffusers will always have the greater head at partload than the stages with the continuous ones under otherwise equal conditions.

These considerations have been empirically examined. Two types of the vanes diffusers were manufactured for this purpose (the continuous vanes diffuser and discontinuous vanes diffuser). They were tested in assembly with a pump and the same impellers at the testing facility.

Figure 7 gives test results; index 1 designates a curve of a pump with the continuous vanes diffuser and index 2 designates a curve of a pump with the discontinuous vanes diffuser.

An analysis of performance curves confirmed the foregoing premises. It should be noted that replacement the continuous vanes diffusers on the continuous ones allowed to correct the shape of the Q-H-curve (to remove the saddle curve) while the maintaining the efficiency level and the BEP position.

IOP Conf. Series: Materials Science and Engineering **233** (2017) 012003 doi:10.1088/1757-899X/233/1/012003

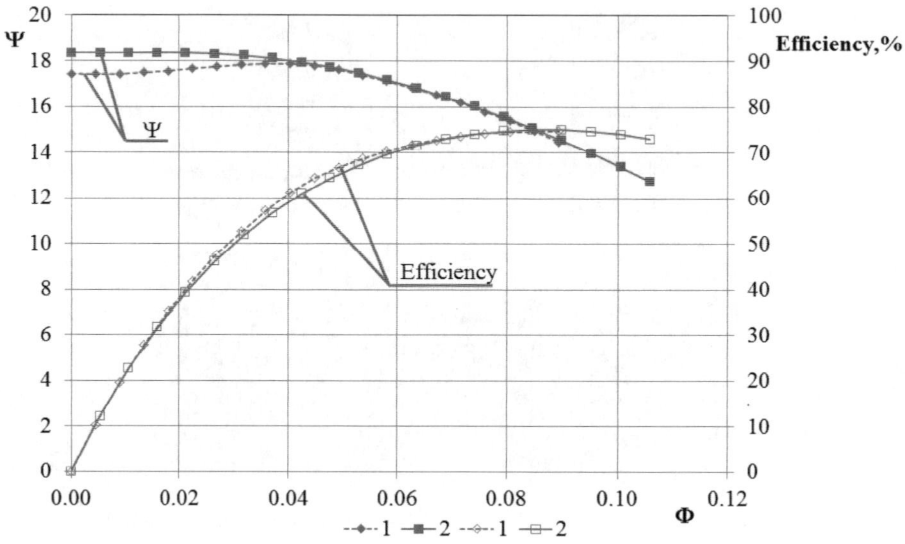

Figure 7. Comparative curve of pumps tested with the vanes diffusers of the different types.

6. Conclusions

1. The vanes diffuser geometrical elements have influence on the shape and steepness of the Q-H-curve.

2. Performed studies at vanes diffuser outlet and impeller inlet have been shown that trimming of return channel vanes at the outlet of the vanes diffuser allowed to decrease the slope of the Q-H-curve while maintaining its stable shape.

3. Performed studies at vanes diffuser inlet and impeller outlet have been shown that trimming of the leading edge of the vanes diffuser diffuser channels allowed to decrease the slope of the Q-H-curve.

4 Replacement the continuous vanes diffuser for the discontinuous one in the stage allows increasing steepness of the Q-H-curve due to a head rise at partload.

References

[1] Mikhailov A K 1977 *Vane type pump. Theory, calculation and design.* (Moscow: Mashinostroyeniye)

[2] *Centrifugal boiler feed pumps. Main characteristics. GOST 22337-77* 1988 (Moscow: Statinform).

[3] *Boiler feed pumps for nuclear power plant supply unit. General technical specifications. GOST 24464-80* 1986 (Moscow: Statinform)

[4] Lugova S O 2008 *Search for multistage pumps performance measuring method: a) in an attempt to optimum flow displacement; b) in attempt to improve the shape of the head curve* NTUU "KPI" pp 279-288.

[5] Lugova S O 2010 Intermediate stage investigation results of high-speed pumps with impellers high hub ratio and suction specific speed ns=85...100 *Journal Technological Systems* **2** (51) pp 80-83

[6] Lugova S O 2015 Addition of the non-model alterations in hydraulic passage of the centrifugal pump in order to change its *characteristics Proceedings of the XVI international scientific and technical conference* (Sumy) p 67

[7] Rudnev S S 1968 Backflows at impeller inlet and its influence on shape of head curve in centrifugal sectional pumps *Proceedings of VNIIGydromash* **37** pp 167-183

[8] Guilich J F 2010 *Centrifugal Pumps (*Springer Heidelberg Dordrecht London New York)

[9] Lugova S O 2011 *Vanes diffuser design of the intermediate stage in centrifugal pumps* (Sumy: Sumy State University) pp 281-288

[10] Ris V F 1964 *Centrifugal compressor unit machines* (Moscow: Mashinostroyeniye) p 336

Increase of economy of torque flow pump with high specific speed

A G Gusak[1], I V Krishtop[1], V F German[1] and V N Baga[2]

[1]Department of Applied Fluid Mechanics, Sumy State University, 2rd Rimsky-Korsakov st., Sumy, 40007, Ukraine
[2]Technical Thermophysics Department, Sumy State University, 2rd Rimsky-Korsakov st., Sumy, 40007, Ukraine

E-mail: gusak-og@ukr.net

Abstract. Torque flow pumps are widely spread types of energy machines, which are used in majority of modern branches of industry for pumping of dirty media. The main task of researchers of torque flow pumps is increase of such pumps effectiveness for higher feed. Hydraulic losses for torque flow pumps are caused by working process of such pumps and are inevitable. Decrease of losses can be obtained by means of optimization of hydraulic flow part geometry.

Modern approach to design of pump outlet introduces new constructive solutions which can increase economy of torque flow pumps.

The aim of this research is increase of economy of torque flow pumps by means of application of spatial outlet and investigation of its geometry on pump characteristics.

Analytical and numerical methods of liquid flow research for hydraulic flow part of torque flow pump were used in this paper. Moreover, influence of hydraulic flow part geometry of different designs of "Turo" type torque flow pumps outlets on pump characteristics was investigated. Numerical research enabled to study process of energy transfer of torque flow pump and evaluate influence of geometrical dimensions of spatial spiral outlet on its characteristics. Besides numerical research confirmed introduced regularity of peripheral velocity distribution in outlet. Velocity moment distribution in outlet was obtained during implementation of numerical research.

Implemented bench tests of torque flow pump prototypes enabled to obtain real characteristics of pump and confirm effectiveness of spatial geometry of outlet application for such pump.

1. Introduction

Torque flow pumps are priority types of pump equipment, which are used in majority of branches of industry for pumping of dirty media.

Opposed to centrifugal pumps, which not always can pump liquids with big solid inclusions, nowadays torque flow pumps are widely used in technological processes of particular production types, such pumps are progressive pumping equipment [1, 2].

Range of optimum values of specific speed for "Turo" type torque flow pumps is well-known and overrun leads to efficiency decrease. Majority of torque flow pumps has maximum energy effectiveness for the range of specific speed values n_S=60–140. Pump operation out of this range is not economically reasonable. However, nowadays non clogging torque flow pumps operating at high specific speed n_S>140 are of great necessity [3, 4].

Experimental material of paper [5] shows that torque flow pump outlet takes part not only in energy transfer process, but simultaneously with impeller forms vortex liquid flow, which is major part in energy transfer process of such pump.

Analysis of earlier researches [6] showed that specific speed and efficiency of torque flow pump are determined by outlet geometry for optimum geometry of impeller. Thereby criterion of achievement of maximum value of efficiency for torque flow pump with high specific speed value turns into optimization of outlet geometry by means of decrease of hydraulic losses.

Taking into account experience of centrifugal pumps designing, expansion of torque flow pump parameters range can be reach by outlet geometry modification. First attempt in this direction was implemented by E. Egger [7] during creation of "Turo" type torque flow pump outlet with internal axial spiral. Side channel in pump body is made in such way that it provides faster medium flow from suction pipe to discharge pipe. However such constructive solution of outlet greatly influences on flow and leads to increase of hydraulic shock losses.

Tests implemented by the author [8] concerning influence of outlet on torque flow pump characteristics showed that its form (annular or spiral) insignificantly influences on pump parameters. However these results differs from experimentally obtained data [9], according to which efficiency of torque flow pump with spiral outlet is 1.5% higher than for pump with annular outlet.

In the paper [4] technical solution was proposed for significant decreasing of metal quantity for torque flow pump and improvement of its characteristics keeping values of other parameters. However the author mentions that considered results can't be stable for "Turo" type torque flow pumps which have different parameters.

For increase of specific speed of torque flow pumps [10] it was proposed to turn into two-stream scheme of liquid flow in pump. Such constructive solution allows moving optimum operation mode of pump to area of higher feed, but significantly complicates its design and consequently pump price.

For creation of torque flow pump with high specific speed authors of the paper [3] decided to modify impeller and outlet design. As a result they proposed "Turo" type torque flow pump outlet elongated in axial direction ("screw-shaped") [11]. However, implemented tests of pump prototype didn't show desired results, because efficiency was too low.

Probing of flow in free chamber of torque flow pump [12] allowed to obtain flow pattern in pump with real velocity distribution. Liquid flow was unsteady. Flow coming from impeller partially is directed backwards to pump inlet and partially to outlet. And meridian velocity in pump has maximum value near impeller decreasing when approaching to outlet side opposite to impeller (Figure 1 (a)).

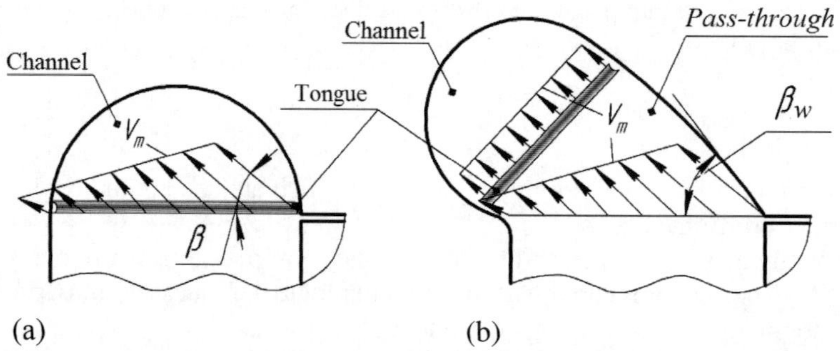

Figure 1. Meridian velocity distribution in free chamber of torque flow pump: (a) annular outlet; (b) spatial spiral outlet.

Obtained meridian velocity distribution leads to increase of hydraulic shock losses because of unconformity of inflow angle and angle of outlet wall setting, especially for torque flow pump with high specific speed value. For decrease of these losses it was proposed to configure outlet wall in such way that it provided free liquid flow. Thereby geometry of outlet should be designed for provision of steady distribution of velocity at outlet entrance.

As a result new design of torque flow pump body [13] with spatial spiral outlet was obtained (Figure 1 (b)). For exclusion forming of stagnant vortex area at outlet, angle of wall, conjugated with pump body β_w, must be close to angle of meridian velocity of flow β. In such case steady distribution of velocity at outlet entrance provides minimum losses and absence of detached flow in its channel.

2. Influence of outlet form on liquid flow forming in hydraulic flow part of torque flow pump

Complex process of energy transfer and turbulent character of liquid flow do not allow identifying optimum design of torque flow pump outlet by means of experiment. However modern numerical methods of research allow to determine influence of outlet design on liquid flow structure forming in free chamber of torque flow pump [14], and indicate optimum design of outlet for pumps with high specific speed $n_s > 140$.

Existing methods of hydraulic flow part design calculation for "Turo" type torque flow pumps were approved for range of specific speed values $n_s = 60–140$ [15], therefore investigated designs of pumps were determined for range of n_s with maximum effectiveness – $n_s = 100$.

Pump designs with different types of outlets were chosen for research, these types of outlets were used earlier or proposed for the first time: with annular No.A, spiral No.B, double-volute spiral No.C, spatial No.D, chamber-free No.E and with reverse spatial spiral No.F (Figure 2)

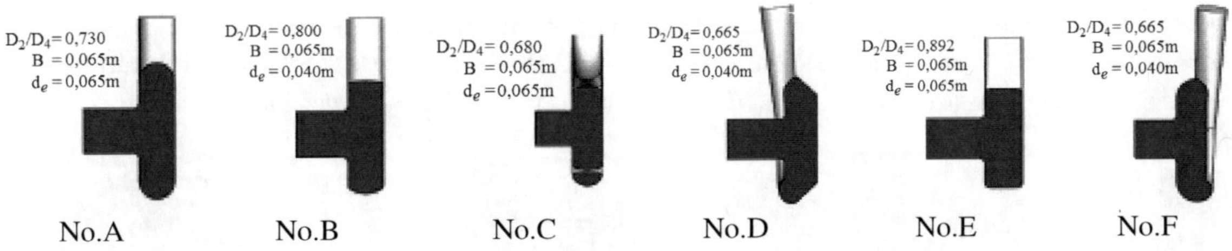

Figure 2. Configuration of investigated outlets of torque flow pump.

Application of spatial spiral outlet (No.D) for torque flow pump allowed increasing pump efficiency in operative mode of feed, where its maximum value is 56% (Figure 3). Satisfactory results showed pump with spiral outlet (No.B), its efficiency value for rated mode is 52 %. Concerning pumps with double-volute spiral, chamber-free and reverse spatial outlets (No.C, E and F), efficiency is in the level of 46%, which indicates their ineffectiveness.

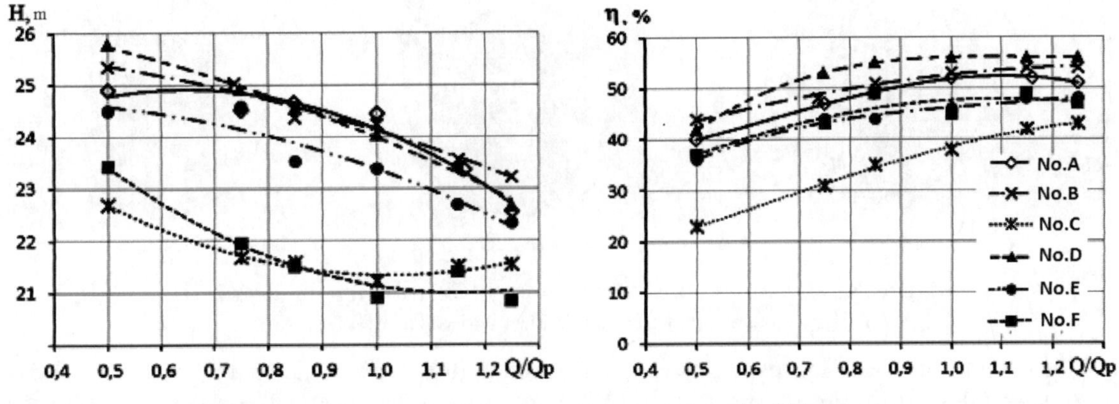

Figure 3. Characteristics of torque flow pump with different types of outlets.

Level of efficiency for investigated outlets is caused by level of hydraulic losses. The lowest value of hydraulic losses was obtained for pump with spatial spiral outlet No.D (Figure 4). It should be mentioned that curve of losses for pump with outlet No.D is almost linear. There is no optimum value on the curve of hydraulic losses for investigated range of pump feed values.

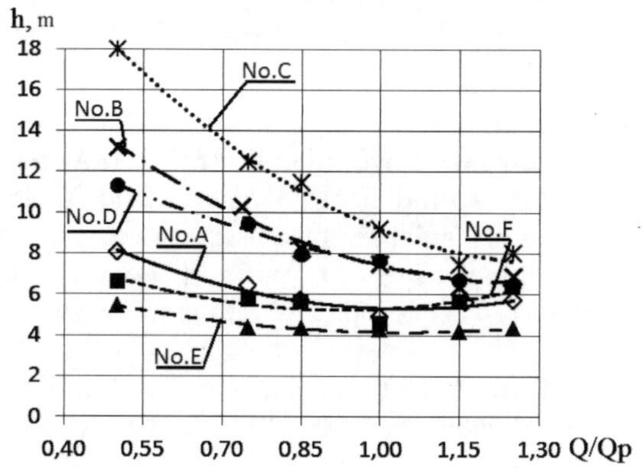

Figure 4. Hydraulic losses *h* for torque flow pumps with different types of outlets.

Analysis of velocity moment distribution patterns (Figure 5) showed, that for design of torque flow pump with spatial spiral outlet No.D more steady velocity moment distribution is observed, it is presented in concentric circles and very close to the law $V_u \cdot r = const$. Presence of local areas of velocity moment increase and its unsteady distribution is typical for other designs of outlets.

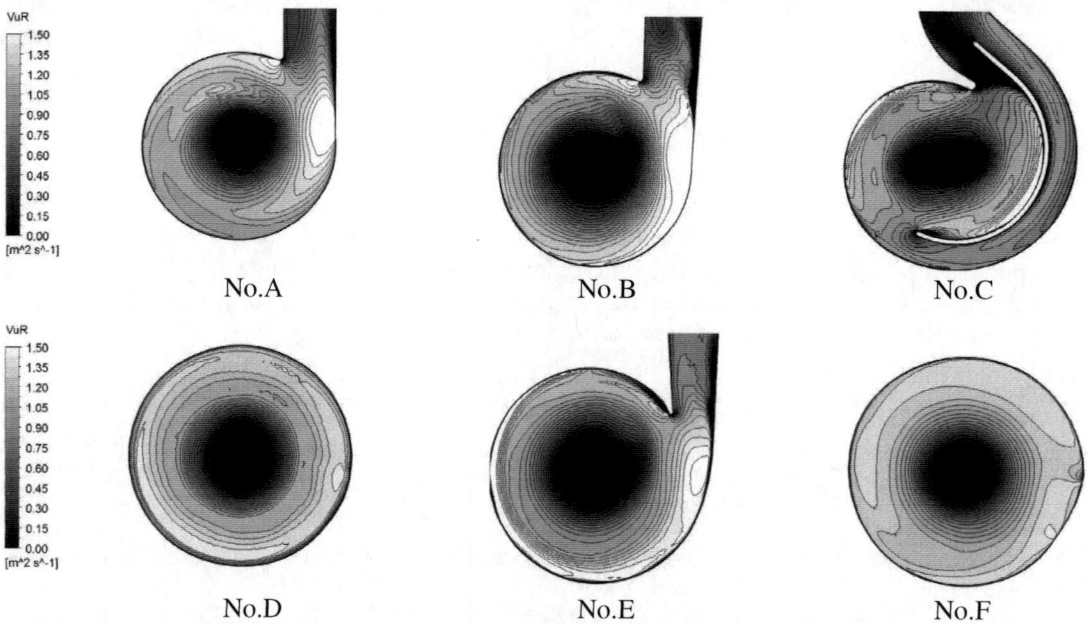

Figure 5. Isoprofiles of velocity moment distribution in middle section of free chamber of torque flow pump with different outlets.

Given results of numerical research showed that torque flow pumps with spatial outlet No.D have more steady flow structure and pressure distribution in free chamber and consequently advanced energy indexes.

3. Research of spatial outlet geometry influence on characteristics of torque flow pump with high specific speed (n_s > 140)

As is known [15], outlet geometry greatly influences on torque flow pump parameters. Modelling of liquid flow in torque flow pump with spatial spiral outlet was implemented for research of this issue.

Influence of outlet design section area on torque flow pump parameters was investigated, this area changed for constant width of free chamber B (Figure 6a). Increase of relative design section area ε from 0.23 to 0.57 ($\varepsilon = \frac{S_e}{S_2} = \frac{1}{D_2} \cdot \left(\frac{4F_c}{\pi} \right)^{1/2}$, where F_c - area of outlet design section) leads to decrease of pump efficiency from 55 to 42%, in this case head of pump decreases by 18%, but specific speed of torque flow pump increases from 100 to 180. Obtained results confirm conclusions of papers [5, 16] about the fact that value of exit section area of outlet significantly influences on torque flow pump parameters.

When relative width of free chamber of pump changes in the range $\overline{B_0} = B / B_2$ =0.29 – 0.4 (D_2 – impeller diameter) pump head increases by 5 % (Figure 6 (b)).

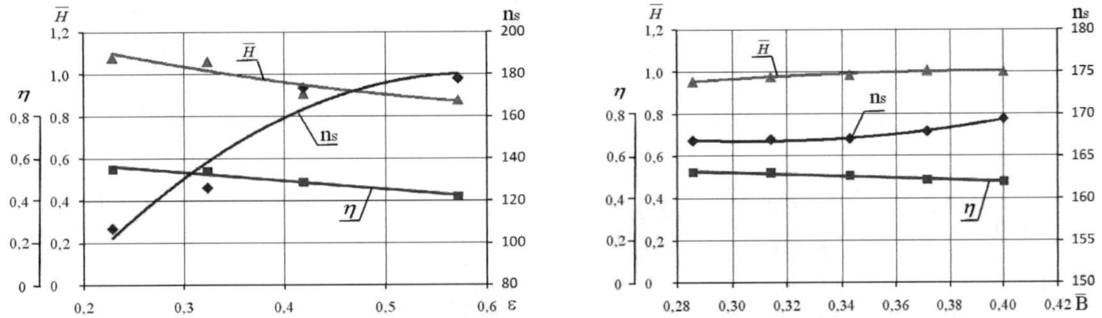

(a) (b)

Figure 6. Relation of optimum parameters of torque flow pump: (a) depending on relative area of exit section with ε ; (b) depending on relative width of free chamber $\overline{B}$ ($\overline{H} = \frac{7200}{\pi^2} \frac{gH}{n^2 D_2^2}$).

Efficiency of torque flow pump will decrease by 4.5% with increase of free chamber width, and specific speed will increase by 2% in this case. Obtained results indicate no effectiveness of specific speed growth by means of free chamber width increase for torque flow pumps. Optimum value of relative width of free chamber is in the range 0.3 – 0.35 for torque flow pump with spatial spiral outlet.

Modification of angle of wall inclination β_{cm} of outlet at the exit of impeller in the range of 30-50 degrees leads to growth of pump head by 11% (Figure 7 (a)). Maximum value of efficiency was obtained for angle of wall inclination equal to 45 degrees.

Modification of suction pipe diameter influences in head and efficiency of pump. And for increase of relative value of inlet diameter $\overline{D_0} = D_0 / D_2$ from 0.4 to 0.62 head decreases by 2%, and pump efficiency – by 2.5% (Figure 7 (b)).

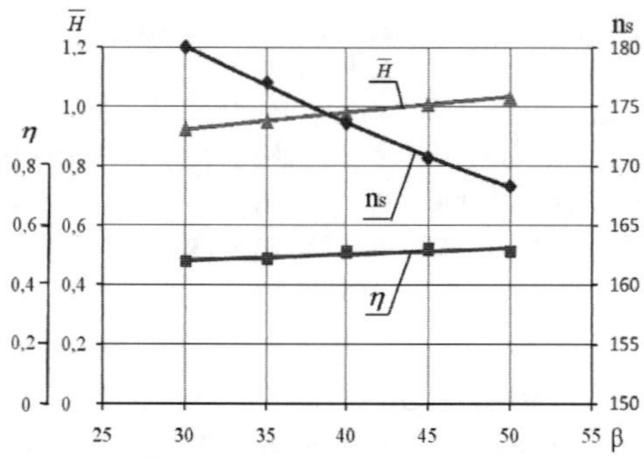

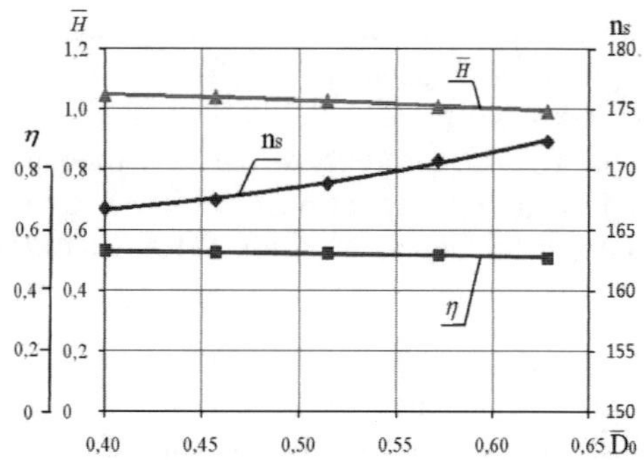

Figure 7. Relation of optimum parameters of torque flow pump:
(a) depending on inclination angle free chamber wall β_{cm} ;
(b) depending on suction pipe diameter D_o.

Taking into account obtained results series of torque flow pumps with spatial spiral outlet was designed and investigated for the range of specific speed n_s=60–200. Significant increase of efficiency for this pump was obtained comparing to pump with annular outlet (Figure 8). Recommended range of application for torque flow pumps with spatial spiral outlet is $n_s = 60 - 180$, for this case pump efficiency equals 48–55 %.

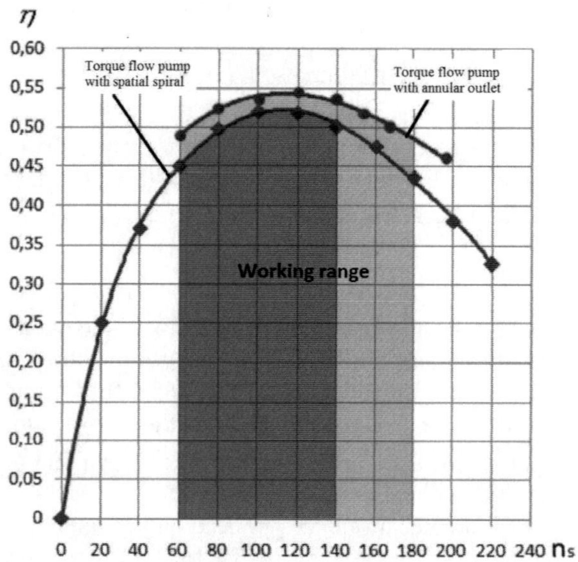

Figure 8. Dependence between torque flow pump efficiency and specific speed.

4. Results confirmation at experimental bench

For verification of operation effectiveness of torque flow pump with spatial spiral outlet and confirmation results of numerical calculation tests of pump prototypes with outlets No. A, No. B and No. D were implemented at the experimental bench (Figure 9 (a)). All pumps had specific speed equal to n_s = 168.

Tests showed that efficiency of pump with annular outlet equals $\eta = 43.5\,\%$. Optimum operation mode of pump with such outlet didn't correspond to its rated value (Figure 9 (b)).

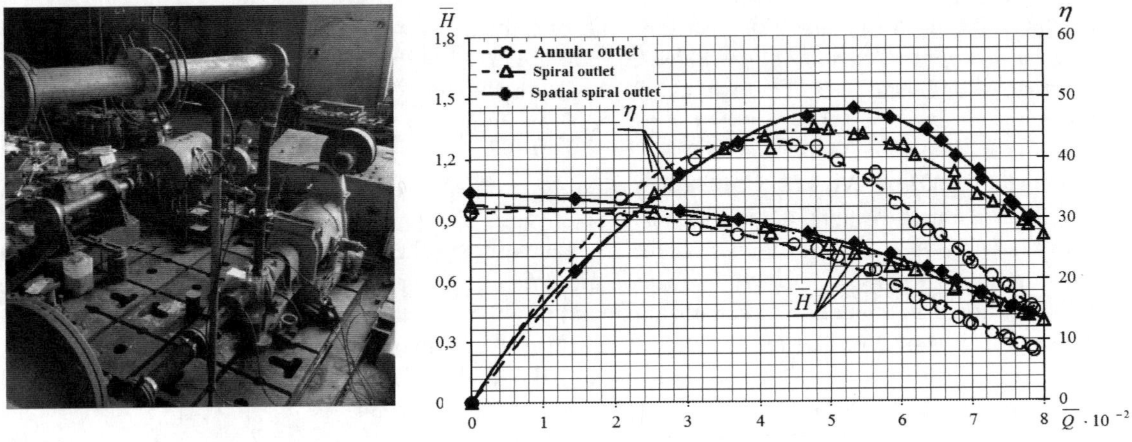

(a) (b)

Figure 9. Experimental researches of torque flow pump: (a) experimental bench; (b) characteristics of torque flow pump with different types of outlets ($\overline{Q}=\frac{240}{\pi^2}\frac{Q}{nD_2^3}$).

Concerning torque flow pump with spiral outlet, its efficiency for optimum operation mode Q=120 m^3/h equals η=45 %. No correspondence of optimum modes of these pumps is caused by the fact that outlet geometry modification leads to flow structure modification and consequently to increase of hydraulic losses and decrease of pump efficiency.

For tests of torque flow pump with spatial spiral outlet, value of efficiency increased by 48% which is much higher than efficiency of pump with annular outlet. Growth of efficiency comparing with pump with spiral outlet is 3% and comparing with pump with annular outlet – 4.5%. Optimum mode of this pump corresponds to rated value area for specific speed n_s = 168.

5. Conclusion

Implemented set of researches showed that application field of torque flow pumps with high specific speed values can be expanded by means of modification of outlet geometry.

Calculations of geometrical parameters of pattern hydraulic flow parts were implemented and tests were implemented for n_s = 100 – 180.

Conclusion concerning significant influence of free chamber width on pump characteristics among all dimensions of hydraulic flow part was experimentally confirmed for invariable diameter of torque flow pump impeller.

Radial and annular irregularity of flow in spatial spiral outlet was eliminated, which allows to create advanced outlet with improved hydraulic indexes, for example efficiency of "Turo" type torque flow pump with high specific speed (n_s =168) can be increased by 3-4% by applying this outlet.

It was proved by physical experiment that torque flow pump with spatial spiral outlet is more energy efficient because its efficiency is 3% higher than efficiency of pump with spiral outlet and 4.5% higher than pump with annular outlet.

Hydraulic flow part of outlet, which can be used for creation of more economical "Turo" type torque flow pumps, was obtained as a result of researches [13].

IOP Conf. Series: Materials Science and Engineering **233** (2017) 012004 doi:10.1088/1757-899X/233/1/012004

References

[1] Gulich J F 2010 *Centrifugal Pumps* (*second edition*) Springer Heidelberg Dordrecht (London, New York) p 957

[2] Krishtop I V 2015 Designing of hydraulic part of torque flow pump with uprated energy efficiency *Eastern-European Journal of enterprise technologies* **2/7** (74) pp 31-37

[3] German V F, Gusak A G, Yevtushenko V O and Panchenko V O 2011 Search for ways to expand the range of operating parameters for torque flow pumps of the type «Turo» *Eastern European Journal of Advanced Technology* **4/8** (52) pp 33-37

[4] Solyanik V O 1999 Workflow and energy pumps such as torque flow pumps type «Turo» Sumy: the dissertation of the candidate of technical sciences: 05.05.17 p 217

[5] Korbutovskiy A A 1977 Influence of the tap geometry on the operating parameters of the torque flow pump *Research, calculation and manufacturing technology of hydraulic machines* (VNIIGidromash: collection of scientific papers) pp 40-52

[6] Presman L S 1967 Vortex (torque flow) pumps (VNIIGidromash: release 44) pp 46-65

[7] United States Patent No 4475868 Switzerland, IPS F04D 29/44. Free Flow Pump / E. Egger. – Appl. No 445971 Filled: Dec. 1. 1982, Date of Patent: Oct. 9. 1984.

[8] Yahnenko S M 2003 Hydrodynamic aspects of block-modular design of dynamic pumps (Sumy: the dissertation of the candidate of technical sciences: 05.05.17) p 210

[9] German V F 1984 Creation and research of mass-flow torque flow pumps of increased economy] (Sumy: the dissertation of the candidate of technical sciences: 05.05.17) p 154

[10] Ukraine Patent No 99588 Ukraine, IPS F04D 7/04. Torque flow pumps/ Krishtop I, German V, Gusak A No u201500091 Filled: Jan. 6. 2015, Date of Patent: June. 10. 2015

[11] Ukraine Patent No 56039 Ukraine, IPS F04D 7/00. Torque flow pumps/ Panchenko V, Yevtushenko V, Solyanik V and Morgal A No u201006394 Filled: May. 25. 2010, Date of Patent: Dec. 27. 2010

[12] German V F 1994 Investigation of the flow structure in a torque flow pump *Hydraulic machines and hydropneumatic units: theory, calculation, design* (Kiev : ISIS) pp 67-81

[13] Ukraine Patent No 84940 Ukraine, IPS F04D 7/04. Torque flow pumps/ Krishtop I, German V, Gusak O and Lugova S No u201303946 Filled: Apr. 1. 2013, Date of Patent: Nov. 11. 2013

[14] Krishtop I, German V, Gusak O, Lugova S and Kochevsky A 2014 Numerical Approach for Simulation of Fluid Flow in Torque Flow Pumps *Trans Tech Publicatoins Inc.: "Applied Mechanics and Materials"* **440** pp 43-51

[15] German V F, Kovalev I A, Kotenko A I and Gusak A G 2013 Torqoe flow pumps (Sumy, Sumy State University) p 159

[16] German V F, Kochevsky A N, Schelyaev A E 2006 Analysis of the flow structure in a torque flow pump *Industrial hydraulics and pneumatics* **3** (13) pp 82-88

IOP Conf. Series: Materials Science and Engineering **233** (2017) 012005 doi:10.1088/1757-899X/233/1/012005

Calculation and designing of volutes of rotodynamic pumps

I V Krishtop[1], P M Kalinichenko[2] and A G Gusak[1]

[1]Department of Applied Fluid Mechanics, Sumy State University, 2rd Rimsky-Korsakov st., Sumy, 40007, Ukraine
[2]Department of General Mechanics and Dynamics of Machines, Sumy State University, 2rd Rimsky-Korsakov st., Sumy, 40007, Ukraine

E-mail: kryshtopihor@gmail.com

Abstract. This work deals with improvements of the calculation and design methods of rotodynamic pump volutes taking into account viscous liquid flow. The research was performed with torque flow pump of "Turo" type. The analytical solution for the problem within the one-dimensional model of viscous liquid allowed to obtain a viscous liquid flow pattern in the volute. The law of velocity variation of viscous liquid flowing through the volute was obtained. There are numerical solutions of the problem concerning of viscous liquid flow in the pump with volute designed for traditional model of non viscous and viscous fluid.
The research results are given by means of head-capacity and power input curves. To estimate the results of numerical study the tests were performed on the test stand.

1. Introduction

Rotodynamic pumps are energy machines. They are widely used in many branches of the industries and sectors of the national economy. Today's tendency of the development of rotodymanic pumps is connected with the increasing of their efficiency without any change in their mass and dimension. Following the energy balance the part of the hydraulic losses occurs in the volute chamber. The decrease of such losses within the total losses balance of a pump needs the improving of the calculation and design methods of pump volute especially in case of viscous liquid.

One of the most effective methods with regard to minimum energy losses is a volute. It consists of a volute chamber casted integrally with a pump casing and a diffuser nozzle.

Volute shall provide axially symmetric flow downstream of an impeller and thus producing steady flow in the impeller, convert part of the kinetic energy of the flow downstream of the impeller into the pressure energy and direct flow to the discharge nozzle or to the next pump stage.

Volute designing is carried out for the rated conditions of the pump. Therefore, a relatively small circumferential non-uniformity of the flow is observed only within a rather short operating range, close to the optimum condition. Beyond the above mentioned range the non-uniformity of the flow around the impeller increases greatly. If this occurs the hydraulic losses as well as pressure pulsations and radial forces also increase in the volute and impeller. The usage of the volute is economically feasible only at the rated conditions.

The existing methods for calculation of the volutes are based on a number of assumptions which simplify the physics of the liquid flow passing through the volutes. These assumptions include [1]:

- the liquid flow in the volute at the rated condition of the pump is assumed as steady flow;

- the liquid flow at the inlet of the volute is axially symmetrical i.e. flow rate throughout any section of the volute is proportional to angle of center φ;
- kinetic energy is transformed into potential energy due to the change of the circumferential velocity V_u (radial velocity V_r may be neglected as its value is low relative to the circumferential velocity).

The flow in the volute is determined by constant velocity moment $V_u r=const$ and proportionality of its flow rate to angle of section: $Q_\varphi = Q\frac{\varphi}{360^0}$ [1,2].

The principle of the geometry designing of the volute consists of uniformly reducing of its capacity from the design section to the "zero" (section through the cut water of the volute) [2]. The law of circumferential velocity distribution in such volute is parabolic rather than linear, as it is treated in the calculations.

Following [3], two approaches to the designing of the volute geometry are considered: the friction in the volute is not taken into account and it is taken into account. In the equation of the slope angle of the volute, which describes its geometry, the correction for liquid friction against the surface of the volute channel is applied. And upon that we have made an assumption that the additional friction surfaces affect the flow as well as completely closed shell. However, the losses in the volute are characterized not only by the wall friction of the liquid but local losses as well (smooth enlargement in section of the outlet, flow separation and vortex formation, etc.) which are not included in the present calculation.

In the work [4] the flow in the volute is treated as plane flow induced by straight vortex source with infinite length. As the result of this assumption we have received equation for flow line in the polar coordinates, which has a form of the logarithmic spiral and is treated as the wall outlines of the volute. And two approaches for volute designing is considered: the method according which the velocity moment is constant and the method consisting of determining of mean velocity of the flow in the volute. The present calculations do not show the pattern of the velocity distribution for viscous flow in the outlet channel as they are based on the flow of non viscous liquid.

Flow area of the volute chamber designed for models of non viscous liquid will differ from that designed for models of viscous flow. Subsequently, the volume of losses will be different in these volutes. That is why to calculate and study the volutes taking into account the influence of viscosity and determination of decreasing of loss volumes while changing the non viscous liquid into viscous liquid is still a topical problem. It is considered in the present work.

2. Theory of calculation and designing of volutes

The liquid flow in the volute is treated as spiral flow. The flow velocity in the circumferential direction is V_u and in radial direction is V_m. The volute shall provide the axisymmetric liquid flow at the inlet of volute which means that the inlet angle of flow entering the volute α_3 in circumferential direction shall be constant and equal angle [5,7] (Figure 1):

$$tg\alpha_3 = \frac{V_{m3}}{V_{u3}} \qquad (1)$$

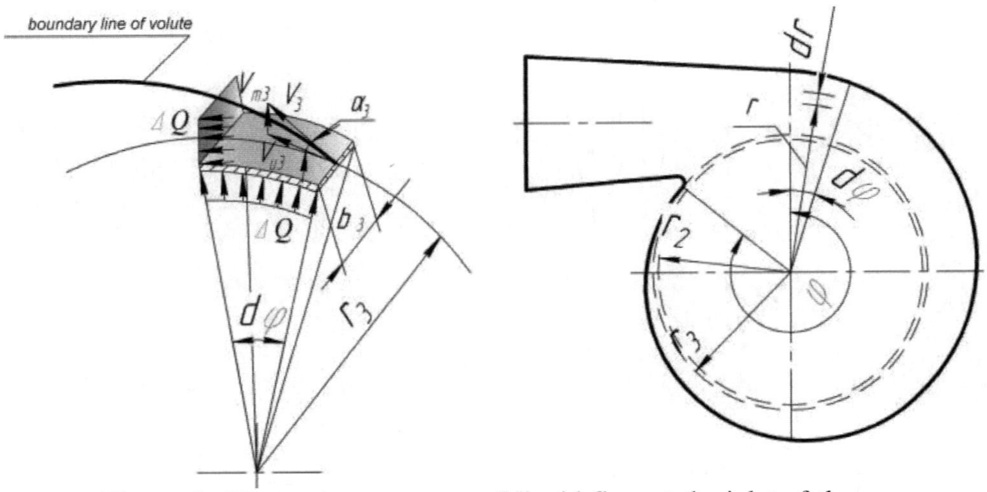

Figure 1. Kinematic parameters of liquid flow at the inlet of the pump volute.

In the first approximation the liquid is assumed as non viscous. Determining the geometry of the volute instead of the traditional approach consisted in building of a finite and therefore intermediate sections we determine the line of volute assuming it as flow line of liquid particle which starts near the bottom of the volute cut water. So we can write the differential equation of the flow line in the form of:

$$\frac{r\mathrm{d}\varphi}{V_u} = \frac{\mathrm{d}r}{V_r} \tag{2}$$

The circumferential velocity of non viscous liquid flow follows the law of potential vortex $V_u r = K = const$ [2]. The radial velocity is determined by the flow rate equation. For the radial velocity following figure 1, we can write:

$$\mathrm{d}Q = V_{r3} b r_3 \mathrm{d}\varphi = V_r b r \mathrm{d}\varphi \tag{3}$$

Taking into account the forgoing, we obtain from the equation (2) the dependency for building of the flow line:

$$\mathrm{d}r = \frac{r}{bK}\mathrm{d}Q \tag{4}$$

Equation (4) determines the value of volute line ascent in the circumferential direction which corresponds to the angle of bend φ (Figure 2 (a)).

For outlets with eccentric volute (Figure 2 (b)) the equation takes the form:

$$\mathrm{d}l = \frac{r}{bK}\mathrm{d}Q \tag{5}$$

where

$\mathrm{d}l$ – element of the flow line length of meridian section of the volute.

IOP Conf. Series: Materials Science and Engineering **233** (2017) 012005 doi:10.1088/1757-899X/233/1/012005

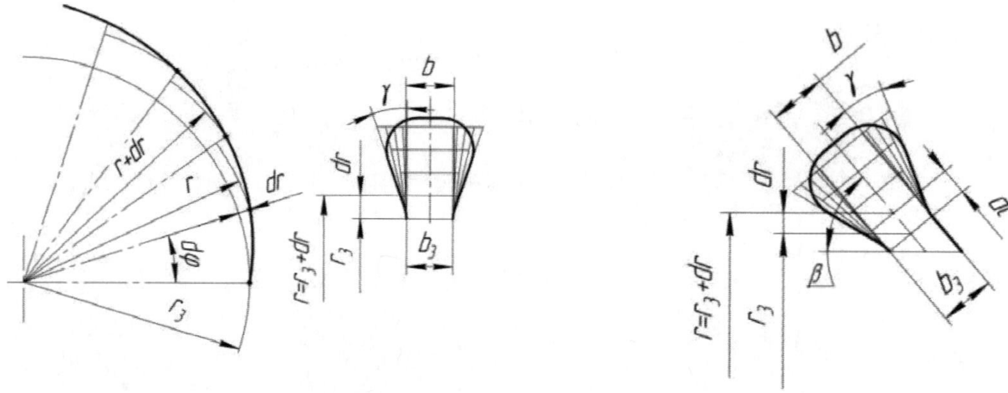

(a) (b)

Figure 2. Calculation model of volute: (a) construction of the volute with coaxial cross-section; (b) construction of the volute with eccentric cross-section.

We shall consider the flow of viscous liquid in the volute chamber of the pump.

The motion of liquid particle can be conceived of as a group of stream tubes flowing into volute and moving along spiral flow lines (Figure 3).

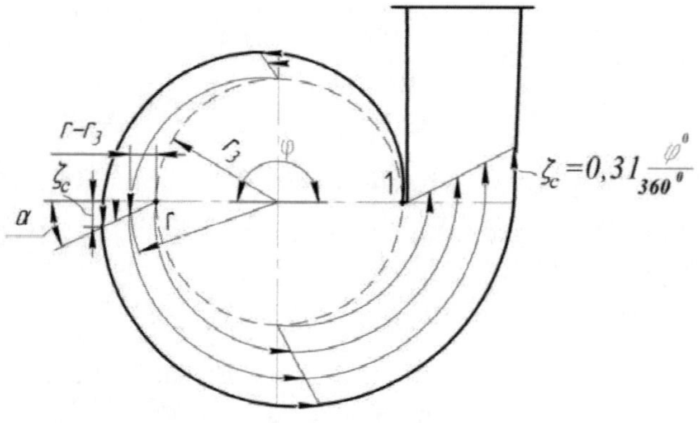

Figure 3. Liquid flow in the pump volute.

For inlet of volute, section 1 and optional section at the angle φ we write Bernoulli equation [6]:

$$z_1 + \frac{P_1}{\gamma} + \frac{V_1^2}{2g} = z + \frac{P}{\gamma} + \frac{V^2}{2g} + h_{1-\varphi} \tag{6}$$

where

$h_{1-\varphi}$ - energy losses along the stream tubes;

$z_1 = z = 0$ - energy of position of liquid particles is neglected.

Energy losses $h_{1-\varphi}$ are assumed as a group of local losses h_M including bend, flow expansion and other as well as losses due to the length h_D:

$$h_{1-\varphi} = h_M + h_D = (\xi_M + \xi_D)\cdot\frac{V^2}{2g} = \xi_c\frac{V^2}{2g} \tag{7}$$

where

ξ_M, ξ_D and $\xi_c = \xi_M + \xi_D$ are coefficients of resistance.

We replace the stream tubes in the form of spiral lines with the stream tubes in the form of concentric circles. The coefficient of resistance is calculated by Welsbach formula [8], obtained empirically:

$$\xi_c = \left[0.31 + 0.16\left(\frac{d}{r}\right)^{3,5}\right] \cdot \frac{\varphi^0}{90^0} \quad (8)$$

where

d - pipe diameter;

r - bending radius of pipe;

φ - bend angle of pipe.

According to equation (8), the coefficients of resistance of the pipe for stream tube with $d \ll r$ equals: $\xi_c = 0.31\frac{\varphi^0}{90^0}$.

As a first approximation we assume the law of variation of the coefficient of the resistance along the section of volute as linear (see Figure 3):

$$\xi_c = \frac{tg\alpha(r - r_3)}{r_0} \quad (9)$$

where

$r_0 = 1$ - a single linear size.

Then calculating the equation (6) with provision for (9), we obtain the dependence of the flow circumferential velocity in the volute on radius:

$$V_u \cdot \left[(a + b \cdot r)^{\frac{a-2}{2a}} \cdot r^{\frac{1}{a}}\right] = const \quad (10)$$

where

$b = tg\alpha$; $a = 1 - b \cdot r_3$ - are coefficients.

From equation (10) it fallows that when the viscous liquid flows through the volute chamber the velocity moment is lower than the velocity moment of non viscous liquid, flowing through the volute, it can be explained by energy losses. It means that the sections of volute calculated for viscous liquid by equation (10) shall be large than those calculated for non viscous liquid by the condition $V_u r = const$.

3. Research methods

The check of calculation and designing of volute were performed with the testing torque flow pump 125/28 with eccentric volute [9]. It is overhung pump designed for Q = 125 m^3/h, H = 28 m. The hydraulic flow part is made according to the method [12]. It is shown in Figure 4.

IOP Conf. Series: Materials Science and Engineering **233** (2017) 012005 doi:10.1088/1757-899X/233/1/012005

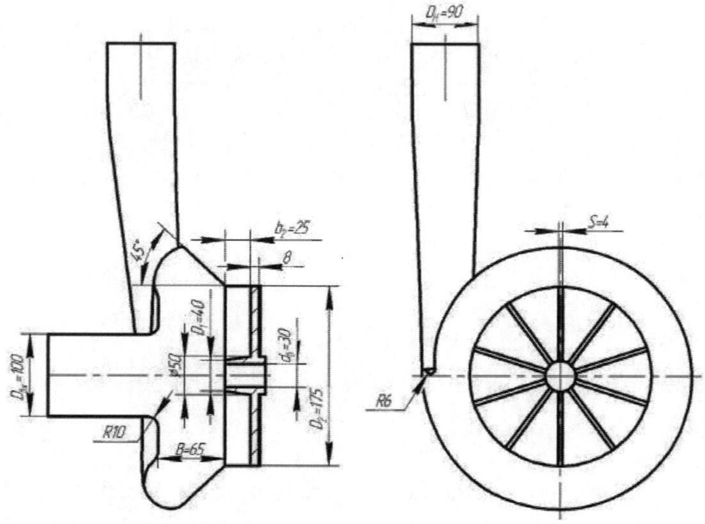

Figure 4. Testing torque flow pump 125-28.

The researches were conducted by means of numeric modeling and physical experimentation.

We used ANSYS CFX version for university to carry out numerical modeling of the liquid flow in the hydraulic flow part of the examined variants of the torque flow pumps. While modeling the operation conditions of the torque flow pump we applied computational domain consisting of pump casing including volute chamber and impeller. Computational mesh of impeller had about 1.63 millions prismatic cells, pump casing – 1.86 millions of cells. Prior to research mesh the independence check was made for each element of the computational domain. The value of the variable Y+ was in the range from 1 to 20 units, it was complied with the recommendations given in the user manual [10].

Typical Menter SST model [11] was used for turbulence modeling.

Numerical calculations were carried out in nonstationary problem. The outlet boundary of the calculation domain was located at 6 diameters away from the inlet boundary of diffuser nozzle. As boundary condition at outlet of calculation domain we set static pressure. As we assumed the presence of back flow at the outlet of the calculation domain the type of the boundary condition was set as "opening".

To conduct this research we designed two variants of casing of the torque flow pump with dimensioned volute according [13] (Figure 5):

variant No.1 – the absent of losses in the volute, according equation (5);

variant No.2 – the losses in the volute channel is taken into account, according equation (10).

The calculation showed that the finite section of the volute No.2 was higher by $\Delta = 7 \div 10\%$ than the finite section of the volute No.1 due to viscous.

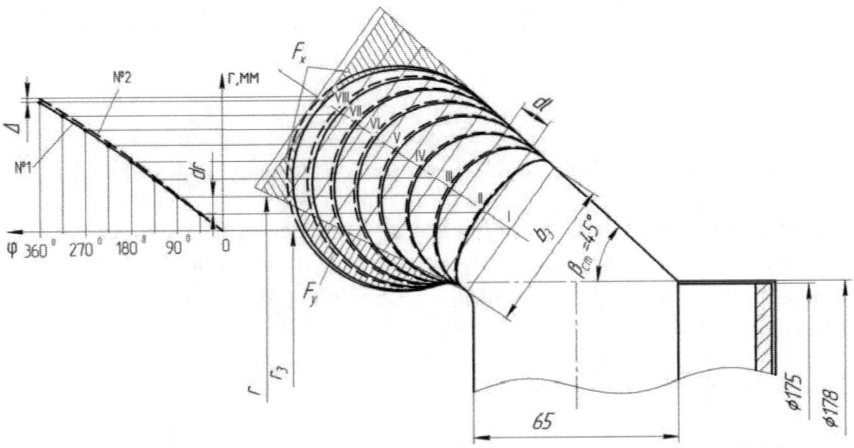

Figure 5. The calculation of casing of torque flow pump 125-28 with eccentric volute.

4. The study results of the volute geometry of the torque flow pump

By means of the numerical study we received a head-capacity and power input curves for volutes of the torque flow pumps (Figure 6). The curves for model hydraulic flow parts were calculated in a capacity range from 0.5 Q_{opt} to 1.25 Q_{opt}.

According to the research results the optimal operating range of the pump with volute No.1 with the highest efficiency 46.4% shifted from the rated capacities towards the area of smaller capacities with the value of Q=120 m^3/h with the head H= 31.1m. The optimal operating range of the pump with volute No.2 corresponds to the rated operating range with efficiency 48.6%: Q=125 m^3/h, H= 30.9m. Flow area of the volute No.2 is greater than the flow area of the volute No.1. The optimal operating range of the pump with volute No.2 corresponds to the rated operating range at the lower losses in the volute than the losses in the volute No.2 (Figure 7).

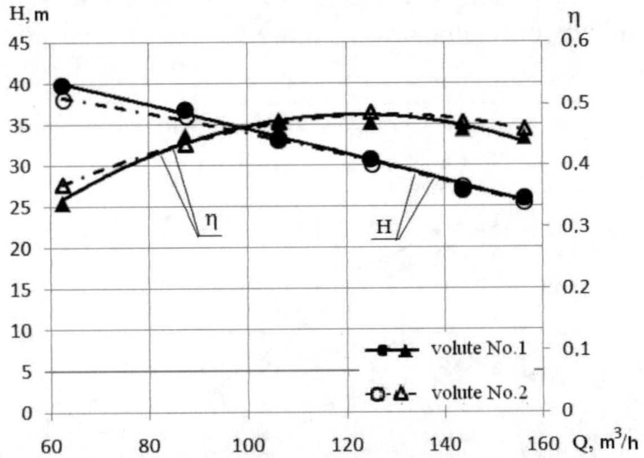

Figure 6. Head-capacity and power input curves for casings of the torque flow pumps.

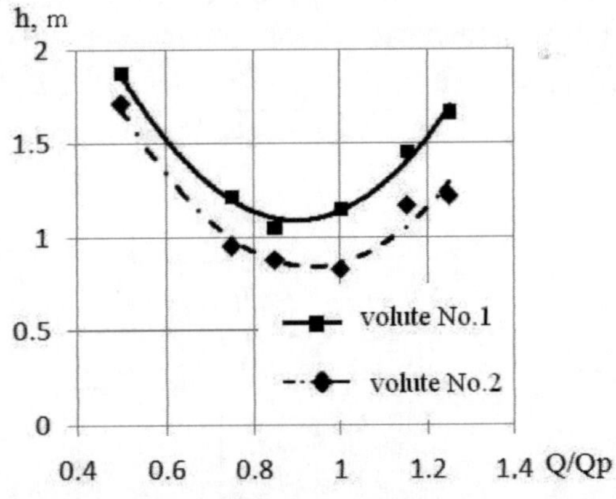

Figure 7. Distribution of the losses in the volutes of the torque flow pumps.

The shift of the optimal operating range in the volute No.1 with decreased flow areas and increased losses can be explained by axially symmetrical distortion of the flow at the inlet of the volute chamber. The volute No.2 is characterized by minimal losses at the design point. Therefore, taking into account the viscous of the liquid during the designing of the volute leads to the axially symmetric flow at the inlet of the volute chamber providing the steady-state flow near the pump impeller thus reducing the hydraulic losses. According to the research results (see Figure 6) efficiency of the pump with volute No.2 calculated with an allowance for viscosity exceeds by 1-2% the efficiency of the pump with volute No.1 calculated for non-viscous liquid. A good example of ensuring axial symmetry of the liquid flow at the inlet of the volute is the dependence of the flow rate $\frac{Q}{K_2}$ of the volute chamber on the contact angle φ (Figure 8).

The account of the viscosity in the calculation of the volute leads to a linear dependency $\frac{Q}{K_2}(\varphi)$, and hence it provides axially symmetric flow at the inlet of the volute chamber. The reducing of the velocity moment K_2 in the volute No.1 below the calculated value leads to the shift of the rated conditions towards the area of smaller capacities. Volute No.1 is characterized by deviations of the analytical and calculation values of 11%, for volute No.2 – 3.5%.

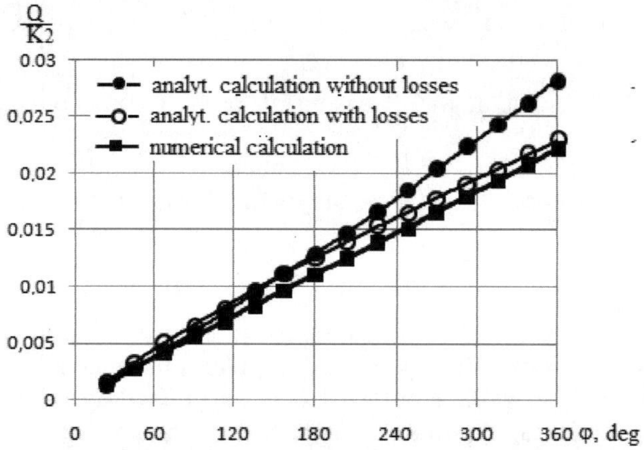

Figure 8. Flow rate of the volute of the torque flow pumps.

In order to confirm the results of the numerical calculation we tested the volute No.2. The testing results are presented by means of head-capacity and power input curves (Figure 9).

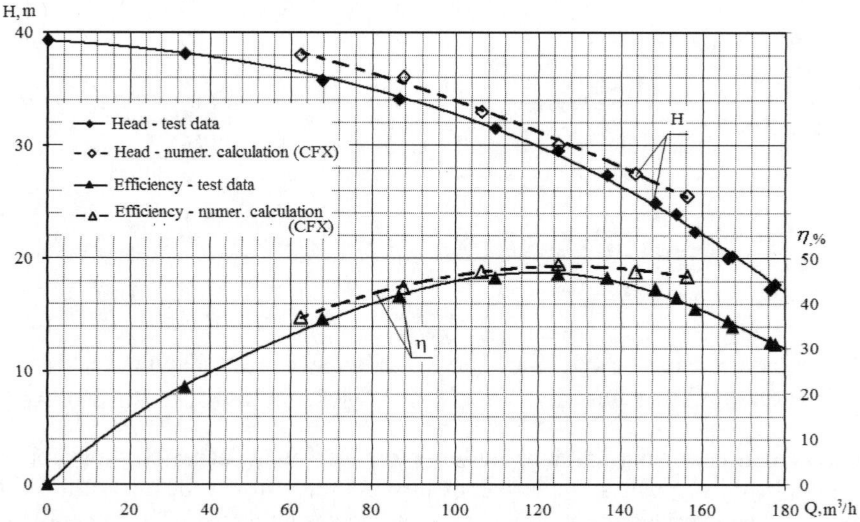

Figure 9. Comparative characteristics of the torque flow pump with dimensioned volute received after numerical calculations and tests.

According to the test results the optimal operating range of the torque flow pump with dimensioned volute No.2 corresponds to it calculated operating range with efficiency 47.8% and head H= 28m. The comparison of the characteristic curves obtained by testing with the results of numerical calculation shows that the head discrepancy is not more than 3% and efficiency discrepancy is about 2%. The most likely cause of such discrepancy of characteristic curves of the torque flow pump is the lack of information about the operation of the torque flow pump.

5. Conclusions

The method for calculating of volute based on the differential equation is presented.

The usage of one-dimensional model for viscous liquid flow allowed to specify the law of circumferential velocity distribution and therefore the velocity moment along the section of the volute.

The volute calculated on the assumption that the velocity moment is constant along the section of the volute chamber $V_u r = const$ leads to the shift of the optimal operating range from the calculated value towards the area of smaller capacities. The discrepancy of the velocity moment obtained by numerical calculation of the viscous liquid problem and non-viscous problem for such volute is 11%. The velocity moment discrepancy obtained by calculation performed by proposed dependency with account for viscous allows reduce the velocity moment discrepancy up to 3%.

According to the numerical calculations the volutes of the torque flow pumps calculated by the method presented in this work improved the efficiency of the pump up to 1-1.5%.

References

[1] Miftahov A A and Zyikov V I 1996 Inlets and outlets of centrifugal compressors (Kazan: Feng (Science)) p 198

[2] Mashin A N 1980 Calculation and designing of volute outlet and semi-volute inlet of centrifugal pump (Moscow: MPEI Publ.) p 44

[3] Pfleiderer K 1960 *Die Kreiselpumpen für Flüssigkeiten und Gase. Wasserpumpen, Ventilatoren, Turbogebläse, Turbokompressoren* (Moscow: GNTI mashinostroitelnoy literatury) p 683

[4] Kostornoy S D and Chaplygin A A Designing of centrifulal pump volutes of trapezoidal shape with rounded corners *Bulletin of Sumy State University. Series: Engineering* **1** (3) pp 105-115

[5] Gulich J F 2010 Centrifugal Pumps (second edition) *Springer Heidelberg Dordrecht* (London, New York) p 957

[6] Barlit V V 1977 Gidravlicheskie turbiny *Hydraulic turbine (*Vyshcha shkola) p 360

[7] Rudnev S S 1967 Calculation of volute chambers of centrifugal pumps (Moscow)

[8] Frenkel N Z 1956 Hydraulics State *Energy Publishing* (Mostka)

[9] Krishtop I V 2015 Designing of hydraulic part of torque flow pump with uprated energy efficiency *Eastern-European Journal of enterprise technologies* **2/7(74)** pp 31-37

[10] ANSYS CFX 10.0 Solver Models. Release 10.0 [Digital source] –Available at: \ www/ URL: http:// ansys.com/ - 2008 p 549

[11] Krishtop I, German V, Gusak O, Lugova S and Kochevsky A 2014 Numerical Approach for Simulation of Fluid Flow in Torque Flow Pumps *Trans Tech Publicatoins Inc.: "Applied Mechanics and Materials"* **440** pp 43-51

[12] German V F, Kovalev I A, Kotenko A I and Gusak A G 2013 Torque flow pumps (Sumy: Sumy State University) p 159

IOP Conf. Series: Materials Science and Engineering **233** (2017) 012006 doi:10.1088/1757-899X/233/1/012006

Selection of design and determination of quality targets for VS6 pump stages

I Tverdokhleb[1], A Yelin[1]

[1] HMS Group Management Company LLC, 7, Chayanova st., Moscow, 125047, Russia

E-mail: tib@hms.ru

Abstract. A comparative analysis of VS6 pump type in the range of specific speed n_q = 25...55 was carried out basing on information about analogous pumps produced by the world's leading manufacturers. The focus was made on mass-dimensional characteristics of the pump stages. It was found that rational design of a diffuser with radial vanes provides lower material consumption at similar hydraulic characteristics compared to using a semi-axial vanes diffuser in the stage.

Centrifugal vertical semi-submersible pumps of a double casing design are widely used in the mainline oil transportation, processing of hydrocarbons as well as condensate pumps at power stations. The pumps of VS6 type with in-line suction and discharge nozzles are the most popular in the world practice [1,2].

Perfection of the pump hydraulics largely determines the pump competitiveness in general and first of all by the criteria of efficiency and minimum weight and size. Two types of stages are used in VS6 pump type: with radial diffuser (hereinafter – type 1) and semi-axial diffuser (hereinafter – type 2). The type 1 is traditionally preferred by European manufacturers, the type 2 – by the North American ones. These types of hydraulic have identical hydrodynamic parameters in the range of specific speed n_q = 25...55 [2]. Hence the choice between type 1 and type 2 for this specified speed range should be based on the possibility to get the minimum weight and size. Obviously, the stage type 1 has relatively smaller axial and larger radial dimensions in comparison with the stage type 2. Therefore, the comparison of the stages of both types should be carried out according to the mass of the stage, and also by the mass of the corresponding section of the outer pump casing.

In view of the foregoing, the development of methodological recommendations for the selection of optimal design for VS6 pump stages in the range of the specific speed n_q = 25...55 has considerable scientific and practical interest.

To solve this problem we collected, summarized, processed and analyzed the available statistical information on analogous pumps manufactured by the world's leading manufacturers. As information sources there were catalogs [3,4], programs for selection of pumps with authorized access on the manufacturers websites [5,6], as well as information bases with open access [7,8].

At the first stage of the work the upper limit of the stage specific speed value $n_{q.BEP}$ corresponding to the maximum energy efficiency was determined. Analysis of the tender specifications for the purchase of VS6 type pumps showed that presently almost all customers limit the pump head rise from the design point to zero flow by value not above 20 ... 25%. In view of this the upper limit of the VS6 pump stage specific speed was defined as $n_{q.BEP} = 55$ (Figure 1).

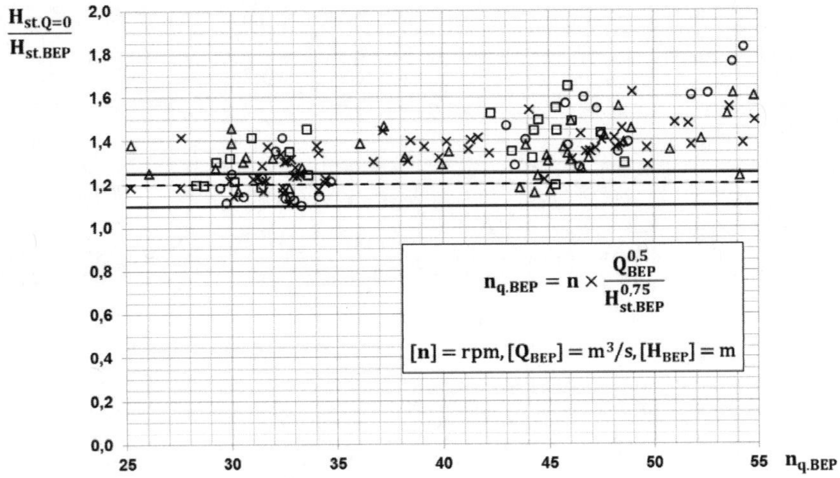

Figure 1. Dependence of head rise of stage type 2 on $n_{q.BEP}$.

When estimating the mass-dimension characteristics, the volume of the outer contour of a stage of type 1 was defined as a volume of a conditional cylinder with a diameter equal to the outer diameter of the diffuser D_{5R} and a length equal to the length of the stage $L_{st.R}$ (Figure 2).

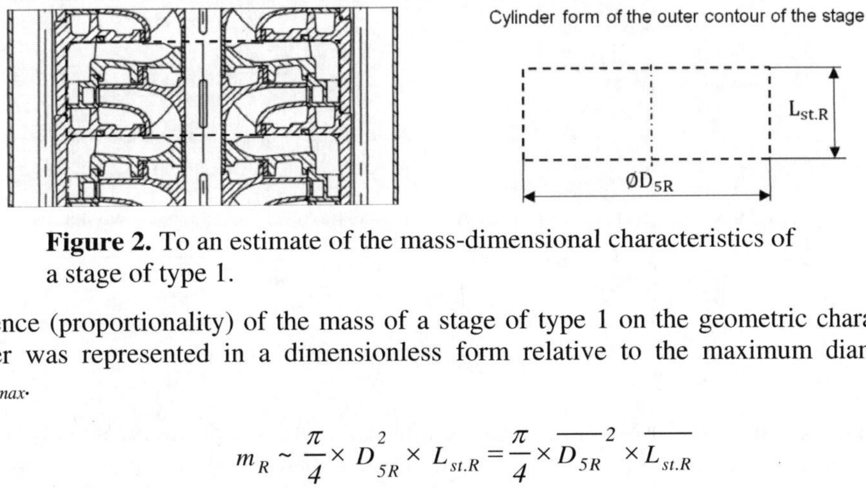

Figure 2. To an estimate of the mass-dimensional characteristics of a stage of type 1.

Dependence (proportionality) of the mass of a stage of type 1 on the geometric characteristics of this cylinder was represented in a dimensionless form relative to the maximum diameter of the impeller D_{2max}.

$$m_R \sim \frac{\pi}{4} \times D_{5R}^2 \times L_{st.R} = \frac{\pi}{4} \times \overline{D_{5R}}^2 \times \overline{L_{st.R}} \tag{1}$$

In turn, dimensionless criteria $\overline{D_{5R}}$ and $\overline{L_{st.R}}$ were determined empirically from the analysis as the functions of $n_{q.BEP}$:

$$\overline{D_{5R}} = D_{5R}/D_{2max} = f_1(n_{q.BEP}),\ L_{st.R} = L_{st.R}/D_{2max} = f_2(n_{q.BEP}) \tag{2}$$

To find the relative diameter of the diffuser of type 1 averaging of the known dependence from the monograph [2] was used:

$$\overline{D_{5R}} = (1{,}05 \div 1{,}15) + n_{q.BEP}/100 \approx 1{,}1 + n_{q.BEP}/100 \tag{3}$$

The statistical dependence of the relative length $\overline{L_{st.R}}$ of the stage of type 1 on $n_{q.BEP}$, obtained with the help of the one of the pump selection programs, is shown in Figure 3.

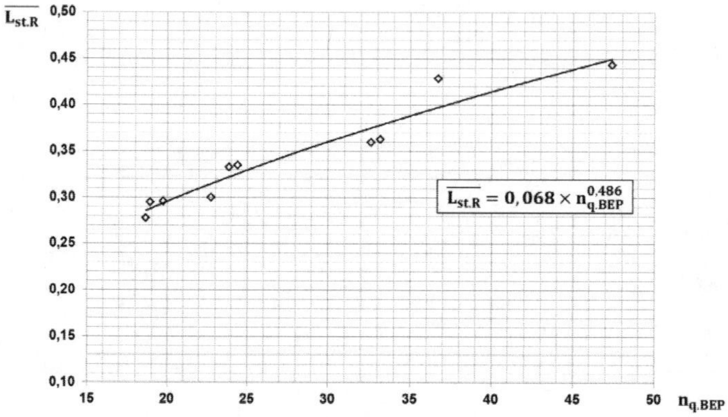

Figure 3. Dependence of the relative length $\overline{L_{st.R}}$ of a stage of type 1 on $n_{q.BEP}$.

An analysis of the design of stages with semi-axial diffuser with respect to pumps of type VS6 showed that in considered range of values of $n_{q.BEP}$, the stages of most manufactures have an outer contour in the conditional form of a cylinder and its associated truncated cone (Figure 4).

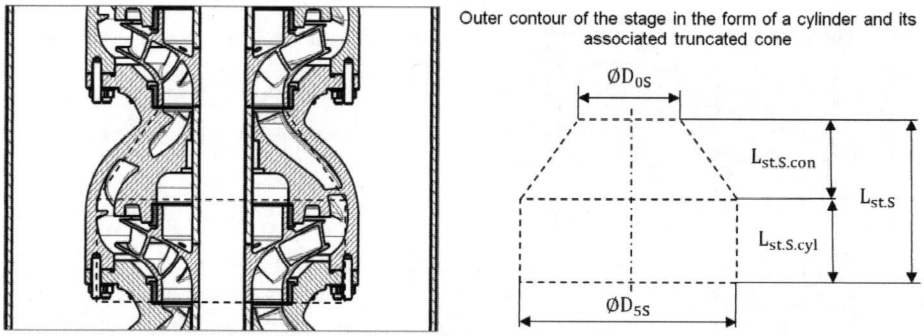

Figure 4. To an estimate of the mass-dimensional characteristics of a stage of type 2.

When estimating the mass-dimensional characteristics, the volume of the outer counter of the stage of type 2 was determined as the sum of the volume of the cylinder with a diameter equal to the outer diameter of the diffuser D_{5S} and the length equal to the length of the cylindrical stage section $L_{st.S.cyl}$ and the volume of the truncated cone with a diameter of a large base equal to the outer diameter of the diffuser D_{5S}, diameter of a smaller base equal to the diameter of the impeller eye D_{0S} and length equal to the length of the conical stage section $L_{st.S.con}$.

Dependence (proportionality) of the mass of a stage of type 2 on the geometric characteristics of its outer contour was represented in a dimensionless form relative to the maximum diameter of the impeller D_{2max}:

$$m_S \sim \frac{\pi}{4} \times D_{5S}^2 \times L_{st.S.cyl} + \frac{\pi}{12} \times L_{st.S.con} \times \left(D_{5S}^2 + D_{5S} \times D_{0S} + D_{0S}^2\right) = \\ = \frac{\pi}{4} \times \overline{D_{5S}}^2 \times \overline{L_{st.S.cyl}} + \frac{\pi}{12} \times \overline{L_{st.S.con}} \times \left(\overline{D_{5S}}^2 + \overline{D_{5S}} \times \overline{D_{0S}} + \overline{D_{0S}}^2\right) \quad (4)$$

The value of $L_{st.S.con}$ is close to the relative axial length of the blade system of the semi-axial diffuser and was estimated from the known dependence from the monograph [2]:

$$\overline{L_{st.S.con}} \approx 0{,}72 \times \left(n_{q.BEP}/200\right)^{0,19} \quad (5)$$

Taking this into account, the value of $L_{st.S.cyl}$ was calculated as:

$$\overline{L_{st.S.cyl}} = \overline{L_{st.S}} - 0{,}72 \times \left(n_{q.BEP}/200\right)^{0,19} \quad (6)$$

Dimensionless criteria $\overline{D_{5S}}$, $\overline{D_{0S}}$ and $\overline{L_{st.S}}$ were determined empirically from the analysis as the functions of $n_{q.BEP}$:

$$\overline{D_{5S}} = D_{5S}/D_{2max} = f_3\left(n_{q.BEP}\right), \overline{D_{0S}} = D_{0S}/D_{2max} = f_4\left(n_{q.BEP}\right), \\ \overline{L_{st.S}} = L_{st.S}/D_{2max} = f_5\left(n_{q.BEP}\right) \quad (7)$$

The empirical dependences of the relative diffuser outer diameter $\overline{D_{5S}}$ on $n_{q.BEP}$ (Figure 5), the relative stage length $\overline{L_{st.S}}$ on $n_{q.BEP}$ (Figure 6) and the relative diameter of the impeller eye $\overline{D_{0S}}$ on $n_{q.BEP}$ (Figure 7) were obtained as a result of analysis and generalization of information about the stages of type 2 of analogous pumps produced by the world's leading manufacturers.

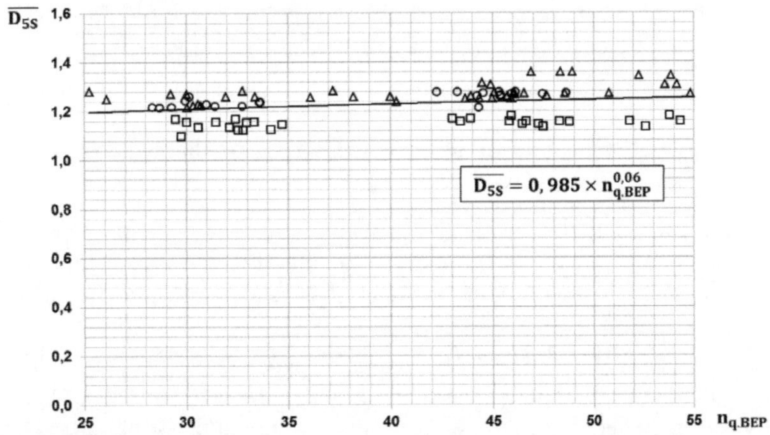

Figure 5. Dependence of the relative diffuser outer diameter of a stage of type 2 on $n_{q.BEP}$.

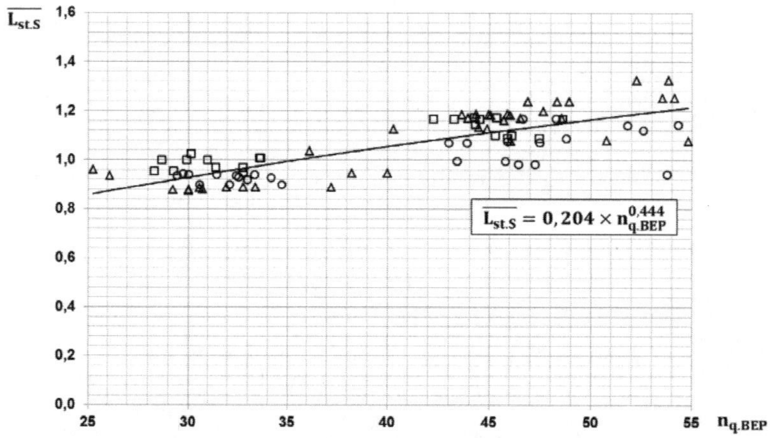

Figure 6. Dependence of the relative length of the stage of type 2 on $n_{q.BEP}$.

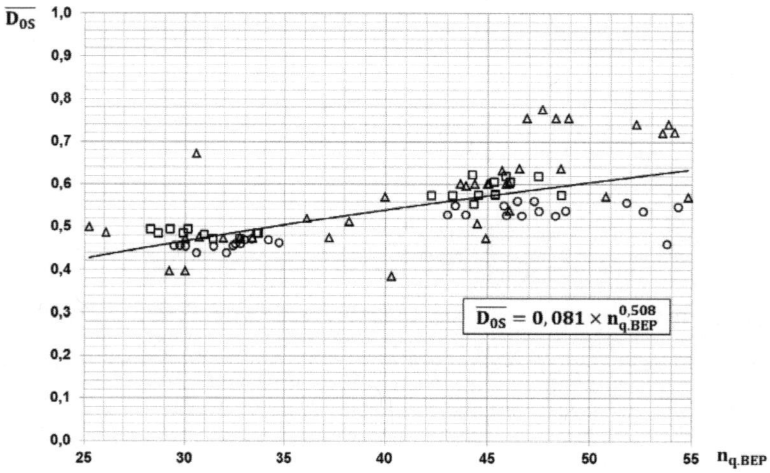

Figure 7. Dependence of relative impeller eye diameter of the stage of type 2 on $n_{q.BEP}$.

On the basis of formulas 1 and 4, using the obtained empirical relationships, were estimated the ratio of the mass of a stage of type 1 to the mass of a stage of type 2 as a function of $n_{q.BEP}$. In this case, the value of criteria $\overline{L_{st.R}}$, $\overline{D_{5S}}$, $\overline{L_{st.S}}$ and $\overline{D_{OS}}$ were determined according to Figure 3, Figures 5 – 7. The values of criteria $\overline{D_{5R}}$, $\overline{L_{st.S.con}}$ and $\overline{L_{st.S.cyl}}$ were calculated using formulas (3), (5) and (6).

As a result, it was established that in the whole range of the values of $n_{q.BEP}$ considered, the mass of a stage with the semi-axial diffuser will be large than the mass of a stage of the type 1 (Figure 8).

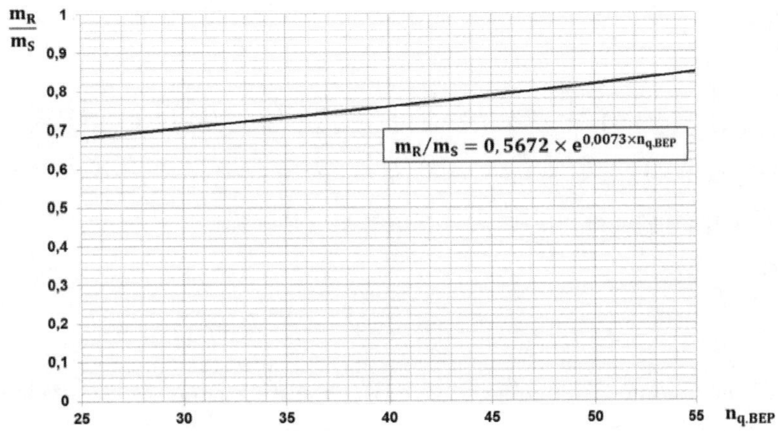

Figure 8. Dependences of the ratio of the mass of a stage of type 1 to the mass of a stage of type 2 on $n_{q.BEP}$.

It should be noted that this statement is valid if the stator elements of the stage of type 1 are designed rationally from the point of view filling its outer contour with metal (Figure 9, a). Otherwise, the mass of a stage of type 1 may exceed the mass of a stage of type 2 (Figure 9, b).

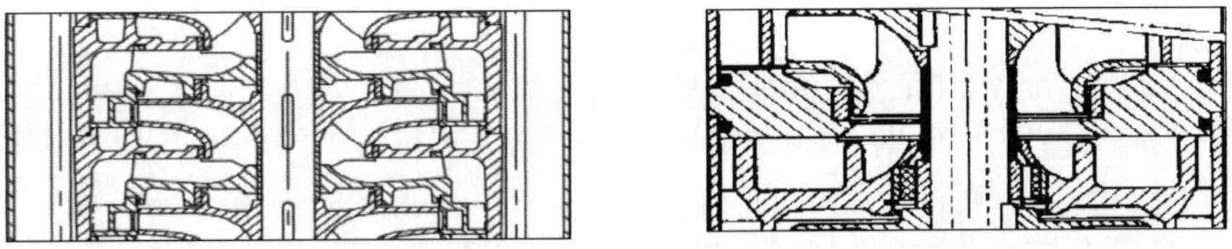

(a) (b)

Figure 9. Examples of rational (a) and irrational (b) design of stator elements of a stage of type 1 from the point of view of metal consumption.

Recommendation of the Hydraulic Institute of the USA [9] on the velocity of liquid between the inner and outer casing of VS6 pump (no more than 1,5 m/s) allows to move from the maximum outer diameter of the diffuser to the diameter of the outer casing of the pump during the design process (Figure 10).

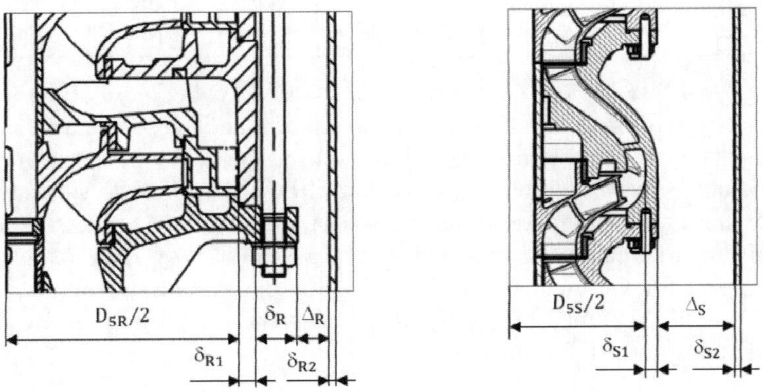

(a) (b)

Figure 10. To an estimate of the mass of outer casing section of VS type pump with a stage of types 1 (a) and 2 (b) per one stage.

For a stage of type 1 the distance between the flange for fastening the pack of stages and the outer casing of the pump w determined as:

$$\Delta_R = \left[Q_{BEP}/(\pi \times v) + (D_{5R}/2 + \delta_{1R} + \delta_R)^2\right]^{0,5} - D_{5R}/2 - \delta_{1R} - \delta_R \quad (8)$$

where δ_{1R} is the wall thickness of the section of the stage of type 1;

δ_R is width of flange for fastening the pack of stages of type 1.

Similarly for a stage of type 2:

$$\Delta_S = \left[Q_{BEP}/(\pi \times v) + (D_{5S}/2 + \delta_{1S})^2\right]^{0,5} - D_{5S}/2 - \delta_{1S} \quad (9)$$

Dependences of the volume of the section of the outer pump casing per one stage of types 1 and 2 were represented as:

$$V_R = \pi \times L_{st.R} \times \left[(D_{5R}/2 + \delta_{1R} + \delta_R + \Delta_R + \delta_{2R})^2 - (D_{5R}/2 + \delta_{1R} + \delta_R + \Delta_R)^2\right] \quad (10)$$

$$V_S = \pi \times L_{st.S} \times \left[(D_{5S}/2 + \delta_{1S} + \Delta_S + \delta_{2S})^2 - (D_{5S}/2 + \delta_{1S} + \Delta_S)^2\right] \quad (11)$$

where δ_{1R}, δ_{1S} are the wall thickness of the outer pump casing with stages of types 1 and 2, correspondingly.

Using formula (3), as well as Figures 3, 5 and 6, a series of preliminary evaluation construction drawings of pumps of type VS6 with rotational speed n = 1500 rpm were made (Table 1).

Table 1. Parameters of the series of preliminary evaluation construction drawings of pumps of type VS6 with rotational speed n = 1500 rpm.

BEP		D_{2max} [2], mm	δ_{1R}, δ_{1S}, mm	δ_{2R}, δ_{2S}, mm	δ_R, mm	Δ_R, mm	Δ_S, mm	$V_R\times10^{-3}$, mm^3	$V_S\times10^{-3}$, mm^3	V_R/V_S
Q, m^3/h	n_q									
200	31,5	290	12	12	50	21	29	2324	4603	0,50
315	34,2	320	16	16	60	29	40	4157	8170	0,51
500	38,4	350	20	20	60	41	56	6825	13635	0,50
800	43,8	385	20	22	70	57	79	10243	20271	0,51
1250	43,8	450	20	24	100	73	105	15863	30822	0,51
1800	52,1	465	20	26	100	97	140	21253	41524	0,51

As can be seen from Table 1, the volume and, correspondingly, the mass of the section of the outer pump casing per one stage for stages of type 2 is much larger than in case of stages of type 1.

The obtained results were used by the HMS Group during the development of its own new product line of the NMV pumps type VS6 with in-line suction and discharge nozzles during the formation of the range chart for the entire pump series and also for the creation and assessment of the actual competitiveness of the hydraulics of this pump type. The use of data of the world's leading manufacturers as targets for the quality parameters of the pump hydraulics predetermined the high competitiveness of the NMV pumps for both energy efficiency and mass-dimension characteristics. The last is clearly demonstrated in Figure 11.

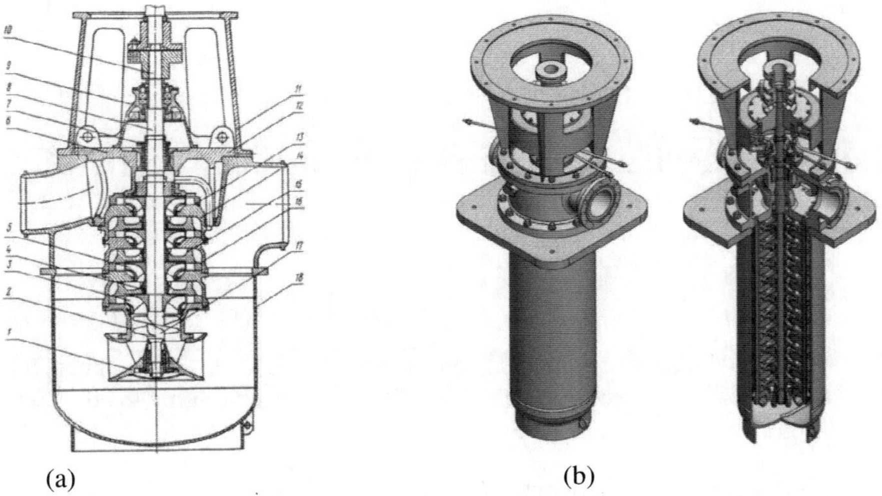

(a) (b)

Figure 11. Pumps of type VS6 with in-line suction and discharge nozzles produced by the HMS Group: (a) the first generation (pump KsV 1000-220); (b) the new generation (pump type NMV 500-350).

Conclusions

1 To use of hydraulics with a specific speed $n_{q.BEP} < 55$ ensures the demand for pumps of type VS6 in practice by the criterion of the limit head rise.

2 With rational design the use of stages with radial diffusers in pumps of VS6 type ensures a lower pump material consumption with similar hydraulic parameters compared to the hydraulics based on semi-axial diffusers. In each case the final choice of the stage type should be performed taking into account the comparative evaluation of the prime cost of machining its stator elements.

3 Effective use of the results of the available information analysis on analogous pumps produced by the world's leading manufacturers at the design stage of the pump hydraulics ensures a high level of competitiveness of the new pumping equipment.

References

[1] American petroleum institute, ANSI/API Standard 610, 11th edition, September 2010 (ISO 13709:2009 (Identical). Centrifugal pumps for petroleum, petrochemical and natural gas industries, pp 218
[2] Guilich J F 2010 *Centrifugal pumps* (Springer) pp 963
[3] Floway pumps 2005 *Vertical turbine pumps for the petroleum industry* PIBI 02/05 p 12
[4] Ebara. Vertical multistage canned pumps. Models VPCS/VPCH. CS2510EK p 7
[5] http://www.sulzer.com
[6] http://www.flowserve.com.
[7] http://www.wcc.nrcs.usda.gov
[8] http://www.powerzone.com
[9] American national standard for pump intake design 1998 (ANSI/HI 9.8, New Jersey) p 83

Cavitation in centrifugal pump with rotating walls of axial inlet device

O Moloshnyi[1], M Sotnyk[1]

[1]Applied Hydro- and Aeromechanics department, Faculty of Technical Systems and Energy Efficient Technologies, Sumy State University, Sumy 40007, Ukraine

E-mail: o.moloshnij@pgm.sumdu.edu.ua

Abstract. The article deals with the analysis of cavitation processes in the flowing part of the double entry centrifugal pump. The analysis is conducted using numerical modeling of the centrifugal pump operating process in the software environment ANSYS CFX. Two models of the axial inlet device is researched. It is shaped by a cylindrical section and diffuser section in front of the impeller, which includes fairing. The walls of the axial inlet device rotate with the same speed as the pump rotor. The numerical experiment is conducted under the condition of the flow rate change and absolute pressure at the inlet. The analysis shows that the pump has the average statistical cavitation performance. The occurrence of the cavitation in the axial inlet device is after narrowing the cross-section of flow channel and at the beginning of the diffuser section. Additional sudden expansion at the outlet of the axial inlet diffuser section does not affect the cavitation characteristics of the impeller, however, improves cavitation characteristics of the axial inlet device. For considered geometric parameters of the axial inlet device the cavitation in the impeller begins earlier than in the axial inlet device. That is, the considered design of the axial inlet device will not be subjected to destruction at the ensuring operation without cavitation in the impeller.

1. Introduction

Cavitation is a negative phenomenon that can occur in a pump. It leads to the destruction of the material, the occurrence of vibration and noise, reduction of the head and the efficiency of the pump. Scientists have devoted much attention to researching the cavitation and ways to reduce the probability of its occurrence in the pumps. One of the methods to fight against the occurrence of the cavitation in a centrifugal pump impeller is an additional inducer in front of it. Guo et.al [1] analyzed the influence of inducer with different designs on the cavitation characteristics of the centrifugal pump. The authors determined by the numerical simulation and confirmed experimentally that long inducer has better anti-cavitation characteristics than a short one. Although, the use of an inducer is considered an effective way to fight against the cavitation, however, there are many other ways. Xie et.al [2] investigated the influence of slots with different configurations in the impeller blades near the leading edge on the distribution of vapour volume fraction in the impeller and cavitation characteristics of the pump. The authors found that the slots with a large diameter and a smaller angle to the tangent of the impeller eye improve cavitation characteristics and does not affect the efficiency of the pump. Si et.al [3] found improved cavitation characteristics of the centrifugal pump with an additional jetting pipe in the pump inlet nozzle. However, cavitation characteristics get worse with increase of the flow rate. Zhuang et.al [4] got worse pump cavitation characteristics of the double entry pump with the radial blades of the impeller than the profiled. Tan et.al [5] found deterioration of the cavitation characteristics in the case of the prewhirl regulation by inlet guide vanes in front of the impeller. Hergt et.al [6] described the best known ways to improve the cavitation performance. The authors noted the impact of a diffuser on the shortening the recirculation zone length and the reducing the risk of cavitation. In addition, Gülich [7] noted the efficiency of using the diaphragm to reduce the cavitation. Kozubkováa et.al [8] researched the cavitation in the diffuser using different models of the cavitation and found a little difference in the results.

2. Models and Boundary Conditions

2.1. Pump geometry mode

The main pump parameters are as follows: nominal flow rate Q_{nom} = 25 m^3/h, head H=12 m, rotational speed n = 1450 r.p.m, specific speed n_q = 13 rpm, impeller eye diameter D_0 = 75 mm impeller outer diameter D_2 = 202 mm, the number of impeller blades z = 7.

The pump axial inlet device is a hollow shaft, which is also simultaneously the pump shaft. It is shaped by cylindrical and diffuser area in front of the impeller, which includes fairing (Figure 1). In addition, an axial cylindrical section of the casing is situated before the axial inlet device.

The numerical simulation of the fluid flow is conducted for 2 axial inlet device models (Figure 1). The design of Model 2 additionally comprises a sudden expansion at the exit of diffuser section. The diameter of the cylindrical section d and the inlet D are 45 mm and 65 mm respectively. The length of rotational wall of the axial inlet device is L = 227 mm and the stationary section of the casing is l = 40 mm. The length of the fairing (l_f) and the diffuser section of the axial inlet device (l_{dif}) are 37 mm and 50 mm respectively, an external (D_{out}) diameter of the outlet of the diffuser section in the model №1 and №2 are 75 mm and 71 mm respectively. The internal (d_{out}) diameter of the outlet of the diffuser section is 40 mm. Axial dimensions of the pump housing limit the axial inlet device dimensions.

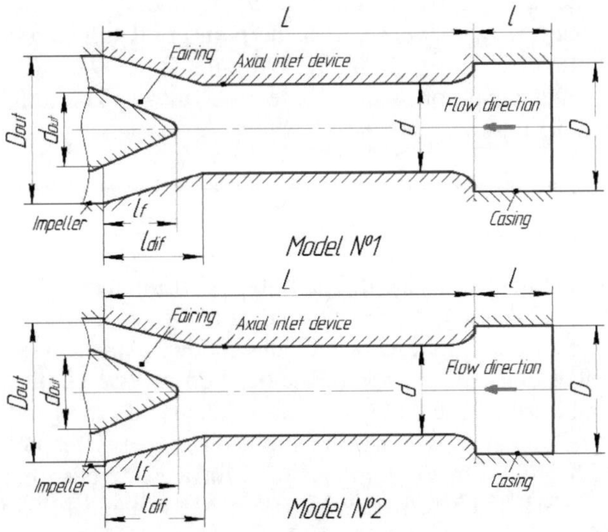

Figure 1. Scheme of the axial inlet device.

2.2. Mesh generation and boundary conditions

A solid model of a fluid computational domain of the double entry centrifugal pump includes the inlet pipe, the outlet pipe, the axial inlet device, the impeller and the annular casing with guide vanes (Figure 3). A geometry of the flowing part was simplified by assumptions about the absence of a sinus pump. Also, symmetry flow relative to the impeller of the double entry centrifugal pump was considered. This reduces complexity and increases the speed of calculation.

An unstructured mesh is generated using software ICEM-CFD (Figure 2). Elements size is selected via mesh independence research. Layers of prismatic elements were created near solid walls in the boundary layer (Figure 3(b)). In addition, elements of the mesh near the leading edge and the trailing edge are smaller. The total number of elements of the pump flowing part is 5 million. The inlet device, the impeller and the annular casing contain 0.97 mln, 2.4 mln and 1.65 mln of elements respectively.

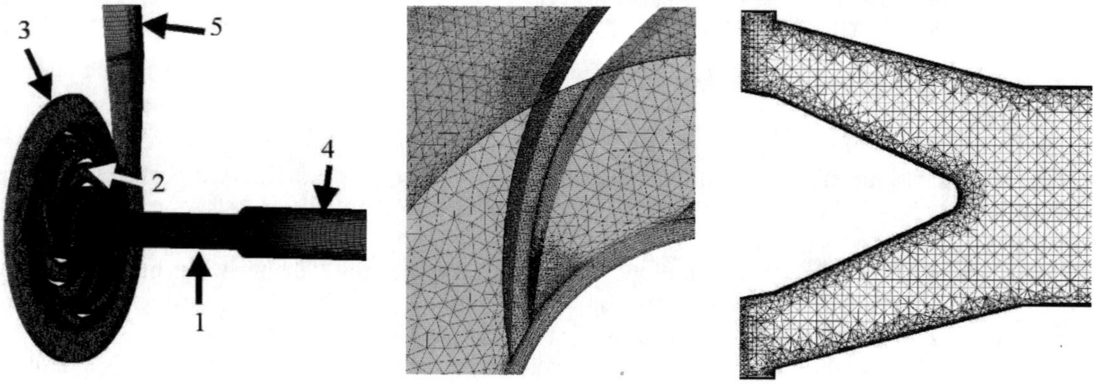

(a) (b) (c)

Figure 2. Calculation mesh: (a) computational domains: 1 – axial inlet device, 2 – impeller, 3 – annular casing, 4 – inlet pipe, 5 – outlet pipe; (b) leading edge; (c) diffuser section of the inlet device.

The numerical simulation of fluid flow in the flowing part of the pump was conducted using the software ANSYS CFX. To get cavitation characteristics was used the Rayleigh-Plesset equation, the standard k-ε turbulence model and Reynolds equation [9]. Boundary conditions was set as: the total pressure at the pump inlet and the mass flow rate at the pump outlet. The numerical calculation was carried out at the flow rate (0.7; 1.0; 1.2) Q_{nom}. Working fluid was water and water-vapor at temperature 25°C. The saturated steam pressure was set as 3167 Pa. The roughness of the axial inlet device surfaces was adopted 6.3 microns.

3. Results analysis

3.1. HPSH analysis

The results of the numerical simulation define the cavitation performance of the pump at the flow rates (0.7; 1.0; 1.2) Q_{nom} (Figure 3). The cavitation performance of Model 1 and Model 2 are qualitatively and quantitatively similar despite the introduction of the sudden expansion at the outlet of the axial inlet device. Minor differences in the magnitude of the head drop between the models observed. Also, there is a slight head drop and rise before its rapid decline. The head drop in Model 2 for the flow rate 1.2 Q_{nom} at NPSH more than 3 m hardly occurs. The above discussed differences do not affect $NPSH_3$, defined head drop by 3%. It is 0.15 m, 0.31 m and 0.45 m respectively for the flow rate (0.7; 1.0; 1.2) Q_{nom} for both models. $NPSH_A$ [7] is respectively 0.75 m, 0.91 m and 1.05 m.

An important criterion for cavitation in a pump is the suction specific speed n_{ss} [7]. It is defined as:

$$n_{ss} = n\frac{\sqrt{Q_{nom}/f_q}}{NPSH_3^{0.75}}$$

where, n is rotation speed; f_q is impeller eyes per impeller.

The values of the suction specific speed at the flow rate Q_{nom} for both models are 206. Standard impellers for axial inlet or between-bearing pumps have a value in the range of 160 - 220 [7]. That is, the indicator of the studied pump is above the average.

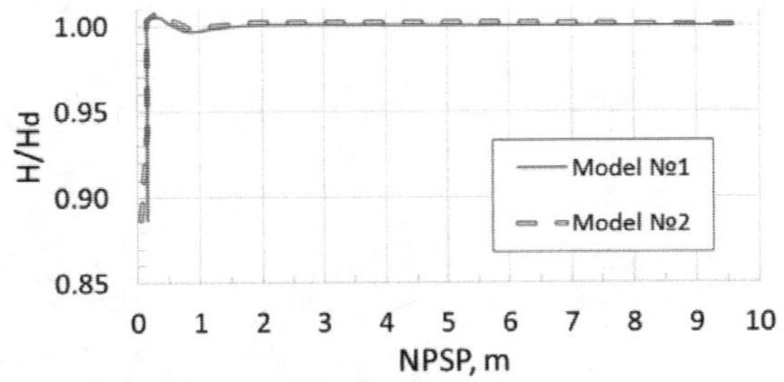

(a)

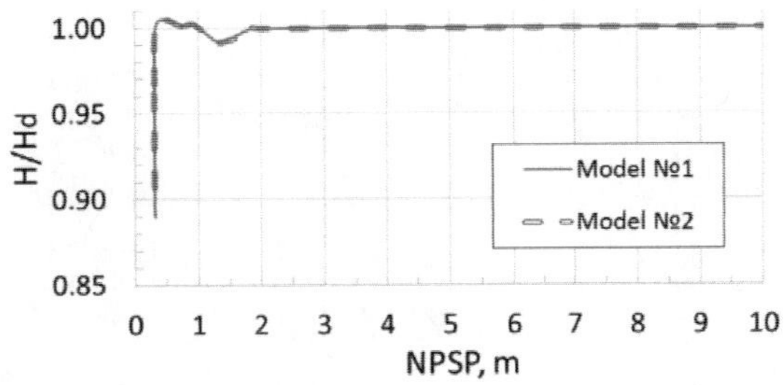

(b)

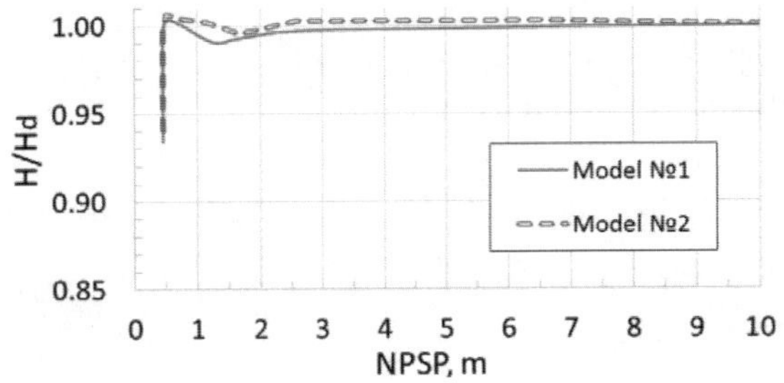

(c)

Figure 3. Cavitation performance curves: (a) 0.7 Q_{nom}; (b) Q_{nom}; (c) 1.2 Q_{nom}.

3.2. Cavitation structure and vapour volume fraction distribution

The occurrence of the cavitation in the impeller at the flow rate Q_{nom} is observed at the NPSP = 2 m (Figure 4(a)). It appears on the suction surface of the blades and almost does not effect the discharge characteristics of the pump. The cavitation zones are increased by reducing the NPSH (Figure 4(b)). At the NPSH3 there is a significant cavitation near the suction surface of the blades, which occupies a significant part of the passage between the blades (Figure 4 (c)). Distribution cavitation zones in the impeller for both models are almost identical.

The size of cavitation zones while reducing cavitation pressure of 3% is different for different flow rate. The cavitation zones and vapour fraction decrease under the condition of increasing the flow rate from 0.7 Q_{nom} to 1.2 Q_{nom} (Figure 5). The same pattern appears with other values of the head drop. Also, there is a slight increase of the cavitation zones at the constant value of the NPSH and increasing the flow rate (Figure 6). This is the result of the changing the approach flow angle, which depends on the value of the velocity and direction of the flow at the impeller inlet. It should be noted that the value of the circumferential component of the absolute velocity is also affected by the rotation of the walls.

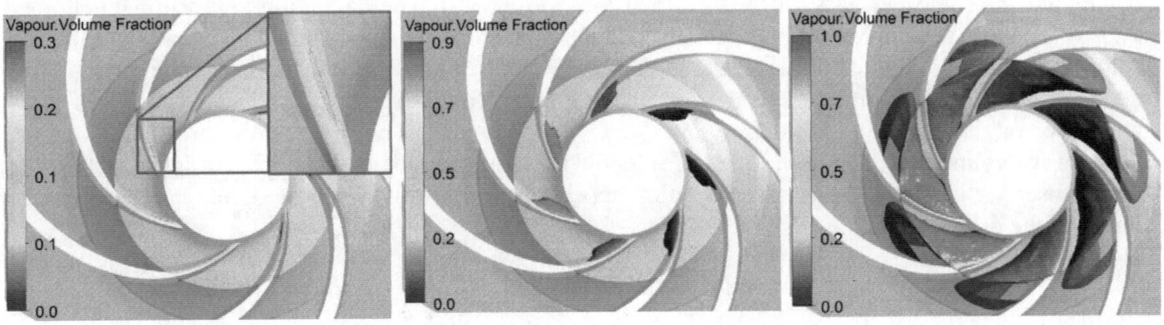

(a) (b) (c)

Figure 4. Cavitation structure in the impeller at Q_{nom}: (a) NPSP = 2 m; (b) NPSP = 1 m; (c) $NPSP_3$ = 0.31 m.

IOP Conf. Series: Materials Science and Engineering **233** (2017) 012007 doi:10.1088/1757-899X/233/1/012007

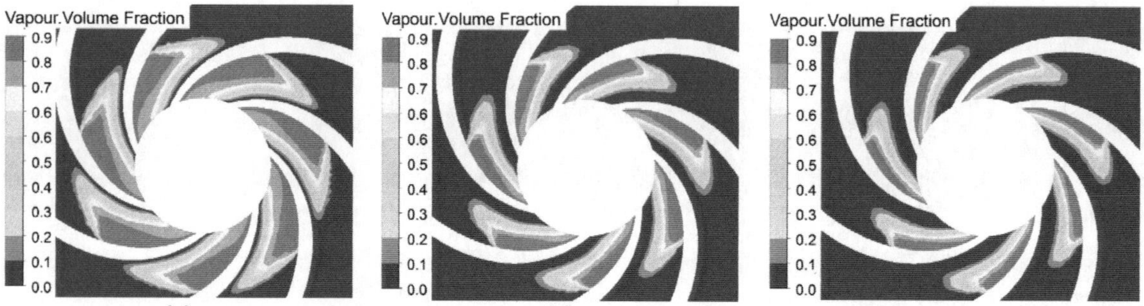

(a) (b) (c)

Figure 5. Vapour volume fraction distribution in the impeller: (a) $NPSP_3$ = 0.15 m, 0.7 Q_{nom}; (b) $NPSP_3$ = 0.31 m, Q_{nom}; (c) $NPSP_3$ = 0.45 m, 1.2 Q_{nom}.

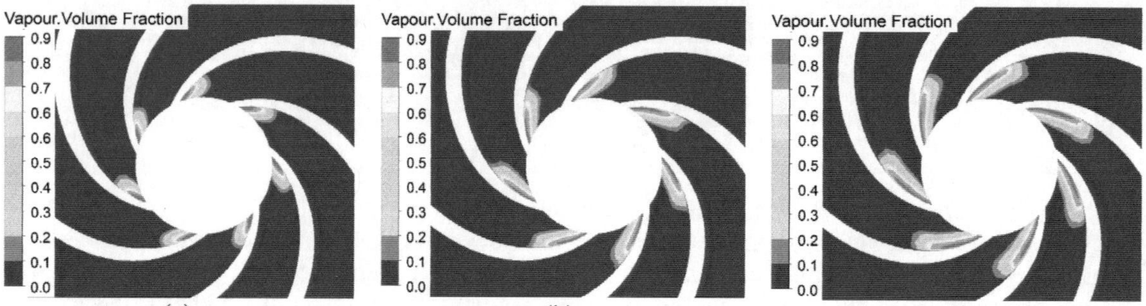

(a) (b) (c)

Figure 6. Vapour volume fraction distribution in the impeller at NPSP = 0.5 m: (a) 0.7 Q_{nom}; (b) Q_{nom}; (c) 1.2 Q_{nom}.

The cavitation occurs in the axial inlet device at the NPSP is 0.17 m, 0.35 m and 0.5 m respectively at the flow rate (0.7; 1.0; 1.2) Q_{nom} in both models. There are two cavitation zones (Figure 7). The first zone is after narrowing the cross-section of flow channel. The second zone is at the beginning of diffuser section of the axial inlet device. It should be noted that cavitation in the impeller starts earlier than in the axial inlet device. $NPSH_A$ for the impeller is higher than NPSP where there are first signs of cavitation in the axial inlet device. That is, the first destruction must be occur in the impeller.

The occurrence of the cavitation zones for different flow rate and the models of the axial inlet device are the same but have different dimensions (Figure 8). The cavitation zones in Model 1 are relatively higher than in Model 2, especially it is manifested at the flow rate 0.7 Q_{nom} and 1.2 Q_{nom}. To some extent this is due to the decrease of the angle diffuser disclosure. The difference value of the zone volume is greater under the condition of the greater flow rate. There are major cavitation zones and vapor particles in them at the flow rate 1.2 Q_{nom} in Model 1. The first zone is stretched along the walls of the cylindrical section of the axial inlet device. The second zone occupies a significant part of the diffuser section. In addition, the cavitation appears in the third zone, located near the fairing at the outlet of the diffuser section.

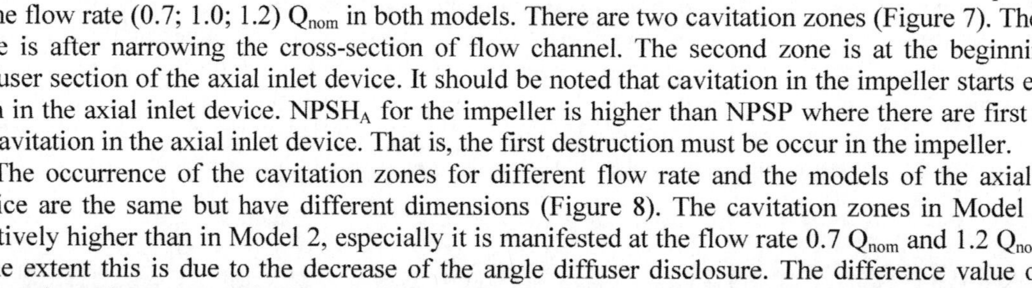

(a) (b)

Figure 7. Cavitation structure in the axial inlet device at Q_{nom}: (a) NPSP = 0.35 m; (b) $NPSP_{3\%}$ = 0.31 m;

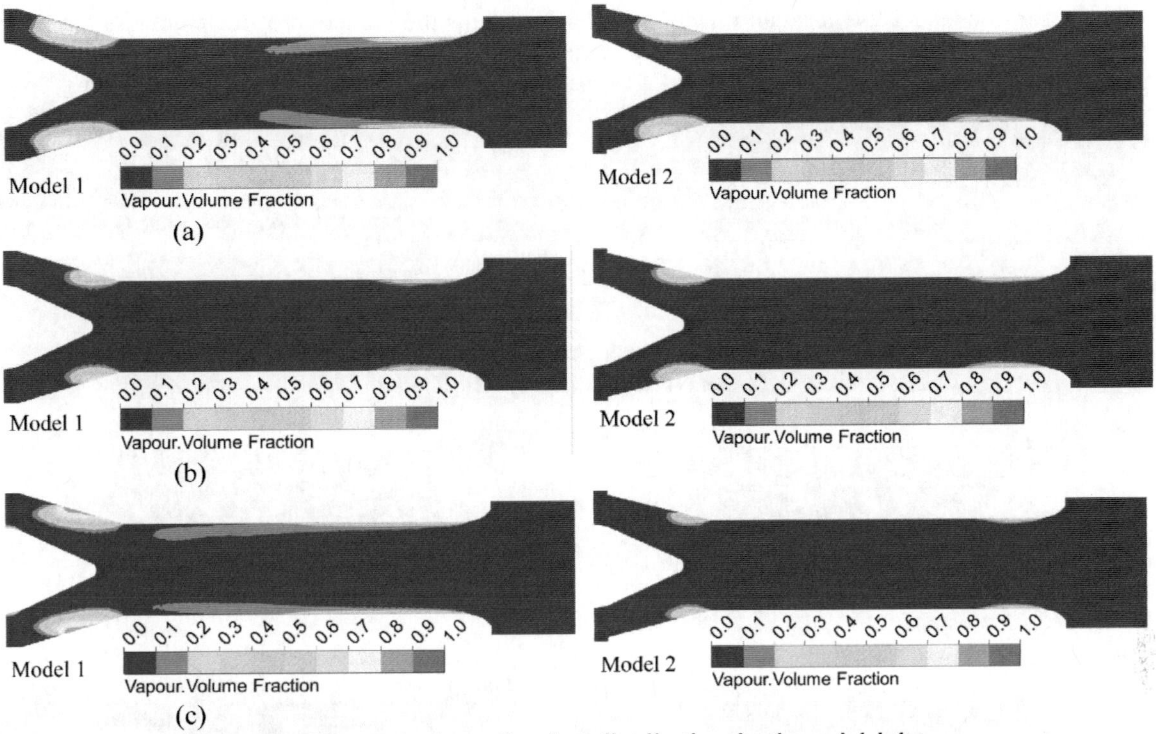

Figure 8. Vapour volume fraction distribution in the axial inlet device: (a) $NPSP_3$ = 0.15 m, 0.7 Q_{nom}; (b) $NPSP_3$ = 0.31 m, Q_{nom}; (c) $NPSP_3$ = 0.45 m, 1.2 Q_{nom}.

3.3. Pressure distribution

The absolute pressure distribution in the impeller for a different NPSP is similar (Figure 9). It should be noted that at the NPSP = 2 m the minimum absolute pressure in the impeller is much more than the saturation pressure, however, the cavitation at the suction surface of the blades already happens.

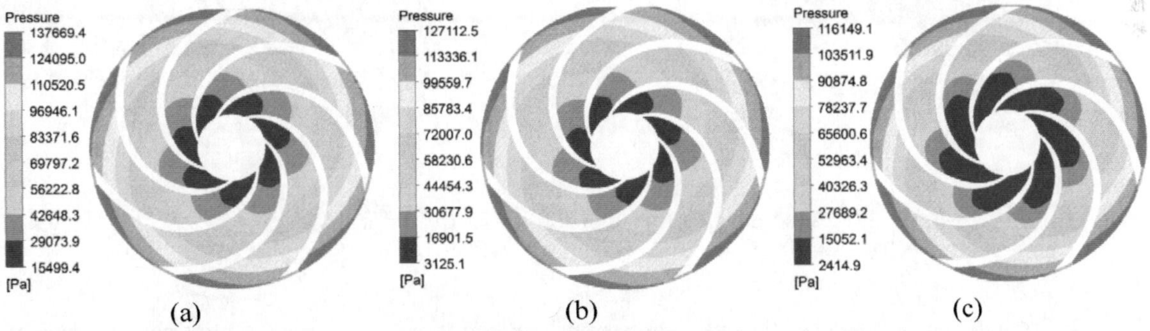

Figure 9. Absolute pressure distribution in the impeller at Q_{nom}: (a) NPSP = 2 m; (b) NPSP = 1 m; (c) $NPSP_3$ = 0.31 m.

The absolute pressure distribution in the axial inlet device demonstrates the zones of the reduced pressure (Figure 10, 11). The first zone is after narrowing cross-section, and the second zone is at the beginning of the diffuser section. The increasing of the flow velocity and the whirls occur in these zones. With the passage of fluid zones there is a significant decrease in pressure, which causes the cavitation. The absolute pressure distribution for both models is similar with the corresponding values of $NPSP_3$ for the considered flow rates (Figure 11). However, the minimum value of the absolute pressure in Model 2 is bigger than in Model 1. This explains the reduction of the cavitation zones in Model 2. In the larger flow rate the pressure difference is also greater.

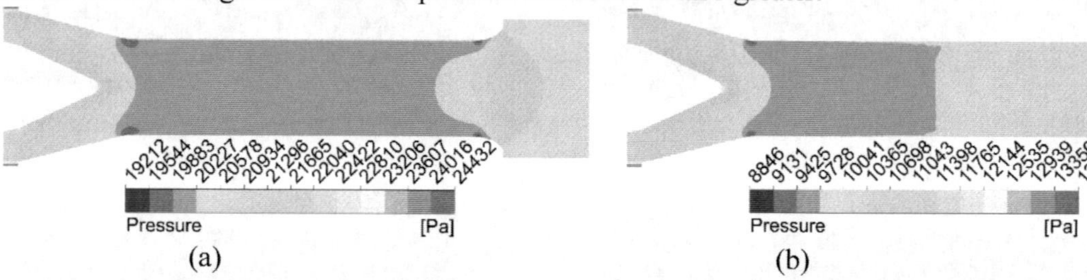

Figure 10. Absolute pressure distribution in Model 1 of the axial inlet device at Q_{nom}: (a) NPSP = 2 m; (b) NPSP = 1 m.

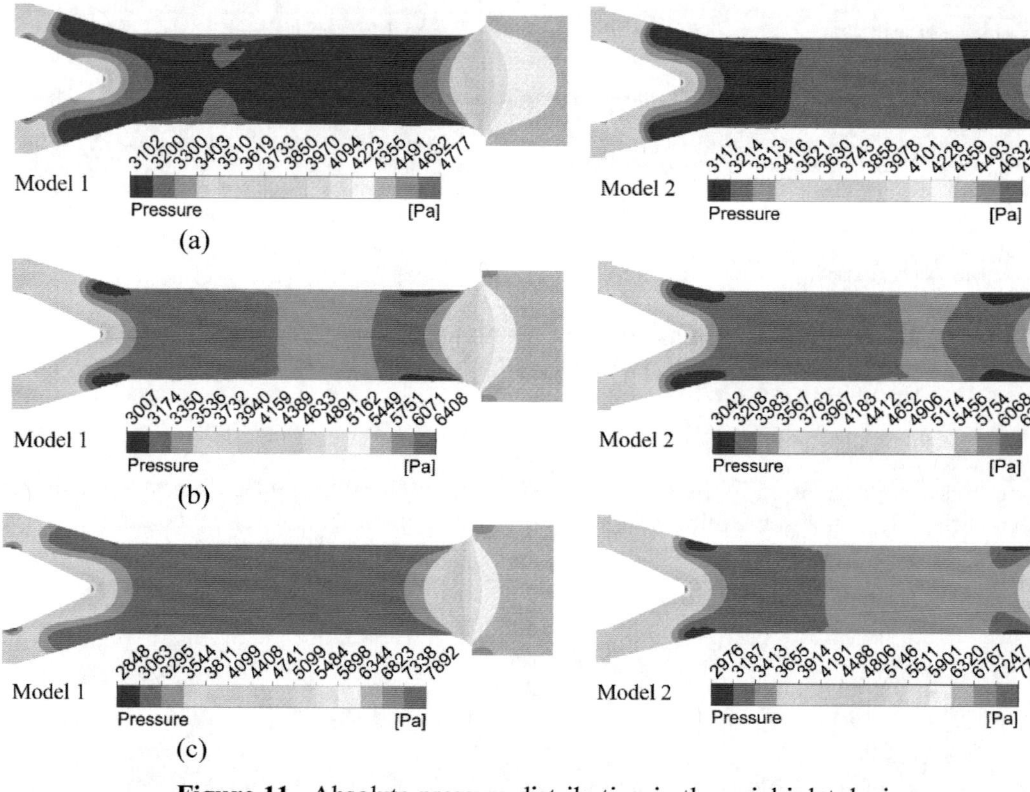

Figure 11. Absolute pressure distribution in the axial inlet device: (a) $NPSP_3$ = 0.15 m, 0.7 Q_{nom}; (b) $NPSP_3$ = 0.31 m, Q_{nom}; (c) $NPSP_3$ = 0.45 m, 1.2 Q_{nom}.

To eliminate the first zone of the absolute pressure reduction it is necessary to eliminate the change of the diameters at the inlet of the axial inlet device. That means, it is necessary to reduce the diameter of the pump inlet to the size of the cylindrical section diameter of the axial inlet device. To eliminate the second zone is necessary to eliminate abrupt change from the cylindrical to the diffuser section.

4. Conclusion

Analysis of the numerical simulation results of the operating process in the double entry centrifugal pump provided changes of the geometric parameters of the diffuser section of the axial inlet device with rotating walls at the flow rate (0.7; 1.0; 1.2) Q_{nom} enables the following conclusions:

1) Both models $NPSH_3$ is 0.15 m, 0.31 m and 0.45 m respectively for the flow rate (0.7; 1.0; 1.2) Q_{nom}. $NPSH_A$ is respectively 0.75 m, 0.91 m and 1.05 m. In this case, for exploitation of the pump without cavitation NPSH must be more than 2 m. The suction specific speed at Q_{nom} is 206 for both models.

2) When changing the flow rate the distribution of the cavitation zones in the impeller is similar, however the size of these zones is reduced with increasing the flow rate from 0.7 Q_{nom} to 1.2 Q_{nom}.

3) The cavitation in the impeller beginning earlier than in the axial inlet device. That is, the considered design of the axial inlet device will not be subjected to destruction at the ensuring operation without cavitation in the impeller.

4) Additional sudden expansion at the outlet of the axial inlet diffuser section does not affect the cavitation characteristics of the impeller, however, improves cavitation characteristics of the axial inlet device due to the decrease of the angle diffuser disclosure

5) To increase the cavitation characteristics of the axial inlet device it is necessary to reduce the pump inlet diameter to the size of the cylindrical section diameter and to eliminate abrupt change from the cylindrical to the diffuser section of the axial inlet device.

References

[1] Guo X, Zhu Z, Cui B and Li Y 2012 Analysis of cavitation performance of inducers, Centrifugal pumps (Dr. Dimitris Papantonis (Ed.), InTech)

[2] Xie S F, Wangl Y, Liu Z C, Zhu Z T, Ning C and Zhao L F 2015 Optimization of centrifugal pump cavitation performance based on CFD. *International Symposium of Cavitation and Multiphase Flow (ISCM 2014) IOP Conf. Series: Materials Science and Engineering* **72** 032023

[3] Si Q, Yuan S, Yuan J, Bois G 2016 Investigation on the influence of jetting equipment on the characteristics of centrifugal pump *Advances in Mechanical Eng.* **8** (8) pp 1–11

[4] Zhuang B, Luo X, Zhu L, Wang X and Xu H 2011 Cavitation in a shaft-less double suction centrifugal miniature pump *International J. of Fluid Machinery and Systems* **4** (1)

[5] Tan L, Zha L, Caol S L, Wang Y C and Gui S B 2015 Cavitation performance and flow characteristic in a centrifugal pump with inlet guide vanes *International Symposium of Cavitation and Multiphase Flow (ISCM 2014) IOP Conf. Series: Materials Sci. and Eng.* **72** 032028

[6] Hergt P, Nicklas A, Mollenkopf G and Brodersen S 1996 The suction performance of centrifugal pumps possibilities and limits of improvements *Proceedings of the international pump users symposium, Texas A&M University system* 13-26

[7] Gülich J F 2014 Centrifugal Pumps 3rd Edition (Springer Berlin, Heidelberg, New York)

[8] Kozubkováa M, Rautováb J, Bojkoc M 2012 Mathematical model of cavitation and modelling of fluid flow in cone *Procedia Engineering* **39** pp 9 - 18

[9] ANSYS CFX Reference Guide, Release 15.0. 2013. http://www.ansys.com

The CFD calculations as a main tool for the mixed–flow pump modernization

J Skrzypacz[1], P Szulc[1]

[1] Faculty of Mechanical and Power Engineering, Department of Design Fundamentals and Fluid–Flow Machinery, Wroclaw University of Science and Technology, ul. Wybrzeze Wyspianskiego 27, 50–370 Wroclaw, Poland

E-mail: janusz.szkrzypacz@pwr.edu.pl

Abstract. The primary aim of the project was to increase cavitation performance of a big mixed–flow pump. The CFD software was used as a main research tool. The two kinds of numerical models were used to solve this problem – both were presented. The results were verified on the basis of comparison with experimental results obtained for real and model pumps. The interesting way of estimating cavitation performance was presented. The pump characteristics before and after the modernisation were shown.

1. Introduction

Pumps are the biggest (after electrical motors) consumers of power produced globally. It is said that the range of this consumption could reach even 30%. In accordance with this estimation the efficiency of the pump should be as high as possible [3], of course, in compliance with BAT (Best Available Technology). The literature [2, 6–9] describes the design method that meets this significant requirement. However, there is another important parameter which should be taken into consideration – suction performance of the pump. This crucial aspect should be analysed, especially for a high specific speed pump.

When a designer designs or modernizes flow elements of turbo machinery there is always a dilemma if a new structure fulfills all requirements. The answer could be obtained by performing a test in the laboratory. A problem arises when it is not possible to realize the experimental investigation due to, e.g. power, flow rate or the pipeline connector dimensions. In such a situation the law of hydrodynamic affinity can be applied. It means that the model machine has to be designed (according to the affinity law), next an experimental test should be carried out and after wards the results of the test should be recalculated for real machine dimensions [2, 9]. This way is very expensive and time–consuming.

Owing to CFD a designer has new possibilities in the area of the turbo machinery design. He can analyze many variants in a „virtual laboratory" and choose the best solution. His actions are not limited by the machine dimensions or operating parameters.

The results shown in this article are part of a bigger project aiming at modernization of the large mixed–flow pump. It was decided that CFD would be the primary research method. The main goal of the project was to increase the cavitation performance with no changes in the level of efficiency.

IOP Conf. Series: Materials Science and Engineering **233** (2017) 012008 doi:10.1088/1757-899X/233/1/012008

2. Research object

The research object was a mixed–flow pump (Figure 1) applied, e.g. in a cooling tower. The pump dimensions and operating parameters are shown in Table 1. Presented parameters exceed the measurement possibilities of the testing station.

From the hydraulic point of view this pump consists of: the inlet element, the impeller and the diffuser.

Table 1. Parameters of the pump.

	Parameter	Symbol	Value	Unit
1	Flow rate	Q	23000	m^3/h
2	Head	H	23	m
3	Rotational speed	n	370	rpm
4	Power	P	1.7	MW
5	Impeller external diameter	d_2	1.412	m
6	Diameter of pump discharge connector	d_d	1.6	m

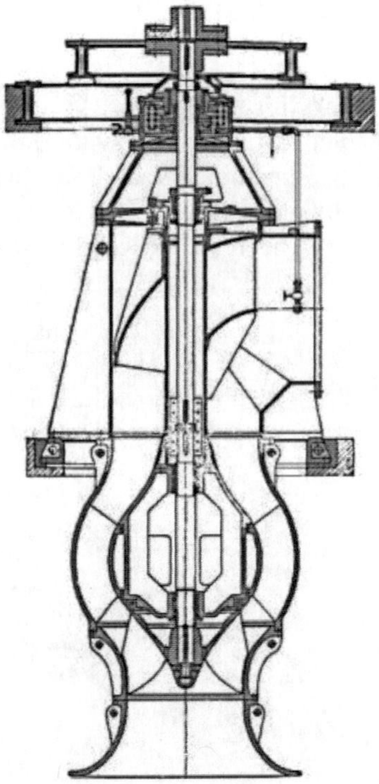

Figure 1. The analysed mixed–flow pump.

3. Numerical model

Considering complexity of the problem numerical calculations were performed in two stages:

- STAGE I – The main aim of this stage was to determine: the characteristics of the performance, efficiency of different kinds of the inlet elements and identify flow phenomena inside the hydraulic channels of the pump. At this stage the numerical model consisted of the inlet element, impeller, diffuser and the auxiliary outlet volume (see Figure. 2). All impeller and stator channels were modelled due to the asymmetric variants of the inlet element. The Fluent [4] software was applied for the calculation. The pump volumetric delivery was taken into account in calculations.

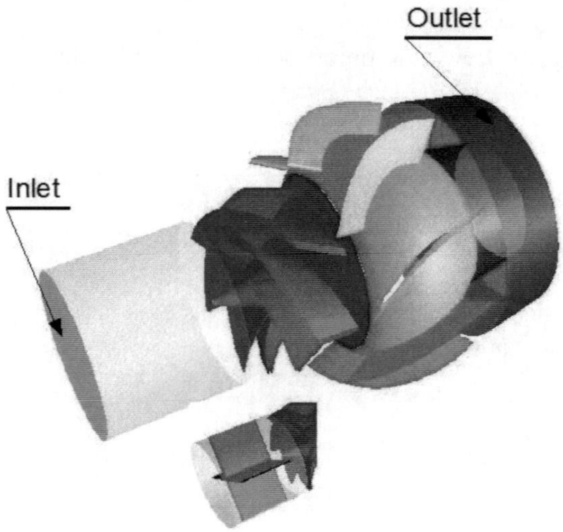

Figure 2. Numerical model. Model for stage I with different inlet elements.

- STAGE II – The main aim of this stage was to determine the influence of individual geometrical features of the impeller inlet on the pump cavitation performance. At this stage numerical model consisted of a single channel of the impeller and auxiliary the inlet and the outlet volumes (Figure 3). The pump volumetric delivery was taken into account in calculations. The Fluent [4] software was used for the calculation. The research object was the mixed–flow pump (Figure 1) applied, e.g. in a cooling tower.

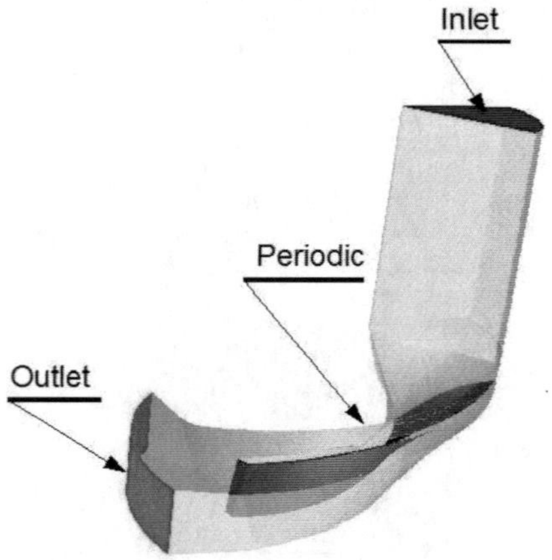

Figure 3. Numerical model. Model for stage II.

In both stages the calculations were carried out as stationary with the second order discretization scheme. The boundary conditions in both stages were defined as velocity on the inlet surface and the static pressure on the outlet surface (Figure 2). Grid in both stages was based on the hexa elements and concentrated next to the blades. The grid on the blades is regular (Figure 4). The number of cells equals, for:

- Stage I – 2340000,
- Stage II – 195000.

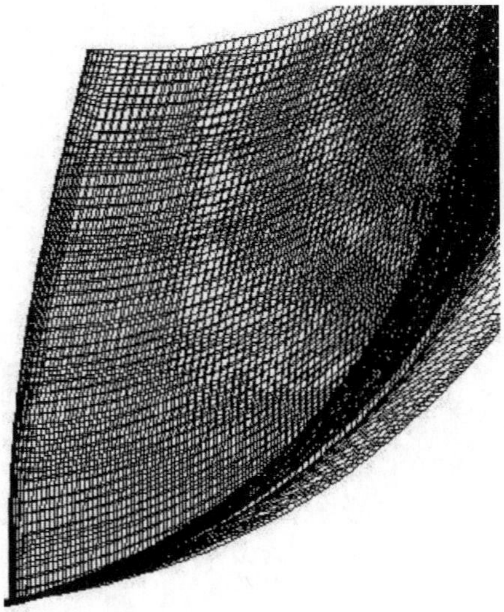

Figure 4. Grid on the blade surface.

To obtain the cavitation resistance of the pump a multiphase cavitation model was applied. The cavitation performance was calculated in accordance with the equation (1), which was verified in [1]. This is a simple and fast way to determine the cavitation performance based on a CFD calculation.

$$NPSH_i = \frac{p_{sin} - p_{s\,min}}{\rho g} + \frac{c_0^2 - c_{s\,min}^2}{2g} \tag{1}$$

where:
$NPSH_i$ – NPSH for inception,
p_{sin} – static pressure on the impeller inlet surface,
p_{smin} – minimal static pressure into the impeller channel,
c_o – velocity on the impeller inlet surface,
c_{min} – velocity at the point where the static pressure is minimal.

4. Numerical model verification

Whenever the results of numerical calculations are analysed their quality raises doubts. Conducted measurements (according to [5]) of the real and the model pump allows to compare physical results with the ones obtained from the CFD analysis. The comparison is shown in Figure 5. In the vicinity of an optimum operating point the average error of calculations is less than 4%. This divergence of the results was mainly caused by the fact that the numerical model did not include the element carrying the fluid from the stator outlet to the pipeline connector of the pump.

It is worth to mention that the difference between the real pump and the model pump characteristics is similar to the difference between the real pump characteristic and the virtual pump characteristic. It means the CFD analyses are a good alternative to experimental tests for both constructions: the real and the model pumps.

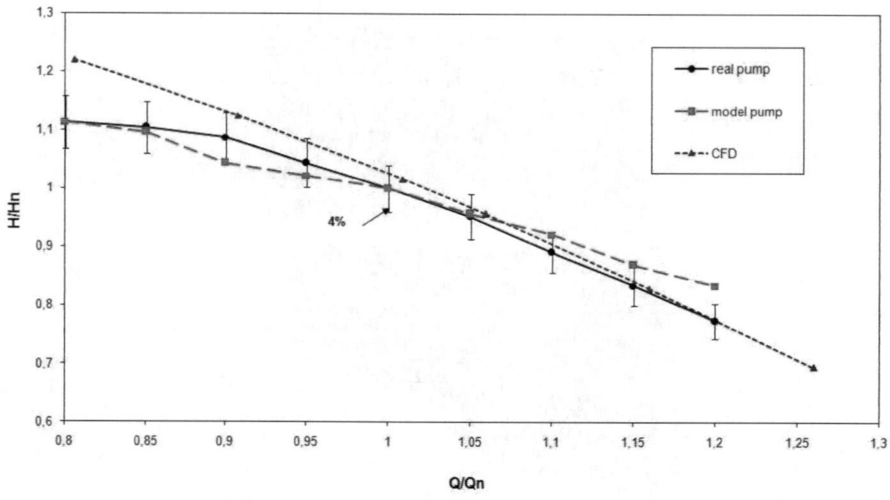

Figure 5. Comparison of dimensionless characteristics of the pump obtained in a different way.

5. Results

The examples of calculation results of the stage I are shown in Figures 6 and 7.

The analysis show, that in the best efficiency point the pump operates without any swirls, nevertheless, the backward flows and other disturbing flow phenomena occur in the other operation points. However, the main goal of the research was to optimise the impeller inlet structure to obtain a better cavitation performance.

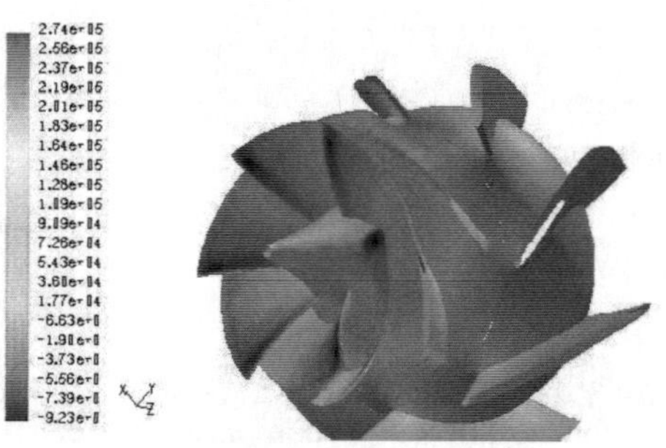

Figure 6. Examples of I–st stage calculation. Static pressure distribution.

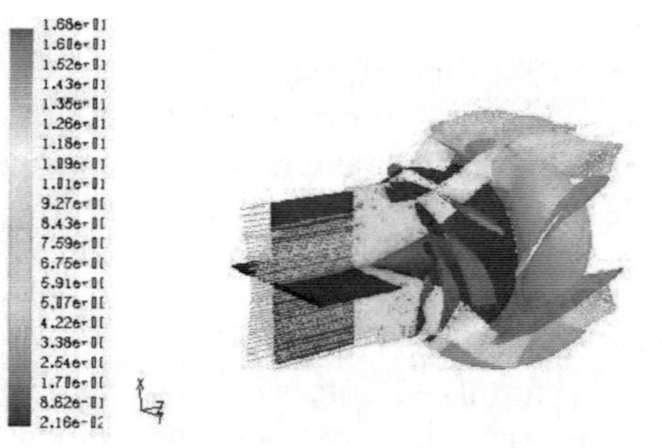

Figure 7. Examples of I–st stage calculation. Velocity distribution.

In course of the project the following geometrical features of the impeller inlet were tested:

- The location of the leading edge in a meridional plane.
- The inlet attack angle (inlet blade angle).
- The numbers of blade.
- The splitters.
- Mix of the above.

Figure 8 shows the comparison of the performance curves calculated for the pump with the impeller before the modernisation and for the best version of the impeller after the modernisation. The model of the stage I was used in both characteristic calculations. It must be emphasised that the NPSH values shown in figure 8 should not be considered as the crucial ones. These values correspond to the initial stage of the cavitation (formula (1)), therefore, the crucial values are considerably lower. Figure

8 shows that the applied modifications have not resulted in the change of optimum operating point parameters (BEP) (the pump head increased slightly), however, the dramatic increase of the cavitation performance of the pump was observed.

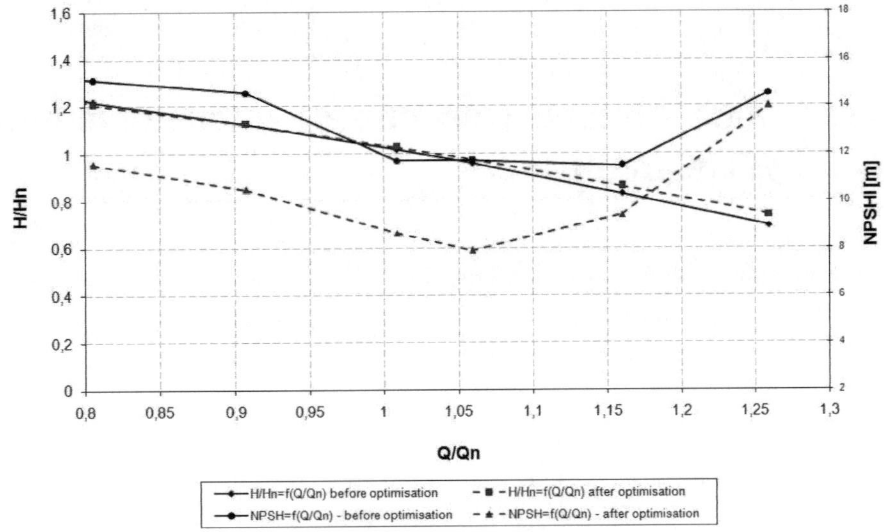

Figure 8. The flow and the cavitation characteristics before and after the impeller modernization (for the best impeller solution).

6. Conclusions

The presented research proved that the CFD software is a very useful tool for industrial applications. Due to numerical calculations realized for many variants the new design of the mixed–flow pump impeller of a much better cavitation performance and a little better performance curves was developed. The new impeller made and it has been operating well in one of the European power plants. Thanks to the calculations some contradictory ideas presented in professional literature were verified, for instance, the ideas of influence of the inlet attack angle on the mixed flow pump cavitation performance.

References

[1] Misiewicz A and Skrzypacz J 2006 Obraz zachowań kawitacyjnych wirników pomp wolnobieżnych projektowanych wg zasady tzw *"ciasnego wlotu" International conference Hydroforum 2005* pp 291–299

[2] Lazarkiewicz S and Troskolanski A 1965 *Impeller pumps* (Warsaw: Wydawnictwo Naukowo–Techniczne)

[3] Shankar V, Umashankar S, Paramasivam S and Hanigovszki N 2016 A comprehensive review on energy efficiency enhancement initiatives in centrifugal pumping system *Applied Energy* **181** pp 495–513

[4] Ansys Fluent documentation

[5] The European Standard EN ISO 9906:2012 2012 *Rotodynamic pumps – Hydraulic performance acceptance tests – Grades 1, 2 and 3*

[6] Varchola M and Hlbocan P 2014 Interaction between impeller and axial vane diffuser of mixed flow pump *Applied Mechanics and Materials* **360**

[7] Jin–Hyuk K and Kwang–Yong K 2011 Optimization of Vane Diffuser in a Mixed–Flow Pump for High Efficiency Design, *International Journal of Fluid Machinery and Systems*

[8] Urban G and Varchola M 2006 Hydraulic relation between the impeller and the diffuser of the mixed–flow pump *Mechanical Engineering (Bratislava)*

[9] Gulich J F 2008 *Centrifugal pumps* (Berlin Heidelberg: Springer–Verlag)

IOP Conf. Series: Materials Science and Engineering **233** (2017) 012009 doi:10.1088/1757-899X/233/1/012009

Centrifugal pump's impeller optimization using methods of calculation hydrodynamics

S Grigoriev[1], S Mayorov[2], R Polyakov[2]

[1] LLC «STC Sirius», Moscovskiy prospect, 11-203, Voroneg, Russia, 394026
[2] Department of Mechatronics and International Engineering, Orel State University, Naugorskoe sh. 29, Orel, Russia, 302020

E-mail: romanpolak@mail.ru

Abstract. The paper features the results of the fluid flow calculation in the channels of varying geometry of the centrifugal pump for the service water in the methanol production chain. Modeling of the flow in ANSYS CFX allowed developing recommendations on adjusting the impeller's profile, significantly decrease the cavitation wear and increase the lifetime by several times.

1. Introduction

The rotational machinery (pumps, compressors, expanders, ventilators, etc.) is the most widespread type of the technical systems, on which the reliability, lifetime expectancy, various technical and economical characteristics, and energy efficiency of the systems in question depend. Given that presently energy, oil, chemical and metallurgical branches of industry are the basic components of the economy of the Russian Federation, the solution of the problem of energy efficient pumps development with active control and high operational characteristics is a quite urgent problem. In order to solve it, either new technologies and techniques have to appear or the existing have to be modernized, which in its turn requires development of the new generation pumps. The present political situation leads to a necessity of replacing the imported pumping equipment used in oil and gas, chemical, metallurgical, pharmaceutical industries. Untimely shortage of imported equipment and its parts can lead to a rapid increase in the number of emergencies and to a decrease in the industrial power. Impossibility of buying new parts from abroad leads to the necessity of import substitution in order to timely repair and replace the parts, which can no longer be bought from the foreign suppliers. Using the example of a pump by 'Egger' (Switzerland) which is installed at the LLC 'Schekinazot' as a part of the drainage cycle of methanol production, a comparative analysis of the technical and economic prospects of import substitution will be given. The price of an imported pump is 5 million RUB; average lifetime is 3 months due to the cavitation wear of the impeller (Figure 1). The price of the parts, i.e. impeller itself, fluid-film bearing and the sealing, is around 3 million RUB. Downtime is unacceptable because its cost in terms of methanol production per day is 6.0 million RUB. It is suggested to optimize the impeller's profile in order to increase the cavitation margin and accordingly the lifetime expectancy up to 10-12 months.

(a) (b)

Figure 1. General view of the impeller of the Egger's pump before (a) and after 3 months of operation (b).

2. Flow modelling of the original impeller

For the studied model of the centrifugal impeller a 3D model of the flow part was developed (Figure 2).

Figure 2. Model of the impeller's flow part.

The fluid flow modeling in the impeller was implemented using the numerical solution of the Navier-Stokes differential equations [1]. The method of finite elements was used, which is realized in the CFD software ANSYS CFX. The first step was to analyze flow of the fluid with the following parameters through the impeller:

– fluid density 997 kg/m^3;
– dynamic viscosity 0,0008899 Pa·s;
– angular velocity 1450 rev/min.

The following boundary conditions were used during the calculations:

– input relative overpressure (Figure 3(a)) 1 kPa;
– output mass flow (Figure 3 (b)) 1000 kg/s.

The standard k-ε turbulence model was used in the calculations. As a result of the calculation, the pressure in the impeller was 52.38 m. To evaluate the possibility of cavitation excitation the pressure distribution was obtained (Figure 4), the analysis of which showed, that in the area of the blades the cavitation is indeed expected, since these are areas of under pressure.

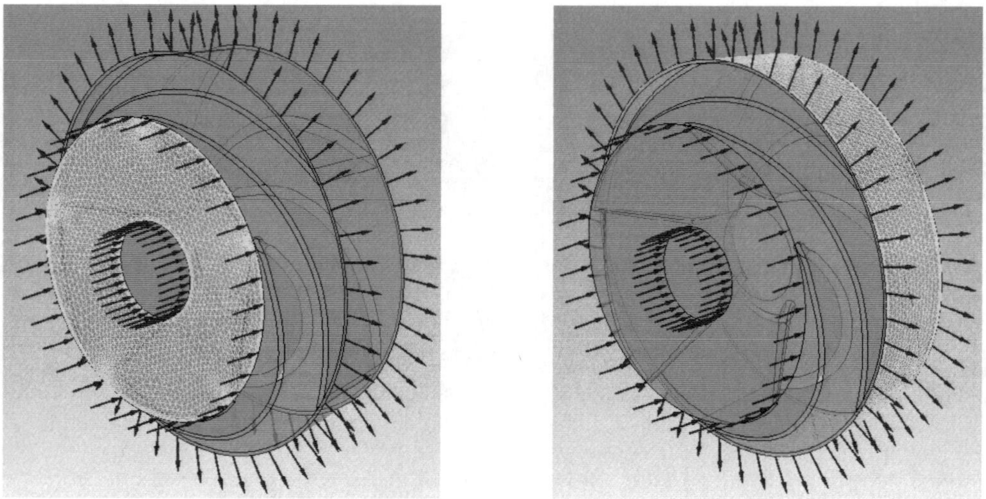

(a) (b)

Figure 3. Boundary condition: (a) input boundary condition; (b) output boundary condition.

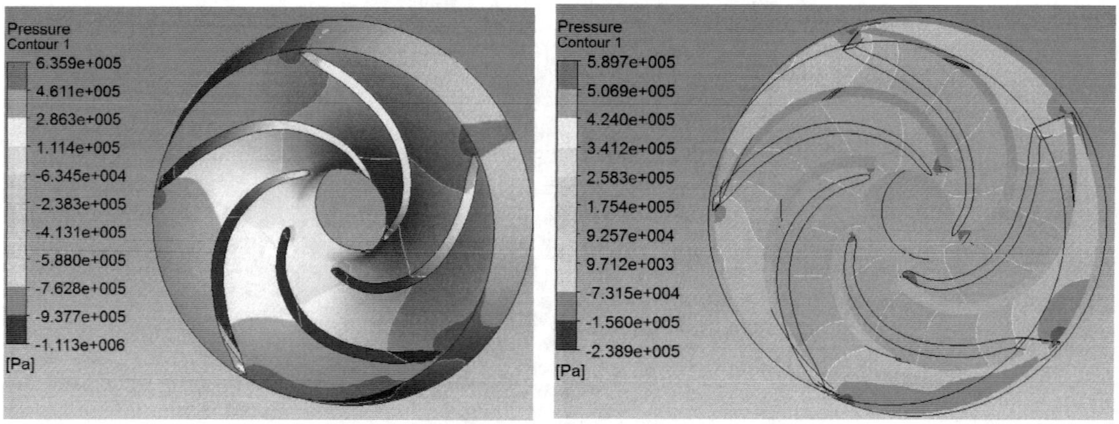

(a) (b)

Figure 4. Pressure distribution on the impeller: (a) impeller's model; (b) cross-sectional area.

In order to study the possibility of cavitation excitation a two-phase fluid flow was analyzed. With the saturated vapor pressure of 3.2 kPa and relative overpressure of 0.1 MPa, average pressure on the impeller is 51.18 m, which indicates the absence of cavitation. This assumption is based on the distribution of volumetric gas content (Figure 5 (a)). However, with the decrease of the pressure to 0.08 MPa the average pressure in the impeller would be 30.11 m, which indicates a cavitational disruption. This assumption is also proven by the distribution of volumetric gas content (Figure 5 (b)). On the basis of the calculation experiments the conclusion can be made that the original impeller of the pump provides the acceptable operation characteristics only in case of the cavitational margin of 10 atm., but with smaller values the impeller is exposed to a significant wear of the impeller's blades.

3. Flow modelling in the modified impeller

The modified profile of the blades of the impeller (Figure 6) is different from the original in oncoming angles, number of blades and the closed design of the impeller. The parameters and the procedures of

IOP Conf. Series: Materials Science and Engineering **233** (2017) 012009 doi:10.1088/1757-899X/233/1/012009

the calculation experiments are similar to the calculations for the original impeller. According to the results, the average pressure in the modified impeller is 56.44 meters, which is comparable to the original impeller.

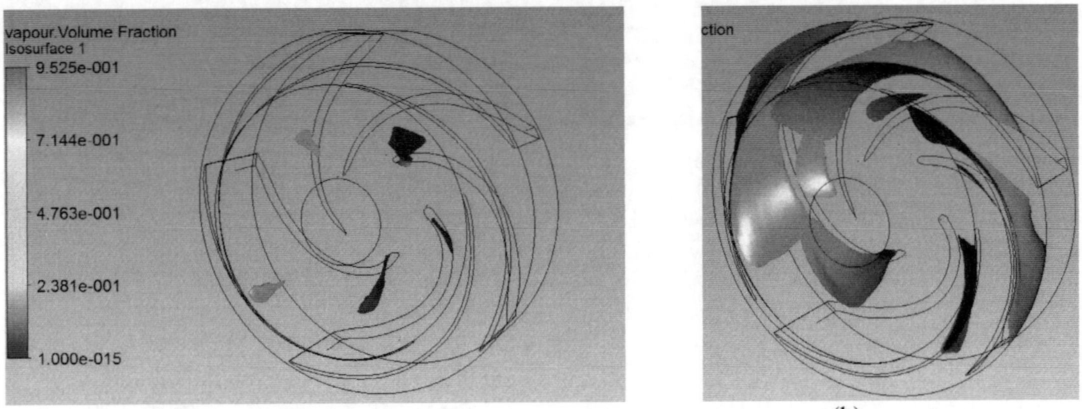

(a) (b)

Figure 5. Isosurfaces of the volumetric gas content.

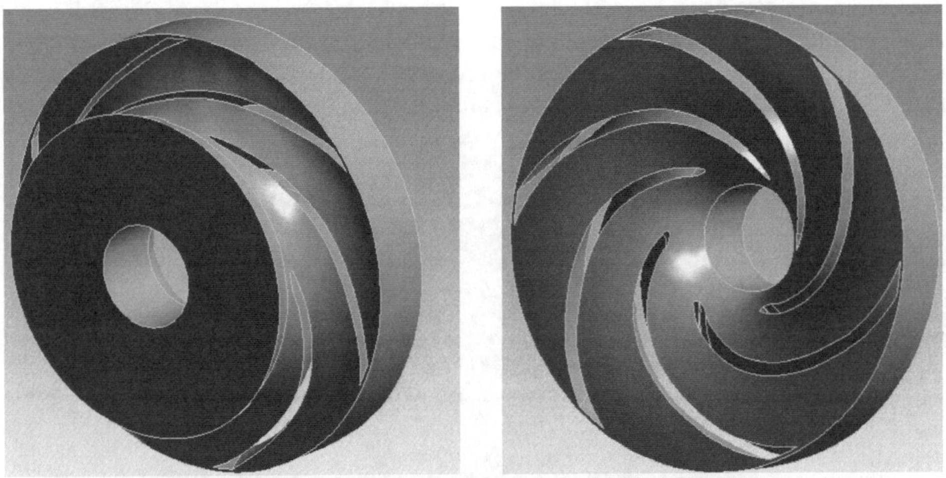

Figure 6. Geometry of the flow area of the modified impeller.

In the Figure 7 the pressure distribution is shown, and it is clear that in a small area on the beginning of the blades there are areas with high disruption and possible cavitation.

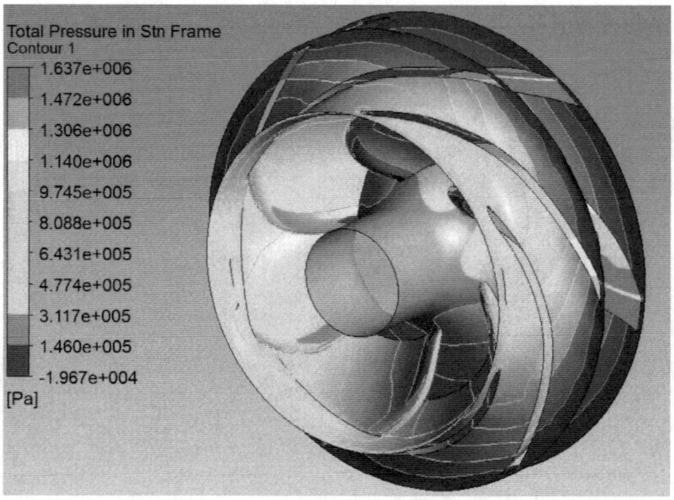

Figure 7. Pressure distribution on the modified impeller.

When analyzing the two-phase fluid flow in the modified impeller it was obtained, that the average absolute pressure on the input is 2.47831 atm., and the average pressure in the impeller is 56.42 m. This indicates the absence of the significant cavitational occurrences, as the drop of the pressure is less than 3% (3% is a widely used value to determine the cavitation tear). This assumption is proved with the distributions of pressure and volumetric gas content (Figure 8).

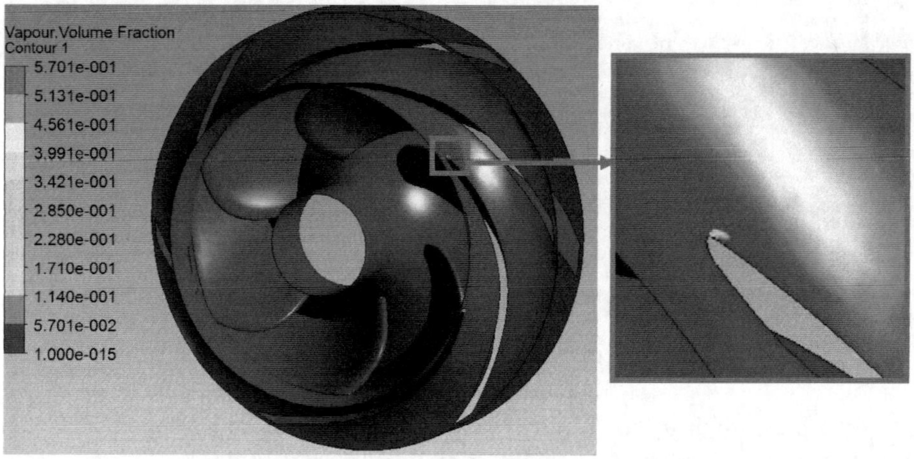

Figure 8. Volumetric gas content distribution.

When the output absolute pressure on the exit drops to 4.8 atm. The absolute pressure on the inlet is 73742 atm., and the average pressure is 55.80 m, which indicates the absence of significant cavitation, as the pressure drop is also less than 3%. This assumption is based on the pressure and volumetric gas content distributions (Figures 9 and 10).

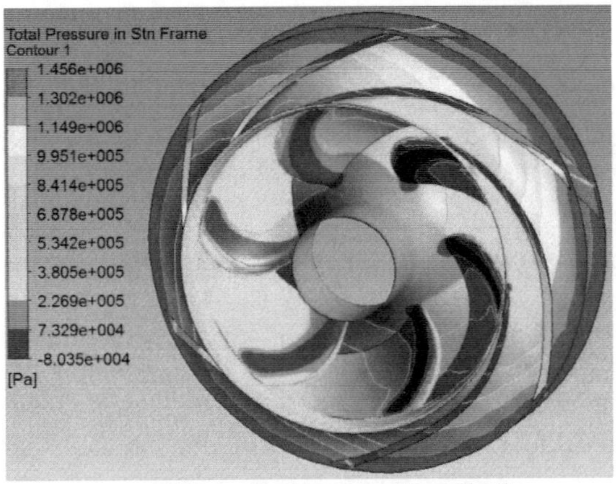

Figure 9. Pressure distribution.

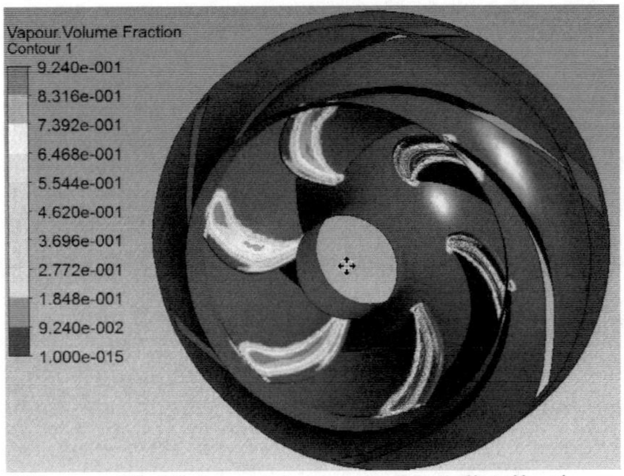

Figure 10. Volumetric gas content distribution.

4. Conclusions

So, it can be concluded, that the modified impeller provides better operational characteristics even with the cavitational margin around -0.26 atm., i.e. with the possibility of suction from the depth 2.6 m. Generally the modified impeller outraces the original design from the point of view of the operational parameters and cavitational margin.

References

[1] S Pope 2000 *Turbulent Flows* (Cambridge University Press)
[2] J Tu, G Teoh, C Liu 2012 Computational Fluid Dynamics, Second Edition: A Practical Approach (Butterworth-Heinemann, Elseiver Ltd)

IOP Conf. Series: Materials Science and Engineering **233** (2017) 012010 doi:10.1088/1757-899X/233/1/012010

Simulation of Vortex Flow in Centrifugal Pump by Method of Hydrodynamic Features

Davydenko A K[1], Kostornoi S D[2], Mieshkova N D[1] and Khatuntsev A Yu[1]

[1] JSC "VNIIAEN", 2, 2nd Zheleznodorozhnaya str., Sumy, 40003, Ukraine
[2] Sumy State University, 2, Rimskogo-Korsakova str., Sumy, 40000, Ukraine

E-mail: admin@vniiaen.sumy.ua

Abstract. A numerical experiment performed by ECM based on different mathematic models and algorithms permits to investigate and evaluate internal and external dynamics of centrifugal pumps being operated in regimes of a flow without separation and separation one around bodies with excessive degree of reliability. This paper researches simulation of the flow within a centrifugal pump impeller by the method of hydrodynamic features. Streamlines and vortex lines were developed. Their relative positions were studied. Friction losses and losses induced by vortex dissipation have been evaluated in dependence to a flow rate.

1. Introduction

When simulating of a vortex flow within different channels or confined spaces, vortex distribution is usually considered to be final and distributed one at the internal boundary on which there is a tangential discontinuity of a velocity vector. A magnitude of a tangential velocity jump is defined according to the integral equation which provides the condition of a flow lack through the body surface.

The method of discrete vortices developed to simulate an ideal liquid [1] is the basic vortex method. This method used successfully for numerical calculations of an unsteady flow around bodies with a tail of a vortex wake at sharp edges [2]. The method of discrete vortices in conjunction with a boundary layer model was used to simulate the separation flow at a smooth surface [3, 4]. Schemes that simulate generation of vorticity at the surface of streamlined bodies are shown in [5, 7].

2. The research purpose

Friction losses and losses induced by dissipation of a vortex in dependence to the flow rate are to be evaluated according to the results of 3D simulation of the vortex flow of an ideal incompressible liquid.

For research purposes the centrifugal pump impeller for parameters of Q=130 m^3/h, H=80 m, n=2979 rpm was simulated by inversion proprietary software.

The design technique that was used to simulate an impeller blade according to defined flow pattern is set out in studies [7, 8].

Hereinafter, Figure 1 gives a software-generated line drawing of the impeller being investigated.

IOP Conf. Series: Materials Science and Engineering **233** (2017) 012010 doi:10.1088/1757-899X/233/1/012010

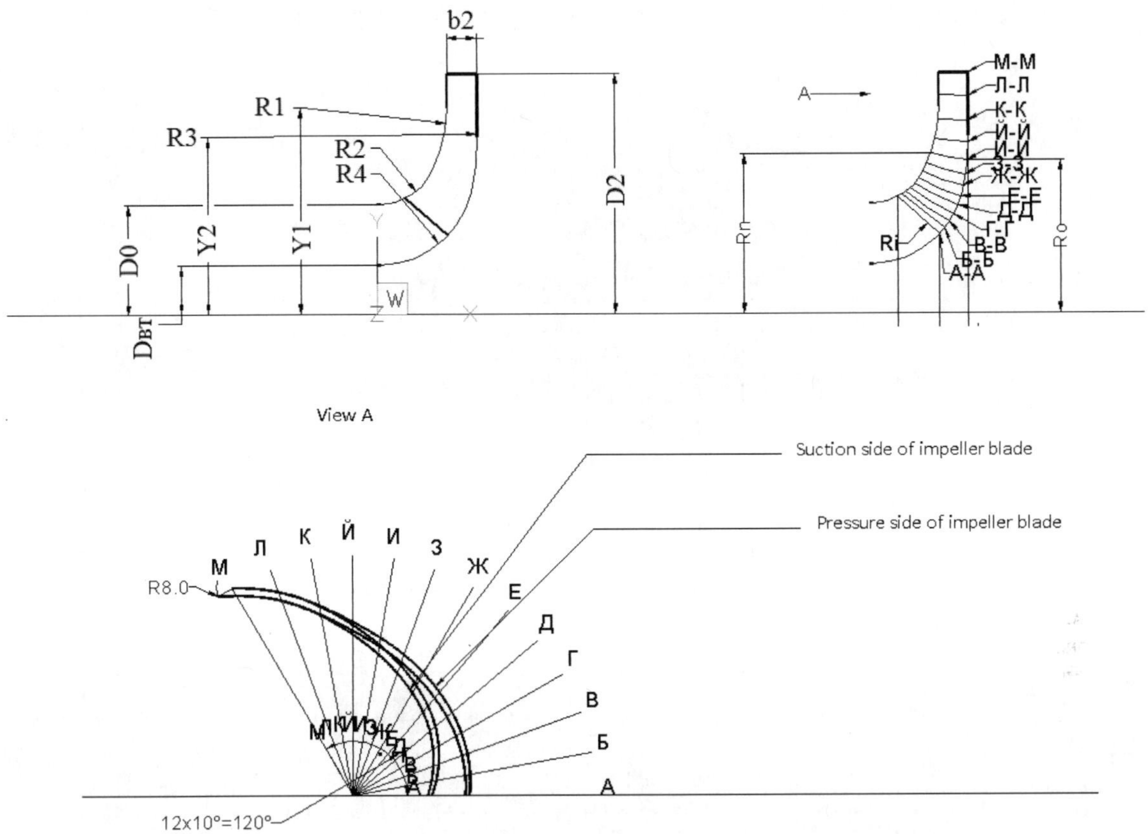

Figure 1. Software-generated line drawing of impeller being investigated.

3. Direct problem

A vortex model of the ideal incompressible liquid based on studies [10-12] has been used for simulation of 3D-flow.

A vortex frame and point dipoles were the main hydrodynamic features being in use, their descriptions are given in [13].

The flow within centrifugal pump impeller with a vaneless diffuser was under consideration.

Boundary conditions.

The uniform velocity distribution corresponding to the required flow rate was specified at the impeller inlet.

Since a backflow was supposed a condition of an open flow (opening) was specified at the impeller outlet.

The impermeability boundary condition was specified at solid surface rotating with specified revolutions. The no-slip boundary condition was not specified.

Figures 2, 3 give reference streamlines in the longitudinal section throughout the height of b2/2. It is properly observed the effect of “no-slipping” of streamlines to the suction side of impeller blade and “repulsing” from the pressure side within interblade channels, that is also simulated by CFD.

Figures 4, 5 give distribution diagrams of absolute and relative velocities within the interblade channels. It is observed significant areas of the backflow that is not realized as a part of the potential flow only.

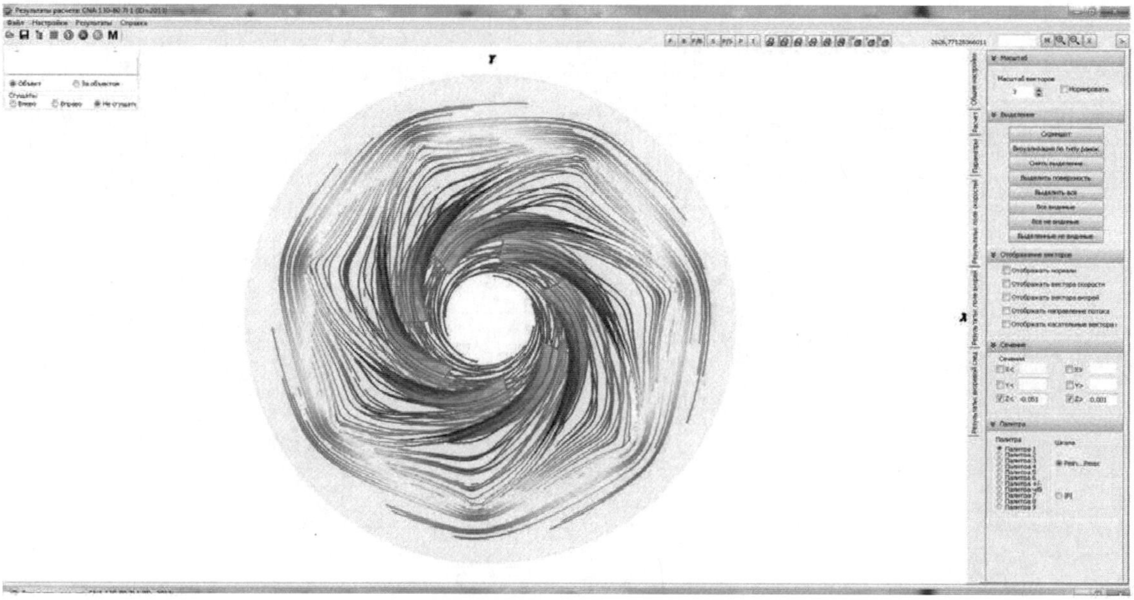

Figure 2. Reference streamlines at rated flow regime.

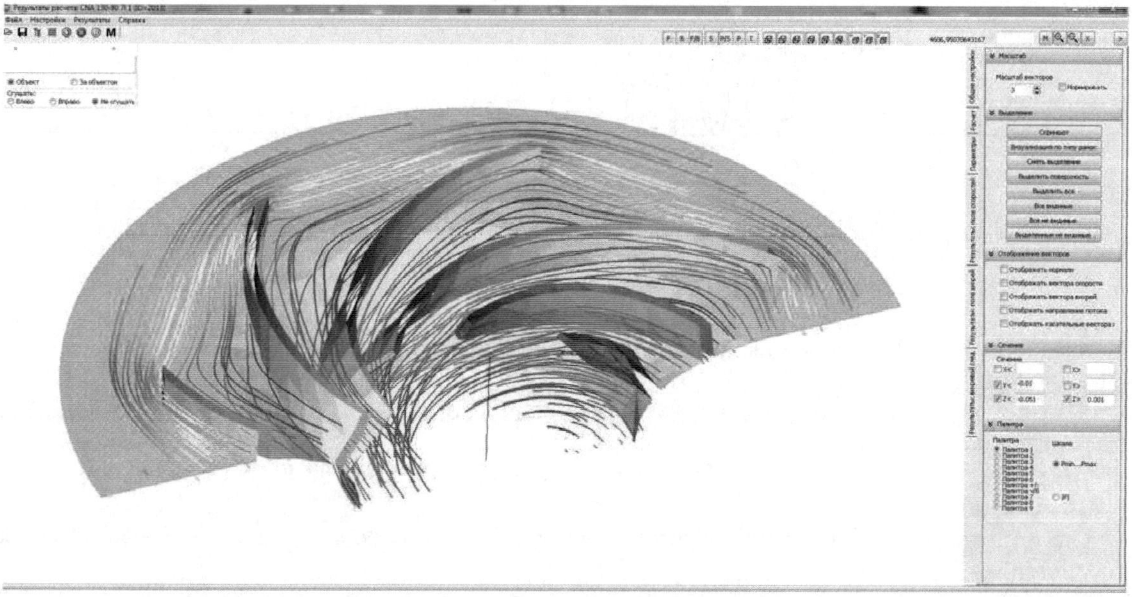

Figure 3. Reference streamlines at rated flow regime (a fragment).

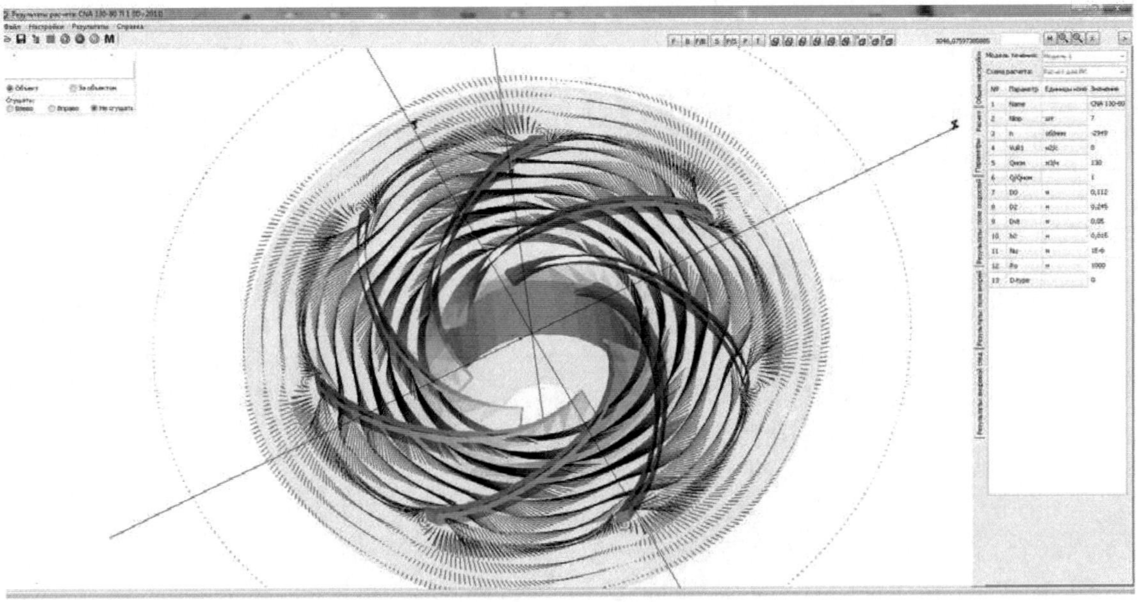

Figure 4. Distribution diagram of absolute velocity vectors within interblade channels.

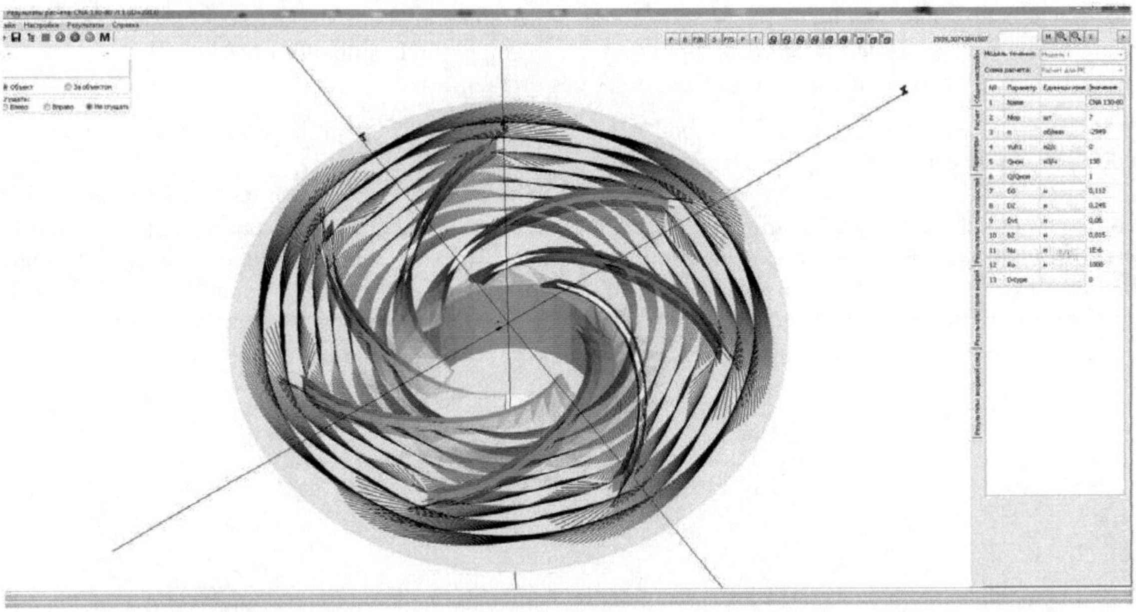

Figure 5. Distribution diagram of relative velocity vectors within interblade channels.

Hereinafter, Figure 6 gives some separately depicted streamlines which run from the impeller inlet to outlet to be viewed more conveniently without impeller geometry. At every computational point of streamlines a vortex vector at this point was calculated and shown by a segment being oriented in direction of vortex vector. The segment length is proportional to a modulus of the vortex.

Surges of vortices at the impeller outlet correspond to positions of impeller blade trailing edges (at the design moment of time), where strong trailing vortices form. It is also observed that the particle rotation direction reverses the direction of motion.

IOP Conf. Series: Materials Science and Engineering **233** (2017) 012010 doi:10.1088/1757-899X/233/1/012010

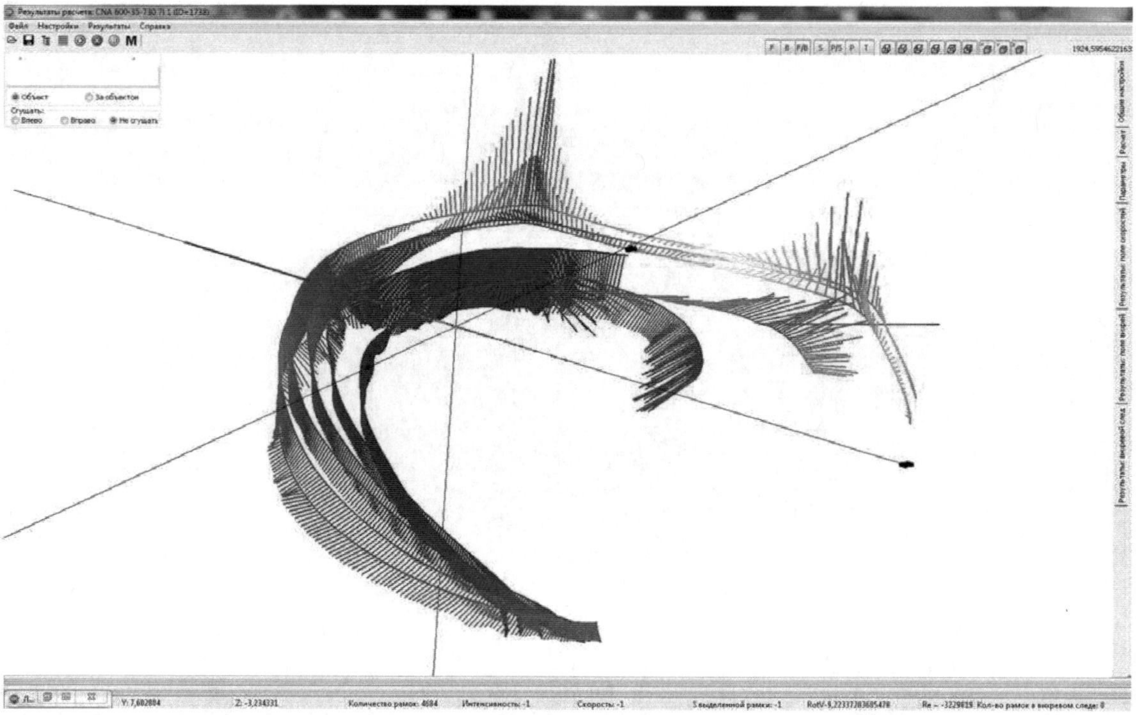

Figure 6. Streamlines directed from inlet section to impeller outlet with relative positions of vortex lines on them.

Figure 7 gives the graph of velocity modulus variance along the streamlines, Figure 8 - the graph of vortex modulus variance along the streamlines.

Figure 9 shows the graph of variance of angle cosine between the streamlines and corresponding vortex lines along the streamlines. Energy transfer in the centrifugal pump impeller occurs proportionally to the value of the dot product V*RotV. Therefore at the sections, where an angle between the direction of the vortex vector (streamline) and relative vortex vector is the smaller, the smaller amount of energy could be taken by liquid from the impeller. A swirling flow (a vortex extends in velocity vector direction) doesn't give and take energy from outside. The swirling flow only transfers energy.

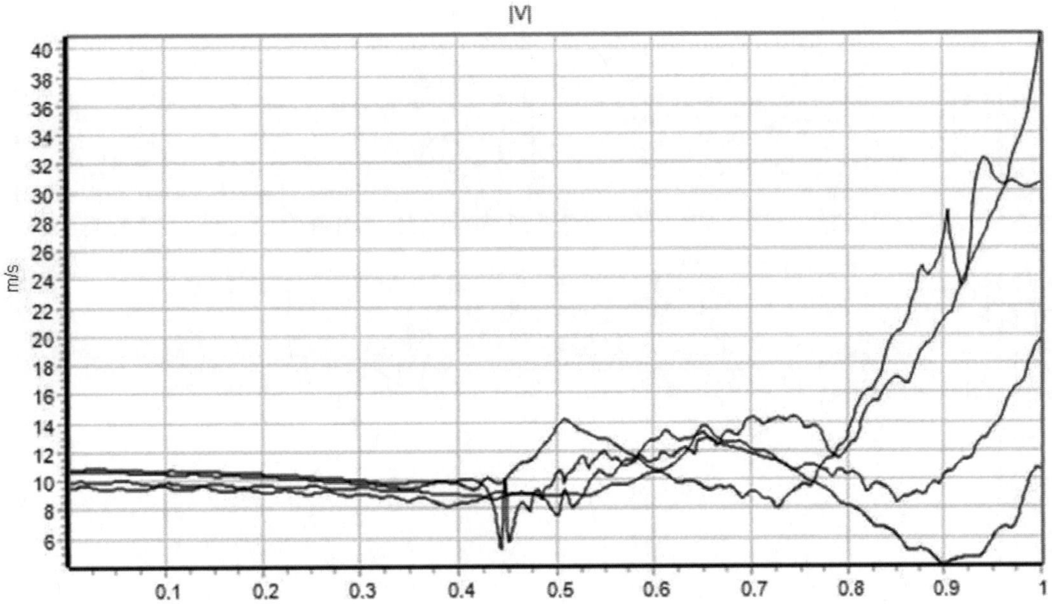

Figure 7. Graph of velocity modulus variance along streamlines.

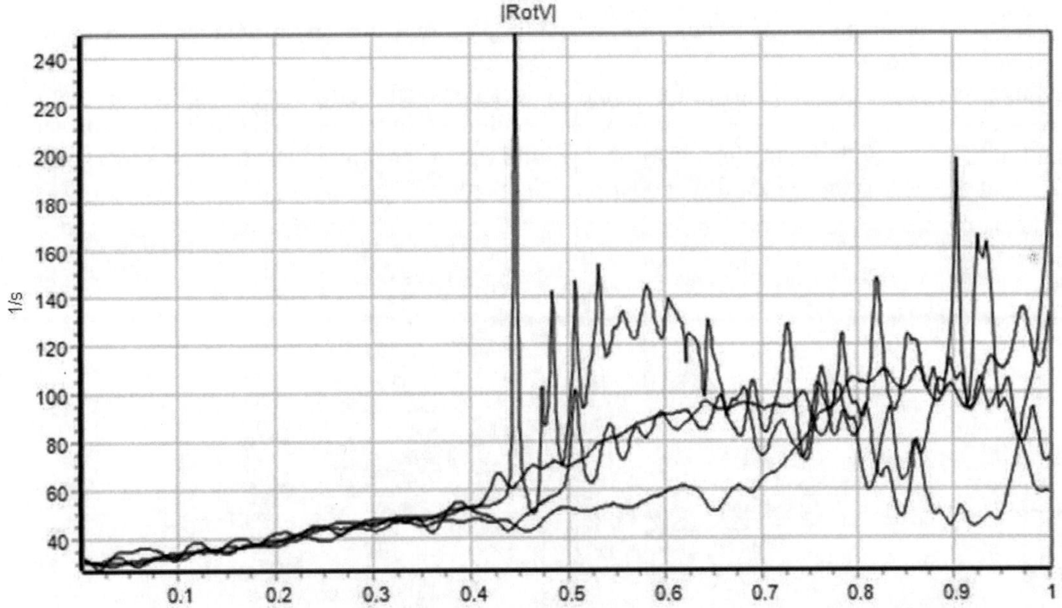

Figure 8. Graph of vortex modulus variance along streamlines.

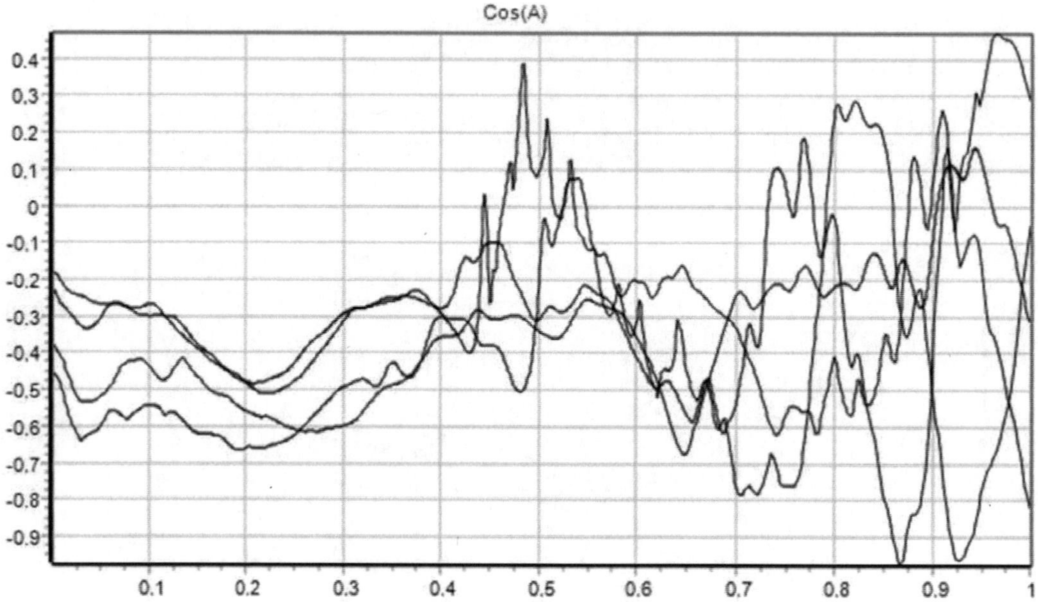

Figure 9.- Graph of variance of angle cosine between streamlines and corresponding vortex lines along streamlines.

Figure 10 shows a pattern of vortex lines in the longitudinal section throughout the height of b2/2 at rated flow regime.

Awareness of a vortex field within a domain under consideration gives the following application results – evaluation of losses due to friction on a solid surface (including a smooth surface) and evaluation of losses with dissipation of vortices within the internal region of the flow emerging due to the fact that the flow is the vortex flow rather than the potential one.

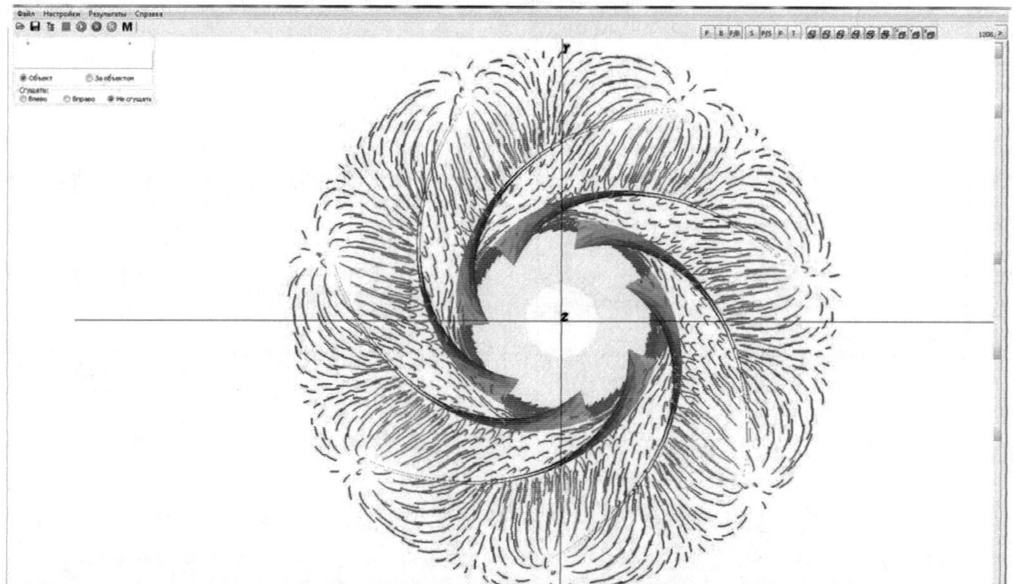

Figure 10. Reference vortex lines in longitudinal section at rated flow regime.

IOP Conf. Series: Materials Science and Engineering **233** (2017) 012010 doi:10.1088/1757-899X/233/1/012010

4. Friction stress analysis

Studies [14, 15] show that force F acting on a streamlined surface without other liquid-acted forces is related with hydrodynamic momentum I by the formula

$$F = -\frac{dI}{dt} - \rho\frac{d}{dt}\oint r\times(V\times n)dl \tag{1}$$

where hydrodynamic momentum

$$I = \rho\int_{S} \rho\times\Omega ds \tag{2}$$

If we differentiate (1) when body moves uniform under the no-slip boundary conditions, we should deduce

$$F = -\rho\int_{l} r\times Jdl - \rho\int_{\Gamma} V_d\times dG \tag{3}$$

The first item represents a resultant of a pressure force, the addend represents a resultant of friction forces. Friction stress τ_w is expressed by the formula:

The first item represents a resultant of a pressure force, the addend represents a resultant of friction forces. Friction stress τ_w is expressed by the formula:

$$\tau_w(l) = -\rho\int_{\Gamma} I_w r((l), r(G))\times dG \tag{4}$$

Figure 11 shows a change of friction force total energy (resistance forces) per a time unit, that acts on an impeller depending on the flow rate.

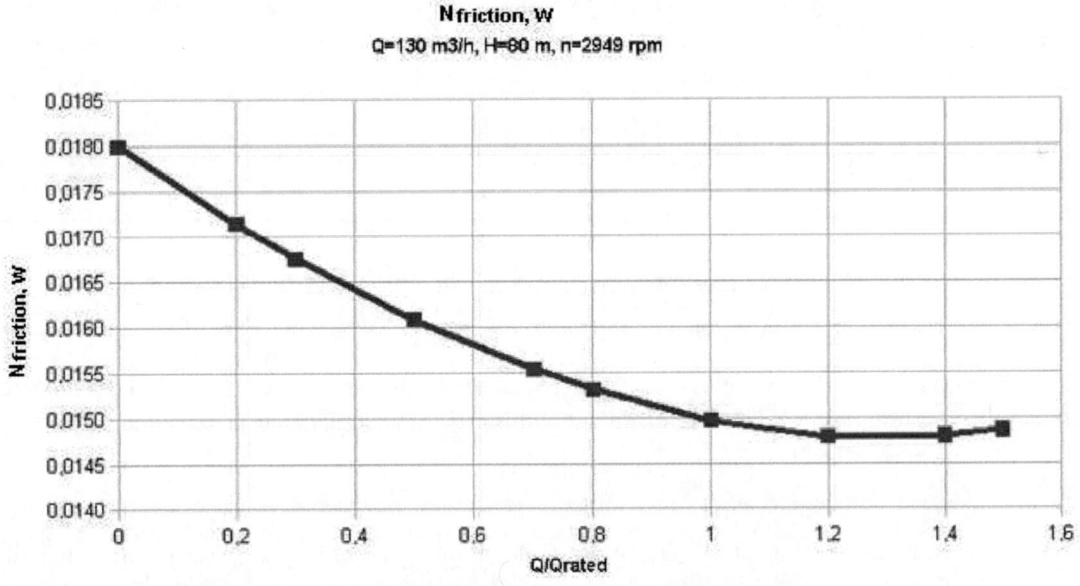

Figure 11. Change of friction force total energy per a time unit, that acts on an impeller in dependence to the flow rate.

5. Energy Dissipation

The work of all forces acting on mass the liquid in finite volume being defined by dot product of the stress tensor and strain velocity tensor for incompressible liquid is determined by the equation [16]

IOP Conf. Series: Materials Science and Engineering **233** (2017) 012010 doi:10.1088/1757-899X/233/1/012010

$$N_{\partial uc}=\mu\begin{bmatrix} 2\left(\frac{\partial V_x}{\partial x}\right)^2+2\left(\frac{\partial V_y}{\partial y}\right)^2+2\left(\frac{\partial V_z}{\partial z}\right)^2+ \\ \left(\frac{\partial V_y}{\partial x}+\frac{\partial V_x}{\partial y}\right)^2+\left(\frac{\partial V_z}{\partial y}+\frac{\partial V_y}{\partial z}\right)^2+\left(\frac{\partial V_x}{\partial z}+\frac{\partial V_z}{\partial x}\right)^2 \end{bmatrix} \quad (5)$$

The right-hand side of the equation (5) is always positive except for the case when all derivatives of velocities become zero in the coordinates. Therefore, a viscous incompressible liquid flows without dissipation of mechanical energy only in the case of absence of particle deformations, i.e. when a liquid moves like a solid body. At all other reasons the losses of mechanical energy occur when the liquid is flowing. When subtracted the term

$$2\mu\left(\frac{\partial V_x}{\partial x}+\frac{\partial V_y}{\partial y}+\frac{\partial V_z}{\partial z}\right)^2=0$$

from the right-side and left-side of equation (5) and added components of the vortex we should deduce:

$$N_{\partial uc}=4\mu\left(\omega_x^2+\omega_y^2+\omega_z^2\right)-4\mu\begin{pmatrix} \frac{\partial V_y}{\partial y}\frac{\partial V_z}{\partial z}-\frac{\partial V_y}{\partial z}\frac{\partial V_z}{\partial y}+\frac{\partial V_z}{\partial z}\frac{\partial V_x}{\partial x}- \\ \frac{\partial V_z}{\partial x}\frac{\partial V_x}{\partial z}+\frac{\partial V_x}{\partial x}\frac{\partial V_y}{\partial y}-\frac{\partial V_x}{\partial y}\frac{\partial V_y}{\partial x} \end{pmatrix} \quad (6)$$

When multiplied the right-side and left-side by volume element $d\tau$ and performed the integration of the whole volume we should obtain an amount of mechanical energy being dissipated per a time unit in finite volume τ.

$$\iiint_\tau N_{\partial uc}d\tau=4\mu\iiint_\tau\left(\omega_x^2+\omega_y^2+\omega_z^2\right)d\tau-$$
$$-4\mu\iiint_\tau\begin{pmatrix} \frac{\partial V_y}{\partial y}\frac{\partial V_z}{\partial z}-\frac{\partial V_y}{\partial z}\frac{\partial V_z}{\partial y}+\frac{\partial V_z}{\partial z}\frac{\partial V_x}{\partial x}- \\ \frac{\partial V_z}{\partial x}\frac{\partial V_x}{\partial z}+\frac{\partial V_x}{\partial x}\frac{\partial V_y}{\partial y}-\frac{\partial V_x}{\partial y}\frac{\partial V_y}{\partial x} \end{pmatrix}d\tau \quad (7)$$

If boundaries of the volume τ are fixed, solid boundaries whereon the velocities will become zero due to no-slip condition for a vector projection upon integration by parts we deduce:

$$\iiint_\tau N_{\partial uc}d\tau=4\mu\iiint_L\left(\omega_x^2+\omega_y^2+\omega_z^2\right)d\tau \quad (8)$$

Therefore, with the motion of the incompressible liquid enclosed in a fixed volume, the total amount of mechanical energy being dissipated per second will depend only on vortex intensity within the volume. This total amount of mechanical energy will represent in the form (8).

Figure 12 shows the graph of variance of energy losses per a time unit within an impeller under consideration depending on the flow rate due to dissipation of vortices. Energy losses decrease with increasing of the flow rate.

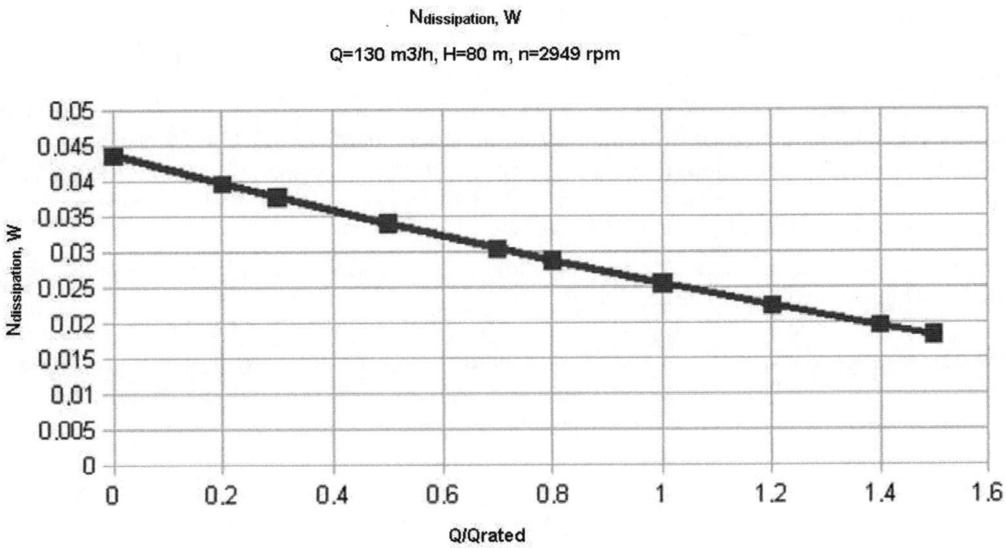

Figure 12. Change of energy losses per a time unit in the impeller under consideration in dependence to the flow rate due to vortex dissipation.

6. Conclusions

- 3D-simulation of the vortex flow of incompressible ideal fluid was performed within centrifugal pump impeller designed for specified flow pattern.
- The velocity fields, streamlines, vortex fields have been calculated.
- Losses induced by friction and vortex dissipation have been evaluated versus the flow rate.
- Obtained results and procedure could be used for target retrofit of hydrodynamic qualities of various hydraulic machines.

References

[1] Rosenhead L 1931 Formation of Vortices from a Surface of *Discontinuity Proc: Roy. Soc. – Series A* **134** p 170–192

[2] Belotserkovskii S M 1978 *Separation and Without Separation Flows of Ideal Liquid Around Thin Wings* (Moscow: Nauka)

[3] Belotserkovskii S M 1988 *Mathematical Modeling of Plane-Parallel Flow Around Bodies* (Moscow: Nauka)

[4] Belotserkovskii S M 1995 *Turbulent Streams and Wakes Based on the Method of Discrete Vortices* (Moscow: Phys.-mat.lit.)

[5] Pavlovets G A 1974 About a Possible Pattern of Calculation of Separation Flow Around Bodies *Trudy TSAGI*

[6] Marchevskii I K 2005 About an Approach to Airfoil Characteristics in Ideal Liquid by the Method of Discrete Vortices *Vestnik NTU "HPI"* pp 182-191

[7] Shcheglov G A 2003 About a Vortex Formation Theory for a Calculation of Hydrodynamic Load acting on Two-Spring Elastic Aircraft *Vestnik NTU "HPI"* pp 244-248

[8] Kostornoi S D 1968 *Designing of Impeller Blade of Radial-Axial Turbine for Orthogonality of Streamlines and Vortex Lines in Absolute Motion* (HGU) p 118

[9] Kostornoi S D 2012 Computer-Aided Design of Centrifugal Pump Hydraulic Part *Kyiv: Modern Science: Researches, Ideas, Results, Technologies* pp 305-311

[10] Kostornoi S D 2015 Flow Model of Ideal Fluid Including Special Features of Boundary Conditions of Real Fluid *Harkiv-Sumy: Proceedings of XVII International Symposium " Discrete*

IOP Conf. Series: Materials Science and Engineering **233** (2017) 012010 doi:10.1088/1757-899X/233/1/012010

Singularity Method in Mathematical Physics Aims pp 126-129

[11] Davidenko AK 2004 Numerical Calculation of Direct Aim of Liquid Flow in Centrifugal Pump Hydraulic Part *Nasosy I oborudovanie* **2** (25) pp 32-35

[12] Dynnikova G Ya 2004 Simulation of Viscous Liquid Flows by Modified Method of Discrete Vortices *Report No. 4698, Institute of Mechanics, Lomonosov Moscow State University*

[13] Dynnikova G Ya 2011 Numerical Calculation of 3D-Flows of Incompressible Liquid based on Dipole Representation of Vorticity *DAN* **437** (1) pp 35-38

[14] Seffmen F J 2000 *Vortex Dynamic* (Moscow: Nauchnyy Mir)

[15] Andronov P R 2006 *Vortex Method of Calculation of Unsteady Dynamic Loads* (Moscow: Publishing House of Moscow University)

[16] Kostornoi S D 2004 Development of Locked Mathematical Models of Calculation of Mechanical Energy Losses Within Hydraulic Part of Hydrodynamic Machines *Vestnik of SumGU* pp 5-13

IOP Conf. Series: Materials Science and Engineering **233** (2017) 012011 doi:10.1088/1757-899X/233/1/012011

Application of the similarity theory for vortex chamber superchargers

A Rogovyi[1], S Khovanskyy[2]

[1] Kharkiv National Automobile and Highway University, 25 Yaroslava Mudrogo st, Kharkov, Ukraine
[2] Sumy State University, 2 Rymskogo-Korsakova st., Sumy, Ukraine

E-mail: asrogovoy@ukr.net

Abstract. On the basis of similarity in vortex chamber superchargers criteria of similarity are defined and their check by numerical researches is made. Vortex chamber superchargers have the big power efficiency of pumping, in comparison with classical jet pumps, at the expense of association of advantages of centrifugal and jet pumps on the basis of the vortex chamber. The concept of vortex chamber superchargers is based on a new principle of energy transfer in jet superchargers at the expense of hydrodynamic effects of rotating flows use. In consequence of pressure decrease near to an axis of the vortex chamber occurs intaking particles of a pumped over flow. Having got to the vortex chamber, particles get tangential velocity interact with a rotating flow at the expense of a momentum exchange. The received criteria take into account of three major factors of similarity: the geometrical sizes, pressure of an active stream on an input in the device, working medium density. Validity of the received criteria is confirmed by a finding of pressure-flow rate characteristics of similar operating modes of a supercharger and their further combination on one characteristic.

1. Introduction

Application of jet superchargers in adverse service conditions, and also at pumping liquids and gases with the big solid abrasive particles content is dictated by rapid wear of mechanical mobile operative parts of dynamic and displacement pumps [1]. Jet devices, unlike classical superchargers possess low power efficiency, but high indicators of reliability and durability [2]. At work of direct-flow jet devices in hydraulic and pneumatic transport of bulk solids their use is limited to low indicators of ejection factor and high pressure values of an active stream [3]. In some way solve these problems in special service conditions [4, 5] can vortex chamber superchargers, possessing all advantages of jet technics: high reliability and durability, work possibility on any structure and concentration of phases, small overall dimensions. Besides, vortex chamber superchargers (VCS), owing to use of centrifugal force in pumping process, possess the best, in comparison with jet direct-flow pumps, characteristics at bulk solids transportation [6].

Works on studying of working capacity and power characteristics of vortex chamber superchargers are spent throughout last fifteen years [4-6]. But, for today, there are no the publications devoted to possibility of preliminary definition of design parameters of the projected device. Besides, there is no possibility of recalculation of characteristics of a supercharger on any standard size.

To solve this problem it is possible by use of the theory of similarity and dimension [7]. Designing of vortex chamber superchargers, as well as any another, is based on preliminary extensive researches. Among them experimental researches [4] have importance. In the dimension and similarity theory it is necessary to define conditions which should be observed in experiences with models, and then allocate the characteristic and convenient parameters defining the main effects and process modes [7].

Complicated movement character of a pumped over liquid in vortex chamber superchargers leads to that the problem of new devices designing decides along with design-theoretical engineering of a design of their flowing part by experimental researches in vitro, and also with the help of CFD-models and in natural conditions [8, 9].

Preliminary definition of design parameters of a projected supercharger, research of working processes on models and propagation of the received results on full-scale superchargers demands fulfillment of geometrical, kinematic and dynamic conditions of similarity [10].

Therefore the purpose of the given work is definition of similarity criteria of vortex chamber superchargers and their check by numerical researches.

2. Results of research

Investigations of vortex chamber superchargers similarity it was spent on the model presented in drawing 1. The VCS [6] works as follows: the basic stream with the flow rate Q_s and pressure p_s comes through the input tangential channel in the vortex mixing chamber and leaves it through the exit tangential channel. A working stream, having mixed up with the pumped over stream arriving through two input axial channels with flow rates Q_{in1} and Q_{in2}, with pressure p_{in1} and p_{in2}, comes in the exit tangential channel with the flow rate Q_e and pressure p_e.

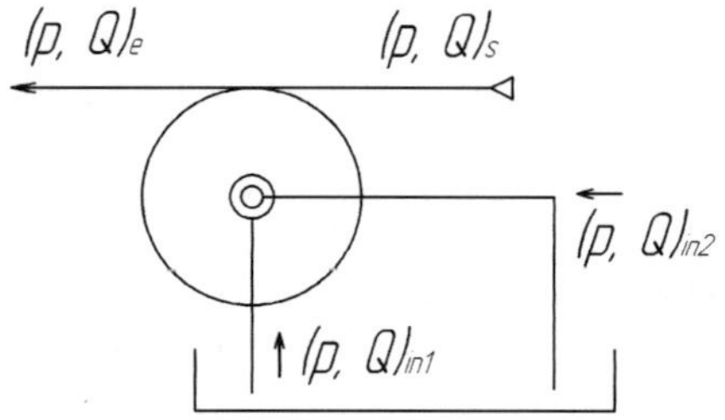

Figure 1. Investigated vortex chamber supercharger.

The supercharger can work in two modes [6] which are characterized by two various working processes. Realization of the first working process leads to occurrence of losses in the drainage channel (the channel in1 in drawing 1) and allows to provide relatively high pressure at the small flow rate of the pumped over medium. Realization of the second working process leads to sucking a pumped over stream through both axial channels, i.e. is fulfill undrainage variant of a supercharger work with relatively low pressure at the high flow rate on the device exit. Both working process of energy transfer in the field of centrifugal force unites, but in the first case occurs at preservation of the angular moment (tangential velocity circulation). And, in the second case energy transfer take place at a momentum exchange between interacting streams due to an eddy motion.

The concept of vortex chamber superchargers is based on a new principle of energy transfer in jet superchargers at the expense of hydrodynamic effects of rotating flows use – vacuum near a rotation axis and overpressure on periphery of the vortex chamber. In consequence of pressure decrease near to an axis of the vortex chamber occurs intaking particles of a pumped over flow. Having got to the vortex chamber, particles get tangential velocity interact with a rotating flow at the expense of a momentum exchange (Figure. 2). Further under the influence of superficial (pressure forces) and mass (dominating gravity) in a potential field move on chamber periphery. Near the vortex chamber walls the particles of a pumped over liquid get in exit tangential channel of a supercharger getting kinetic energy of the overpressure in periphery chamber.

IOP Conf. Series: Materials Science and Engineering **233** (2017) 012011 doi:10.1088/1757-899X/233/1/012011

In classical jet superchargers the only method of energy transfer at the expense of an momentum exchange of interact collision flows is used that is accompanied by essential dissipative process [11]. In vortex chamber superchargers, unlike classical jet devices, the particles get the basic energy mainly in a conservative field at the expense of moving on periphery of the vortex chamber under the influence of centrifugal force. The active flow on periphery has high values of potential energy which can reach 90 % from the spent energy of a supply flow.

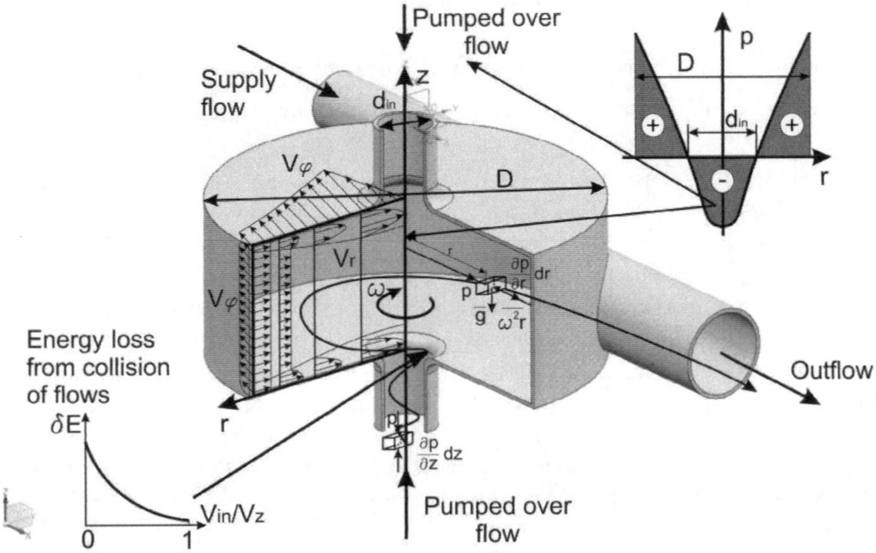

Figure 2. Justification of a pumping over process in the vortex chamber supercharger.

At physical modeling of hydromechanical processes it is accepted to distinguish geometrical, kinematic and dynamic similarity [10]. At the same time use analogous points – points of geometrically similar figures equally located in position to borders of these figures. Hydromechanical systems name geometrically similar if them analogous the sizes are proportional, and corners in analogous points are identical [7].

Geometrical similarity of superchargers is defined by a constancy of similarity linear factor:

$$K_L = \frac{D_H}{D_M} = \frac{d_{inH}}{d_{inM}} = \frac{H_H}{H_M} = ... = const \tag{1}$$

where D_H, d_{inH}, H_H and D_M, d_{inM}, H_M – diameter of the vortex chamber, diameter of the inflow channel and height of the chamber of modelling and full-scale superchargers accordingly. Geometrical similarity demands a proportion constancy of any other sizes of modeling and full-scale superchargers $D_M / H_M = D_H / H_H = const$.

Strictly speaking, for full geometrical similarity it is necessary, that also roughness of vortex chamber surfaces also were similar, but fulfillment of this condition is seldom possible.

The kinematic similarity of superchargers says that dimensionless velocity fields in them should be equal. Besides, mechanical trajectories of liquid particles should be geometrically similar also. The conditions of the kinematic similarity are expressed in the form of ratios:

$$K_V = \frac{V_{rH}}{V_{rM}} = \frac{V_{\tau H}}{V_{\tau M}} = \frac{V_{zH}}{V_{zM}} = const \tag{2}$$

where $V_{rH}, V_{\tau H}, V_{zH}$ and $V_{rM}, V_{\tau M}, V_{zM}$ – the radial, tangential and axial components of points velocity of model and full-scale superchargers.

Using geometrical similarity from which follows, as $V_s \sim Q_s / d_s^2$, and also $V_e \sim Q_e / d_e^2$ similarity of flow rates in a supercharger, we gain:

$$\frac{\left(Q_s / d_s^2\right)_H}{\left(Q_s / d_s^2\right)_M} = \frac{\left(Q_e / d_e^2\right)_H}{\left(Q_e / d_e^2\right)_M} = const, \qquad \frac{Q_{eH}}{Q_{eM}} = \frac{Q_{sH}}{Q_{sM}} = \frac{d_{sH}^2 \, p_{sH}^{1/2} \, \rho_H^{-1/2}}{d_{sM}^2 \, p_{sM}^{1/2} \, \rho_M^{-1/2}}. \tag{3}$$

As at an outflow from a supercharger in an aerosphere the total pressure is defined by formula $p_e = 0.5\rho V_e^2$ and in the equation (3) it is necessary to use a ratio $\left(p_{sH} / p_{sM}\right)^{1/2}$ as a ratio of dynamic pressures.

Dynamic similarity, except observance of geometrical and kinematic similarity conditions, demands similarity of the forces acting in matching points of a flow. Presence of dynamic similarity is a sufficient condition for existence of the kinematic and geometrical similarity. The solution of the flow equations hydrodinamically similar streams among themselves differ from each other only a scale factor. Therefore, choosing scale for quantities entering into such equations, it is possible to convert them to an aspect when factors near each member of the equation become dimensionless. The fundamentals for turbomachinery are criteria (numbers) of Reynolds, Strouhal and Euler. By flows consideration in which force of liquid weight plays the important role, necessary there is Froude number use. At achievement near- and supersonic velocity values of a flow the important role is played by a Mach number. In practice of hydro-machinery modeling very large value has Euler's similarity parameter. With reference to observed conditions and vortex chamber superchargers it can be expressed as follows:

$$Eu_H = \left(\frac{p}{\rho V^2}\right)_H = Eu_M, \qquad Eu_H = \left(\frac{p}{\rho V^2}\right)_H = \left(\frac{p_e \pi^2 d_e^4}{16\rho Q_e^2}\right)_H = \left(\frac{p_e \pi^2 d_e^4}{16\rho Q_e^2}\right)_M \tag{4}$$

$$V_e = \frac{4Q_e}{\pi d_e^2} \tag{5}$$

Using (3) and (4) it is possible to express a similarity parameters for VCS:

$$\frac{Q_{eH}}{Q_{eM}} = \frac{d_{sH}^2 \, p_{sH}^{1/2} \, \rho_H^{-1/2}}{d_{sM}^2 \, p_{sM}^{1/2} \, \rho_M^{-1/2}} \tag{6}$$

$$\frac{p_{eH}}{p_{eM}} = \frac{p_{sH}}{p_{sM}} \tag{7}$$

$$\frac{N_{eH}}{N_{eM}} = \frac{d_{sH}^2 \, p_{sH}^{3/2} \, \rho_H^{-1/2}}{d_{sM}^2 \, p_{sM}^{3/2} \, \rho_M^{-1/2}} \tag{8}$$

The wide experience of hydraulic modeling and turbomachinery modeling shows that at pump work in self-similarity region change of Reynolds number does not render noticeable influence on hydraulic efficiency [10].

For confirmation of choice correctness of vortex chamber superchargers similarity criteria in wide range of pressure, density and the geometrical sizes on the basis of the solution of equations RANS (Reynolds-Averaged Navier-Stokes) numerical research has been made for an incompressible liquid with use SST-model of turbulence with rotation-curvature correction.

Set of equations SST model looks as follows [12]:

$$\frac{\partial(\rho k)}{\partial t}+\frac{\partial}{\partial x_j}(\rho u_j k)=\frac{\partial}{\partial x_j}\left(\mu_{ef}\frac{\partial k}{\partial x_j}\right)+P_k-\beta^*\rho k\omega \tag{9}$$

$$\frac{\partial(\rho\omega)}{\partial t}+\frac{\partial}{\partial x_j}(\rho u_j\omega)=\frac{\partial}{\partial x_j}\left(\mu_{ef}\frac{\partial \omega}{\partial x_j}\right)-\rho\beta\omega^2+Cd_\omega+\alpha\frac{\rho}{\mu_t}P_k \tag{10}$$

where: ρ – density; k – kinetic energy of turbulent pulsation; x_j – Cartesian coordinates; u_j – components of the mean velocity vector; $\mu_{ef}=\mu+\mu_t$ – effective viscosity; μ_t – turbulent viscosity; μ – molecular viscosity; ω – turbulence eddy frequency; P_k – production of turbulence kinetic energy; Cd_ω – cross-diffusion term in SST model. Constants and the specification statement of the equations (9) - (10) can be found in [13].

Rotation-curvature correction in SST-model is realized by multiplication of production of turbulence kinetic energy in the equations (9) - (10) on function [13]:

$$f_{r_1}=\max\left\{\min\left(f_{rotation},1.25\right),0.0\right\} \tag{11}$$

where

$$f_{rotation}=\left(1+c_{r1}\right)\frac{2r^*}{1+r^*}\left[1-c_{r3}\tan^{-1}\left(c_{r2}\tilde{r}\right)\right]-c_{r1} \tag{12}$$

The constants c_{r1}, c_{r2}, c_{r3} are equal 1.0, 2.0 and 1.0 respectively [13, 14]. Terms r^* and $\tilde{r}$ are calculated as follows:

$$r^*=\frac{S}{\Omega}, \tag{13}$$

$$\tilde{r}=2\Omega_{ik}S_{ik}\left[\frac{DS_{ij}}{Dt}+\left(\varepsilon_{imn}S_{jn}+\varepsilon_{jmn}S_{in}\right)\Omega_m^{rot}\right]\frac{1}{\Omega D^3} \tag{14}$$

Components of the mean strain tensor:

$$S_{ij}=\frac{1}{2}\left(\frac{\partial u_i}{\partial x_j}+\frac{\partial u_j}{\partial x_i}\right) \tag{15}$$

Components of the vorticity tensor – $\Omega_{ij}=\frac{1}{2}\left(\left(\frac{\partial u_i}{\partial x_j}-\frac{\partial u_j}{\partial x_i}\right)+2\varepsilon_{mji}\Omega_m^{rot}\right)$, where ε_{mji} – tensor of Levi-Civita.

$$S^2=2S_{ij}S_{ij};\quad \Omega^2=2\Omega_{ij}\Omega_{ij};\quad D^2=\max\left(S^2,0.09\omega^2\right) \tag{16}$$

Adequacy of calculations by means of the chosen model confirmed by comparison with experimental data on integrated parameters (pressures and flow rates in all channels) and on distribution of pressure along radius of the vortex chamber on the basis of a Student's and Fisher's tests [8]. In [8, 14] results of verification and selection optimum on errors of mathematical model are presented.

Experimental researches of the vortex device (Figure 3) are made for a finding of the best mathematical model on errors. The error estimation of pressure distribution along radius of the vortex chamber is made at the closed outlet tangential channel of a supercharger. Thus the top end cover of the vortex chamber was drained sixteen apertures. Experimental facility included the vortex device, blower, receiver and measuring equipment. At different stages of experimental researches of the vortex chamber superchargers characteristics standard instruments and equipment were used. Their relative error did not exceed 1 %. Pressure in channels was measured cup manometers, ambient temperature – mercury thermometers, the flow rates in a channels - flowmeters. The model of a supercharger with diameter of the vortex chamber 60 mm was used. Pressure differences on flowmeter devices it was measured by micromanometers with the maximum error in 8 Pa. The total error between computed and experimental values on pressure did not exceed 12 % in central zone of vortex chamber.

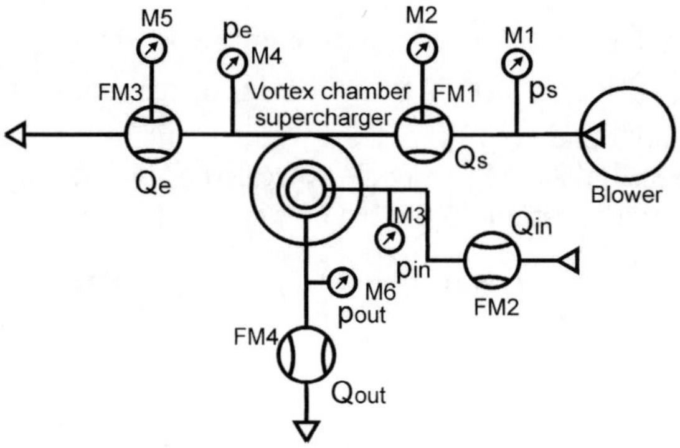

Figure 3. Experimental facility of the vortex chamber supercharger.

Calculations are made in program complex OpenFOAM (OpenCFD Ltd) [15, 16] at following boundary conditions: on a wall – no slip wall, at inflow of the supply channel was set a stagnation pressure, in outflow channels – equality to null of static pressure. The mesh made about 5 million elements for fulfillment of parameter Y+ <2.

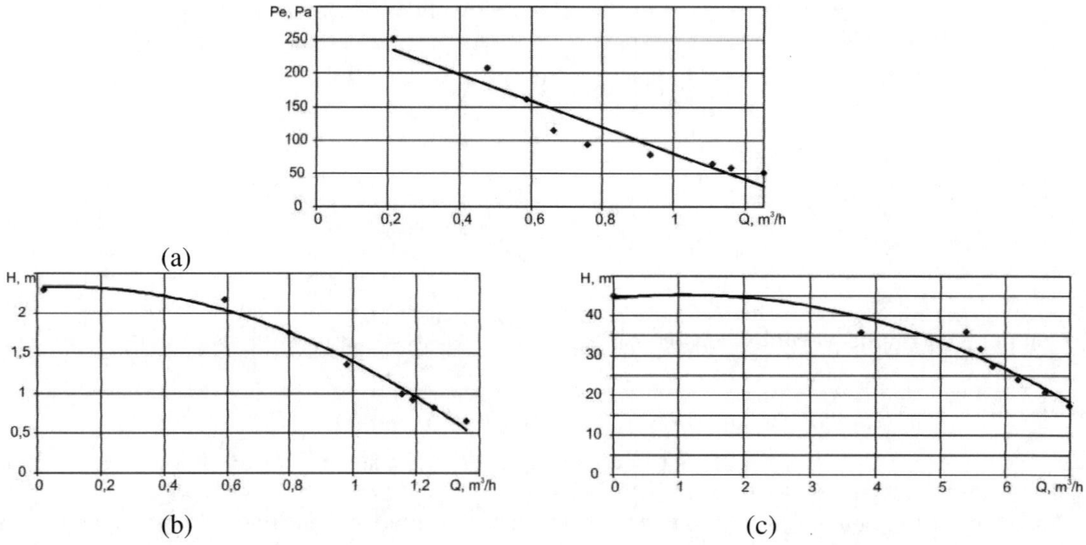

Figure 4. Pressure-flow rate characteristics of vortex chamber superchargers: (a) variant 1 (air); (b) variant 2; (c) variant 3.

For research following parameters of a supercharger work have been chosen: 1) actuating medium - air with density 1.185 kg/m^3, $D = 0.03$ m, a total pressure at supercharger tangential inlet $p_s^{tot} = 500$ Pa; 2) actuating medium – water with density 997 kg/m^3, $D = 0.05$ m, a total pressure at supercharger tangential inlet of the $p_s^{tot} = 50000$ Pa; 3) actuating medium - water with density

997 kg/m^3, $D = 0.05$ m, a total pressure at supercharger tangential inlet $p_s^{tot} = 1.25$ MPa. Thus, wide ranges of a vortex chamber supercharger use on various actuating mediums - gas and liquid, with different levels of pressure from 0.5 kPa to 1.25 MPa are chosen.

In Figure 4 pressure-flow rate characteristics of VCS for the observed three cases are presented. For all three variants of superchargers work parameters have been selected so that conditions of geometrical, kinematic and dynamic similarity were observed.

Observance of dynamic similarity can be checked by combination of pressure-flow rate characteristics of all three observed cases on one dimensionless characteristic taking into account recalculation on the basis of the (6). I.e. for combination, for example, the first and second variants, the characteristic first needs to be converted using linear factor of similarity $K_L = D_H / D_M = 0.05/0.03 = 1.67$. Values of a supercharger pressure and flow rate are similarly computed. The result of characteristics combination is presented in Figure 5.

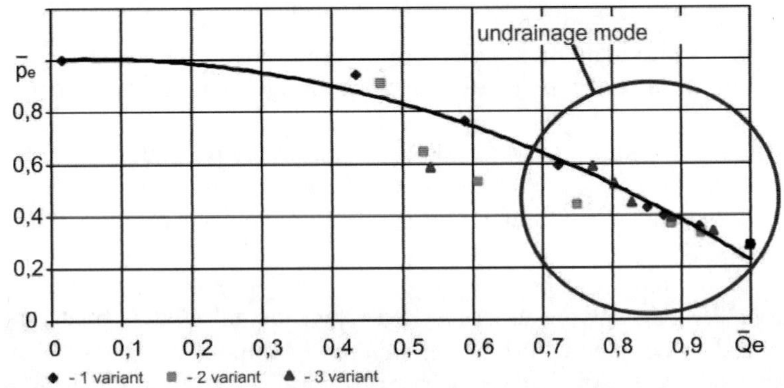

Figure 5. The combined characteristics for three cases of a supercharger work.

In Figure 5 characteristics of a VCS work are rated to the maximum pressure and to the maximum flow rate of the pressure-flow rate characteristic of the second variant. The points corresponding to different operating modes and transporting of different medium closely lie to the pressure curve of the second variant. It speaks about justice of the statement about similarity of operating modes of all three variants. Especially it is become apparent in the undrainage mode of a supercharger at which deviations of the majority points make no more than 10 %.

Fulfillment of the kinematic similarity demands is illustrated in Figure 6.

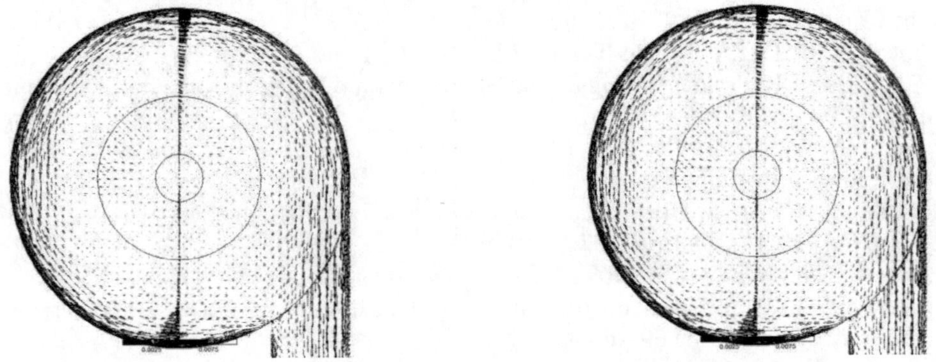

(a) (b)

Figure 6. The velocity vector field in a supercharger: (a) variant 1 (air); (b) variant 3 (water).

IOP Conf. Series: Materials Science and Engineering **233** (2017) 012011 doi:10.1088/1757-899X/233/1/012011

In Figure 6 it is visible that the kinematic similarity consisting in equality of dimensionless velocity field in the device, working on similar regimes is carried out.

For comparison of characteristics hydrodinamically similar pumps inter se often use, formed on the basis of mathematical expressions of similarity rules, the so-called, reduced values – dimensionless or dimensional complexes which are defined from physical parameters of hydraulic machine (flow rate, head, power and others). Equality of the last provides presence of partial dynamic working processes similarity of model and full-scale pump.

At observance of similarity all conditions, on the basis of the equations (6) - (8), equality for similar regimes of efficiency follows. However, influence of the roughness on hydraulic losses in a supercharger will lead to some difference in values of efficiency.

3. Conclusion

Similarity parameters of the vortex chamber superchargers is defined and their check by numerical way is made.

1. Criteria of geometrical, kinematic and dynamic similarity on the basis of ratios of three major factors of similarity are gained: geometrical sizes, actuating medium inlet pressure in the device, actuating medium density.
2. Validity of the gained criteria is confirmed by a finding of pressure-flow rate characteristics of three similar operating modes of a supercharger and their further combination on one characteristic. At the undrainage operating mode a deviation of the majority points make no more than 10 %.

References

[1] Xu J, Liu X and Pang M 2016 Numerical and experimental studies on transport properties of powder ejector based on double venturi effect *Vacuum.* **134** pp 92-98

[2] Besagni G, Mereu R and Inzoli F 2016 Ejector refrigeration: a comprehensive review *Renewable and Sustainable Energy Reviews* **53** pp 373-407

[3] Mills D 2004 *Pneumatic conveying design guide* (Oxford: Elsevier) p 636

[4] Syomin D and Rogovoy A 2010 Power characteristics of superchargers with vortex work chamber *Teka Komisji Motoryzacji i Energetyki Rolnictwa* **19** pp 232-240

[5] Beck J L. 2014 Underwater oil and gas leak containment systems and methods. *US 8887813 B2*

[6] Syomin D and Rogovyi A. 2012 Features of a working process and characteristics of irrotational centrifugal pumps *Procedia Engineering.* **39** pp 231–237

[7] Sedov L I 1993 *Similarity and dimensional methods in mechanics.* (Boca Raton:CRC press) p 496

[8] Rogovyi A 2016 Use of detached-eddy simulation method (DES) in calculations of the swirled flows in vortex apparatuses *Teka Komisji Motoryzacji i Energetyki Rolnictwa* **16** (3) pp 57-62

[9] Karapuzova M, Lugova S and Tverdokhleb I 2012 Flow Structure Investigation in the Lateral Inlet Branches of Hydraulic Machines and Some Recommendations on Their Designing *Procedia Engineering* **39** pp 140-147

[10] Dixon S L and Hall C 2013 *Fluid mechanics and thermodynamics of turbomachinery* (Oxford:Butterworth-Heinemann)

[11] Sokolov E Y, Zinger N M 1989 *Jet devices* (Moscow:Energoatomizdat) p 352

[12] Menter F R 1994 *AIAA Journal.* **32** 1598-1605

[13] Smirnov P E and Menter F R 2009 *Journal of Turbomachinery* **131** (4) pp 041010

[14] Rogovyi A S 2016 Verification of Fluid Flow Calculation in Vortex Chamber Superchargers *Automobile Transport* **39** pp 39-46

[15] Peyret R and Taylor T D 2012 *Computational methods for fluid flow.* (New York: Springer Science & Business Media)

[16] Jasak H 2009 *Int. J. of Naval Architecture and Ocean Engineering* **1** (2) pp 89-94

IOP Conf. Series: Materials Science and Engineering **233** (2017) 012012 doi:10.1088/1757-899X/233/1/012012

Influence of Geometric Parameters of Inducer Bush Design on Cavitation Erosion Characteristics of Centrifugal Inducer Stage of Pump

P Y Tkach[1]

[1]Research Department of Hydraulic Passages, JSC "VNIIAEN",
2nd Zheleznodorozhnaya str., 2, Sumy, 40003, Ukraine

E-mail: admin@vniiaen.sumy.ua

Abstract. The article represents the research results of a centrifugal inducer stage with inducer bush. We determined the optimal inducer bush design that would improve the cavitation erosion characteristics without deteriorating the energy levels and preserving overall dimensions of the centrifugal inducer stage at the same time.

1. Introduction

Currently, one of the topic issue in the field of pumpbuilding is the assurance of centrifugal pump service life of at least 40,000 hours without the critical consequences of cavitation wear in accordance with modern requirements for feed water pumps operated in nuclear and thermal power plants [1]. A guaranteed solution of this problem is the use of double suction fist stage and two inducers. However, this method will lead to a significant complication of the design, as well as an increase in overall dimensions and mass, respectively. The analysis of the literature sources [2-5], as well as experience of the centrifugal inducer stage design suggests that cavitation erosion resistance increase due to only the geometry of the inducer vane system has already been exhausted, therefore, it seems reasonable to find a solution of the problem through others elements of the centrifugal inducer stage. It is known that inducer peripheral areas are subjected to the greatest damage caused by cavitation erosion, therefore attention was paid to this place.

Proceeding from the abovementioned, the idea to solve the problem of the cavitation erosion wear using the centrifugal inducer stage with the inducer bush was proposed. This will permit to improve the cavitation erosion characteristics and will not impair the energy properties at the same time. Figure 1 shows a schematic diagram of traditional design of centrifugal inducer stage, where a smooth bush (Figure 2 (a)) is placed over the inducer (Figure 1, item 2) and idea is to replace this smooth bush with a bush with axial grooves – (Figure 2 (b)).

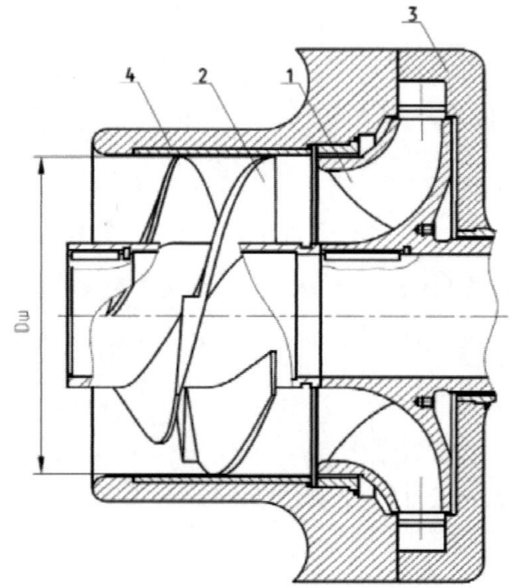

Figure 1. Principal scheme of traditional centrifugal inducer stage design: 1 - impeller; 2 – inducer; 3 – diffuser; 4 – smooth inducer bush.

The advantage of this technical solution is that the improvement of the cavitation erosion characteristics of the centrifugal inducer stage can be obtained with the current hydraulic passage while maintaining the overall dimensions of the stage. Presumably, the positive effect of the implementation of the inducer bush with grooves is achieved due to the occurrence of vortexes in bush grooves which positively influence on reverse flows at the inducer inlet and also eliminate cavitation cavities located on periphery thus protecting inducer from damage.

(a) (b)

Figure 2. Design of bush being under research: (a) – smooth bush; (b) - bush with grooves.

2. Research Task

It is known that first traces of cavitation erosion in pumps appear only after the incubation period which can last dozens of hours [6, 7]. Such a long duration of incubation period greatly complicates the possibility of systematic tests for pump cavitation wear and determine its cavitation erosion characteristics. Therefore, in order to confirm the expediency of inducer bush a method for assessing the cavitation erosion characteristics of the centrifugal inducer stage was developed [8]. This method will allow to estimate the cavitation wear as well as save time for research as much as possible. According to this method, a threshold value of the resistance to cavitation erosion K_{er} is chosen as a value of the cavitation erosion characteristics of the centrifugal inducer stage with grooved bush being under research [2].

$$K_{er\,(ex)} = K_{er\,(b)} \left(w_{c\,(b)} / w_{c\,(ex)}\right)^{1/3} \quad (1)$$

where $w_{k\,(ex)}$ and $w_{k\,(b)}$ – cavitation vibration acceleration value for the centrifugal inducer stage with the grooved inducer bush being under research and the centrifugal inducer stage with the smooth inducer bush (reference value) that are tested under identical conditions, respectively;

$K_{er}\,(b)$ – threshold value of cavitation erosion parameter for centrifugal inducer stage with smooth inducer bush (reference value).

The research first step is to carry out tests on the model test stand using the test procedures where the following geometric parameters of the inducer bush are used as the main influence factors: *z* - number of grooves, b - width of grooves, l1 - length upstream and l2 – length downstream the leading edge of inducer vanes. The 16 variants of the inducer bush with different geometric parameters have been designed, where combinations of the geometric parameters for inducer bush are combined within the variability interval of the factor level.

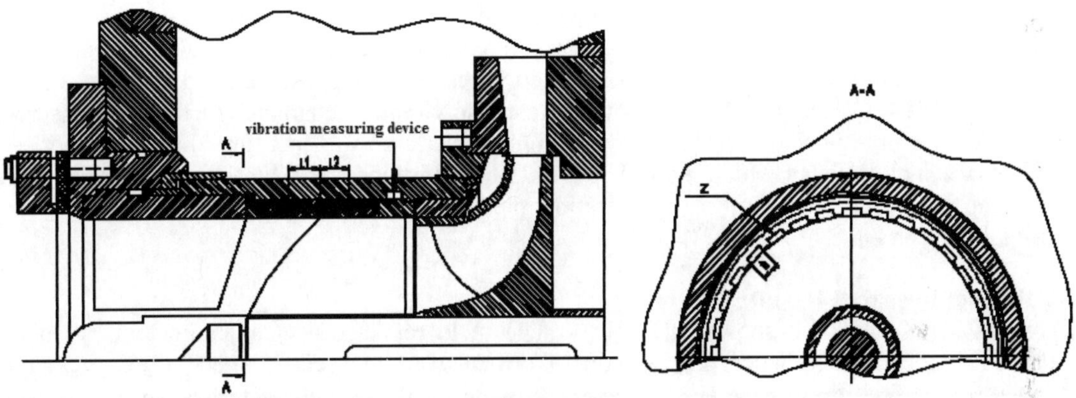

Figure 3. Constructive diagram of the experimental device.

Figure 3 shows a schematic diagram of the experimental device in which the centrifugal inducer stage is placed. It consists of inducer, impeller, diffuser and axial grid at the inlet.

3. Description of Physical Experiment

The tests were conducted on the model test stand, the basic diagram of which is shown in Figure 4. The test stand operates as a closed loop and uses service water.

The main element of the test stand is the experimental device (Figure 3) - a single-stage axially split pump with a transfer shaft, which permits the operative replacement of the working parts without complete disassembly of the pump. The experimental device is driven by a balanced motor which provides the possibility of testing at speed range of 1000-3000 rpm.

IOP Conf. Series: Materials Science and Engineering **233** (2017) 012012 doi:10.1088/1757-899X/233/1/012012

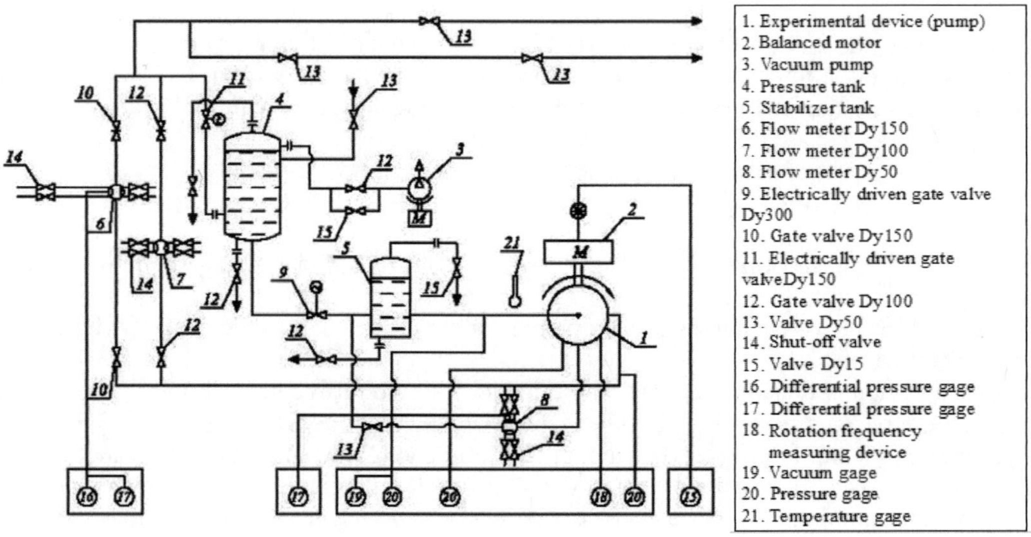

Figure 4. Principal scheme of the experimental stand.

The tests were conducted on the experimental model test stand at the rotor speed of n=2000 rpm. The method of recording and plotting energy and cavitation characteristics curves during the tests meets the requirements of the Standard [9]. Cavitation tests were carried out at 5 flow rates: 0.3 Q_n, 0.5 Q_n, 0.75 Q_n, 1.0 Q_n and 1.2 Q_n. During these tests the vibration characteristics were measured. In determining the vibration characteristics the vibration acceleration was measured by waterproof sensor located on the inducer bush which made it possible to determine the vibration characteristics more precisely and to refuse of vibration acceleration measurements on other surfaces of the examine centrifugal inducer stage.

4. Analysis of Research Results

The results of the tests are shown in Table 1. The table represents such parameters as: index of cavitation erosion characteristics $K_{er(ex)}$, ratio of critical suction specific speed $C_{cr}/C_{cr\ sm}$, ratio of efficiency η/η_{sm} and ratio of head H/H_{sm}, where the index $_{sm}$ denotes characteristics of the centrifugal inducer stage with a smooth inducer bush, and without an index - characteristics of the centrifugal inducer stage with a inducer bush with grooves and corresponding number of assembly variant from first column of the table.

Table 1a. Research results.

Assembly variant No.	Parameter	0.3 Q_n	0.5 Q_n	0.75 Q_n	1.0 Q_n	1.2 Q_n
1	$K_{er\ (ex)}$	18.42	20.83	20.37	19.93	21.05
	$C_{cr}/C_{cr\ sm}$	1.46	1.20	1.08	0.97	0.84
	η/η_{sm}	0.91	0.92	0.93	0.95	0.95
	H/H_{sm}	1.00	1.00	1.00	0.99	0.97
2	$K_{er\ (ex)}$	20.26	21.66	20.09	20.21	19.7
	$C_{cr}/C_{cr\ sm}$	1.16	1.20	1.17	1.00	0.91
	η/η_{sm}	0.96	0.96	0.97	0.97	0.97
	H/H_{sm}	1.02	1.01	1.02	1.00	0.98

Table 1b. Research results.

Assembly variant №	Parameter	0.3 Q_n	0.5 Q_n	0.75 Q_n	1.0 Q_n	1.2 Q_n
3	$K_{er\ (ex)}$	18.96	21.08	20.26	20.57	22.37
	$C_{cr}/C_{cr\ sm}$	1.10	1.10	1.24	1.06	0.93
	η/η_{sm}	0.97	0.98	0.96	0.97	0.97
	H/H_{sm}	1.02	1.01	1.01	0.99	0.98
4	$K_{er\ (ex)}$	20.03	21.28	20.43	20.81	23.16
	$C_{cr}/C_{cr\ sm}$	1.07	1.20	1.20	1.09	0.91
	η/η_{sm}	0.98	0.98	0.99	0.98	0.98
	H/H_{sm}	1.02	1.01	1.01	0.98	0.96
5	$K_{er\ (ex)}$	18.78	21.04	21.31	21.88	23.23
	$C_{cr}/C_{cr\ sm}$	1.46	1.31	1.11	0.97	0.80
	η/η_{sm}	0.92	0.92	0.93	0.93	0.93
	H/H_{sm}	1.01	1.01	1.01	0.99	0.97
6	$K_{er\ (ex)}$	34.65	20.73	20.71	19.90	19.63
	$C_{cr}/C_{cr\ sm}$	1.13	1.40	1.05	1.00	0.72
	η/η_{sm}	0.96	0.96	0.97	0.96	0.96
	H/H_{sm}	1.01	1.01	1.01	0.99	0.97
7	$K_{er\ tlv(ex)}$	18.72	21.39	21.28	22.53	24.91
	$C_{cr}/C_{cr\ sm}$	1.10	1.23	1.32	1.12	0.88
	η/η_{sm}	0.96	0.96	0.97	0.97	0.96
	H/H_{sm}	1.01	1.01	1.00	0.98	0.96
8	$K_{er\ (ex)}$	19.75	21.57	20.55	20.48	19.7
	$C_{cr}/C_{cr\ sm}$	1.10	1.13	1.20	1.12	0.89
	η/η_{sm}	0.97	0.97	0.98	0.97	0.97
	H/H_{sm}	1.01	1.01	1.00	0.98	0.96
9	$K_{er\ (ex)}$	23.71	23.32	22.11	21.74	20.90
	$C_{cr}/C_{cr\ sm}$	1.07	1.36	1.32	1.12	0.91
	η/η_{sm}	0.94	0.93	0.94	0.95	0.96
	H/H_{sm}	1.02	1.02	1.02	1.00	0.97
10	$K_{er\ (ex)}$	21.31	21.73	20.80	20.44	19.58
	$C_{cr}/C_{cr\ sm}$	1.40	1.51	1.20	1.12	0.95
	η/η_{sm}	0.99	0.99	0.99	0.97	0.97
	H/H_{sm}	1.04	1.02	1.01	1.00	0.98
11	$K_{er\ (ex)}$	24.64	21.73	21.97	21.60	20.81
	$C_{cr}/C_{cr\ sm}$	1.23	1.51	1.24	1.09	0.89
	η/η_{sm}	0.98	0.99	0.99	1.00	1.02
	H/H_{sm}	1.02	1.01	0.99	1.00	1.00
12	$K_{er\ (ex)}$	18.29	21.15	21.22	21.65	19.06
	$C_{cr}/C_{cr\ sm}$	1.20	1.27	1.28	1.06	0.92
	η/η_{sm}	1.00	1.00	1.00	1,01	1.04
	H/H_{sm}	1.01	1.01	1.01	1.01	1.01
13	$K_{er\ (ex)}$	34.65	33.23	21.97	21.53	19.48
	$C_{cr}/C_{cr\ sm}$	1.40	1.31	1.32	1.03	0.88
	η/η_{sm}	0.95	0.96	0.97	0.98	1.01
	H/H_{sm}	1.02	1.01	1.03	1.03	1.02

Table 1c. Research results.

Assembly variant №	Parameter	0.3 Q_n	0.5 Q_n	0.75 Q_n	1.0 Q_n	1.2 Q_n
14	$K_{er\ (ex)}$	32.61	23.35	30.63	21.91	19.61
	$C_{cr}/C_{cr\ sm}$	1.07	0.98	1.24	1.06	0.81
	η/η_{sm}	0.98	0.98	0.99	1.00	1.02
	H/H_{sm}	1.01	1.01	1.02	1.02	1.02
15	$K_{er\ (ex)}$	26.25	22.92	21.45	21.40	23.92
	$C_{cr}/C_{cr\ sm}$	0.98	1.10	1.17	1.09	0.88
	η/η_{sm}	0.99	0.98	0.99	1.01	1.03
	H/H_{sm}	1.01	1.00	1.01	1.01	1.01
16	$K_{er\ (ex)}$	41.08	21.41	22.30	20.30	19.8
	$C_{cr}/C_{cr\ sm}$	1.20	1.13	1.11	1.06	0.89
	η/η_{sm}	0.99	0.98	1.00	1.00	1.02
	H/H_{sm}	1.01	1.00	1.01	1.01	1.01

Since this article is devoted to research the influence of the inducer bush on the cavitation erosion characteristics of the centrifugal inducer stage, it is necessary to pay attention to threshold value of the cavitation erosion parameter $K_{er\ (ex)}$ and ratio of the critical suction specific speed $C_{cr}/C_{cr\ sm}$. According to the method described in [8] and experimental value of $K_{er(b)}$, depending on the inducer design and type of handled medium [2], the confirmation of improvement of the cavitation erosion characteristics of examine inducer centrifugal stage will be the accomplishment of the inequation $K_{er\ (ex)}>20$. As can be seen from Table 1, the value of $K_{er(ex)}$ exceeded the required threshold value for most assembly variants, but not stable within full flow rate range, only some assemblies gave a significant increase of the cavitation erosion parameter within full flow rate range.

According to Table 1 the ratio of the critical suction specific speed $C_{cr}/C_{cr\ sm}$ which characterizes the cavitation characteristics of the centrifugal inducer stage indicates that the use of the inducer bush improves the cavitation characteristics of the centrifugal inducer stage during operation at partload flow rates for all assembly variants and nominal flow rates for most assembly variants. During operation at overload flow rate all assembly variants with the inducer grooved bush had a negative effect on the cavitation characteristics.

As for the energy curve of the centrifugal inducer stage, the use of the inducer bush leads to efficiency decreasing in the most assembly variants. It can be said that only in some variants with a smooth inducer bush the efficiency remains at the same level and only the one variant (assembly No.12) showed a positive result – the efficiency was slightly increased and improvement of the cavitation performance has been achieved.

The head curve of the centrifugal inducer stage with inducer bush is practically unchanged throughout the full flow rate range.

Analyzing the physical experiment research results, we found that the best values for both energy and cavitation erosion characteristics were shown by inducer bushes which grooves have a square cross section. This fact confirms the theory that vortices occurred in the bush grooves reduce the influence of reverse flows at the inducer inlet and also eliminate the cavitation cavities located on its periphery thus protecting inducer from damage, since exactly this form of grooves provide more stable vortices in comparison with grooves that has rectangle cross section.

Also, there was observed an ambiguous influence of the position of inducer bush relatively to the inducer vane leading edge on the characteristics of the centrifugal inducer stage as the assemblies that have a similar shape and number of grooves but differ only in configuration have differences in energy and cavitation erosion characteristics. The general recommendation for obtaining better values is that the length of inducer bushes should be not more than half length of the inducer vane system starting

IOP Conf. Series: Materials Science and Engineering **233** (2017) 012012 doi:10.1088/1757-899X/233/1/012012

from the leading edge and lengthen outside the inducer vane system by 0.4-0.7 of the length of the inducer vane system.

5. Conclusions

1. We experimentally proved the general possibility of using the inducer bush for improving cavitation erosion characteristics of the centrifugal inducer stage as well as maintaining the economic qualities and overall dimensions of the stage.

2. The influence analysis of the geometric parameters of the inducer bush on cavitation erosion and energy characteristics of the centrifugal inducer stage and the selection guideline of the inducer bush optimal design are given.

3. In order to confirm obtained results of erosion wear reduction by means of evaluating the cavitation erosion characteristics of the stages based on the cavitation vibration acceleration it will be necessary to perform their selective qualitative inspection using the method of evaluation of cavitation resistance by applying fragile lacquer coatings for stages with the best variants of inducer bush design and their further comparison with similar results of a smooth inducer bush. After that it is possible to develop a design procedure of inducer bushes of centrifugal inducer stages.

References

[1] Chumachenko B 2002 Theoretical Foundations and Experimental Studies Aimed to Create Hydraulic Parts of Vane Type Pumps, Providing a Combination of High Efficiency, Suction Capacity and Low Vibration Level *Doctoral thesis Moscow* p 411

[2] Rzhebaeva N 1990 Centrifugal Inducer Stage of Pump: Recommended Practices for Course Work and Diploma Plaining *Department of Hydraulic Machines Sumy Unit of Kharkiv Technical University "Kharkiv Polytechnic Institute" Kharkov* p 40

[3] Vizenkov G, Tverdokhleb I, Kutsenko V 2008 *Pumps of Special and General Industrial Purpose with Inducers. Review of Research Experience, Development and Operation of Inducer Pumps* 3 pp 46-50

[4] Kowalik M 1993 *World pumps Inducer – state of the art* pp 32-35

[5] Gulich J 2010 *Centrifugal Pumps* (Springer Heidelberg Dordrecht, London) p 964

[6] Karelin V 1970 *Vaned Hydraulic Machines Wear Caused by Cavitation and Pumps* (Moscow)

[7] Knepp R, Deyli J, Khemmit F 1974 *Cavitation* (Moscow)

[8] Elin O, Tkach P 2014 Ukrainian scientific and technical journal Industrial hydraulics and pneumatics *Experimental Setup and Methods of Research Influence of Inducer Bush Design on the Cavitation Erosion Characteristics of Inducer Centrifugal Stage of the Pump* **3** (45) pp 16-21

[9] DSTU6134-2009 (ISO 9906:1999) *Rotodynamic pumps. Method of testing* State Standart of Ukraine Kiev p 105

IOP Conf. Series: Materials Science and Engineering **233** (2017) 012013 doi:10.1088/1757-899X/233/1/012013

Experimental Measurements and Mathematical Modeling of Static and Dynamic Characteristics of Water Flow in a Long Pipe

J Jablonska[1], M Kozubkova [1]

[1]Departament of Hydromechanics and Hydraulic Equipment, VSB Technical University of Ostrava, 17. listopadu 15/2172, CZ- 70830 Ostrava

E-mail: jana.jablonska@vsb.cz

Adstract. Static and dynamic characteristics of flow in technical practice are very important and serious problem and can be solved by experimental measurement or mathematical modeling. Unsteady flow presents time changes of the flow and water hammer can be an example of this phenomenon. Water hammer is caused by rapid changes in the water flow by means the closure or opening of the control valve. The authors deal with by hydraulic hammer at the multiphase flow (water and air), its one-dimensional modeling (Matlab SimHydraulics) and modeling with the use of the finite volume method (Ansys Fluent) in article. The circuit elements are defined by static and dynamic characteristics. The results are verified with measurements. The article evaluates different approaches, their advantages, disadvantages and specifics in solving of water hammer.

1. Introducton

The dynamics of hydraulic systems is often accompanied by a hydraulic hammer, and it has been investigated experimentally and mathematically by a number of authors who have used in the past many partially simplified approaches. It has to be said that the quality of results always depended on obtaining high quality data regarding the physical properties of liquids, such as the modulus of elasticity and, of course, the viscosity and density of the flowing liquid. The article focuses on the possibilities of solution of the hydraulic hammer, the creation of more precise mathematical models and the presentation of the solution approach.

A typical example of a non-stationary isothermal flow of compressible fluid is a hydraulic hammer which is typical by the pulsations of pressure and noise. In experimental hydraulic hammer measurements, it is necessary to record the time series of pressures at various points of the circuit, respectively flow rate to the computer and to devote their evaluation.

Numerical simulation is used to examine the dynamic properties of the system. Computer processing and simulation of mathematical models to describe variables of fluid flow has a number of advantages. With the massive development of computational capacity, it is useful to use mathematical modeling in designing and constructing of hydraulic devices in many engineering applications. However, some caution should be exercised when using the simulation results, as the issue of turbulent flow is not yet fully resolved. Therefore, the simulation can be used as an exploration tool, which must be further verified and tested. A great emphasis must be put on the flow prediction as a result of the simulation.

For the experimental investigation of the hydraulic hammer, the hydraulic circuit was created where the water was as the flowing medium. By quick closing or opening the valve at the end of the long pipeline, a hydraulic hammer was simulated. Because this is a dynamic task, pressure sensors were used to record to a computer [1]. The results were used as boundary conditions for mathematical simulations and subsequently for their verification.

Due to the electrohydraulic analogy, hydraulic elements and systems can be examined by the theory of electrical circuits that have similar transient properties and the circuit is defined as a one-dimensional mathematical model. For hydraulic circuits, the following hydraulic resistances have been specified [4, 7]:

- The local and frictional resistances: $R = \dfrac{d(\Delta p)}{dQ}$

- The resistance to acceleration (inductance): $L = \dfrac{\Delta p}{\dfrac{dQ}{dt}}$

- The resistance to deformation (capacity): $C = \dfrac{Q}{\dfrac{d(\Delta p)}{dt}}$

Resistance sorting (serial, parallel) is the same in fluid systems as in electrical systems (the most famous is so called RLC T-cell). For combined sorting, this is a resistance network, the solution of which is described by the node law and the circuit law.

The software Matlab-Simulink-SimHydraulics was used to solve the hydraulic circuit by electro-hydraulic analogy. The program allows to model systems consisting from hydraulic and mechanical elements, called multi-domain systems. It uses a direct analogy with real system elements (SimHydraulics blocks correspond to the real elements in the circuit). Each single connection between Matlab blocks corresponds to the power transmission as in the actual system. This makes it possible to simulate systems by describing their physical structure and it is not necessary to derive mathematical relationships between controlled quantities [1, 3, 5].

A more precise model is defined as a multidimensional mathematical model or a multidimensional multiphase model. The basic equations describing the flow of real liquids, expressing the basic physical laws of conservation of mass, momentum and energy, physically demonstrate the balance of a certain variable ζ in the control volume as a continuous environment and in certain coordinate system. According to this balance, the time change (accumulation) of the variable ζ in the given control volume V is equal to the flow of this variable by the control surface S limiting this volume and its production within the volume V [7].

$$\frac{\partial}{\partial t}\int_V (\rho\zeta)dV + \int_S \left(\rho \vec{F}(\zeta).\vec{n} \right) dS = \int_V P(\zeta)dV \quad (1)$$

where

ζ – the balancing variable,

$\vec{F}(\zeta)$ - the flow density vector of variable ζ by area dS,

$\vec{n}$ - the vector normal to the area dS,

$P(\zeta)$ - the density of production of variable ζ (production per unit of time in volume V).

The ANSYS Fluent software enables the solution of the above equation system by the finite volumes method from its setting, through stationary and transient calculation, to subsequent display of

IOP Conf. Series: Materials Science and Engineering **233** (2017) 012013 doi:10.1088/1757-899X/233/1/012013

results in two-dimensional, axially symmetrical and generally three-dimensional space. It is important to choose the correct mathematical model of laminar or turbulent flow, or multiphase flow with or without admixture. Furthermore, the physical properties of used media, initial and boundary conditions are set. The following is a numerical calculation and evaluation of results [1, 2].

2. Experiment

Figure 1 illustrates a hydraulic circuit on which a number of static and dynamic measurements have been made. The result was the transition characteristics that served to define boundary conditions and verify mathematical modeling, see [1].

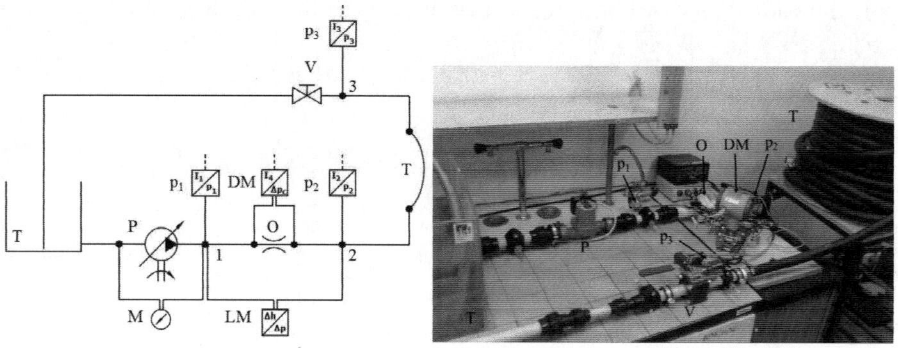

Figure 1. Experimental equipment for measuring the hydraulic hammer and circuit diagram.

The circuit is composed of following elements: water tank (T), pump (P), digital manometer (M), orifice (O) with a differential manometer of gauge for measuring the flow rate (DM), liquid manometer (LM), ball valve (V), long tube (T), three electrical pressure sensors p_1, p_2 and p_3 (connected to the sampling points 1 (behind the pump), 2 (behind the orifice) and 3 (before the closing ball valve)) and the connecting elements of the pipeline.

On the circuit for measuring the hydraulic hammer, the static characteristics of the hydrodynamic pump for the maximum speed of 2 000, the orifice and the ball valve at full opening, and their loss / flow coefficients were measured. Subsequently, the hydraulic hammer was measured for comparison with the mathematical model.

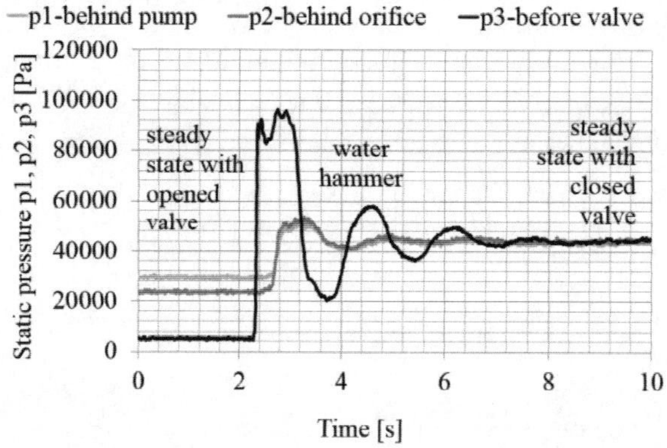

Figure 2. Static pressure behind pump, behind orifice, before valve.

The circuit can be considered a low-pressure system. Figure 2 evaluates the measured transient characteristics where the curves present steady states in the left and right parts of the graph and the hydraulic hammer is evident in the middle [4, 7]. The results are summarized in Table 1.

Table 1. Measured and calculated results.

flow rate Q	m^3/hour	1.120
wave run time T	s	0.86
real velocity of sound a_s	m/s	111.63
real bulk modulus of elasticity K_s	Pa	12 438 360
closing time of the valve t_u	s	0.11
mean pressure behind the pump p_1	Pa	29599
mean pressure behind the orifice p_2	Pa	23783
mean pressure before the valve p_3	Pa	4985

3. One-dimensional model (electro-hydraulic analogy)

For hydraulic hammer modeling the hydraulic circuit was created in the Matlab-SimHydraulics, see Figure 3. The tank is defined by the given level of the water, the energy source is the pump given by the static characteristic, the orifice is like the resistance to flowing, the long pipe is given as segmented pipe, which is presented approximately by number of T-cells, the ball valve is controlled by the closing signal and the short pipe feeds the water back to the tank. There are four pressure sensors and a flowmeter in the circuit. An important element is the custom hydraulic fluid defining the parameters such as density and kinematic viscosity of the fluid, the measured or the theoretical modulus of water elasticity and air content [6]. The numerical parameters and the solution method are entered in the configuration and simulation parameters.

The first step was to verify the stationary solution, i.e. the conformity of measured and calculated values of pressure and flow rate, and then to resolve and verify transient solution of hydraulic hammer.

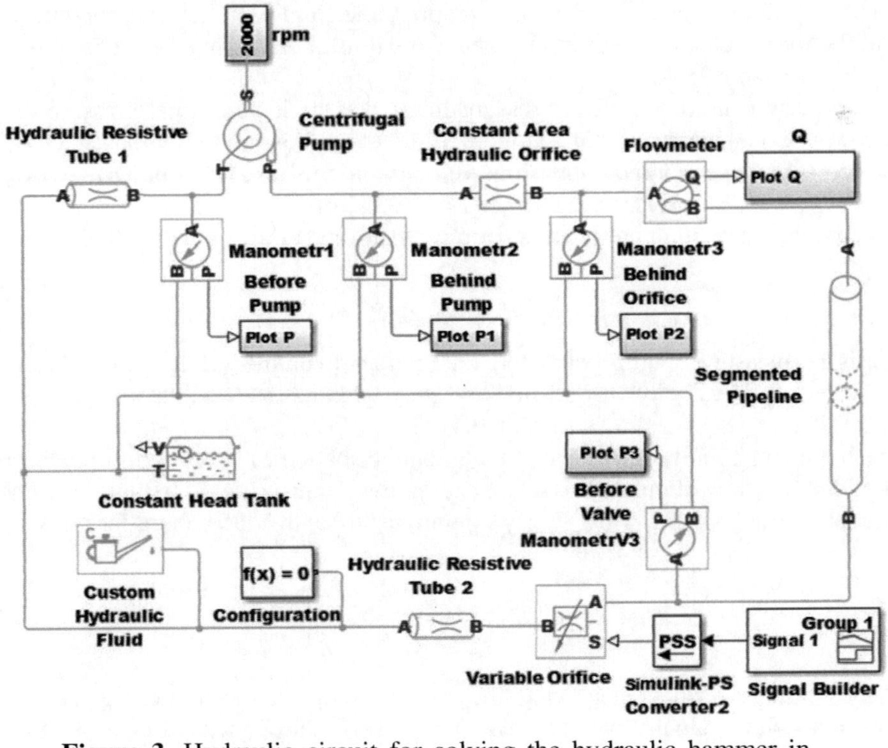

Figure 3. Hydraulic circuit for solving the hydraulic hammer in Matlab-SimHydraulics.

Table 2. The parameters for definition the fluid.

	Variant 1	Variant 2	Units
Fluid density	998	998	$kg.m^{-3}$
Kinematic viscosity	0,000001	0,000001	$m^2.s^{-1}$
Bulk modulus	12 401 859	2 300 000 000	Pa
Relative amount of trapped air	0	0,02	1

4. Two-dimensional modeling of final volum method

For two-dimensional modeling, the geometry, physical properties and boundary conditions were defined based on the experiment. Pipeline geometry has been simplified as axially symmetrical region for simulation in software Ansys Fluent. The computational mesh was refined near the wall and contained a smaller number of cells (300 970). Then the calculation was not time consuming. The area diagram and seven created boundary conditions are shown in Figure 4.

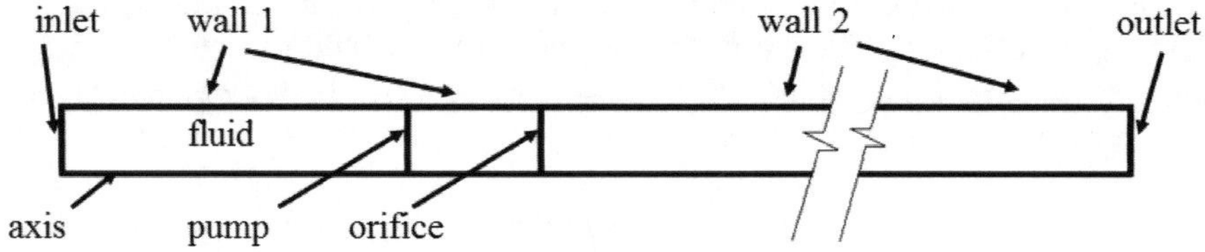

Figure 4. The area diagram and boundary conditions.

The task was solved as axially symmetrical and time-dependent with the selected time step of 0.001 s. Due to the lower Reynolds numbers, a two-equation k-ε RNG turbulent model was chosen as the computational model, the wall function was setting as "Standard Wall Function" [2]. The flowing medium (water) was defined as a single-phase compressible fluid with the bulk modulus of elasticity obtained from the measurement and as a multiphase mixture of water and air (density is given by the equation of state), see table 2.

The inlet boundary condition called "Pressure-Inlet" was defined as total pressure of 2 693.8 Pa (static pressure was calculated as a hydrostatic pressure, caused by a column of water in the tank and the dynamic pressure depending on the flow velocity in the tube). Turbulence was defined by: Turbulent Intensity of 1% and Hydraulic Diameter of 0.0232 m.

The pump boundary condition has been defined as "Fan" and allows input of static characteristic by polynomial, i.e.

$$\Delta p = a_2 v^2 + a_1 v + a_0 \tag{2}$$

On the basis of measured pump characteristic for the revolution 2000 min-1, the graph of the pressure drop versus flow velocity was created and the regression method determined the coefficients as a_0=41 053.24422, a_1=-20 942.39698, a_2=2 099.41419.

The orifice boundary condition was specified by simplification as "Porous-Jump" as it represents a pressure drop due to flow around obstacle. The porous thin-walled medium of final thickness represents the pressure loss given by Darcy's relationship and added inert loss according to the relationship

$$p = -\left(\frac{\mu}{\alpha} v + C_2 \frac{1}{2} \rho v^2\right) t \tag{3}$$

From the values obtained when measuring the orifice static characteristic, the regression of the pressure loss versus flow velocity was created $\Delta p = c_2 v^2 + c_1 v$. Subsequently, face permeability $1/\alpha$=2 694 205 m^2 and pressure-jump coefficient C_2=46,87922 m^{-1} at porous medium thickness t=0,49 m was determined [2, 8].

The output boundary condition was set as "Velocity-Inlet". The size of velocity was calculated from the measured volume flow rate Q_v and tube internal diameter (v = 0,633 ms^{-1}). Turbulence was defined by turbulent intensity by value of 1% and a hydraulic diameter of 0,0232 m. The velocity value was set to 0 for closed valve. The closing time is assumed to be 0,11 s. The boundary condition was specified using the table as instructed [1, 2]. On the walls, the roughness has been estimated so that the pressure drop of the stationary result matches the measurement.

5. Comparison of results

The results of the mathematical modeling of the two variants with respect to the modulus of elasticity obtained from simulations using Matlab-SimHydraulics and Ansys-Fluent were compared with the experimental results with the same specification of boundary conditions, element characteristics and wall roughness. Figure 5 compares the pressure values before the valve, which is the most interesting for the hydraulic hammer and the simulation of the flow rate is evaluated in Figure 6. The dynamic flow rate was not measured with the use of existing measuring devices.

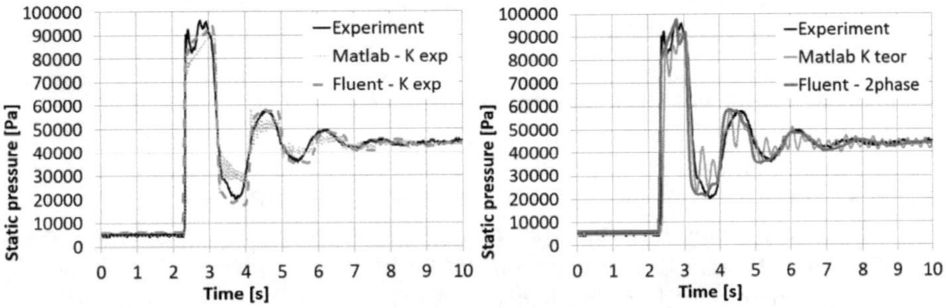

Figure 5. Comparison of static pressure versus time between simulation results and experiment result.

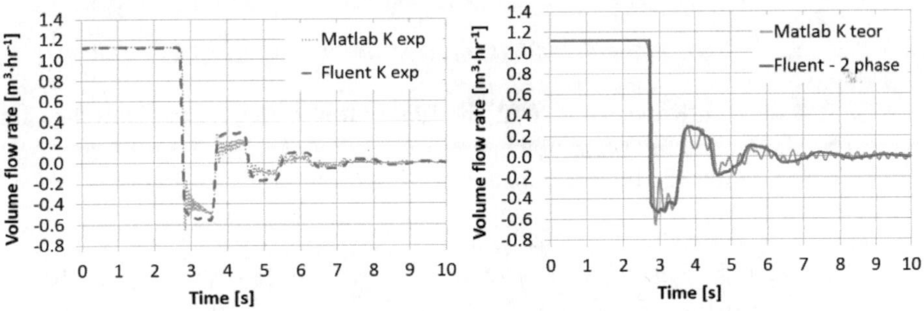

Figure 6. Comparison of volume flow versus time from simulation results.

6. Conclusion

This work was devoted to hydraulic hammer simulation in the low-pressure hydraulic system. Two time-dependent mathematical models were tested by simulation methods and verified by experimental measurement. Two variants of the solution were made. The first variant supposed the fluid with zero air content and the volume modulus of elasticity obtained from hydraulic hammer measurement. The second variant solved the fluid defined by the theoretical volume modulus of elasticity (2 300 MPa) and the volume fraction of air α = 0,02.

The first one-dimensional flow model was compiled in the Matlab-SimHydraulics software. Circuit design in this program is simple and intuitive, the resulting model is very similar to a real hydraulic scheme. It is important to choose the correct numerical solver in setting. The number of segments in the segmented pipeline also has a significant influence on the accuracy of the calculation. The more

IOP Conf. Series: Materials Science and Engineering **233** (2017) 012013 doi:10.1088/1757-899X/233/1/012013

segments you choose, the more accurate the calculation and more the calculation time is longer. The second mathematical model, allowing simulation of multi-dimensional flow, was created in Ansys Fluent software under the same boundary and initial conditions. For reasons of shortening calculation times, geometry was defined as 2D and axially symmetrical. Simulated pressures were compared with the measured values and the results are evaluated. It can be said that the amplitudes and periods of static pressure are almost the same and in good accordance with experiment. Multi-dimensional modeling also enables to solve problems with complex geometries.

An important benefit of this work is that simulation of the hydraulic hammer by a one-dimensional model (SimHydraulics) and a multi-dimensional model (Fluent) are nearly of the same quality. Both programs model the real circuit, inlet parameters of the hydraulic elements are determined by static characteristics, theoretical modulus of elasticity and content of air, no auxiliary parameters are estimated or calculated. This allows to model the dynamics of the newly designed hydraulic circuit and to predict the behaviour of the pressure and flow rate at various points in the circuit, to optimize the circuit elements and subsequently prepare the design of the circuit for implementation. Further research will be focused on solving the dynamics of oil flow in hydraulic circuits.

References

[1] Fajka M 2017 Measurement of losses in the low-pressure hydraulic system with hydrostatic pressure *Diplomová práce* (Ostrava)

[2] *ANSYS Fluent User's Guide* [online]. Release 15.0. 2013, 2692 s. [cit. 2017-05-05] Dostupné z: http://148.204.81.206/Ansys/150/ANSYS%20Fluent%20Users%20Guide.pdf

[3] Wang C, Nilsson H, Yang J, Petit O 2017 1D–3D coupling for hydraulic system transient simulations *Computer Physics Communications* **210** pp 1-9

[4] Noskievič J 1995 Dynamics of fluid mechanisms (VŠB - TU Ostrava, Ostrava)

[5] Urbanowicz K, Firkowski M, Zarzycki Z 2016 Modelling water hammer in viscoelastic pipelines: short brief. *In: Journal of Physics: Conference Series* (IOP Publishing) p 012037

[6] Jablonská J 2014 Compressibility of the fluid *In EPJ Web of Conferences* (EDP Sciences) **67** p 02048

[7] Kozubková M 2009 *Mathematical models of cavitation and hydraulic hammer* (VŠB - TU Ostrava, Ostrava)

[8] Štigler J, Klas R, Kotek M, Kopecký V 2012 The Fluid Flow in the T-Junction. The Comparison of the Numerical Modeling and Piv Measurement *Procedia Engineering* **39** pp 19-27

[9] Brennen E 1995 Cavitation and bubble dynamics *Cambridge University Press*

Acknowledgment

The work presented in this paper was supported by a grant SGS „Výzkum v oblasti dynamiky tekutinových systémů“ SP2017/103.

This work was supported by The Ministry of Education, Youth and Sports from the Large Infrastructures for Research, Experimental Development and Innovations project "Innovations National Supercomputing Centre – LM2015070".

Application of vortex three-phase separators for improving the reliability of pump and compressor stations of hydrocarbon processing plants

A E Artyukhov[1], V I Sklabinskiy[1]

[1]Processes and Equipment of Chemical and Petroleum-Refineries Department, Sumy State University, 2 Rymskogo-Korsakova st., 40007, Sumy, Ukraine

E-mail: artyukhov@pohnp.sumdu.edu.ua

Abstract. The article is devoted to the description of the theoretical foundations of the work of a new type of vortex three-phase separators for hydrocarbon processing plants. The expediency of using vortex flows in the technology of three-phase separation to improve the reliability of pumping and compressor stations is substantiated. The principle of operation of separators with a variable cross-sectional area of the working space and an inter-ring drain of the liquid is presented. A theoretical description of the hydrodynamic conditions for the separation of multiphase flows is given. A scheme for the installation of hydrocarbon feedstock processing using a three-phase vortex separator is proposed.

1. Introduction

The majority of the largest oil and gas fields in the stages of primary oil separation, hydrocarbon gas preparation, transportation and compression face problems of qualitative separation of liquid and gaseous hydrocarbon feedstocks. An additional problem is the presence of solid inclusions in the hydrocarbon feedstock. This fact significantly narrows the range of equipment for further transportation of hydrocarbon raw materials and its compression.

Currently three-phase separators [1-5] are used, which are are characterized by sufficient efficiency, however:

- they have large mass-dimensional characteristics;
- the design of the separator involves a complex internal arrangement (cyclone elements, inertial and filter sections, partitions, etc.), which leads to hydraulic resistance increasing;
- when the composition of raw materials changes (which can occur over time), structural change in the internal elements of the separator is necessary.

The actual problem of chemical engineering in this industry is the creation of new types of separators with relatively simple construction, which can work on different types of raw materials (they do not need to be modified in a constructive way, further operation of the separator is possible due to adjustment of process parameters).

The introduction of new methods of intensification of heat and mass exchange processes will increase the specific productivity of equipment and ensure high indicators of products quality. The use of new technologies makes it possible to improve the mass-dimensional characteristics of separation equipment and to increase the productivity and efficiency of the separation process.

Authors [6-9] classify the methods of heat-mass transfer processes intensification into two categories:

1. Active methods: mechanical action on the surface of heat-mass transfer (rotation, surface vibration, mixing etc.); the impact on the flow by electric, magnetic or acoustic field, pressure pulsations; blowing or suctioning of working environment through porous surfaces etc.

2. Passive methods, that are based on influence principle on the flows by the surface form of individual structural elements of equipment: application of inserted intensifiers (swirling flow units), ribbing surfaces, increasing surface area of heat-mass transfer from the working environment with less heat-mass transfer coefficients; intensification of heat-mass transfer during phase transformations (surface treatment, use of surface tension effect, drop condensation), additive in liquid solid particles or gas bubbles etc.

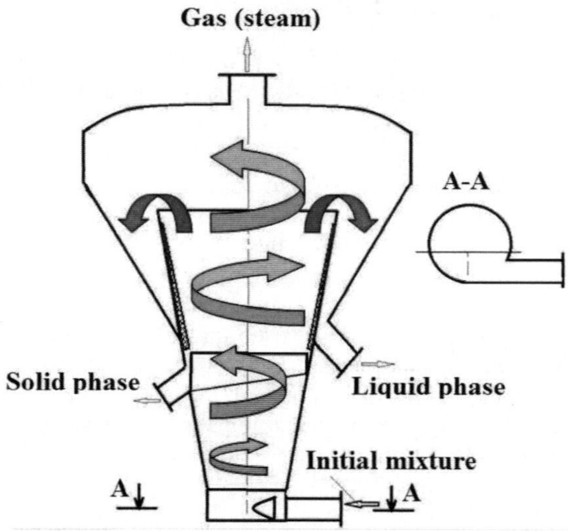

Figure 1. Schematic diagram of a three-phase separator.

Among the passive methods of heat-mass transfer processes intensification the method of inserted intensifiers application, that create vortex flow in the workspace of device, is one of the most promising.

Using of vortex flow as a method of intensifying heat exchangers or heat-mass transfer processes has spread in processes of combustion [10], absorption [11], rectification [12], in turbines [13], ejecting devices [14], reactors [15], granulators [16,17] (theoretical description and experimental study of certain aspects of granulation in vortex granulators also is considered in series of papers describing the hydrodynamics of flow motion [18, 19] and the thermodynamic conditions of granulators operation [20-23], processes of granules classification and separation [24-26], ecological aspects of block developing of vortex recycling modules for production wastes [27,28]) etc., but the use of this method in separation process is not widespread.

In this paper, we offer a description and theoretical basis of three-phase vortex separators operation (Figure 1).

2. Physical model of flows movement in vortex separator

In the separator workspace it is necessary to allocate three areas of gas flow movement in height (Figure 2). In each of the areas the intensity of gas flow movement is determined by velocity components and preferred direction of total velocity.

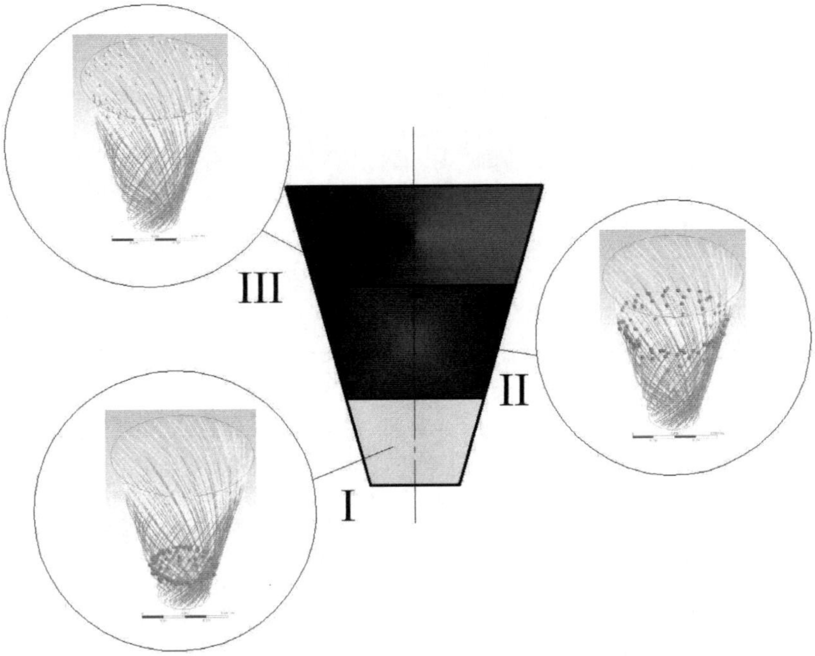

Figure 2. The main zones of gas flow movement in vortex separator in height of working space: I – zone of preferred gas flow vortex movement; II – zone of combined vortex and upward gas flow movement; III – zone of preferred seating upward gas flow movement.

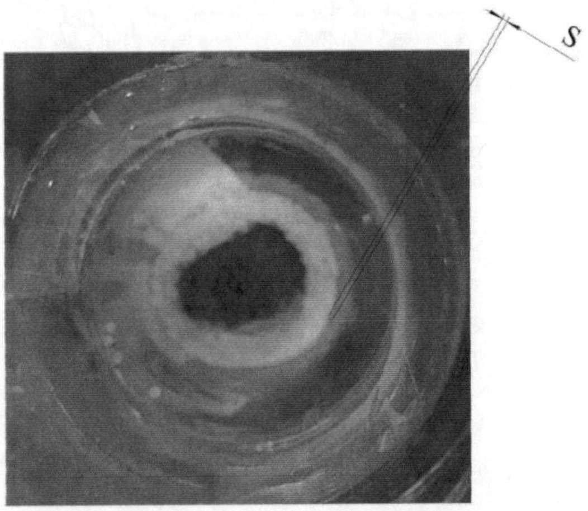

Figure 3. The phenomenon of disperse phase movement in the dense ring at some distance S from the wall of separator.

During the study of disperse phase trajectories we have found that in zone I in the intensive initial gas flow twisting we see the phenomenon of disperse phase movement in dense ring at some distance S from the wall of separator (Figure 3). This phenomenon can be explained by the presence in the vortex separator workspace significant pressure difference at the center and the periphery (Figure 4). This pressure difference is caused by the opposition to the centrifugal force. Further this phenomenon can be used to avoid contact of disperse phase with the wall of separator.

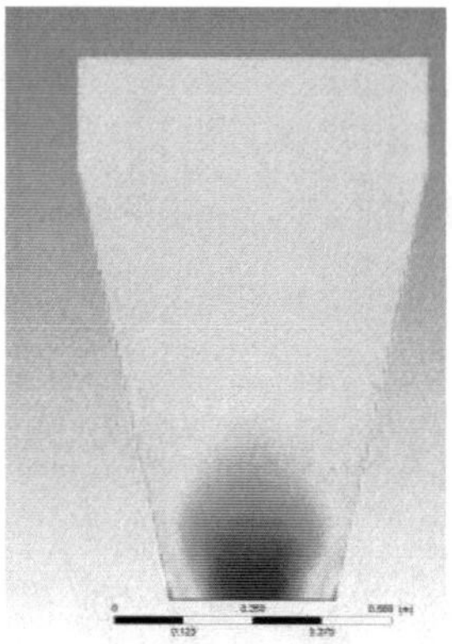

Figure 4. Pressure field in working space of vortex separator.

The feature of separator, that is investigated, is a variable height of cross sectional area of workspace. Devices with the constant cross-sectional area don`t provide the full process of classification and separation of disperse phase with a different mass (density, size) in working volume. This can be explained by the fact, that in the workspace of vortex separator consistency upward gas flow rate remains, that corresponds to the velocity of disperse phase. To carry out classifications processes in device with a constant cross-sectional area is possible with the introduction of gas to the unit in several streams with a different injection height marks. This method of classification is quite energy intensive and is not widely used.

Using separators with variable cross-sectional area of working chamber allows to do classification process of polydispersed system of wide fractional composition. Through the creation different hydrodynamic gas flow movement conditions at different height marks of working volume of device it becomes possible to classify of disperse phase to the required number of factions and to carry out separation of small particles (Figure 5).

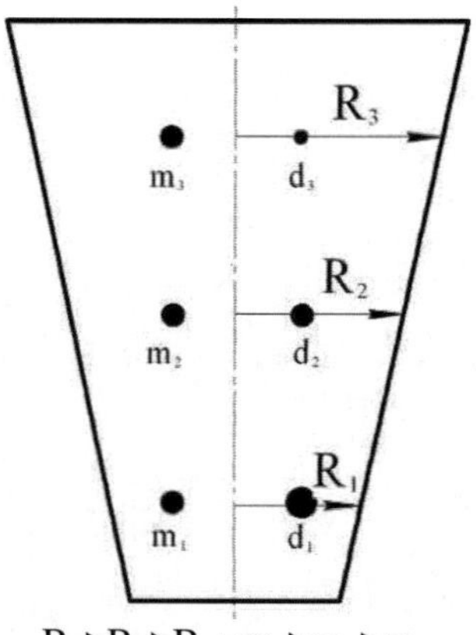

$R_1 > R_2 > R_3$; $m_1 > m_2 > m_3$

Figure 5. Distribution of disperse phase sizes and mass in the working volume of vortex separator.

3. Mathematical model of flows movement in vortex separator

Gas flow

The immediate modeling of turbulent flows by quantitative decision of Navier-Stokes equations, recorded for instant velocities, today are unsolved. However, for specific tasks solving it becomes possible application of this fundamental equation of hydrodynamics when performing of number of conditions:

- working with time-averaged value of the velocity in modeling of turbulent flows;

- selection the coordinate system, which is most satisfying for the specific case and facilitates the recording of equations.

For the simulation the solving of these equations is comfortable when using numerical methods (DEM – Discrete Element Method), defining discrete solution instead of continuous set of values in desired location (cell, grid node) of space (at stationary mode of flow motion). For maximum accuracy of solution it is chosen a way of discrete values representing, which on this occasion sample corresponds analogues of algebraic equations. As a result the mathematical problem solution of differential or integral equations can be reduced to the problem of solving the algebraic equations.

In practice, various models of numerical solution of classical hydrodynamic equations for turbulent flows, which in one way or another way are successfully used in different cases and have their advantages and disadvantages [29]: Direct numerical simulation (DNS), Reynolds Averaged Navier-Stokes (RANS), Large Eddy Simulation (LES).

Given that, in practice, interest is usually deals not with the instantaneous, but the average in time velocity values for the mathematical description of gas flow turbulent swirling motion the Reynolds equation as a modification of Navier-Stokes equations is used [24]:

$$\frac{\partial}{\partial t}\left(\rho\overline{V}_i\right)+\frac{\partial}{\partial q_j}\left(\rho\overline{V_iV_j}\right)+\frac{\partial}{\partial q_j}\left(\rho\overline{V_i'V_j'}\right)=-\frac{\partial p}{\partial q_i}+\frac{\partial}{\partial q_j}\left[\mu\left(\frac{\partial\overline{V}_i}{\partial q_j}+\frac{\partial\overline{V}_j}{\partial q_i}\right)\right]+f_i \qquad (1)$$

where $\overline{V}$ – time averaged velocity values; $\overline{V'}$ – components of velocities pulsation; μ – turbulent viscosity coefficient; t – time; ρ – gas density; p – pressure; f_i – variable, describing the action of mass forces; q_j – coordinate axes (in the case of hydrodynamic modeling in working volume of separator it is more advisable to use curvilinear coordinate system), i, j – 1…3; for the cylindrical coordinate system (Figure 6 (a)) index «1» – axial direction (z), index «2» – radial direction (r), index «3» – circular direction (φ) (Figure 6 (b)).

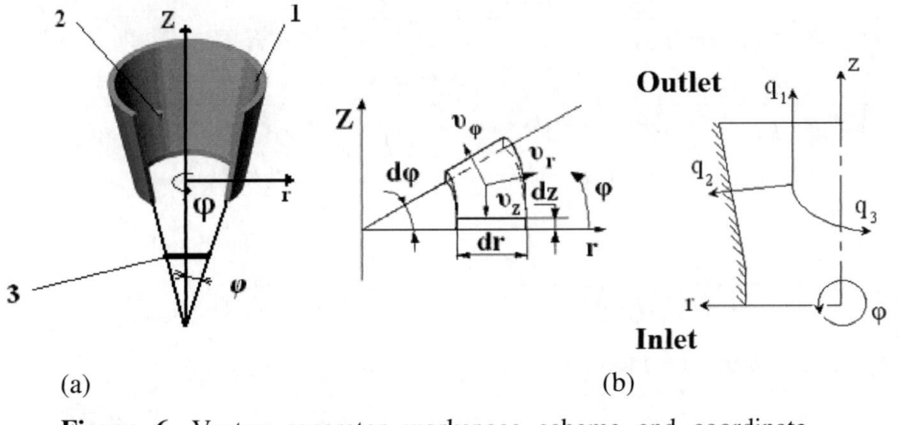

Figure 6. Vortex separator workspace scheme and coordinate system.

The Reynolds equations system is complemented with flow continuity equation

$$\frac{\partial\rho}{\partial t}+\frac{\partial}{\partial q_j}\left(\rho V_j\right)=0. \qquad (2)$$

The main advantage of description and solving method of hydrodynamics problems, based on the numerical solution of complete equations Reynolds, is precision and versatility.

For Reynolds equation (1) Boussinesq hypothesis is used [30]. According to this hypothesis, the members with velocity pulsations $\left(\rho \overline{V_i' V_j'}\right)$ in equation (3) associated with the averaged flow characteristics of such relation:

$$\rho \overline{V_i' V_j'} = -\mu \left(\frac{\partial \overline{V_i}}{\partial q_j} + \frac{\partial \overline{V_j}}{\partial q_i} \right) + \frac{2}{3} \rho \delta_{ij} k \tag{3}$$

where $k = 0,5\left(\overline{V_j' V_j'}\right)$ – kinetic energy of turbulence, $\delta_{ij} = 1$ when $i = j$, $\delta_i = 0$ and $i \neq j$.

The Reynolds equations system is elliptic. It is used to calculate the trends in those cases where flow characteristics at an arbitrary point area depend on the structure of flow both above and downstream, ie when the dominant direction of the fluid is absent or weakly expressed. Ellipticity of equations system means, that to address it it is necessary to set the boundary conditions for all variables in all borders of settlement area.

If axisymmetrical flow modeling equation of motion (1) and flow continuity (2) is significantly simplified. For curved (cylindrical) coordinate system they are follows (with the introduction of equation (1) Reynolds number $Re = V_0 D/\nu$, – where characteristic parameters, D – diameter of input section of calculation area; V_0 – average expenditure rate in the input section; ν – kinematic viscosity):

– Reynolds equations projected on the axial direction q_1:

$$\begin{aligned}
&\frac{V_1}{H_1}\frac{\partial V_1}{\partial q_1} + \frac{V_2}{H_2}\frac{\partial V_1}{\partial q_2} - \frac{V_2^2}{H_1 H_2}\frac{\partial H_2}{\partial q_1} - \frac{V_3^2}{H_1 H_3}\frac{\partial H_3}{\partial q_1} + \frac{1}{H_2}\frac{\partial\left(\overline{V_1' V_2'}\right)}{\partial q_2} + \frac{\left(\overline{V_1' V_2'}\right)}{H_1^2 H_2 H_3}\frac{\partial\left(H_1^2 H_3\right)}{\partial q_2} + \\
&+\frac{1}{H_1}\frac{\partial\left(\overline{V_1' V_1'}\right)}{\partial q_1} + \frac{\left(\overline{V_1' V_1'}\right)}{H_1^2 H_2 H_3}\frac{\partial\left(H_1 H_2 H_3\right)}{\partial q_1} - \frac{\left(\overline{V_2' V_2'}\right)}{H_1 H_2}\frac{\partial H_2}{\partial q_1} - \frac{\left(\overline{V_3' V_3'}\right)}{H_1 H_3}\frac{\partial H_3}{\partial q_1} = -\frac{1}{H_1}\frac{\partial p}{\partial q_1} + \\
&+\frac{1}{\mathrm{Re}}\left(\frac{1}{H_1^2}\frac{\partial^2 V_1}{\partial q_1^2} + \frac{1}{H_2^2}\frac{\partial^2 V_1}{\partial q_2^2} + \frac{1}{H_1 H_2 H_3}\frac{\partial V_1}{\partial q_1}\frac{\partial (H_2 H_3 / H_1)}{\partial q_1} + \frac{1}{H_1 H_2 H_3}\frac{\partial V_1}{\partial q_2}\frac{\partial (H_1 H_3 / H_2)}{\partial q_2} + \right.\\
&-\frac{2}{H_1 H_2^2}\frac{\partial H_2}{\partial q_1}\frac{\partial V_2}{\partial q_2} + \frac{V_1}{H_1}\frac{\partial}{\partial q_1}\left(\frac{1}{H_1 H_2 H_3}\frac{\partial\left(H_2 H_3\right)}{\partial q_1} \right) + \\
&\left.+\frac{V_2}{H_1}\frac{\partial}{\partial q_1}\left(\frac{1}{H_1 H_2 H_3}\frac{\partial\left(H_1 H_3\right)}{\partial q_2} \right) - \frac{V_2}{H_2 H_3}\frac{\partial}{\partial q_2}\left(\frac{H_3}{H_1 H_2}\frac{\partial H_2}{\partial q_1} \right)\right)
\end{aligned} \tag{4}$$

– Reynolds equations in projection on the radial direction q_2:

$$\begin{aligned}
&\frac{V_1}{H_1}\frac{\partial V_2}{\partial q_1} + \frac{V_2}{H_2}\frac{\partial V_2}{\partial q_2} - \frac{V_1 V_2}{H_1 H_2}\frac{\partial H_2}{\partial q_1} - \frac{V_3^2}{H_2 H_3}\frac{\partial H_3}{\partial q_2} + \frac{1}{H_1}\frac{\partial\left(\overline{V_1' V_2'}\right)}{\partial q_1} + \frac{\left(\overline{V_1' V_2'}\right)}{H_1 H_2^2 H_3}\frac{\partial\left(H_2^2 H_3\right)}{\partial q_1} + \\
&+\frac{1}{H_2}\frac{\partial\left(\overline{V_2' V_2'}\right)}{\partial q_2} + \frac{\left(\overline{V_2' V_2'}\right)}{H_1 H_2^2 H_3}\frac{\partial\left(H_1 H_2 H_3\right)}{\partial q_2} - \frac{\left(\overline{V_3' V_3'}\right)}{H_2 H_3}\frac{\partial H_3}{\partial q_2} = -\frac{1}{H_2}\frac{\partial p}{\partial q_2} + \\
&+\frac{1}{\mathrm{Re}}\left(\frac{1}{H_1^2}\frac{\partial^2 V_2}{\partial q_1^2} + \frac{1}{H_2^2}\frac{\partial^2 V_2}{\partial q_2^2} + \frac{1}{H_1 H_2 H_3}\frac{\partial V_2}{\partial q_1}\frac{\partial (H_2 H_3 / H_1)}{\partial q_1} + \frac{1}{H_1 H_2 H_3}\frac{\partial V_2}{\partial q_2}\frac{\partial (H_1 H_3 / H_2)}{\partial q_2} + \right.\\
&\left.+\frac{2}{H_1 H_2^2}\frac{\partial H_2}{\partial q_1}\frac{\partial V_1}{\partial q_2} + \frac{V_2}{H_2}\frac{\partial}{\partial q_2}\left(\frac{1}{H_1 H_2 H_3}\frac{\partial\left(H_1 H_3\right)}{\partial q_2} \right)\right)
\end{aligned} \tag{5}$$

– Reynolds equations projected on circular direction q_3:

$$\frac{V_1}{H_1}\frac{\partial V_3}{\partial q_1}+\frac{V_2}{H_2}\frac{\partial V_3}{\partial q_2}+\frac{V_1V_3}{H_1H_3}\frac{\partial H_3}{\partial q_1}+\frac{V_2V_3}{H_2H_3}\frac{\partial H_3}{\partial q_2}+\frac{1}{H_2}\frac{\partial\left(\overline{V_2'V_3'}\right)}{\partial q_2}+\frac{\left(\overline{V_2'V_3'}\right)}{H_1H_2H_3^2}\frac{\partial\left(H_1H_3^2\right)}{\partial q_2}+$$
$$+\frac{1}{H_1}\frac{\partial\left(\overline{V_1'V_3'}\right)}{\partial q_1}+\frac{\left(\overline{V_1'V_3'}\right)}{H_1H_2H_3^2}\frac{\partial\left(H_2H_3^2\right)}{\partial q_1}=\frac{1}{\mathrm{Re}}\left(\frac{1}{H_1^2}\frac{\partial^2V_3}{\partial q_1^2}+\frac{1}{H_2^2}\frac{\partial^2V_3}{\partial q_2^2}+\right.\tag{6}$$
$$\left.+\frac{1}{H_1H_2H_3}\frac{\partial V_3}{\partial q_1}\frac{\partial(H_2H_3/H_1)}{\partial q_1}+\frac{1}{H_1H_2H_3}\frac{\partial V_3}{\partial q_2}\frac{\partial(H_1H_3/H_2)}{\partial q_2}+\frac{V_3}{H_1H_2}\frac{\partial}{\partial q_2}\left(\frac{H_1}{H_2H_3}\frac{\partial H_3}{\partial q_2}\right)\right)$$

– continuity equation:

$$\frac{1}{H_1H_2H_3}\left(V_1\frac{\partial(H_2H_3)}{\partial q_1}+V_2\frac{\partial(H_3H_1)}{\partial q_2}\right)+\frac{1}{H_1}\frac{\partial V_1}{\partial q_1}+\frac{1}{H_2}\frac{\partial V_2}{\partial q_2}=0\tag{7}$$

where H_1, H_2, H_3 – Lame coefficient [31].

Further simplification of equations system (4)-(7) for gas phase vortex flow simulation in the separator workspace is possible using the following assumptions [32]:

- presence of flow dominant direction, along which the axial component of gas flow velocity is everywhere positive and far exceeds the radial;
- component of gas flow velocity in the axial direction varies considerably slower than in the radial;
- velocity and pressure in every gas flow elementary volume depends only on the downstream conditions and don`t depend on upstream conditions.

These assumptions allow to conduct analysis of components orders in equations (4)-(7) and discard those, that provide significant impact on the result of the calculation.

After accounting assumptions for axisymmetrical gas flow equation (4) – (7) will be written as

$$\frac{V_1}{H_1}\frac{\partial V_1}{\partial q_1}+\frac{V_2}{H_2}\frac{\partial V_1}{\partial q_2}-\frac{V_3^2}{H_1H_3}\frac{\partial H_3}{\partial q_1}+\frac{1}{H_2}\frac{\partial\left(\overline{V_1'V_2'}\right)}{\partial q_2}+\frac{\left(\overline{V_1'V_2'}\right)}{H_1^2H_2H_3}\frac{\partial\left(H_1^2H_3\right)}{\partial q_2}=$$
$$=-\frac{1}{H_1}\frac{\partial p}{\partial q_1}+\frac{1}{\mathrm{Re}}\left(\frac{1}{H_2^2}\frac{\partial^2V_1}{\partial q_2^2}+\frac{1}{H_1H_2H_3}\frac{\partial V_1}{\partial q_2}\frac{\partial(H_1H_3/H_2)}{\partial q_2}\right)\tag{8}$$

$$\frac{\partial p_r}{\partial q_2}=\frac{V_3^2}{H_3}\frac{\partial H_3}{\partial q_2}\tag{9}$$

$$\frac{V_1}{H_1}\frac{\partial V_3}{\partial q_1}+\frac{V_2}{H_2}\frac{\partial V_3}{\partial q_2}+\frac{V_1V_3}{H_1H_3}\frac{\partial H_3}{\partial q_1}+\frac{V_2V_3}{H_2H_3}\frac{\partial H_3}{\partial q_2}+\frac{1}{H_2}\frac{\partial(\overline{V_2'V_3'})}{\partial q_2}+\frac{(\overline{V_2'V_3'})}{H_1H_2H_3^2}\frac{\partial(H_1H_3^2)}{\partial q_2}=$$
$$=\frac{1}{\mathrm{Re}}\left(\frac{1}{H_2^2}\frac{\partial^2V_3}{\partial q_2^2}+\frac{1}{H_1H_2H_3}\frac{\partial V_3}{\partial q_2}\frac{\partial(H_1H_3/H_2)}{\partial q_2}+\frac{V_3}{H_1H_2}\frac{\partial}{\partial q_2}\left(\frac{H_1}{H_2H_3}\frac{\partial H_3}{\partial q_2}\right)\right)\tag{10}$$

$$\frac{1}{H_1H_2H_3}\left(V_1\frac{\partial(H_2H_3)}{\partial q_1}+V_2\frac{\partial(H_3H_1)}{\partial q_2}\right)+\frac{1}{H_1}\frac{\partial V_1}{\partial q_1}+\frac{1}{H_2}\frac{\partial V_2}{\partial q_2}=0\tag{11}$$

This system of equations is closed with the equation of sustainability costs:

$$\int_0^{Q_2} V_1 \, H_2 H_3 dq_2 = const \tag{12}$$

where Q_2 – coordinate q_2 on the wall of the working volume of the vortex separator.

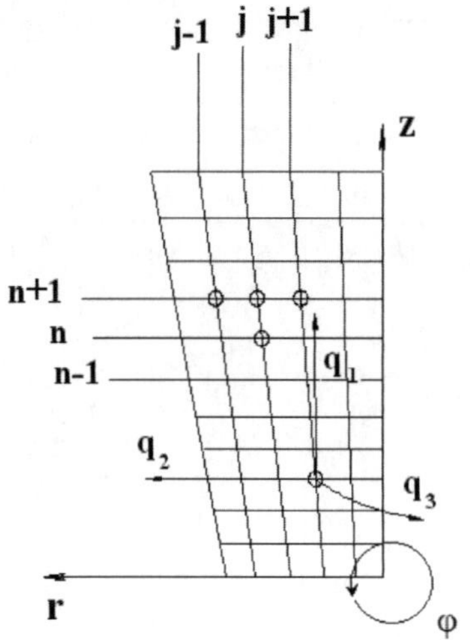

Figure 7. Construction of calculation grid.

The resulting system of equations (8) - (12) is parabolic in nature, and its decision based on the method proposed Patankar and Spalding [33] and realized in the process SIMPLE (Semi-Implicit Method for Pressure-Linked Equations) and its modifications.

Numerical solutions of equations of mathematical models implemented in one marching passage from the input section to the initial working volume using finite volume elements of finite-element approach. Before calculating the estimated net constructed (Figure 7), and unknown values of velocity and pressure are found in the nodes of the grid.

For computer simulation and visualization of research results in this study it was used complex ANSYS CFX, which allows to export obtained using program «Conical channel»©.

Disperse phase (solid particles, liquid)

This approach is used to simulate two-phase flows in which the substance of one of the phases is presented in the form of dispersed particles, and the volume fraction, occupied by these particles, is small (up to 10% of the total volume). The substance that forms the main phase is assumed to be a continuous medium, and its flow is modeled by the Navier-Stokes (or Reynolds) and continuity equations. The substance present in the flow in the form of discrete particles does not form a continuous medium, the individual particles interact with the main phase flow and with each other discretely. To model the motion of the particles of the scattered phase, the Lagrange approach is used, i.e. the movement of separately particles of the scattered phase is monitored under the action of forces on the side of the main phase flow.

The forces acting on the dispersed phase are caused by the difference in the velocity of the disperse phase and the flow velocity of the main phase, as well as the displacement of the main phase by the dispersed phase. The equation of motion of such a particle was derived in [10] and has the form:

$$\begin{aligned} m_p \frac{dv_p}{dt} &= 3\pi\mu d C_{cor}\left(v_f - v_p\right) + \frac{\pi d^3 \rho_f}{6}\frac{dv_f}{dt} + \frac{\pi d^3 \rho_f}{12}\left(\frac{dv_f}{dt} - \frac{dv_p}{dt}\right) + \\ &+ F_e - \frac{\pi d^3}{6}\left(\rho_p - \rho_f\right)\vec{\omega} \times \left(\vec{\omega} \times \vec{r}\right) - \frac{\pi d^3 \rho_p}{3}\left(\vec{\omega} \times v_p\right) \end{aligned} \tag{13}$$

Here m_p - particle`s mass, d - particle`s diameter, v - velocity, μ - dynamic viscosity of the main phase substance, C_{cor} – its viscous drag coefficient; $\vec{\omega}$ - angular velocity of rotation, $\vec{r}$ - radius vector (when considering motion in the relative frame of reference). The index p refers to the disperse phase, index f - to the substance of the main phase.

The left-hand side of equation (13) is the sum of all the forces acting on the particle, expressed in terms of the mass and acceleration of this particle. The first element on the right-hand side expresses the deceleration of the particle as a result of viscous friction against the flux of the main phase according to the Stokes law. The second element is the force applied to the particle, due to the pressure drop in the main phase surrounding the particle caused by the acceleration of the main phase flow. The third term is the force required to accelerate the weight of the main phase in the volume displaced by the particle. These two elements must be taken into account when the density of the main phase exceeds the particle density, for example, when considering air bubbles in a liquid flow. The fourth element (Fe) is an external force acting directly on a particle, for example, gravity or the strength of an electric field. The last two elements are the centrifugal force and the Coriolis force, which take place only when considering motion in a relative frame of reference. In addition, sometimes it is necessary to take into account some additional forces on the right-hand side of (13) (for example, if there is a significant temperature difference in the flow).

Equation (13) is a first-order differential equation in which the only unknown quantity is the particle velocity vp, and the argument is the time t. The velocity of the substance of the main phase vf is assumed to be known throughout the space points. As initial data, in addition to the size and properties of the particle, its position at the initial instant of time is given. It is also indicated what should occur when a particle collides with a wall or with another particle. To carry out the calculation, the terms containing vp are transferred to the left-hand side of equation (15). The velocity and position of the particle at each subsequent moment of time is determined by numerical integration with respect to time with some step Δt of all the remaining terms of equation (15).

The coefficient of viscous resistance C_{cor} at moderate Reynolds numbers 0.01 <Rep <260 can be calculated, for example, using formulas

$$C_{cor} = \begin{cases} 1+0.1315\left(\mathrm{Re}_p\right)^{0.82-0.05\alpha} & npu\ \mathrm{Re}_p < 20 \\ 1+0.1935\left(\mathrm{Re}_p\right)^{0.6305} & npu\ \mathrm{Re}_p > 20 \end{cases} \tag{14}$$

where $Re_p = \rho_f \mid v_f - v_p \mid d / \mu$, $\alpha = \log Re_p$.

4. Conclusions

The proposed design of the vortex separator makes it possible to achieve:

- prolongation of equipment service life;
- reducing the cost of equipment as a result of improving its mass-size characteristics
- reducing the downtime during the modernization of equipment
- reduction of operating and capital costs.

On Figure 8 the basic schemes for the preparation of hydrocarbon gas before its further processing and the compressor section are shown; at the primary separation stage the vortex separator is used. The use of a vortex separator makes it possible to eliminate filtering equipment from the plant and to ensure a better quality of separation of the multicomponent mixture.

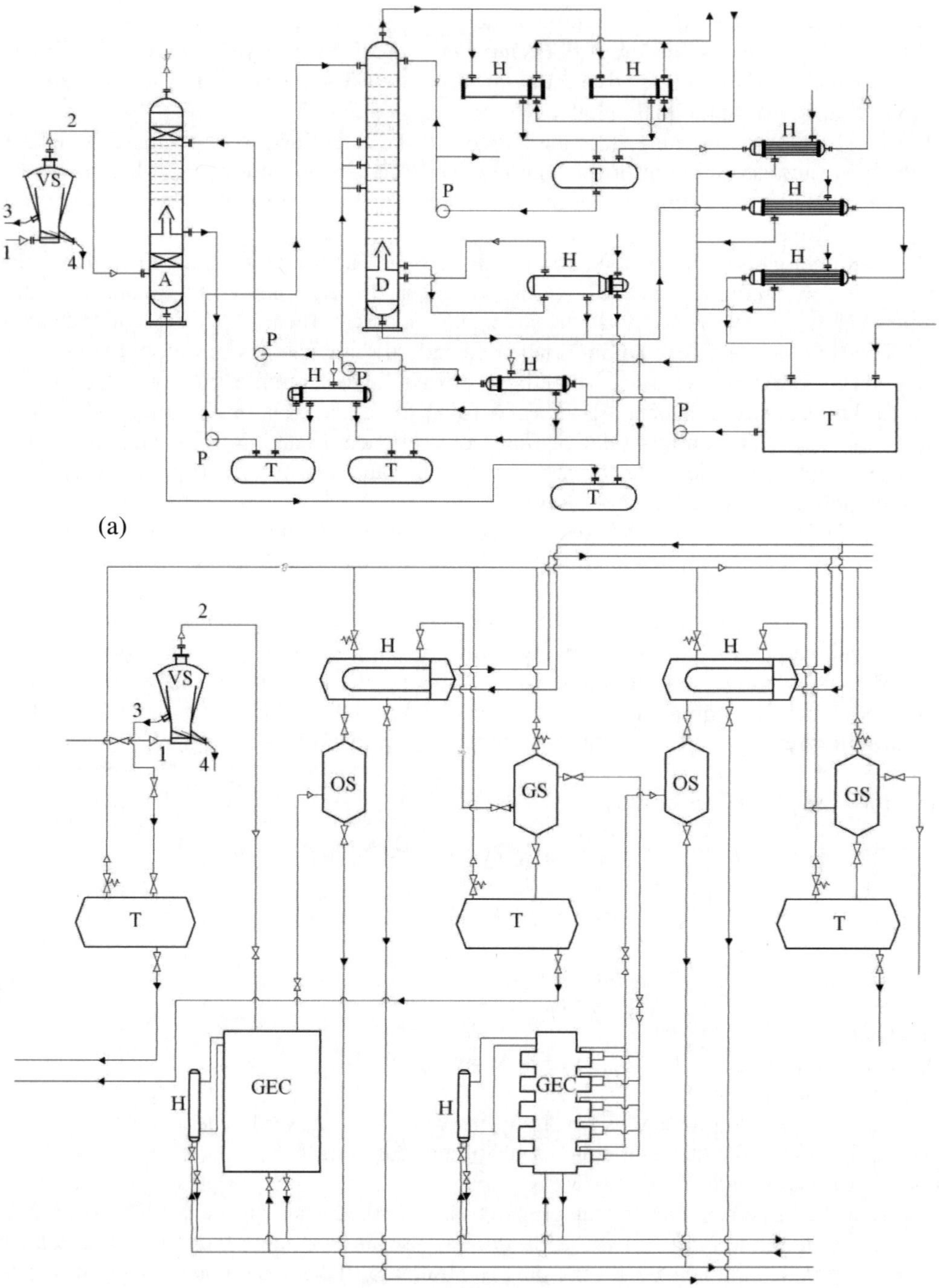

Figure 8. Schematic diagram of the unit for the preparation of hydrocarbon gas (a) and booster compressor station (b): 1- initial mixture (gas-liquid-solid); 2 - hydrocarbon gas; 3 - liquid hydrocarbons + water; 4 - mechanical impurities; VS – vortex separator; A – absorber; D – desorber; T – tank; H – heat exchanger; P – pump; GS – gas separator; OS – oil separator; GEC – gas engine compressor.

References

1. Yucheng L , Qixuan C, Bo Z, Feng T 2015 *Chemical Engineering Research and Design* **100** 554–560
2. Ahmed M, Ibrahim G, Farghaly M 2009 *Int. J. Miner. Process.* **91** pp 34–40
3. Guo G, Zhang F, Deng S, Chen Z 2010 *Chem. Eng. Mach.* **37** pp 128–130
4. Zhao L, Jiang M, Xu B, Zhu B 2012 *Chem. Eng. Res. Des.* **90** pp 2129–2134
5. Kharoua N, Khezzar L, Nemouchi Z 2010 *Petroleum Science and Technology* **28** (7) pp 738-755.
6. Aoune A, Ramshaw C 1999 *International Journal of Heat and Mass Transfer* **14** pp 2543–2556
7. Stankiewicz A, Moulijn J 2002 *Industrial & Engineering Chemistry Research* **41** (8) pp 1920-1924
8. Stankiewicz A I, Moulijn J A 2004 *Re-Engineering the Chemical Processing Plant: Process Intensification* (New York: Marcel Dekker) Keil J F (Ed.) 2007 *Modeling of Process Intensification* (Weinheim: Wiley-VCH Verlag GmbH&Co. KGaA)
9. Moulijn J A ; Stankiewicz A, Grievink J, Gorak A 2008 *Computers and Chemical Engineering* **32** (1-2) pp 3-11
10. Kaewklum R, Kuprianov V I 2010 *Fuel* **89** pp 43-52
11. Volchkov E P, Dvornikov N A, Lukashov V V, Abdrakhmanov R Kh, 2013 *Thermophysics and Aeromechanics* **20** (6) pp 663-669
12. Khalatov A A 2010 *Journal of Engineering Physics and Thermophysics* **83** (4) pp 794-800
13. Levchenko D, Arseniev V, Meleychuk S 2012 *Procedia Engineering* **39** pp 28-34
14. Shiomi Y, Kutsuna H, Akagawa K, Ozawa M 1993 *Nuclear Engineering and Design* **141** pp 27-34
15. Ashcraft RW, Heynderickx G J, Marin G B 2012 *Chemical Engineering Journal* **207-208** pp195-208
16. Artyukhov A E, Sklabinskyi V I 2013 *Naukovyi Visnyk Natsionalnoho Hirnychoho Universytetu* **6** pp 42-48
17. Artyukhova N A, Shandyba A B, Artyukhov A E 2014 *Naukovyi Visnyk Natsionalnoho Hirnychoho Universytetu* **1** pp 92-98
18. Artyukhov A E, Sklabinskyi V I 2015 *Chemistry & chemical technology* **9** (2) pp 175-180
19. Artyukhov A E, Sklabinskyi V I 2015 *Chemistry & chemical technology* **9** (3) pp 337-342
20. Artyukhov A E, Sklabinskyi V I 2016 *Journal of Nano- and Electronic Physics* **8** (4) 04051.
21. Artyukhov A E, Sklabinskyi V I 2016 *Journal of Nano- and Electronic Physics* **8** (4) 04083.
22. Artyukhov A E, Voznyi AA 2016 *6th International Conference Nanomaterials: Application & Properties (NAP-2016)* **5 (2)** 02NEA01.
23. Artyukhov A E, 2016 *6th International Conference Nanomaterials: Application & Properties (NAP-2016)* **5** (2) 02NEA02.
24. Artyukhov A E, Fursa A S, Moskalenko K V 2015 *Chemical and Petroleum Engineering* **51** (5-6) pp 311-318
25. Artyukhov A 2016 *CEUR Workshop Proceedings* **1761** pp 363-373
26. Artyukhov A, Sklabinskiy V , Ivaniia A , Moskalenko K 2016 *CEUR Workshop Proceedings* **1761** pp 374-385
27. Prokopov M G, Levchenko D A, Artyukhov A E 2014 *Applied Mechanics and Materials* **630** pp 109-116.
28. Artyukhov A E 2014 *Chemical and Petroleum Engineering* **49** (11-12) pp 736-740
29. Pope S B 2000 *Turbulent flows* (Cambridge: Cambridge University Press)
30. Schmitt F G 2007 *Comptes Rendus M'ecanique* **335** (9-10) pp 617-627
31. Zou J 1998 *Int. J. Comput. Math.* **70** pp 211–232
32. Hamrock B J, Schmid S R, Jacobson B O 2004 *Fundamental of Fluid Film Lubrication* (New York, Basel: Marcel Dekker, Inc.)
33. Patankar S V, Spalding D B 1972 *Int. J. Heat Mass Transfer* **15** pp 1787-1806

Acknowledgments
This work was carried out under the project «Improving the efficiency of granulators and dryers with active hydrodynamic regimes for obtaining, modification and encapsulation of fertilizers», state registration No. 0116U006812. The authors thank researchers of department "Processes and equipment of chemical and refining industries", Sumy State University for their valuable comments during the article preparation.

IOP Conf. Series: Materials Science and Engineering **233** (2017) 012015 doi:10.1088/1757-899X/233/1/012015

The hydraulic and numerical analyses of the operation of the control valve in the central heating system

J Rogula[1], P Szulc[1]

[1]Wroclaw University of Science and Technology, Faculty of Mechanical and Power Engineering, Department of Design Fundamentals and Fluid–Flow Machinery, ul. Wybrzeze Wyspianskiego 27, 50–370 Wroclaw, Poland

E-mail: janusz.rogula@pwr.edu.pl

Abstract. The article presents the results of experimental and numerical tests of a surface heating flow control system consisting of a distribution rail along with the fixed valve. The components were measured, their solid models were prepared and used to make a flow model. In a flow model geometry the measurements of the test stand elements were considered in order to compare the results.

1. Introduction

In a surface heating the whole area of a floor, wall or ceiling of the room [2] is a heater. It is a kind of a low temperature heating where the medium (water or glycol) flows through the spread on the plane built-in installation and heat exchange takes place predominantly on the basis of radiation. An advantage of this type of heating system is a possibility of spreading out the installation on very large surfaces in a few or several independent branches with the ability to control the heating of each branch from one central point. It allows to simplify the way of regulation.

The efficiency of a surface heating system and the accuracy of temperature regulation in a room depends on the quality of control valves fixed to a distribution rail (Figure 1). Up to several control valves can be fixed on one rail [3].

Warm water under pressure flows through the rail. It is then divided into particular streams flowing through the branches. The quantity of water in the branches is controlled by the valves fixed in a rail. Control with valves causes the increase of local resistance in a stream of liquid.

Knowledge of the flow resistance through the determination of the valve coefficient k_v, which can be defined on a test stand as well as numerically, by means of a CFD software, will help to specify pressure losses in a heating system and choose an appropriate pump.

2. Experimental research

2.1. Object of study

A set consisting of a brass distribution rail and a flow control valve plug were applied for the analysis concerning the operation of control valves of the surface heating. The elements presented in Figure 1 were assembled, connected to the test stand and further tested. The construction of a distribution rail and a valve were analyzed.

IOP Conf. Series: Materials Science and Engineering **233** (2017) 012015 doi:10.1088/1757-899X/233/1/012015

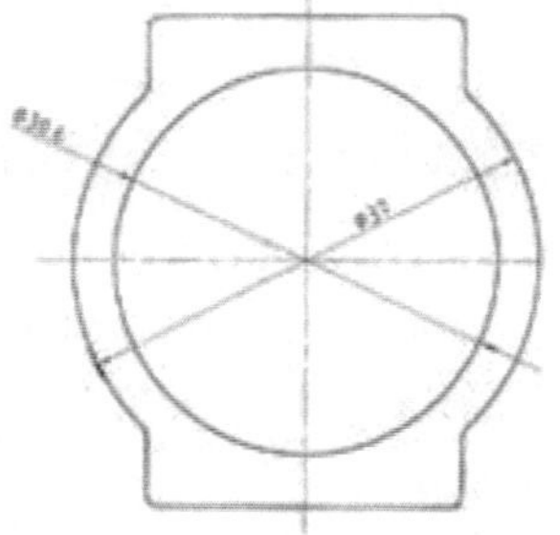

(a)

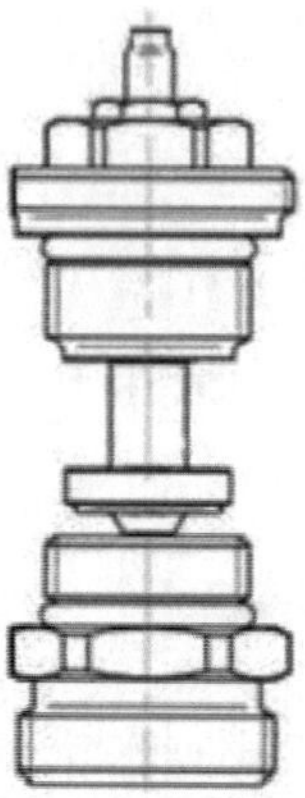

(b)

Figure 1. Object of analysis: (a) distribution rail cross-section, (b) model valve.

A distribution rail 155 mm long, of an inner diameter $\phi30,6$ mm , and outer $\phi37$ mm makes it possible to connect three control valves in a serial configuration. Three threaded holes were made in a rail G1/2 which allow to fix three enabling to put three branches. Moreover, two connectors G1 were made in the rail. One connector is used for fixing a pipeline with the supplied water and the second one allows to join another connecting rail. A valve seat ring was fixed to one side of the rail, on the other side there was the body with a fixed spindle and a valve plug of a diameter $\phi17$ mm (Figure 2). The valve plug (4) was made in a form of a vessel where the sealing was located (5) $\phi15$ mm. The task of the sealing was to eliminate any leakage in a closed position, i.e. when the valve plug touches the seat ring (6). Maximum lift of the valve plug amounts to 2.5 mm which at the same time characterizes the level of the valve regulation. A spindle (1) of $\phi8$ mm was used for flow throttling. The valve spindle for tests was driven by means of a handle connected to the pin sealed with an O-ring in body stuffing box (3). The force deforming the spring (2) regulates the pressure of the valve plug to the seat ring.

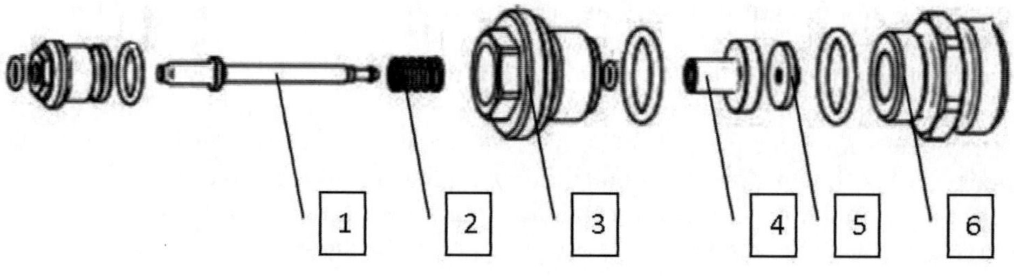

Figure 2. Model valve, dedicated for tests: 1– spindle, 2 – spring, 3 – body, 4 – valve plug, 5 – elastic insertion, 6 – seat ring.

In order to prepare the distribution rail for tests one of the connectors as well as two pairs of the outlet holes were covered.

2.2. Test stand and measurements methodology

The tests of flow resistance through the valve were carried out on a test stand dedicated for testing central heating flow elements. The test stand was shown in Figure 3.

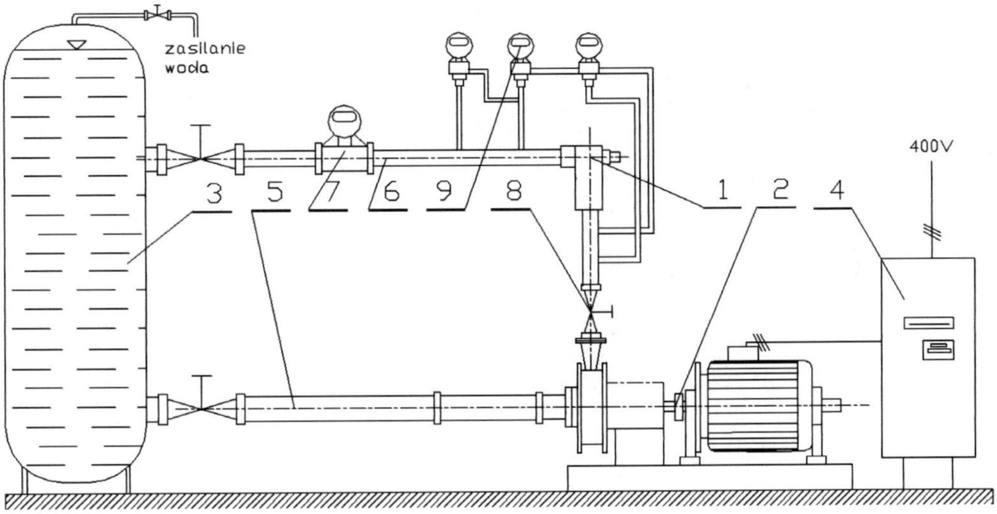

Figure 3. Test stand for testing flow elements of central heating systems: 1 – rail with tested valve, 2 – pump, 3 – tank, 4 – frequency converter, 5 – suction pipe, 6 – discharge pipe, 7 – electromagnetic flow meter, 8 – control valve, 9 – manometers.

Water flow on a test stand took place in a closed system. The measuring system included a rail with one valve (1) and a flow in the system was forced by means of a centrifugal pump (2). Since it was necessary to regulate pump's performance a frequency converter (4) was used in the measuring system. The feeding element was a closed tank (3) from which the pump sucked the medium with a suction pipe (5) and drained it with a discharge pipe (6). The test stand was also equipped in a control valve (8) throttling the flow before the tested valve. The flow through the valve was measured by means of an electromagnetic flowmeter (7) in grade 0,2 ensuring standardized lengths of pipelines before and after the place of measurement; due to flow disorders caused by the valve work the places of measurement were located in the distance of 15 diameters before and after the valve. The measurement of static pressure was realized by means of an electronic manometers (9) in grade 0,2. In order to compensate the pressure losses in straight pipe fractions the loss on equivalent to the length and diameter pipelines fractions was measured: feeding and return. The temperature of the medium was determined by means of a mercury thermometer. The pumped medium was pure water supply of density equal 999 kg/m^3. While taking measurements of parameters necessary to determine basic characteristic of the valve constant feeding conditions were maintained. The measurements were taken in the steady state.

To determine characteristics of the change of valve coefficient k_v as a function of the number of the regulation wheel rotations it was necessary to alter performance of the pump forcing the flow through the tested valve. It was assumed, in accordance with norm [1] that on the valve, regardless of its setting the drop of pressure equal 1 bar would be kept. To enable the realization of pressure drop condition on the valve, rotations of the pump shaft were changed and locally regulated with a control valve (8) for a given setting of the regulation wheel of the tested valve.

To determine the flow coefficient k_v the following formula was used:

$$k_v = \frac{Q}{\sqrt{\Delta p}} \tag{1}$$

where:
Q – volumetric flow,
Δp – pressure drop on the tested valve.

IOP Conf. Series: Materials Science and Engineering **233** (2017) 012015 doi:10.1088/1757-899X/233/1/012015

2.3. Results of the real research
Experimental research included determination of the flow coefficient k_v in course of the valve opening and closing. The results can be found in Figure 4.

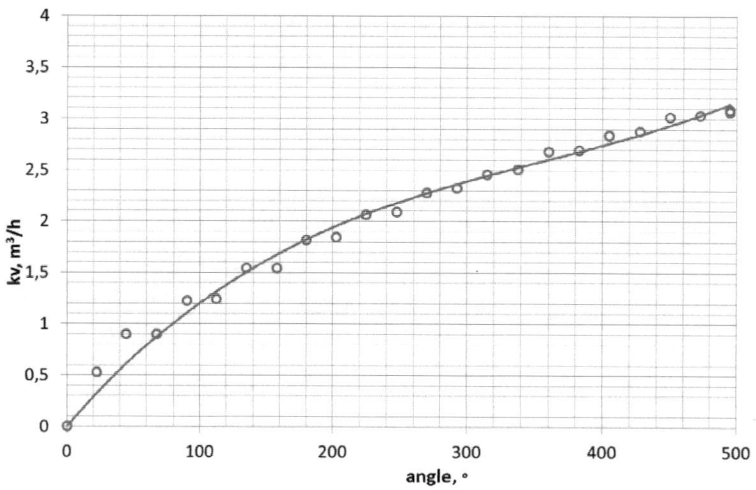

Figure 4. Characteristic of the valve fixed in a rail kv = f (rotation angle).

While analyzing the curve course it should be noticed that for a full opening of the valve the coefficient k_v equals about 3.1 m^3/h The curve of changes k_v in the whole course of changes of the control valve rotation angle is increasing. Due to steepness we can divide the characteristic into two parts. To the opening angle value 150° the increase k_v equals 1.13 m^3/h for 100° of the valve pin rotation, however, more than 150° only for 0.37 m^3/h for 100°.

3. Numerical simulations

3.1. Computational model
Computational Fluid Dynamics makes it possible to conduct numerical tests of the real objects with a view to checking the accuracy of construction calculations as well as explaining and providing the reapicture on accompanying phenomena [4,5]. Due to CFD, reducing investment costs, we are able to explain a lot of phenomena, particularly when the real measurement methods may lead to, e.g. devastation of the studied object.

The three dimensional flow geometry model subjected to simulation was presented in Figure 5. To precisely reflect the physical phenomena the 3D model constituted a carbon copy of a real model. It was decided to carry out flow simulations for three chosen settings of the valve. It was assumed that the valve plug lift would be equal 0.8 mm (Figure 6, 7), 1.6 mm (Figure 8, 9) and 2.4 mm (Figure 10, 11).

The process of model discretization was conducted by means of Ansys Icem CFD software by making a non structural grid (tetrahedral). A prismatic grid was applied close to the walls in areas partil cularly sensitive to flow disorders, i.e. valve areas. This allowed to secure the value y+ during the calculations below 1.5.

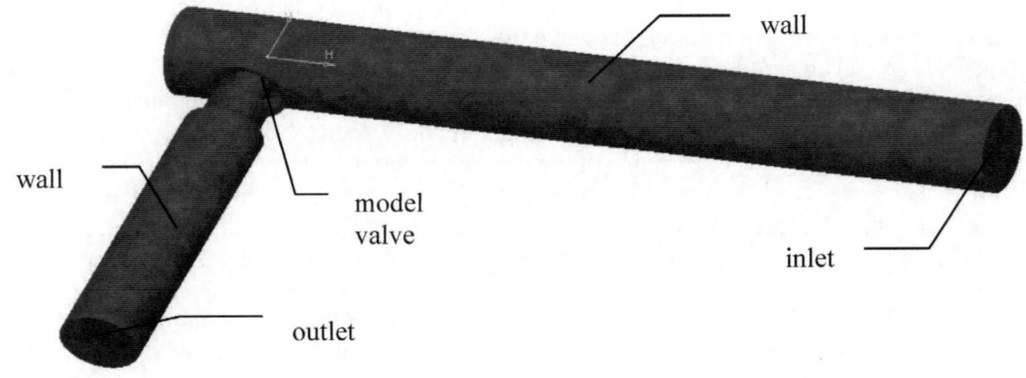

Figure 5. Three dimensional model of the flow through partly open valve fixed in a distribution rail subjected to CFD simulations.

Numerical research of the investigated object was conducted based on commercial software Ansys Fluent CFD. The software enables, with the use of the finite volume method, iterative solution of momentum, energy and mass maintaining equations. Calculations were performed by means of a turbulence model k-w SST and changeable geometric parameters of the valve. Numerical analysis was carried out as stationary, a working medium was pure water. Convergence of the results for all equations was obtained at the level 10^{-5}. In the analyzed configuration the value of static pressure as well as estimated turbulence level were assumed at the inlet and outlet. An output parameter was the flow Q. The discussed numerical model was divided by planes for a better visualization of the phenomena taking place in the channel. The distance between the planes was chosen so that, at the same time, it was possible to show the swirls of the flow. As a result of the carried out numerical simulations data enabling the analysis of a stationary flow through the tested system of the valve fixed in a distribution rail were collected.

3.2. Qualitative results of numerical simulations

The prepared computational models enabled to perform the qualitative identification of the flow in the investigated flow system.

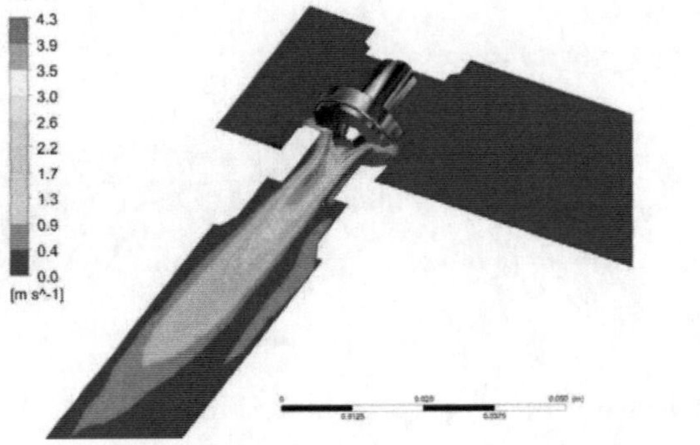

Figure 6. Water velocity distribution in the valve, valve plug lift 0.8 mm.

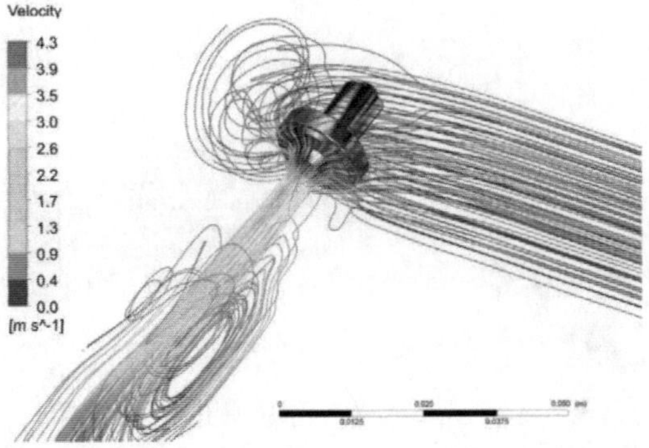

Figure 7. Tracks of water particles in the valve, valve plug lift 0.8 mm.

Figure 6 presents flow velocity distribution in the plane going through the axis of the distribution rail as well as pipe draining the liquid from the valve. Figure 7, on the other hand, presents the lines of liquid particles course for the gap between the valve plug and a seat ring equal 0.8 mm.

While analyzing the obtained results one should notice a zone of the increased velocity which occurs directly between the valve plug and the seat ring. The gap between the flat surface of the seat ring and the flat surface of the valve plug decides about the velocity. The biggest velocity occurs directly under the valve plug. The occurrence of the dead area directly under the elastic insertion of the valve plug and sharp edges of the seat ring influencing the flow decide directly about the localization of the biggest velocity. The received flow picture can be improved by the application of the valve seat ring construction with removed sharp edges on the flow way, e.g. with the rounded edges or of a bell-mouth cross-section. The zone of the non uniform flow behind the valve plug leans left which is the result of the impact on the vortex zone, visualized through the comparison of Figure 7, Figure 9 and Figure 11, where the tracks of water particles were shown at a different location of the valve plug in relation to the seat ring. The vortex zone, which causes the dead zone, fills about a half of the flow channel. It causes the increase of local velocity, therefore, the rise of flow loss. This area is localized on the internal side of the elbow which is a standard picture of the flow while maintaing a constant winding [6]. Its existence is also connected with a violent extension of the pipe, and the flow in the elbow blocks its propagation in a reverse way.

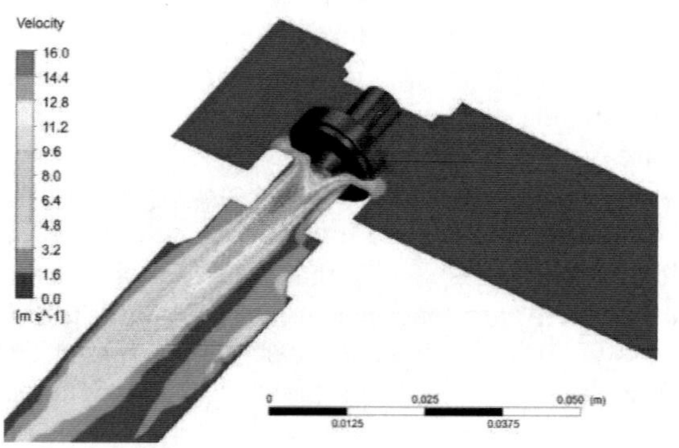

Figure 8. Water velocity distribution in the valve, valve plug lift 1.6 mm.

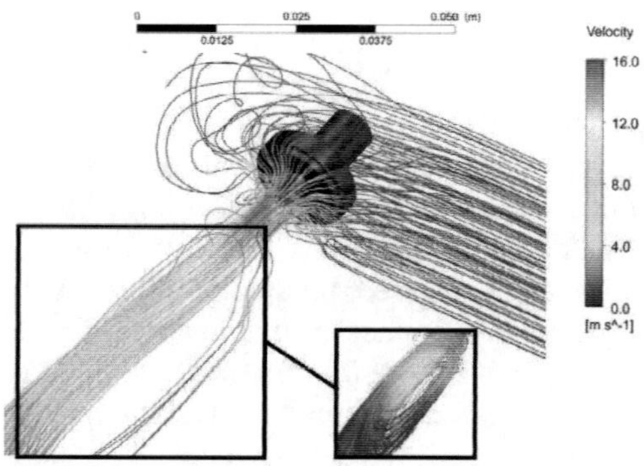

Figure 9. Tracks of water particles in the valve, valve plug lift 1.6 mm.

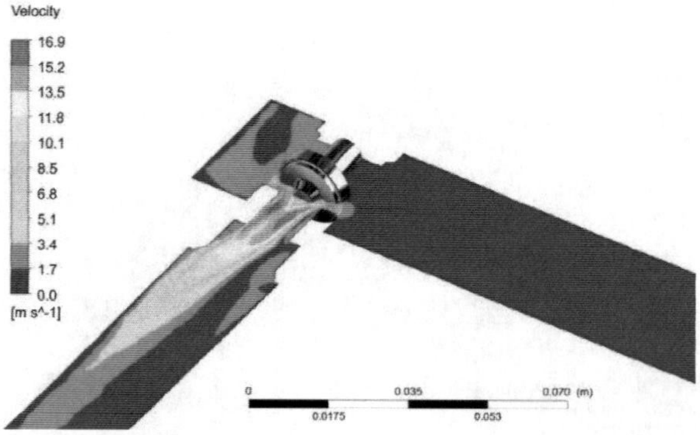

Figure 10. Water velocity distribution in the valve, valve plug lift 2.4 mm.

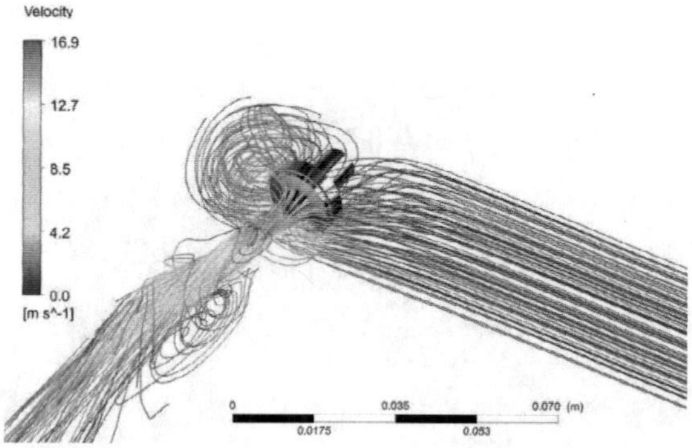

Figure 11. Tracks of water particles in the valve, valve plug lift 2.4 mm.

3.3. Quantitative comparison of the results of real and numerical research

The performed numerical simulations made it possible to conduct a comparative quantitative analysis of the flow through the investigated valve. Therefore, the results of the real research, obtained on a test stand, were compared with the results of CFD calculations. The results can be found in Figure 12.

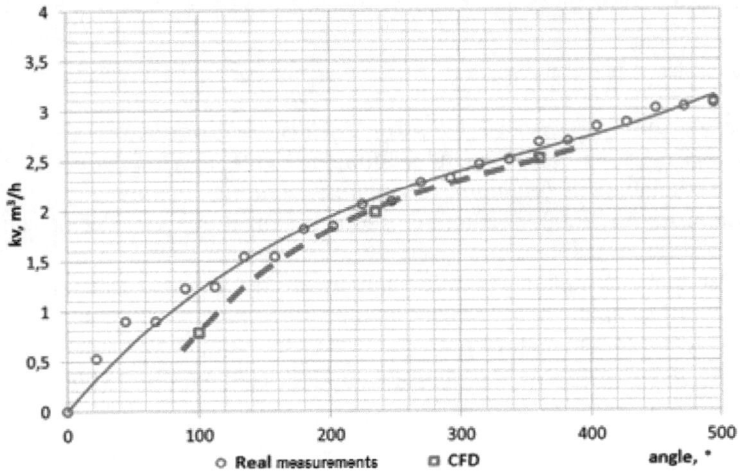

Figure 12. Comparison of numerical simulations results with measurements.

While analyzing the chart one should notice that discrepancy between the two research methods is small. Its maximum value equals 0,4 m^3/h for the rotation angle of a control wheel equal 100° and it decreases along with opening the valve. The minimum value was obtained for the rotation angle 360° and it was about 0,1 m^3/h. The courses of two characteristics are close in the range of a rotation angle exceeding 240°. Below this value the curve obtained through CFD calculations is characterized by a higher steepness, which causes approximately linear increase of the difference between the two research methods. It should be noticed that all result points for numerical simulations lie below the curve defined by the real research. This might be caused by the fact that the computational model does not include the influence of time on the obtained results and it is also important that along with the reduction of the rotation angle the height of the gap between the valve plug and seat ring diminishes and modeling of the flow in the gaps is currently quite problematic and often the obtained results are far from reality.

It should be concluded that the conducted numerical simulations are close enough to the real physical phenomena accompanying the liquid flow through the valve. It allows to implement those methods in a process of the valve construction optimization, variant analysis of the solution of its flow hydraulics as well as conducting the process of its optimization. It should be highlighted that although the accuracy of the CFD method is high each construction must undergo the real research in the final phase.

4. Conclusions

The obtained real research results related to the results of numerical simulations allowed to come to the following conclusions:

- methods of flow simulations can be successfully applied to determine the valve flow coefficient k_v;
- dicrepancy between the CFD simulations and the real research increases along with the reduction of the valve opening angle. This being caused by the change of the solid body height limited by the seat ring and the valve plug surface and the necessity of flow simulations in narrow gaps;
- presented velocity field as well as tracks of water particles showed a considerable non uniformity of the flow behind the valve;

IOP Conf. Series: Materials Science and Engineering **233** (2017) 012015 doi:10.1088/1757-899X/233/1/012015

- the swirl occurring behind the valve is inadvisable;
- for the value of lift 2.4 mm the stream of liquid has swirls after flowing through the area of - valve plug - seat ring;

Currently the authors are working on the experimental and numerical analyses of the cooperation among the three valves fixed on one rail.

References

[1] PN-EN 60534-2-1 *Przemysłowe zawory regulacyjne. Wydajność przepływowa. Równania wymiarowania zaworów do przepływu płynów w warunkach instalacji*
[2] Bean R 2009 Hisory o Radiant Heating and Cooling Systems *ASHRAE Journal*
[3] Chapuran A 2009 Radiant idea *Technology-Hydronic Heating*
[4] Davis J and Steward M 2002 Predicting Globe Control Valve, Performance—Part I: CFD Modeling *ASME*
[5] Sonawane1 V and others 2012 Design and Analysis of Globe Valve as Control Valve Using CFD Software, *Journal of Mechanical and Civil Engineering*
[6] Ferrari J and Leutwyler Z 2008 Fluid flow force measurement under various cavitation state on a globe valve model *Proceedings of 2008* Pressure and Vessels Piping *ASME conference*

Axial pico turbine – construction and experimental research

G. Peczkis[1], Z Goryca[2], A Korczak[1]

[1]Politechnika Śląska, St. Konarski 18 Street, 44-100 Gliwice, Poland
[2]Kielce University of Technology, Aleja Tysiąclecia Państwa Polskiego 7, PL- 25314 Kielce, Poland

E-mail: peczkis@wp.pl

Abstract: The paper concerns axial water turbine of power equal to 1 kW. The example of axial water turbine constructional calculations was provided, as well as turbine rotor construction with NACA profile blades. The laboratory test rig designed and built to perform measurements on pico turbine was described. The turbine drove three-phase electrical generator. On the basis of highest efficiency parameters, pico turbine basic characteristics were elaborated. The experimental research results indicated that pico turbine can achieve maximum efficiency close to the values of larger water turbines.

1. Introduction

The most prevalent natural hydrological conditions in the lowlands, predominating in Central Europe, are characterized by low heads in small and medium watercourses flows rates. Such conditions favor use of water head energy by means of water wheels, Banks - Mitchell turbines and propeller turbines [3]. Rational locations for these turbines possible installations are largely dispersed, so that their powers do not usually exceed several to several dozen kW.

Classification of hydropower plants, according to capacity installed:

- ✓ Large hydropower plants > 10 MW,
- ✓ Small hydropower plants (MEW) < 10 MW,

including:

- ✓ mini < 1 MW,
- ✓ micro < 100 kW,
- ✓ pico < 5 kW.

The paper concerns construction, experimental research and assessment of pico turbine performance, which is designed to work in following conditions:

- water courses with low head and low discharge,
- regeneration,

Moreover, the investigated construction can be concerned as the model machine for the higher capacity turbines.

2. Water turbines efficiency

The decision concerning constructing and building a power machine results, among others, from efficiency that it can achieve. In Figure 1. the chart shows hydraulic losses, which for turbines with impeller diameter <0,46 m grow progressively.

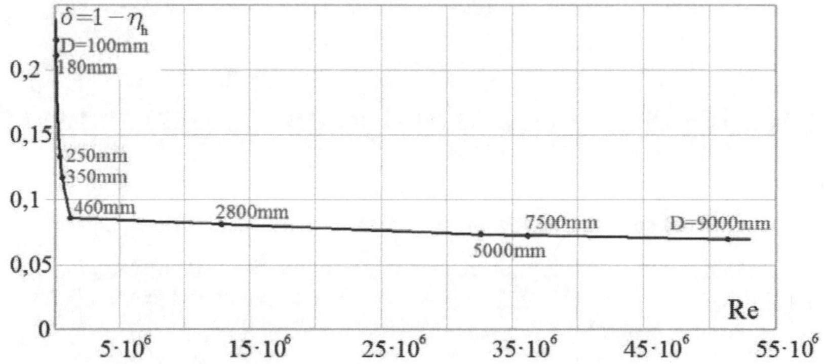

Figure 1. Hydraulic losses in axial turbine as function of Reynolds number [A.M. Chrystjakov].

Reynolds number for propeller turbines is as follows [1]:

$$Re = \frac{d_3\sqrt{2gH}}{\nu} \tag{1}$$

where: d_3 – propeller turbine outlet diameter, m

H – water head, m

ν – kinematic viscosity, $\frac{m}{s^2}$

g – gravitational acceleration, $\frac{m}{s^2}$

According to Figure 1, the pico turbines are characterized by additional hydraulic losses, resulting from low Reynolds numbers. The empirical investigation is justified, as it aims to minimize the total loss during machine performance. The measurements on pico turbines can be carried out on the machine with scaling factor 1. On the basis of dynamic similarity the results of the measurements can be used as the model test in designing process of geometrical similar, higher capacity machines.

For the specific speed $n_b \geq 0{,}33$ and discharge $Q < 0{,}65\frac{m^3}{s}$ total efficiency η_C can be obtained from:

$$\eta_C = \sqrt{\left(\frac{Q}{0{,}048}\right)^{0{,}07} - \left(-0{,}722 - \log n_b\right)^5} - \xi = 0{,}814 \tag{2}$$

where $\xi = 0{,}2$ for one stage turbine.

The total efficiency obtained from experiment investigation was proved by result of calculation conducted according to cited formulas.

The hydraulic efficiencies η_h of a turbine cannot be obtained from experiment results, but they are necessary for constructional calculations. They are determined according to following formulas proposed by:

Wislicenus
$$\eta_h = \sqrt{\eta_C} - (0{,}02 \div 0{,}06) = 0{,}88 \div 0{,}84 \tag{3}$$

Lomakin
$$\eta_h = 1 - \frac{0{,}42}{\left[\log\left(4{,}2 \cdot 10^3 \sqrt[3]{\frac{Q}{60n}}\right) - 0{,}172\right]^2} = 0{,}85 \tag{4}$$

Jekat
$$\eta_h = 1 - \frac{0,8}{(15850,3Q)^{0,25}} = 0,86 \tag{5}$$

Michajlow
$$\eta_h = 0,7 + 0,0835 log\left(4 \cdot 10^3 \sqrt[3]{\frac{Q}{60n}}\right) = 0,84 \tag{6}$$

The research described in this article aims to verify the accuracy of these formulas on the basis of empirical investigation of machine performance.

Due to the diameter of pico turbine $d<0,25m$ for construction calculations the assumption of $\eta_h = 0,75$ was done.

3. Determination of principle dimensions of axial turbine rotor

During construction process the highest efficiency operating conditions were assumed, what was proved in empirical investigation:

flow rate $Q = 4\frac{m^3}{min}$, head $H = 2,m$, rotational speed $n = 850\frac{obr}{min}$

The dynamic specific speed n_s from formula:

$$n_s = 3,65n\frac{\sqrt{Q}}{H^{0,75}} = 476 \tag{7}$$

The specific speed n_b, used in following calculations [3]:

$$n_b = \frac{n_s}{1213,9} = 0,392 \tag{8}$$

The value of specific speed indicates that the investigated turbine is an axial one. The power on turbine shaft is defined as:

$$P = \frac{g\rho QH}{\eta_h} \tag{9}$$

Meridional velocity:

$$c_m = K_m\sqrt{2Y_N} = 2,07\frac{m}{s} \tag{10}$$

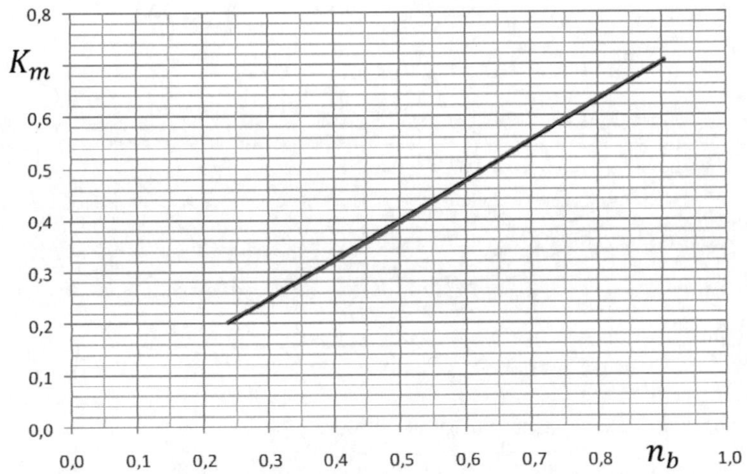

Figure 2. The relation between the specific speed and Km coefficient.

IOP Conf. Series: Materials Science and Engineering **233** (2017) 012016 doi:10.1088/1757-899X/233/1/012016

The coefficient Km for the average meridional velocity can be obtained from the chart (Figure 2) or by the formula:

$$K_m = 0{,}0688 + 0{,}733 n_b^{1,1} = 0{,}331 \tag{11}$$

Turbine rotor diameter ratio can be calculated from:

$$\frac{D_n}{D_A} = 0{,}63 - 0{,}346(n_b - 0{,}25) = 0{,}584 \tag{12}$$

Meridional velocity:

$$c_m = \frac{Q + \Sigma Q_z}{\frac{\pi}{4}\left(D_n^2 - D_A^2\right)} = 2{,}11 \frac{m}{s} \tag{13}$$

The diameter D_A can be determined:

$$D_A = \sqrt{\frac{4(Q + \Sigma Q_z)}{\pi c_m \left[1 - \left(\frac{D_n}{D_A}\right)^2\right]}} = 0{,}228m \tag{13a}$$

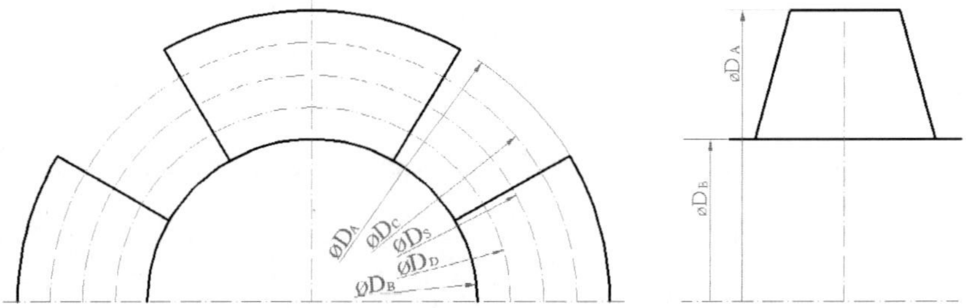

Figure 3. The diameters of designed turbine rotor.

With the assumption of constant meridional velocity and dividing the flow between the diameters $D_{A,\,C,\,S,\,D,\,B}$ shown in Figure 3 equally, these diameters can be calculated from:

$$D_S = \sqrt{\frac{D_A^2 + D_n^2}{2}} = 0{,}187m \qquad D_C = \sqrt{\frac{D_S^2 + D_A^2}{2}} = 0{,}208m \qquad D_D = \sqrt{\frac{D_B^2 + D_S^2}{2}} = 0{,}162m \tag{14 a,b,c}$$

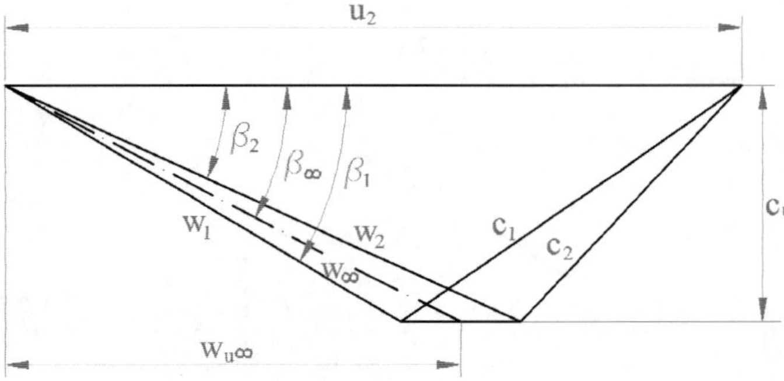

Figure 4. Velocity triangles of axial rotor.

The change of tangential component of rotor velocity on diameter D is equal to:

$$c_{u1} - c_{u2} = \frac{\Gamma}{2\pi r} = 2{,}19 \frac{m}{s} \tag{15}$$

Where Γ- circulation around rotor blade foil, which is defined as:

$$\Gamma = 2\pi r \frac{\eta_h g H}{u} = 1{,}28 \tag{16}$$

The average tangential component of relative velocity (Figure 4) is:

$$w_{u\infty} = \frac{w_{u1} + w_{u2}}{2} = u - \frac{gH}{2\pi u \eta_h} \tag{17}$$

The change of tangential component of relative velocity:

$$\Delta w_u = w_{u2} - w_{u1} \tag{18}$$

Assuming $c_{m\infty} = c_m$ it can be stated:

$$\beta_\infty = arctg \frac{c_{m\infty}}{w_{u\infty}} \tag{19}$$

The theoretical angle β_∞, has to be increased by an additional angle $\Delta\beta = \Delta\delta$ and as a result, the angle of attack $\beta_e = \beta_\infty + \Delta\beta = 15{,}7^\circ$ is obtained for rotor outer diameter. The angle β_e for other streamlines is calculated in the same way.

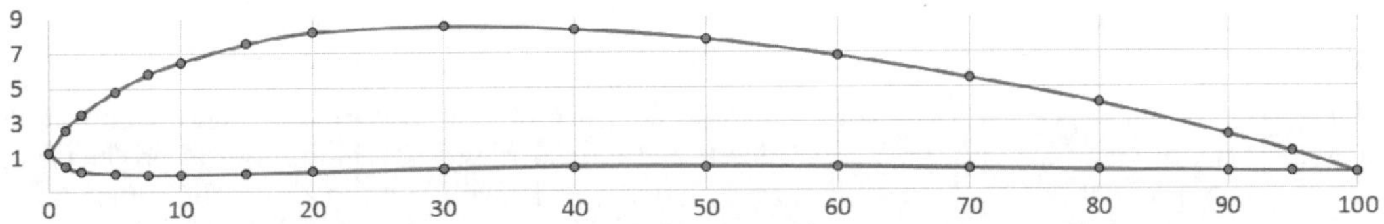

Figure 5. The shape of NACA 2412 profile, used to design the blades.

4. Axial turbine blade design process

The shapes of turbine blades were designed on basis of NACA 2412 foil (Figure 5). The foil is described by characteristic dimensions ratio. The profile sketch was drawn on XY plane. Next, the profile chord was increased, with proportions kept. Then it was decided to change the proportions, so that the thickness of blade was reduced to technologically possible minimum. The obtained blade profile no longer kept the proportions of original NACA 2412 profile, particularly the ratio between the maximum thickness and chord length was interrupted. In Figure 6 the different rotor diameters profiles obtained from this method are shown.

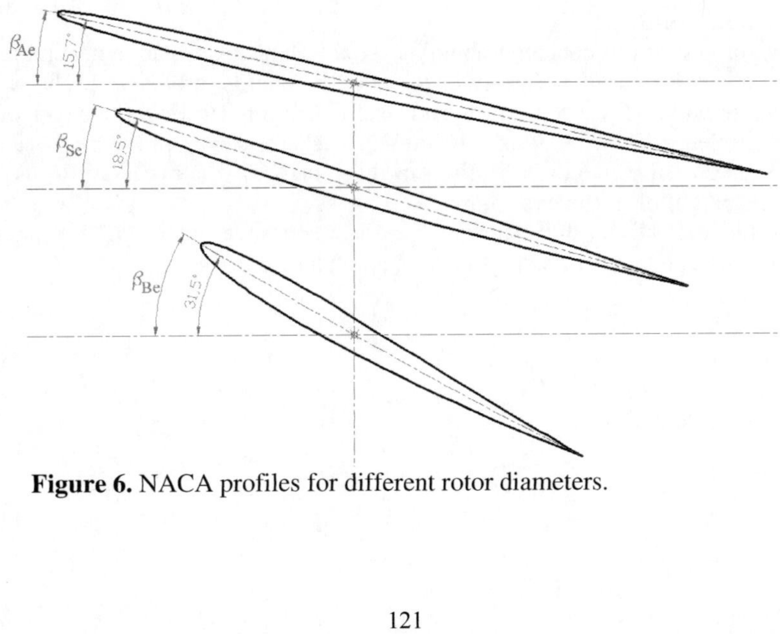

Figure 6. NACA profiles for different rotor diameters.

Next constructional work, performed in XYZ coordination system, included connecting different diameters profiles with one surface and blade root, which enables proper, explicit connection with the rotor hub. The shape of blade in shown in Figure 7. In the investigated construction, the blade is placed on cylinder hub, so that the angle of attack changes are not possible. The blade placing on a hub results in one configuration, shown in Figure 8. The change of blade angles is possible only in case of at least partial spherical hub. Presented turbine rotor blade was used in experimental research. It was made of PA6 aluminum alloy, with use of multi axis milling technique.

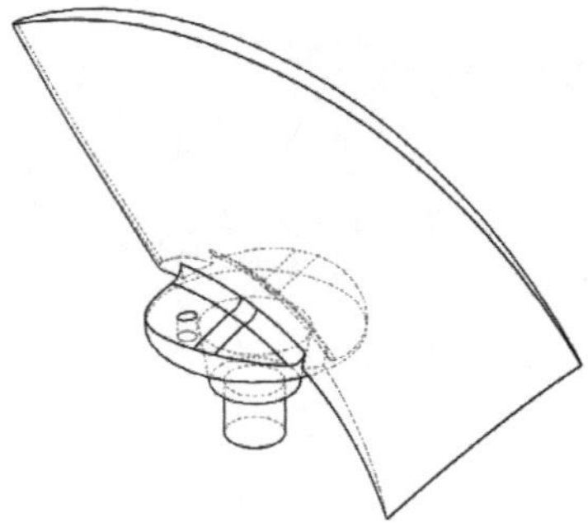

Figure 7. The blade obtained from profiles connection. The blade root embedded on conical fitting.

The processing of these types of elements meets the problem of deflection and vibration of processed material under the cutter pressure (due to the low stiffness). During the serial production it should not be a problem, as the blades would be a casting or a forging, but it was a difficulty during prototype elements production.

Figure 8. The view of the hub with 4 blades.

5. The water turbine prototype

For investigation purposes it was decided to build an axial turbine connected to generator with a clutch (see Figure 9). The ceramic bushing with water passing through the turbine as a lubricating fluid was used. This solution provides very low mechanical losses in bearing. However, this type of bearings has low resistance to resilience. The lack of axial bearing in a scheme shown in Figure 9 is a result of axial forces transfer by means of generator's bearing. In this situation, it is necessary to use an untypical clutch, that rotating part of the turbine is hung on.

At the inlet to the turbine rotor the stator with profiled blades was installed. The stator blades have the regulated angle. The geometry of stator blades is not included in the paper.

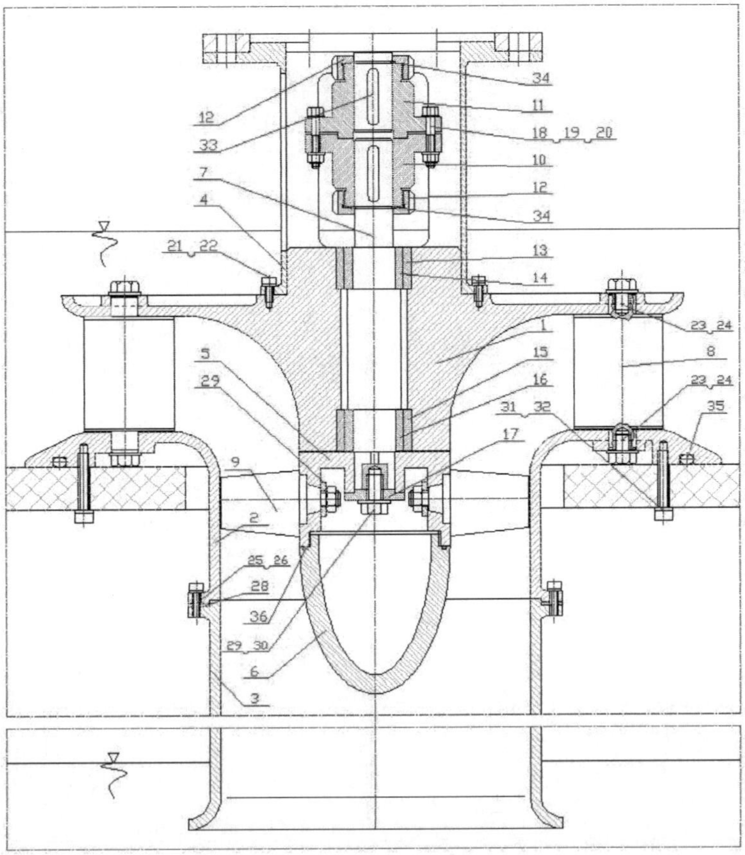

Figure 9. The scheme of the turbine. Four blade rotor version.

6. The turbine test rig

The scheme of pico turbine test rig is shown in Figure 10. Taking into consideration specific speed of the turbine and after literature research, it was decided to build symmetrical inflow channel with one crosswise rib, that prevents water circulation at the turbine inlet. The lack of rib resulted in cyclic vortex appearance, which caused air inflow to the turbine.

The water turbine test rig was designed for only one size of the turbine. The proposed shape of inflow channel can be installed under target conditions (e.g. a river estuary). Water was delivered to the circuit by means of a centrifugal pomp. A draft tube of constant diameter was manufactured for measurement purpose. In future work, it is planned to apply a diffuser with appropriate angle.

IOP Conf. Series: Materials Science and Engineering **233** (2017) 012016 doi:10.1088/1757-899X/233/1/012016

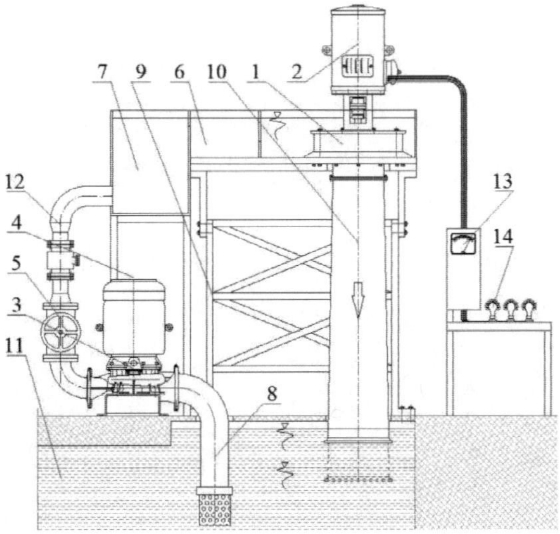

Figure 10. The scheme of water turbine and three – phase generator test rig. 1- turbine, 2- generator, 3-supplying pump, 4- pump motor, 5- valve, 6 – inflow channel, 7- buffer, 8 – suction pipe, 9 – load-bearing construction 10, - draft tube 11 – underfloor tank, 12 – discharge pipe, 13 – electrical meter, 14 – electrical resistor.

6.1 Performing the experiment

To obtain turbine performance curves it is necessary to perform the measurements at constant rotational speed. During the measurements series of points with different operating conditions were determined. The further analysis included highest efficiency operating point.

The view of the test rig is shown in Figure 11. Water was pumped from a tank located under the floor by means of the pump with smooth rotational speed regulation. It flowed evenly to the turbine through the inflow channel. To reduce the flow rate of the water (to lower water velocity), the flow was directed to a tank and then it overflowed to the inflow channel. After passing the turbine water was flowing through the draft tube back to the underfloor tank. Regulation of water flow rate was performed by changing the rotational speed of supplying pump. Regulation of water head was done by changing the water level in the underfloor tank. Electricity produced by three - phase turbine generator supplies set of lightbulbs connected in a wye system. The measurements of electrical parameters were also carried out. To increase the load of each phase the additional lightbulbs were added serially and also in parallel.

Figure 11. View of the measurement system during the experiment.

6.2 Measurements results

The efficiency of water turbine and its generator is derived from flow rate Q, head H and electrical power P_e measurement results. Measurements of flow rate were provided by means of electromagnetic flowmeter. The value of head was defined as a distance between water level in the inflow channel and water level in the underfloor tank. To measure the electrical power an universal meter was used. It indicated voltage, current and phase angles (three-phase system).

6.2.1. Mechanical parameters measurement

The Best Efficiency Point for the investigated pico turbine, at specific head H, was characterized by following parameters:

Head $H = 2{,}18\ m$,
Flow rate: $Q = 4{,}22\ m^3/min$,
Rotational speed: $n = 859\ rev/min$,
The average water axial velocity at turbine outlet:

$$V_{sr} = 1{,}723\frac{m}{s}$$

The measured work parameters indicate the hydraulic power of water flow:

$$P_h = Q\,\rho\,g\,H = 1504\ W \tag{20}$$

They also indicate the specific speed of the turbine [4]:

$$n_s = 1{,}167n\frac{\sqrt{P_h}}{H^{5/4}} = 464{,}1 \tag{21}$$

where: n [rev/min], P_h [kW], H [m].

6.2.2 Measurements of generator's electrical parameters

The measurements were taken at electrical side of the generator driven by the turbine. The generator was loaded by electrical resistors. The three – phase generator, at rotational speed n = 859 rev/min produced electrical current of following parameters:

Voltage U = 458 V,
Current I = 1,65 A,
Electrical power:

$$P_e = \sqrt{3}UI\cos\varphi = 1050W \tag{22}$$

Shaft power :

$$P = \frac{P_e}{\eta_e} = \frac{1050}{0{,}85} = 1123{,}53W \tag{23}$$

The efficiency of turbine - generator system was derived from water hydraulic power and generator's electrical power:

$$\eta_a = \eta_T\eta_G = \frac{P_e}{P_h} = 0{,}698 \tag{24}$$

The efficiency of pico turbine is defined as:

$$\eta_T = \frac{\eta_a}{\eta_G} = 0{,}821 \tag{25}$$

where η_G – generator efficiency was determined on basis of its characteristic shown in Figure 12.

IOP Conf. Series: Materials Science and Engineering **233** (2017) 012016 doi:10.1088/1757-899X/233/1/012016

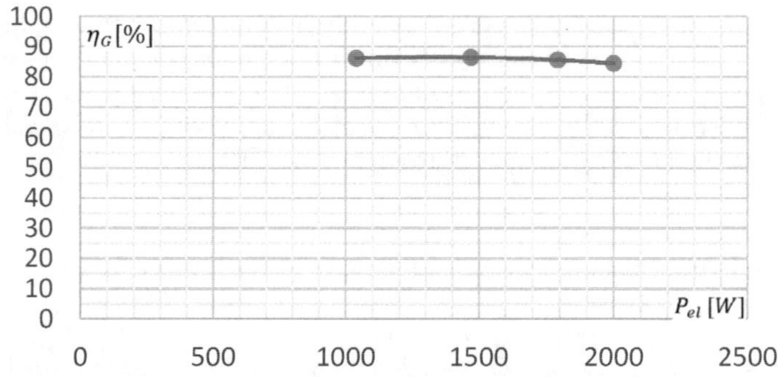

Figure 12. Generator efficiency η_G as a function of load P_{el}.

6.3 The basic performance curves for pico turbine

On the basis of Best Efficiency Point parameters for the investigated turbine and typical performance curves for water turbines the characteristics for pico turbine were elaborated. They are depicted in Figure 13.

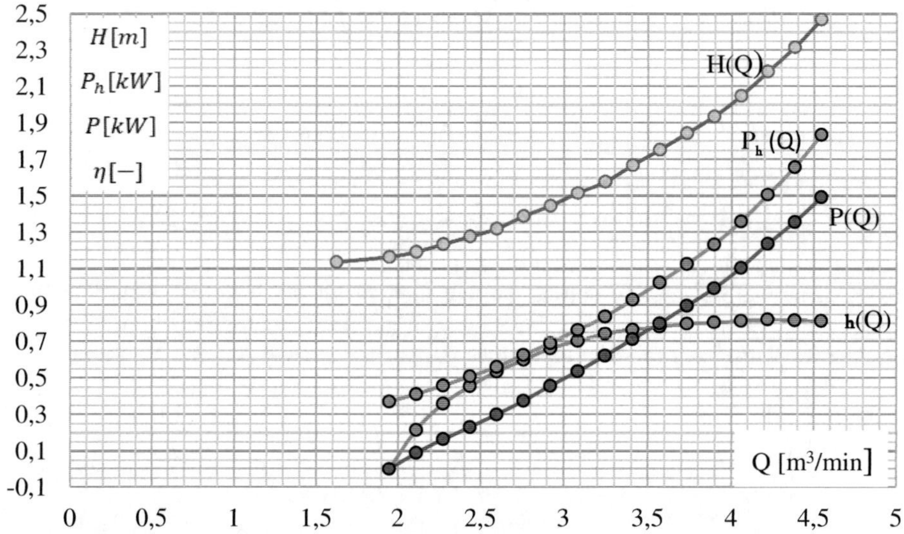

Figure 13. The investigated pico turbine characteristics, elaborated from Best Efficiency Point parameters.

7. Conclusions

1. The experimental research showed that efficiency of a pico turbine of rotor diameter d = 228 mm can be as high as it is indicated in Equation 2 (with no additional losses shown in Figure 1) under following conditions: precise elements production and pico turbine assembly and application of appropriate bearings.
2. The algorithm described in [2] was used in hydraulic and constructional calculations. The calculations results were in good agreement with turbine parameters obtained during the experiment.
3. The application of NACA 2412 as rotor blades profiles was a good choice, which proves the conclusion 1.
4. The use of ceramic bearings, lubricated and cooled by water passing through the turbine was also a good choice. Moreover, they had low rolling resistance.
5. The application of an inflow channel with one crosswise rib (with dimensions recommended in [1]) enables to provide stable flow rate through the turbine.
6. As it was mentioned in Introduction, pico turbine investigation results can be applied during production of this type of turbines and as model machine investigation for higher capacity turbines.

7. Stabilization of parameters after generator load change (and turbine load consequently) was quite time consuming. It was decided to build a different regulation system in further research, so that the turbine load can be changed smoothly, with simultaneous shaft power measurement.

References

[1] W Krzyżanowski 1971 *Turbiny wodne, konstrukcja i zasady regulacji* (Warszawa: WNT)
[2] Varchola M, Hlbočan P 2016 *Hydraulický návrh axiálneho stroja* (Bratislava: SUT)
[3] Korczak A, Rduch J 2009 *Energetyka wodna w Polsce, stan i perspektywy rozwoju* Monografia Komitetu Inżynierii Środowiska PAN, III Kongres Inżynierii Środowiska (Lublin) **3** pp 33-59
[4] Kowaliew N N 1971 *Gidroturbiny* (Leningrad: Maszinostrojenije) p 584

IOP Conf. Series: Materials Science and Engineering **233** (2017) 012017 doi:10.1088/1757-899X/233/1/012017

Influence of design and mode parameters on pump performance curve of heat generating aggregate

O Barykin[1], S Kovalyov[1], M Ovcharenko[1], A Papchenko[1]

[1] Sumy State University, R-Korsakova str., 2, 40007, Sumy, Ukraine

E-mail: barakudapgm@gmail.com

Abstract. Classification of multi-functional heat generating aggregates according to the function is considered in this article. Analysis of operating process mathematical model was implemented and methods for its refinement were proposed. Results of physical investigation of heat generating aggregate design and mode parameters influence on its power and head were presented.

1. Introduction

Multi-functional heat generating aggregates (HGA) – machines of vortex operating principle, where milling process of operating medium solid inclusions, their active mixing, pumping and hydrodynamic heating are implemented in the hydraulic flow part. These machines are widely used as technological equipment of most branches of industry (food, chemical, pharmaceutical, agriculture, municipal engineering and etc.) [1]. Concerning functionality of heat generating aggregate there is possibility of specific process intensification (milling, mixing, heating) by means of some design changes. Classification of heat generating aggregates according to the function is given in Figure 1.

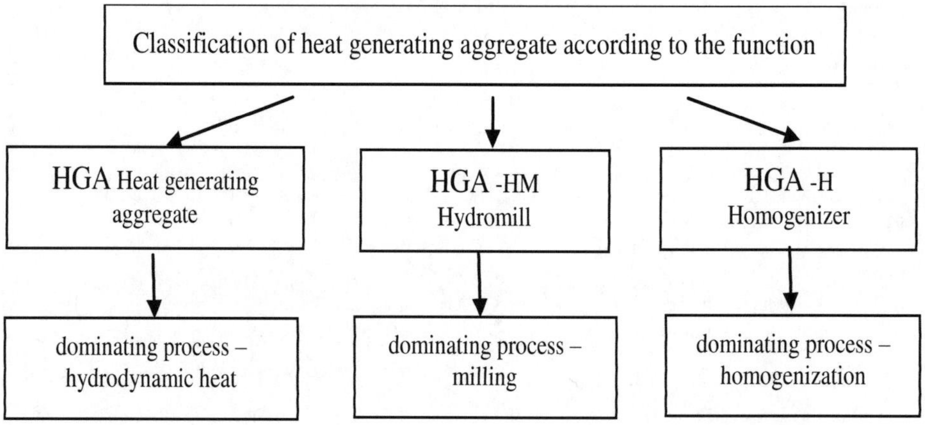

Figure 1. Classification of heat generating aggregate according to the function.

Results of research of heat generating aggregates, hydromills and homogenizers are detailed given in [2-4]. But the most widely spread are heat generating aggregates HGA, where heating of operating medium is dominating process. Previous results of machines research are given in [1].

2. Problem statement

During manufacture of heat generating aggregates and their application it was determined that there is possible to adjust its operation not only by means of alteration of impeller design parameters and its rotating speed but by means of alteration of design parameters of stator apparatus. Moreover in some cases it is necessary to forecast pumping effect of aggregate for the purpose of technological requirements provision concerning operating medium transportation. Earlier obtained mathematical model doesn't permit to include mentioned parameters. For these conditions the decision was taken concerning experimental research of stator apparatus design and diameter influence on power and head performance curve of heat generating aggregate.

3. Description of research implementation

Hydraulic test bench (Figure 2) was developed for research implementation, this bench operates on closed circuit.

Figure 2. Experimental test bench of heat generating aggregate.

Hydraulic principle scheme of hydraulic test bench is given in Figure 3.

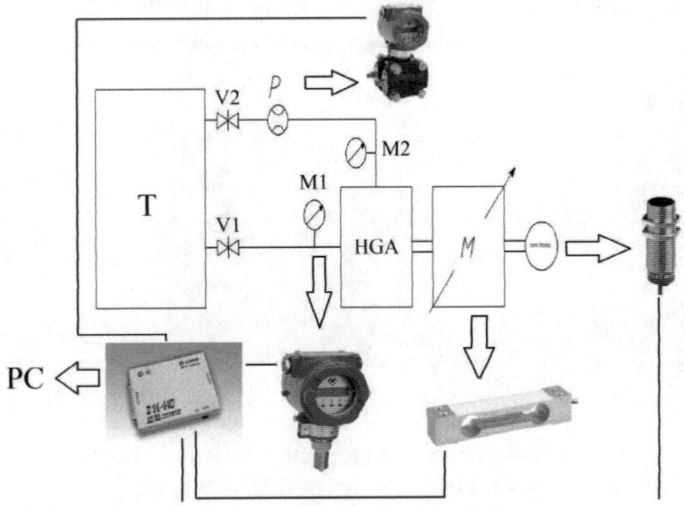

Figure 3. Hydraulic principle scheme of experimental test bench.

The bench consists of tank T (1,5 m^3), to which investigated heat generating aggregate is connected by means of suction and pressure piping. Aggregate drive is implemented by means of pendulum DC motor. Valve V_2 is installed for adjustment of liquid flow rate. Measuring equipment provides possibility to implement power trials of aggregate according to [5]. Set of measuring equipment consists of: sensor of pressure difference «Safir M», sensor of excess pressure «Safir M», tensometric sensor ZEMIC L6D-C3-5kg, inductive tachometer, digital-analogue inverter L-Card E14-140MD for conversion of analogue signals of sensors into digital signals for PC, PC Lenovo V570 for signals collection and analysis.

Design of hydraulic flow part of investigated aggregate is given in Figure 3. Investigated specimen of heat generating aggregate consists of arm on which shaft is installed by means of bearings. Casing is fixed on bearing flange of arm. Impeller is installed on the shaft (Figure 5), which is fixed by means of nut-fairing. Intake is installed at the casing. Stator apparatus are installed at both sides of the impeller (Figure 6). By means of spacer rings there is possibility to adjust gap between impeller and stator apparatus. Mechanical seal is installed for shaft hermetic sealing.

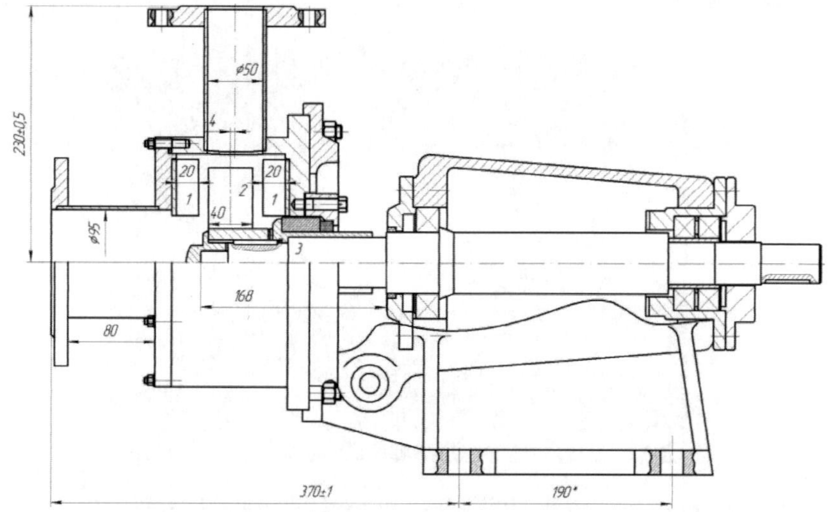

Figure 4. Design scheme of investigated heat generating aggregate.

(a) (b) (c)

Figure 5. Design of operating elements: (a) – impeller; (b) – stator apparatus without dividing ring; (c) – stator apparatus with dividing ring.

Experimental research was implemented for the following parameters: impeller diameter – 170 mm, impeller width - 40 mm, width of stator apparatus – 20 mm, rotating speed - 1500 rev/min, quantity of impeller vanes – 8, quantity of stator apparatus vanes – 9, gap 2.5 mm. Variating parameters – external diameter of stator apparatus – 160-185 mm, liquid flow rate – 0-4.5 m^3/hour.

Defined parameters – power, head, pumping effect (percentage of energy, which is consumed for pumping).

4. Results of experimental research

Results of experimental research are given in form of performance curves in Figure 6.

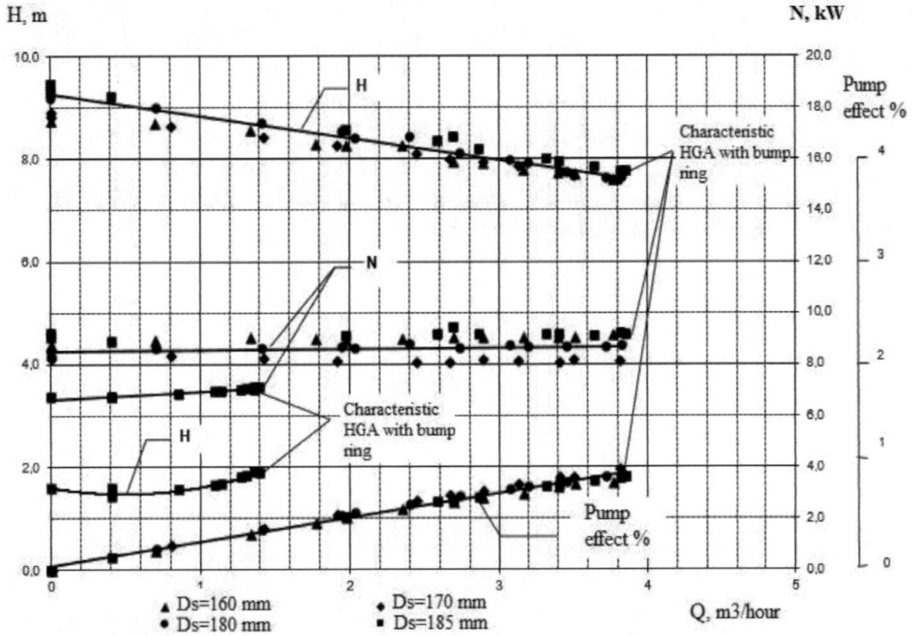

Figure 6. Experimental performance curves of heat generating aggregate.

After analysis of obtained results it can be stated that stator apparatus diameter increase relating to impeller diameter up to 10 % leads to head increase up to 5% and provision of more effective hydraulic interaction of vortex flows, which leads to aggregate power increase up to 5%. Head increase can be proved by flow kinetic energy (angular component of speed) conversion to potential energy. Besides it should be mentioned that application of dividing rings for stator apparatus installed at internal diameter leads to head increase up to 3 times. For this case consumed power increases up to 25%.

5. Conclusions

Implemented experimental research allows producing the following general conclusions:

- it is reasonable to apply stator apparatus with dividing rings at internal diameter of vanes for increase of pumping effect of heat generating aggregate and providing more effective energy interaction in hydraulic flow part. For this case it can be obtained head increase up to 200-300%, and energy effectiveness increase up to 25%;
- variation of external diameter of stator apparatus in range of 0,95-1,08 impeller diameter leads to head variation up to 5% and power – up to 5%.
- it is reasonable to research influence of stator internal diameter and dividing rings installation at stator external diameter on head and power performance curve of heat generating aggregate.

By means of generalization of analytical, experimental and computational research of heat generating aggregate provides possibility to obtain semiempirical mathematical model for forecast of operating characteristics of heat generating aggregates.

IOP Conf. Series: Materials Science and Engineering **233** (2017) 012017 doi:10.1088/1757-899X/233/1/012017

References

[1] Papchenko A, Yevtushenko A and Kovalyov S 2012 *Technological heat generating aggregates*: monograph (Sumy: Sumy state university) p 120

[2] Papchenko A, Kovalev S, Ovcharenko M, Barykin O 2014 Development of mathematical model of working process of multi-functional heat-generating aggregate-hydromill *Applied Mechanics and Materials* **630** pp 124-129

[3] Papchenko A, Kovalev S, Ovcharenko M 2013 *Calculation and design of multi-functional heat generating aggregates – hydromills* (Sumy: Sumy state university) p 128

[4] Papchenko A, Kovalev S, Ovcharenko M 2012 Experimental research of influence of rotor and stator channels quantity of rotor-dynamic aggregate – homogenizer on its energy characteristics *Bulletin of Sumy state university. Series "Technical sciences"* **4** pp 54-57

[5] *Dynamic pumps Methods of trials Project 1st revision* 2002 GOST P 6134-01 ISO 9906:1999Actual since 2001-01-01 (M: Standardinform) p 129

IOP Conf. Series: Materials Science and Engineering **233** (2017) 012018 doi:10.1088/1757-899X/233/1/012018

Effect of hydraulic parts construction arrangement on the overall performance of API 610 VS6 type pumps

S O Lugova[1], I B Tverdokhleb[2], A S Nadtochiy[3] and R I Horovyi[1]

[1] JSC "Nasosenergomash" Sumy, 1 Pryvokzalna square, Sumy 40011, Ukraine
[2] HMS GROUP, 12, Aviakonstructor Mikoyan str., Moscow, 125252, Russia
[3] Moskow Branch Office of HMS Livgidromash, 12, Aviakonstructor Mikoyan str., Moscow, 125252, Russia

E-mail: Lugovaya_SO@nempump.com

Abstract. The article presents the results of computing investigations of various construction arrangements of the VS6 type pumps. Variants of hydraulic parts differing in geometric parameters have been compared. Experimental performance curves, measured during the full-scale pump tests, corroborate the investigation results for two constructions of hydraulic parts.

1. Introduction

At present, a demand for higher level of economical operation of centrifugal pumps becomes more and more vital. In case of pumps with a high power input (over 1000 kW) even a 1 to 2 % increase in the efficiency has an essential impact on their power consumption. Savings of electric energy may reach the figure of 80 thousand kW per year.

Certainly, one of the most common ways of raising the economical operation of centrifugal pumps is to reduce friction losses in the hydraulic passages. Mostly, it implies an enhancement of hydraulic design of impellers and a reduction of friction losses in the outlet hydraulic passage.

This article deals with methods of increasing pump efficiency, which relate mainly not to hydraulic parts themselves, but their mutual influence. Investigations were conducted on the basis of the existing design of a VS6 type pump according to API 610 [1].

2. Theoretical basis

VS6 type pump is a vertically suspended diffuser radially split double casing centrifugal pump. The suction and discharge nozzles of the pump are arranged horizontally and oriented in the opposite directions. The pump consists of a can and a withdrawable element of (multistage) ring section cartridge type, featuring a diffuser downstream of each impeller. In view of high cavitation performance (low NPSH value) of the pump, an inducer is typically arranged immediately upstream of the first stage.

As VS6 type pumps have such design features (vertically suspended, low NPSH pumps), they find their applications as condensate extraction pumps in the power generation industry as well as booster pumps for main pipeline pumps.

While designing new machines, with VS6 type pump design features in mind, it is not economically sound to modify outer casings (cans), as frequently only cartridge (rotating assembly) is supplied for the cans already existing in site.

3. Methods of investigation

In order to improve the economical operation of centrifugal pumps, the design of their hydraulic parts is tried out and developed on a multi-purpose experimental test stand that enables to investigate one, two or three pressure stages of the pump. Such test stands do not replicate in all details design features of the pump, for which hydraulic parts are tried out and developed, and do not allow to take account of mutual influence of design features of a full-scale pump. Making experiments on a real pump is a too costly process. Therefore, in this case a virtual experiment is the most efficient method of investigation. The Nasosenergomash Company applies various software products for pump investigations, the most widely used of them being PumpLinx and ANSYS CFX.

PumpLinx software makes it possible to obtain rather quickly and easily integral characteristics of any particular pump on the whole and even elementwise, but not analyzes in depth the flow in the hydraulic passages with the purpose of revealing potential areas to be optimized. ANSYS CFX software offers in-depth studies of the flow pattern in the elements of hydraulic parts with allowance for their interaction.

Thus within the scope of this research the PumpLinx was used to perform a prompt relative assessment of the effect of one or another modification on the performance characteristics of the pump on the whole, and the flow pattern in the individual elements of hydraulic parts was investigated in ANSYS CFX. The use of ANSYS CFX enabled to obtain a more comprehensive flow pattern in the same and to determine locations where corrections (adjustments) are to be made.

4. Analysis of investigation results

The analysis was carried out for two construction arrangements of hydraulic parts. The following elements, namely: a suction casing, a first stage impeller combined with an upstream inducer, an intermediate stage impeller, a last stage impeller and a flow transfer passage between the first and subsequent pressure stages were optimized in terms of reduction of hydraulic losses.

Suction casing. When analyzing the configuration of its suction chamber, two variants were studied: P1 – the variant, which had been earlier applied in the serially produced pumps of the similar type; P2 – an updated variant, in which the suction chamber is axially more extended and thus offers a smooth approach flow to the inducer. In addition, the P2 suction casing features integral flow straightener ribs having outlet tips tapered off to minimize the length of their downstream vortex wake (see Figure 1).

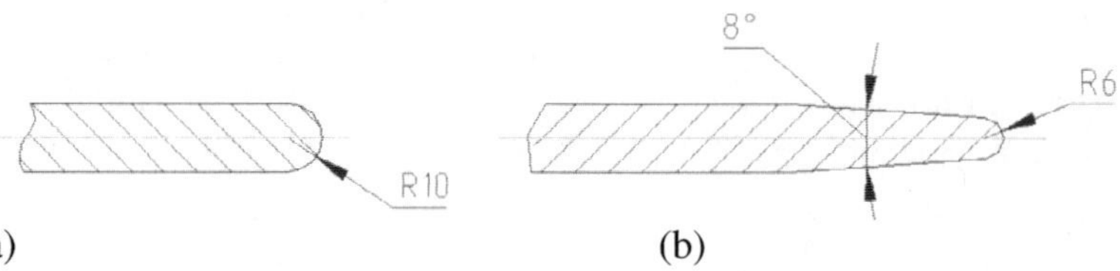

(a) (b)

Figure 1. Rib Cross Section in the P1 (a) and P2 (b) Suction Casings.

Figure 2 shows a flow pattern in the P1 suction casing as against that in the P2 suction casing. For comparison, we may mention that the P2 suction casing has a more uniform flow pattern.

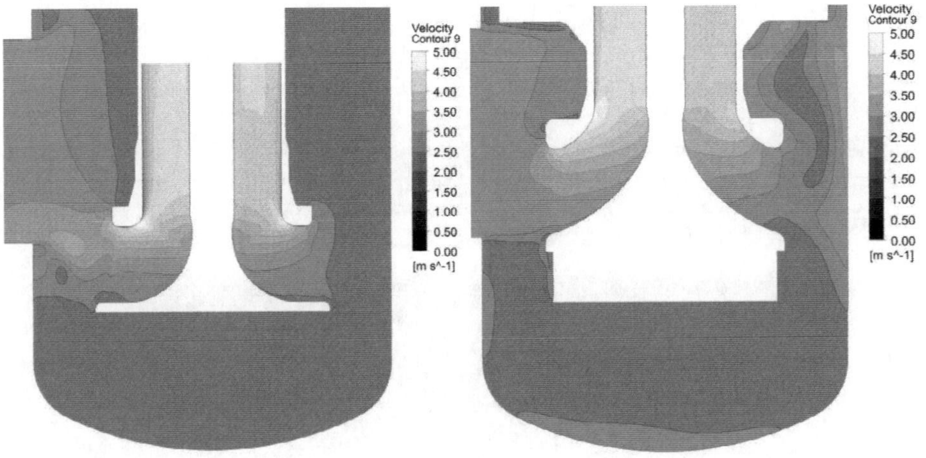

(a) (b)

Figure 2. Flow Pattern and Velocity Distribution in the P1 (a) and P2 (b) Suction Casings.

First stage impeller combined with an upstream inducer. In order to ensure a cavitation-free operation (with a low NPSH required), the first pressure stage of pumps of such type is equipped with a suction impeller in combination with an inducer installed upstream of the same. Besides that, the first stage is frequently lowered, i. e. is arranged at the bottom of the can (caisson) at a considerable distance from the rest of pressure stages. Within the scope of this research two construction arrangements of the first stage were compared: arrangement No. 1 with an inducer located at the bottom of the can and the first stage impeller elevated up to the impellers of subsequent stages, and arrangement No. 2 with all the first stage, comprising an inducer and a suction impeller as well as its diffuser, lowered down to the can bottom. A spacer with flow transfer passage is provided between the first stage diffuser and the impeller of the subsequent stage for maintaining the axial length of the cartridge (a withdrawable element).

A relative position of the inducer and suction impeller close to each other in the arrangement No. 2 allows to apply an inducer generating a lower head. Besides that, losses in the flow transfer passage between the inducer and suction impeller, which occur in case of the arrangement No. 1, are eliminated.

In addition, an inducer without a cuneate ledge was applied in the arrangement No 2 for loss reduction.

Figure 3 shows 3D images of fluid models of the first stage incorporating a suction impeller and an upstream inducer.

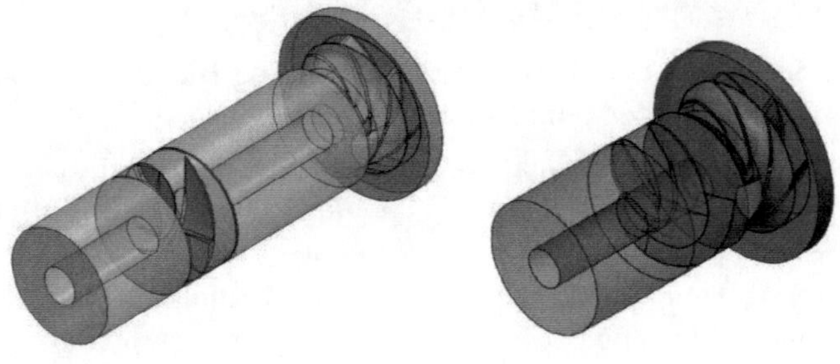

(a) (b)

Figure 3. Fluid Models of the First Stage Incorporating Suction Impeller and Upstream Inducer in the Arrangement No. 1 (a) and in the Arrangement No. 2 (b).

Flow fields of distribution of relative velocities in the blading cascades of the first stage arrangements No. 1 and No. 2 are represented in Figure 4 and Figure 5. It should be mentioned that the arrangement No. 2 has a more uniform distribution of velocities both in the inducer and in the suction impeller.

By the numerical calculation it was established that the application of the combination of the suction impeller and upstream inducer according to the arrangement No. 2 resulted in a 11.5 % gain in efficiency as compared with that produced in case of the arrangement No. 1 due to the reduction of power consumption.

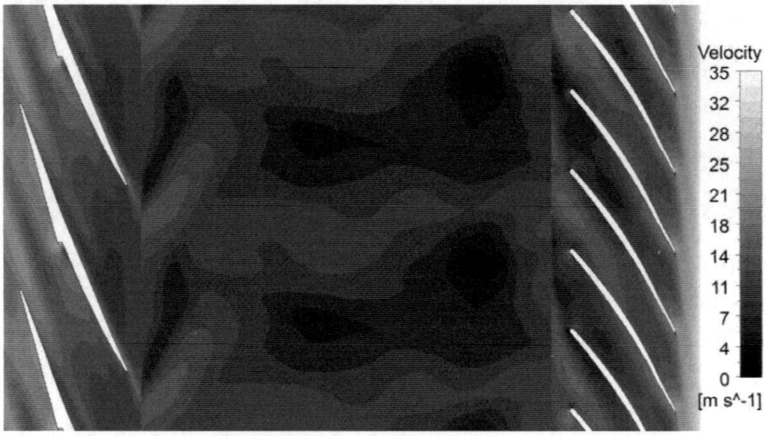

Figure 4. Relative Velocity Distribution in the Blading Cascade in Case of the Arrangement No. 1 First Stage.

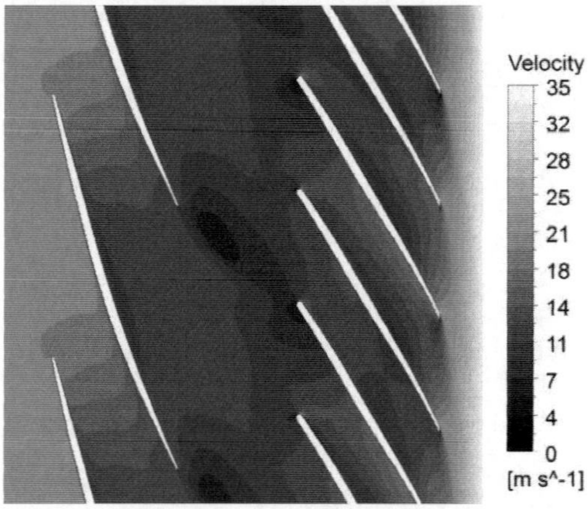

Figure 5. Relative Velocity Distribution in the Blading Cascade in Case of the Arrangement No. 2 First Stage.

Intermediate stage and last stage impellers. Within the scope of this research we studied the flow pattern in the intermediate stage impeller and last stage impeller and optimized their geometry.

We compared two variants of impellers: RK 1 (a standard one) and RK 2 (an upgraded one) with each other. The geometry of the upgraded impeller RK 2 differs in the suction eye diameter as well as in the entrance angle and outlet angle of impeller blades. A comparison table containing major geometric dimensions of the impellers is given below.

Table 1. Major Geometric Dimensions of the Impellers.

	D_h/D_2	D_1/D_2	W_1/W_2	$ß_{2b}/ß_{2f}$, degrees	$ß_{1b}/ß_{1f}$, degrees	$\Delta ß_{1b}/\Delta ß_{1f}$, degrees
RK 1	0.298	0.716	1.396	28.2 / 24.3	35.1 / 17.7	9.1 / 1.1
RK 2	0.298	0.690	1.397	28.5 / 23.5	35.5 / 22.6	6.0 / 3.7

The reduction of the suction eye diameter in the RK 2 impeller permitted to reduce its outer hub diameter at the leakage joints too and thereby to improve the volumetric stage efficiency.

Flow patterns in the channels between blades of the RK 1 and RK 2 impellers are shown in Figure 6 and Figure 7 for a circulation-free inlet flow condition.

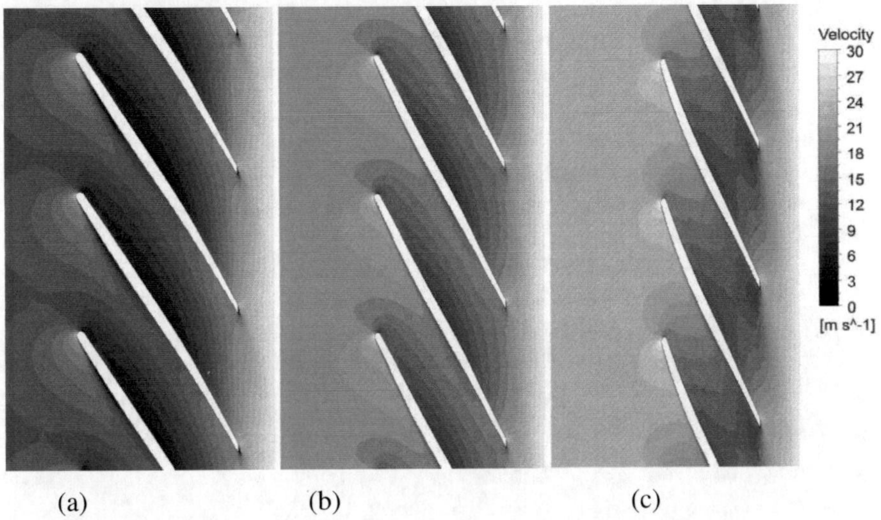

Figure 6. Flow Patterns in the Channels between Blades in Case of the RK 1 Impeller: (a) – near the hub; (b) – in the middle of the channel; (c) – near the front plate.

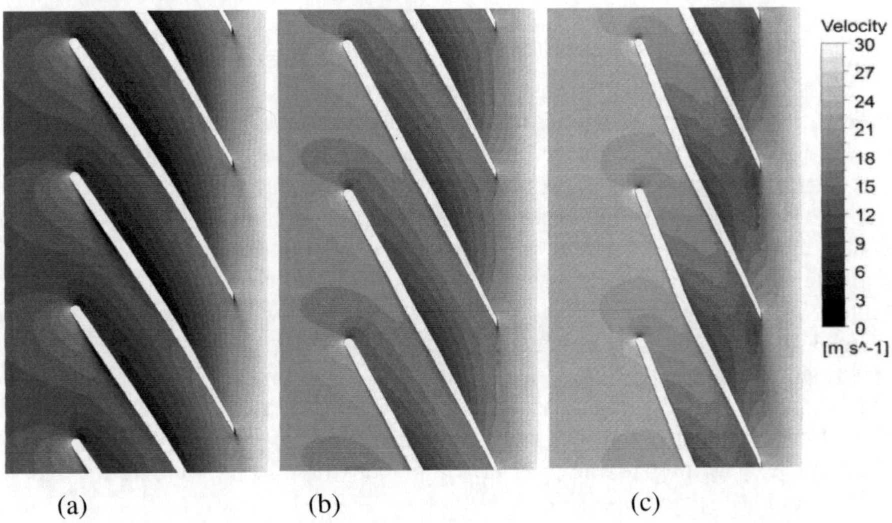

Figure 7. Flow Patterns in the Channels between Blades in Case of the RK 2 Impeller: (a) – near the hub; (b) – in the middle of the channel; c) – near the front plate.

It should be noted that the reduction of angles of incidence in case of the RK 2 impeller resulted in a more uniform velocity distribution in the channels between blades.

Flow transfer passage between the first and subsequent pressure stages. As the first stage in the arrangement No. 2 is lowered down to the bottom of the can (caisson), between the first stage diffuser and impeller of the next stage there is a circular passage where, although the flow downstream of the diffuser being practically straightened, a hydraulic loss occurs. When selecting the best variant, we considered the following two constructions: the 1st one without edge at the diffuser outlet machined to a radius, for flow velocity of 5.1 m/s through the transfer passage and the 2nd one with edge at the diffuser outlet machined to a radius, for flow velocity of 4.0 m/s through the transfer passage.

Flow fields of velocity distribution in the transfer passage for both construction variants are represented for comparison in Figure 8. It shows that with the introduction of the diffuser outlet edge rounding-off (a kind of fairing) the extent and rate of flow separation at the transfer passage inlet, and, in addition, the enlargement of the free cross sectional area of the passage resulted in the reduction of losses therein due to the flow velocity decrease. Besides that, the flow pattern at the inlet of the intermediate stage impeller was improved by providing a convergent zone in the transfer passage upstream of the same.

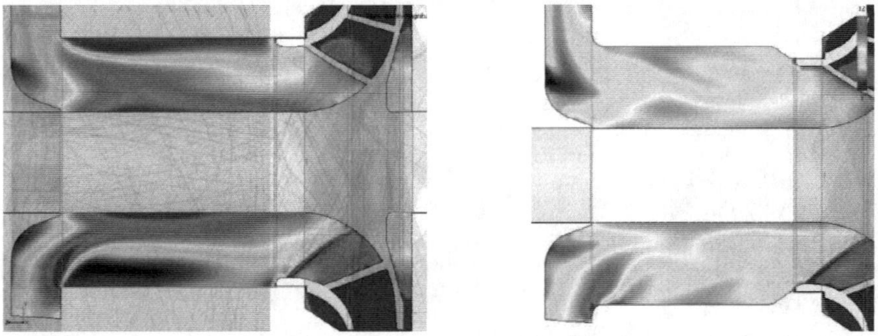

(a) (b)

Figure 8. Flow Fields of Velocity Distribution in the Transfer Passage between the First Stage Diffuser and Impeller of the Next Stage: (a) – 1st variant of construction and b) – 2nd variant of construction.

According to the results of comparison of integral quantities losses calculated for the 2nd variant of construction are 0.8 m less than those in case of the 1st variant.

Investigation of hydraulic parts on the whole. In order to enable a relative assessment of the economical operation of the two suggested construction arrangements of hydraulic parts of the VS6 type pump, numerical computations were performed in PumpLinx for both systems. The numerical computations were conducted with the same grid parameters and boundary conditions. Figure 9 shows computation models of both construction arrangements as well as check cross-sectional areas, which were used for determining values of power input, head, hydraulic loss and volumetric leakage loss. The arrangement No. 1 hydraulic parts comprised the following: the P1 suction casing, the arrangement No. 1 first stage incorporating suction impeller and upstream inducer as well as the RK 1 variant of intermediate stage and last stage impellers. The arrangement No. 2 hydraulic parts comprised the following: the P2 suction casing, the arrangement No. 2 first stage incorporating suction impeller and upstream inducer along with the flow transfer passage between the first and subsequent pressure stages made according to the 2nd variant of construction as well as the RK 2 variant of intermediate stage and last stage impellers.

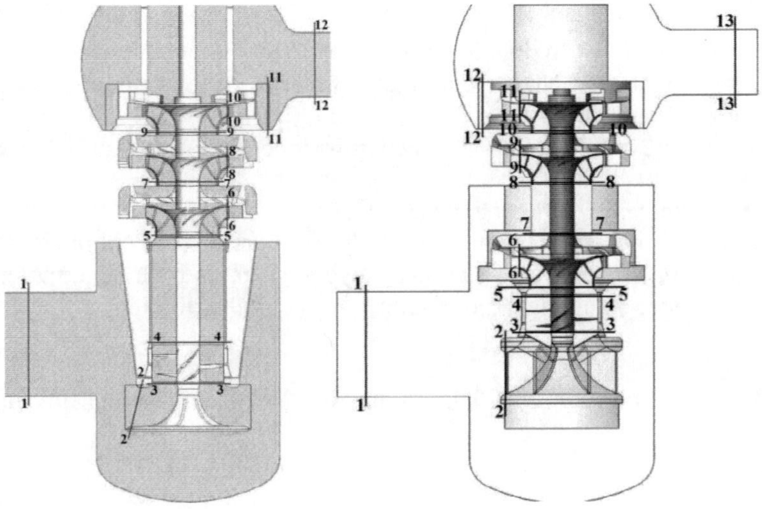

(a) (b)

Figure 9. Locations of Check Cross Sectional Areas in the Pumps under Study: (a) – arrangement No. 1 and b) – arrangement No. 2.

It has been established that on the whole the hydraulic parts in the arrangement No. 2 is more economical than those made according to the arrangement No. 1.

Verification of the computing experiment results. In order to verify the results of computing experiment, the pumps (No. 1, an earlier serially produced pump with hydraulic parts in the arrangement No. 1, and No. 2, a new pump with hydraulic parts according to the arrangement No. 2) were subjected to comparison tests. As it is clear from the comparison of performance curves (refer to Figure 10), the maximum efficiency obtained with the pump No. 2 is 3.5 % higher than that in case of the pump No. 1.

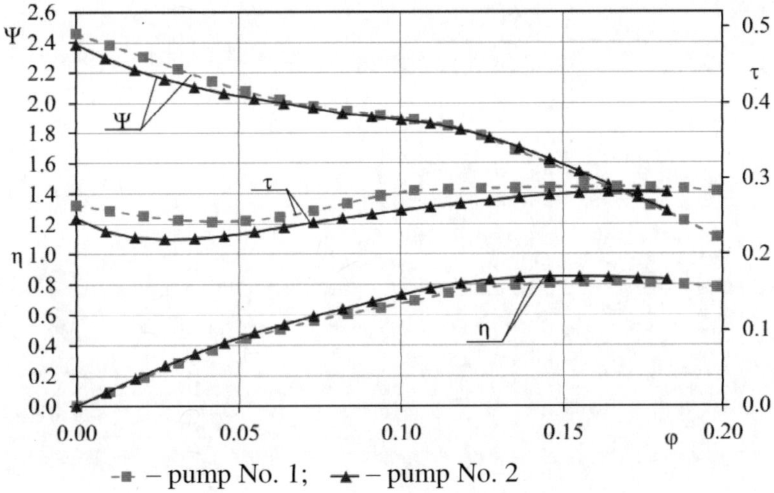

Figure 10. Comparison of Performance Curves of the Pumps with Different Construction Arrangements of Hydraulic Parts.

IOP Conf. Series: Materials Science and Engineering **233** (2017) 012018 doi:10.1088/1757-899X/233/1/012018

5. Conclusions

Two construction arrangements of hydraulic parts of the VS6 type pump have been compared. It was demonstrated that the arrangement No. 2 is preferable in terms of loss minimization. The decrease in losses has been obtained as a result of:

1. Application of the combination of the suction impeller and upstream inducer spaced apart on the minimum distance;
2. Application of an inducer without a cuneate ledge;
3. Application of intermediate stage and last stage impellers with an optimized hydraulic design;
4. Optimization of the geometry of stationary hydraulic parts both in terms of the decrease of flow velocity and in terms of obtaining a more uniform velocity distribution.

References

[1] ANSI/API Std 610/ISO 13709, Centrifugal Pumps for Petroleum, Petrochemical and Natural Gas Industries.

IOP Conf. Series: Materials Science and Engineering **233** (2017) 012019 doi:10.1088/1757-899X/233/1/012019

Effect of spiral outlet hydraulic passage geometrics on the radial thrust

S O Lugova[1] and P I Ignateva[2]

[1] JSC "Nasosenergomash" Sumy, 1 Pryvokzalna square, Sumy 40011, Ukraine
[2] Sumy State University, 2 Rymsky-Korsakov street, Sumy 40007, Ukraine

E-mail: Lugovaya_SO@nempump.com

Abstract. The article presents the results of investigation of the effect of geometrics of a double-volute outlet hydraulic passage on the magnitude and direction of acting of the radial reaction forces in a between-bearings single-stage centrifugal pump with a double entry impeller. The investigation has been performed with the use of a computing experiment. Characteristic curves as well as values and directions of the radial thrust have been compared for three variants of outlet hydraulic passages differing in the width at their entrance.

1. Introduction

Between-bearings double flow volute casing centrifugal pumps are widely used in the various branches of industry. A dependable service of those pumps is conditional primarily on the reliable operation of their bearings.

A non-uniform pressure distribution around the impeller results in arising radial forces that act on the pump rotor. If a resultant radial reaction is calculated correctly, one can select adequate bearings to meet the reliability criterion: mean time between failures (MTBF) of at least 100,000 working hours.

In case of a single-volute pump casing design the radial thrust is the largest, especially at flows other than the best efficiency point. The application of the double-volute casing design enables to reduce the magnitude of the radial thrust, though it makes the pump casing manufacturing technology more sophisticated and such pumps less economical.

At present, there is a great number of calculation and experimental techniques for determination of radial forces in the centrifugal pumps. However, the calculation techniques are empirical and do not take account of actual flow conditions in the pump hydraulic passages, and the experimental ones call for considerable material and time costs as well as availability of expensive measuring instruments. Nowadays, as computing machine capacity increases, computing investigations have emerged into a really new trend in the research.

Geometrics of an outlet hydraulic passage affect directly the pressure distribution around the impeller, and hence, the resultant radial reaction force arising in the pump casing. The current radial thrust calculation techniques [1, 2] do not permit to take into account all the factors, including also outlet hydraulic passage geometrics affecting the magnitude of the radial thrust. Thereupon, it is appropriate to do computing investigations with the application of numerical flow analysis technics, which, in minimizing the material and time costs, allow to evaluate the magnitude of radial thrust as

IOP Conf. Series: Materials Science and Engineering **233** (2017) 012019 doi:10.1088/1757-899X/233/1/012019

well as to carry out a factorial experiment that prevents from being influenced by production deviations that may occur while making variants of hydraulic passages.

2. Theoretical analysis

Radial forces arise in the pumps as a result of a non-uniform distribution of static pressure around the impeller. The very existence itself of a spiral outlet hydraulic passage in the casing design causes a radial thrust because of the axial symmetry breakdown. It goes without saying that in the diffuser-type pumps a set of (5 to 12) spiral channels between the diffuser vanes, axially symmetrized around the impeller, give rise to little if any radial forces, cancelling each other. Therefore, the case in question is a volute-casing type design.

Generally, the radial thrust (F_R) can be estimated from the following formula:

$$F_R = k_R \cdot \rho \cdot g \cdot H \cdot D_2 \cdot b_2 \qquad (1)$$

where:
k_R – radial thrust factor;
ρ – density of handled medium;
H – head per stage;
D_2 – outside diameter of the impeller;
b_2 – impeller outlet width including shrouds.

Stepanoff has proposed the following formula for calculating the magnitude of the radial thrust applicable with single-volute casings [2]:

$$F_R = 0.1 \cdot 0.36 \cdot [1 - (Q / Q_n)^2] \cdot H \cdot D_2 \cdot b_2 \qquad (2)$$

where:
Q – current value of capacity;
Q_n – rated (nominal) capacity.

It should be noted that just the "nominal capacity" (rate of flow) is indicated in the book by Stepanoff. However, in the judgement of the writers, it should be understood that the nominal capacity in this case shall be the same as the capacity at best efficiency point.

A number of other authors [1, 3] have suggested with reference to the investigations [4] to use relations (characteristic curves), experimentally obtained during the investigation of various types of outlet designs of a pump pressure stage with the specific speed n_q = 19, for calculating the magnitude of the radial thrust.

In any case, neither referenced technique represents the effect of outlet hydraulic passage geometrics on the magnitude of the radial thrust. That's why in this paper we bring to your attention a study into the effect of the width of a double-volute outlet hydraulic passage on the magnitude of the radial thrust.

3. Computing experiment

The computing experiment has been undertaken for three variants (designs) of outlet hydraulic passages, designed for the same impeller, which have different volute entrance widths but the same volute throat area. Some impeller and spiral hydraulic passages geometrics and ratios are presented respectively in Tables 1 and 2.

Table 1. Impeller Geometrics.

Description	Symbol	Unit	Value
Outside diameter of the impeller	D_2	m	0.515
Impeller outlet width including shrouds	b_2	m	0.091
Number of impeller blades	Z_{imp}	pieces	7

Table 2. Spiral Hydraulic Passage Geometrics.

Description	Symbol	Unit	Value		
			Outlet No. 1	Outlet No. 2	Outlet No. 3
Diameter of volute pitch circle (cutwater diameter)	D_3	m	0.525		
Volute entrance width	b_3	m	0.105	0.110	0.120
Width ratio of volute entrance to impeller outlet	b_3/b_2	-	1.15	1.21	1.32
Throat height	h	m	0.0725	0.0690	0.0625
Volute throat area	A_{3q}	m^2	0.0248	0.0249	0.0248

The computing experiment has been conducted for hydraulic passages of a volute pump that has a casing with double-volute outlet and semi-spiral inlet hydraulic passages, and a double entry impeller. Assuming that the flow in the pump is symmetric relative to the vertical axis of the outlet hydraulic passage, the calculations have been performed only for the half of the pump hydraulic passages. Besides, as a simplification of the computational grid, the flow in the impeller sidewall gaps and leakage joints has not been simulated. In order to obtain an actual flow pattern at the pump inlet and outlet, adjacent runs of the suction and delivery pipelines have been additionally simulated. A model of the pump hydraulic passages is shown in Figure 1.

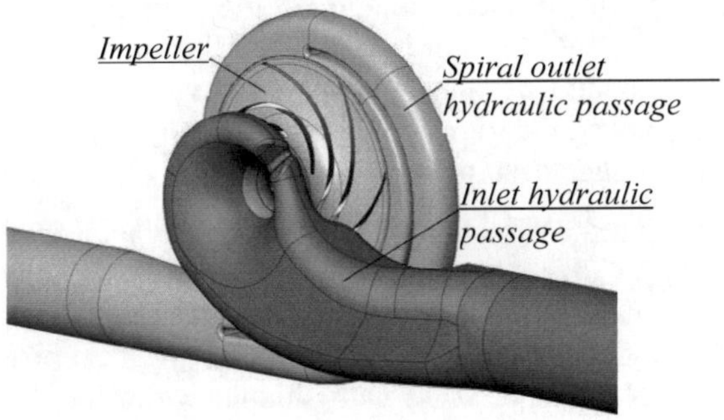

Figure 1. Model of Computation Domain.

To enable computations, unstructured tetrahedral meshes for elements of the computation domain, namely: inlet hydraulic passage, impeller and volute outlet passage have been generated. Block-structured computational grids have been plotted for the adjacent runs of the suction and delivery pipelines. The total number of computational grid elements amounts to 3.6 million cells. The common computational grid comprises 1.4 million nodes in total.

Numerical computations have been conducted for the steady flow conditions. Mass rate of flow and static pressure have been taken as a boundary conditions respectively at the pump inlet and outlet. Leakproofness has been assumed as a boundary condition for all solid walls. To close the RANS (Reynolds-averaged Navier–Stokes) equations we have used a standard k-ε turbulence model with scalable near-wall functions. The computations have been done for four pump duty points: rated flow (1.0 Q_{BEP}), overload (runout) flow (1.25 Q_{BEP}) and two part load flow conditions (0.5 Q_{BEP} and 0.75 Q_{BEP}).

4. Analysis of the computing experiment results

As a result of the completion of the computing experiment we have obtained patterns of pressure distribution around the impeller. Then the region around the impeller has been divided into cross-sections and the magnitude and the vector direction of radial forces have been determined for each

IOP Conf. Series: Materials Science and Engineering **233** (2017) 012019 doi:10.1088/1757-899X/233/1/012019

cross-section. By adding up the vectors, the magnitude and direction of the resultant radial reaction force (thrust) has been determined. Figure 2 shows the results of determination of the magnitude and direction of acting of the radial thrust for each of three variants of outlet hydraulic passage designs at the four operating points of the pump.

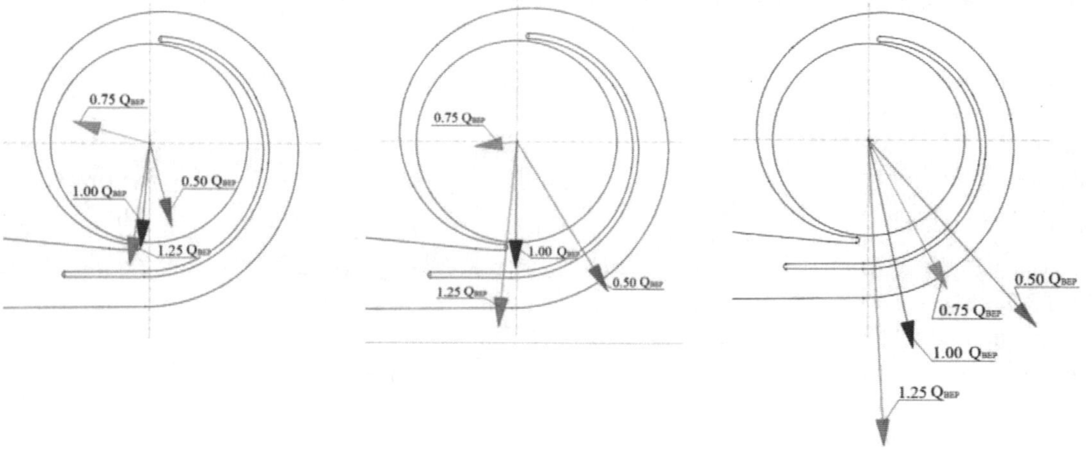

(a) (b) (c)

Figure 2. Results of Determination of the Magnitude and Direction of Acting of the Radial Thrusts: (a) – for the variant No. 1 of outlet hydraulic passage design, (b) – for the variant No. 2 of outlet hydraulic passage design, (c) – for the variant No. 3 of outlet hydraulic passage design).

It is of interest to note that the direction of the maximum radial thrust practically has not changed for various volute entrance widths, though in case of the widest outlet hydraulic passage the thrust vector direction has somewhat shifted towards the smaller cross-sections. The direction of acting of the minimum radial thrust depends substantially on the volute entrance width. In case of outlet hydraulic passage variants No. 1 and No. 2 with smaller width the thrust vector is directed towards smaller cross-sections at an angle of 90 ° with the cutwater. The radial thrust vector is directed towards the cross-sections of larger area for the outlet hydraulic passage design with the maximum volute entrance width.

The computation results are also represented in Figure 3 as a plot of radial thrust magnitude versus pump operating conditions. Here characteristic curves constructed according to the techniques [1, 2] are also presented for comparison.

It should be noted that the minimum value of the radial thrust does not fall at the best efficiency point but it is shifted to the left towards the 0.75 Q_{BEP} duty point. In case of the variant No. 1 with the narrowest outlet hydraulic passage we have obtained the value of the radial thrust for the 0.75 Q_{BEP} duty point, which is greater than that in case of the variant No. 2. However, notice should be taken of the fact that among all the design variants, if their characteristic curves are considered over the entire range of flows, the outlet hydraulic passage variant No. 1 is the most preferable, as its characteristic curve has the minimum values of the radial thrust at the 0.5 Q_{BEP}, 1.0 Q_{BEP} and 1.25 Q_{BEP} duty points, i. e. its graph of radial thrust against pump duty point is more flat. The maximum values of the radial thrust have been obtained for the widest outlet hydraulic passage of the variant No. 3.

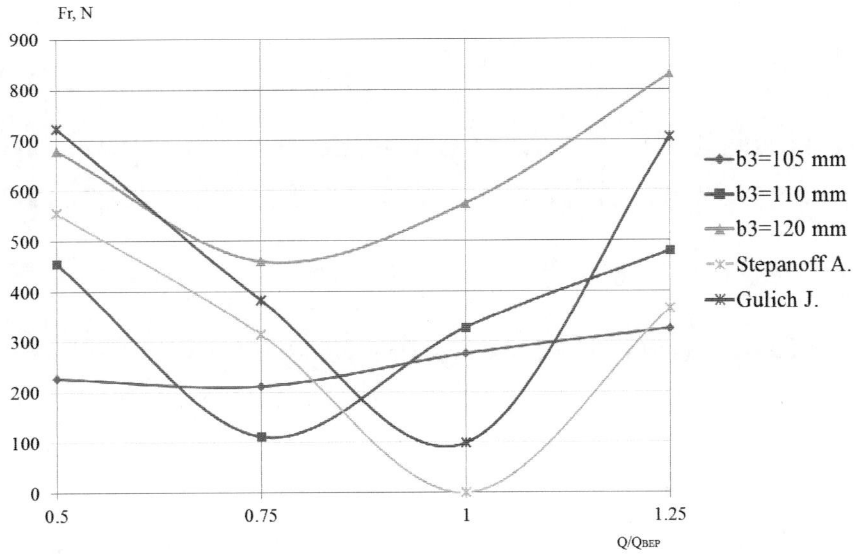

Figure 3. Radial Thrust Magnitude as a Function of Pump Duty Point.

5. Conclusions

Based on the results of the computing experiment, the effect of spiral outlet hydraulic passage width on the magnitude and direction of radial static (reaction) forces acting on the impeller has been analyzed by applying, to this effect, a numerical investigation in the flow through hydraulic passages of a between-bearings single-stage volute-casing centrifugal pump with a double entry impeller.

The behavior of the $F_r(Q/Q_{BEP})$ characteristic curves, on the whole, is similar for all the three investigated outlet hydraulic passage designs and correspond to their presentation described in [1, 2], where the radial thrust has its maximum values at the part load and overload operating conditions, and its minimum value near the best efficiency point. It has been revealed that the radial thrust magnitude is the minimum for all the three investigated variants at the 0.75 Q_{BEP} duty point.

On the grounds of the results obtained during the investigation conducted, we have come to the conclusion that the variant No. 1 featuring a narrow outlet hydraulic passage is the most preferred in terms of minimizing the radial thrust.

In view of the fact that making physical experiments is costly, the application of the computing experiment has enabled us to take account of and to analyze the effect of volute entrance width on the magnitude of radial thrust, which cannot be taken into account when using empirical calculation techniques.

References

[1] Gülich J F 2010 *Centrifugal Pumps.* (Springer Heidelberg Dordrecht London New York)
[2] Stepanoff A J 1957 *Centrifugal and Axial Flow Pumps: Theory, Design, and Application* (New York, Wiley)
[3] TM.P S.p.A. 2004 *Termomeccanica Pompe* (Via del Molo, 3 – 19126 – La Spezia – Italy)
[4] Rebernik B 1979 *Radialkräfte von Kreiselpumpen mit unterschiedlichen Gehäuseformen.* ("25 Jahre ASTRÖ", Aströ, Graz)

IOP Conf. Series: Materials Science and Engineering **233** (2017) 012020 doi:10.1088/1757-899X/233/1/012020

Research of multi-flow and multi-channel flow parts of the vortex expansion machines with the external peripheral channel

S Vaneev[1], D Miroshnichenko[1], S Meleychuk[1], V Baga[1]

[1]Sumy State University, Sumy, R-Korsakova str., 2, Ukraine, 40018

E-mail: 116mirdv@gmail.com

Abstract. At present, work on the use of vortex expansion machines for utilization of the expander installations is being carried out to solve energy saving problems. Significant radial load on the rotor arises at high pressures in the single-flow vortex stages. This problem can be solved by switching to a multi-flow scheme. In this connection, the flow and radial loads on the rotor in single- and multi-flow flow parts of the vortex expansion machine with an external peripheral channel were studied. As the result of applying the theory of experimental planning and optimization studies in the ANSYS software complex, the geometric and gas dynamic parameters of the studied flow parts are determined from the point of view of the efficiency.

1. Problem setting and purpose of the research

Over 1,600 gas distribution stations (GDS), more than 51,000 gas-regulating points (GRP) and 80 compressor stations (CS) with gas turbine gas pumping units are operated in the branched gas transportation system of Ukraine [1]. The main gas pipelines and their outlets feed gas to the gas distribution system with the pressure of 5.5 or 7.4 MPa. At the gas distribution system, the gas pressure is reduced to 1.2 MPa or 0.3 and 0.6 MPa by means of the reducing valves. The gas from the GDS is supplied to the distribution network of inhabited localities and industrial enterprises, where the gas pressure is reduced to 0.3-0.05 MPa by means of the reducing valves. At present, as a rule, the gas pressure reduction on GDS and GRP is carried out by throttling, which is accompanied by the loss of energy.

The wide introduction of turbo-expander utilization units, especially with a capacity of up to 500 kW, is possible only based on the solutions that ensure a quick and inexpensive reconstruction of existing pressure reduction systems. The main problem here is an expansion machine, which should be simple in design, reliable in operation, and should not require complex auxiliary systems.

There are low power turbine sets based on jet-powered turbines [2-4], which are simple in construction but have a high rotor speed.

To utilize the energy of excess pressure at the reduction nodes, it is prospective to use a vortex expansion machine in turbine sets and installations in the power range up to 500 kW. Vortex expansion machines, having all the values of centripetal and axial turbines, also have a number of advantages over them: they are much easier to construct and cheaper to manufacture and maintain, relatively low-speed, which makes it possible to create the turbo-aggregates in gearless performance.

IOP Conf. Series: Materials Science and Engineering **233** (2017) 012020 doi:10.1088/1757-899X/233/1/012020

However, vortex expansion turbo-machines have disadvantages, too. The pressure along the flow part of the vortex single-flow stage varies significantly from input to output (Figure 1). When using a single-flow circuit, it leads to the emergence of radial force, which can be significant. This problem can be solved by switching to a multi-flow scheme. A survey of the known sources showed the absence of systematic studies on the expediency of transition to a multi-flow scheme of the expansion machine. The available experimental values of the efficiency of multi-flow schemes do not exceed 30%. In this connection, a parametric model (Figure 2) of the multi-flow part of the vortex expansion machine with an external peripheral channel was created and the effect of the main geometric and gas dynamic parameters of the dual-flow scheme on its efficiency was studied [5, 6].

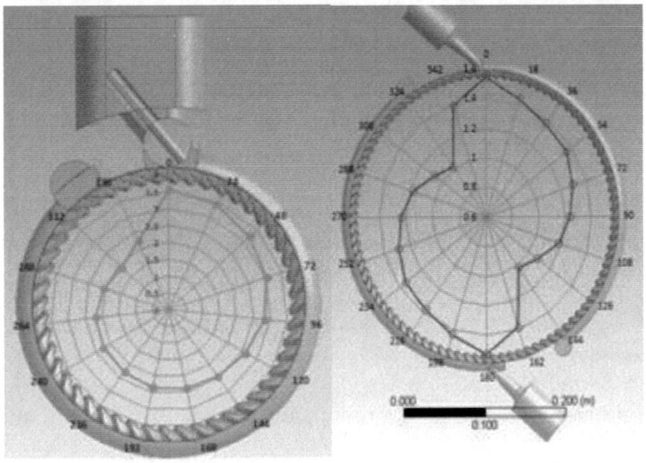

Figure 1. The pressure along the flow part of the vortex single-flow and dual-flow stage.

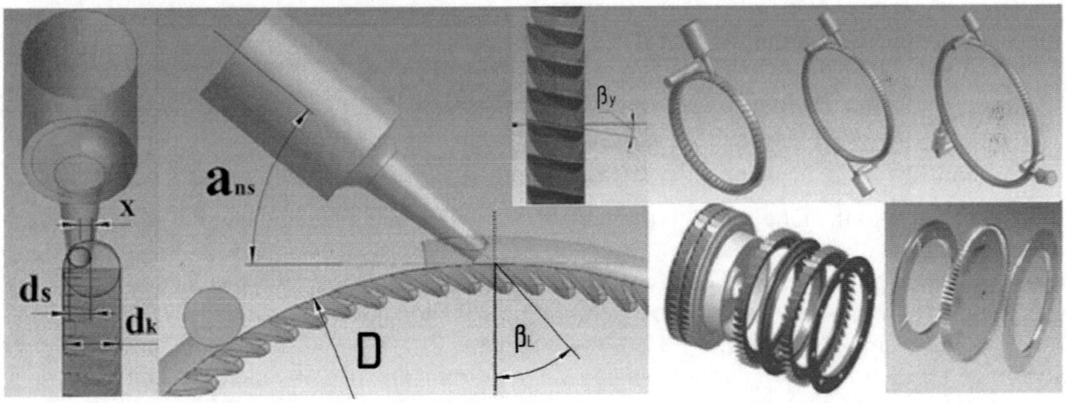

Figure 2. Parametric model of the vortex expansion machine with an external peripheral channel.

The purpose of the research is to determine the optimum geometric and gas dynamic relationships for a single-, dual- and three-flow part of the vortex expansion machine with an external peripheral channel from the point of view of the efficiency.

2. Methods of the research

The optimization task is to find such an admissible, i.e. satisfying the constraints, combination of factors that would give an extreme value to the objective function. To obtain a formal model, the regression analysis apparatus and the theory of experimental design were used [7]. As a functional connection between the geometric parameters of the flow part and the output data, a second-order polynomial was chosen:

$$y = a_0 + \sum_{i=1}^{n} a_{ii} x_i^2 + \sum_{i=1}^{n} \sum_{j\langle 1} a_{ij} x_i x_j \tag{1}$$

To reduce the number of experiments, the most influential parameters were determined and their number was reduced based on the already existing results of the vortex machines studies [8,9]. The reduction of the number of influencing parameters can also be achieved by the transition from several separate parameters to dimensionless complexes, which are formed from them. Influencing parameters were set in a certain range, where it was supposed to obtain the optimum of the objective function (efficiency).

The following most influential factors were distinguished:

- the reduced peripheral speed of the impeller on the outer diameter $\overline{U}$,
- the relative diameter of the flow part, equal to the ratio of the diameter of the meridian section of the flow part dк to the outer diameter of the impeller D, $\overline{d_{ph}} = d_k / D$ (Figure 2);
- the relative diameter of the nozzle, equal to the ratio of the diameter of the nozzle to the diameter of the meridian section of the flow part dк, $\overline{d_s} = d_s / d_k$ (Figure 2);

and the range of their variation was determined [8,9]: $\overline{U} = 0.1...0.2$; $\overline{d_{ph}} = 0.03...0.16$; $\overline{d_s} = 0.3...0.45$; $\alpha_{ns} = 25...50 deg$.

The reduced peripheral speed $\overline{U}$ characterizes the turnover and loading of the expansion machine:

$$\overline{U} = \frac{U}{C_s} = \frac{\pi \cdot D \cdot n}{60 \cdot \sqrt{2h_s}} \tag{2}$$

where

U is the peripheral speed of the impeller at the outer diameter, m/s;
D is the outer diameter of the impeller, m; n is the rotational speed of impeller, rot/min;
Cs is the isentropic flow rate, characterizes the available specific work of the expansion machine, m/s;
h_s is the specific isentropic enthalpy drop (specific available work of the expansion machine), J / kg.

The geometric relationships and parameters of $\overline{d_{ph}}$, $\overline{d_s}$, α_{ns} influence the organization and quality (intensity) of the longitudinal vortex flow in the flow part of the machine.

To construct a quadratic model of the response function, it is necessary to vary the independent factors on at least three levels. In order to study the influence of four factors on three levels, 81 experiments are required. To reduce the number of experiments, symmetric Box-Bencken non-compositional plans are applied, which allow obtaining the values of the coefficients of a quadratic polynomial for the four factors performing only 27 experiments. In this paper, the calculation points of the computational experiment were chosen according to the Box-Bencken plan. At the points of the plan, numerical simulation of the gas flow was carried out using the ANSYS CFX software. The problem was solved on the basis of the Reynolds-averaged Navier-Stokes equations. Modeling of turbulent effects performed by means of the SST model. The total pressure and the total inlet temperature, static pressure at the outlet of the computational domain and the rotor speed were used as the input data for the calculations. The working body is viscous compressible air. According to this technique, single-, dual- and three-flow single-channel flow parts were optimized.

3. Results of the research

As a result of the studies, the optimal values of the influencing factors in the investigated ranges and the regression equations for the single-, dual- and three-flow flow part of the vortex expansion machine were obtained (Figure 2).

Figures 3-5 show the maps of the level lines (response surfaces) for the output parameter (adiabatic efficiency) of the single-, dual- and three-flow flow parts, respectively.

On the basis of the results obtained, certain ratios of the geometric and gas dynamic parameters can be recommended, providing a relatively high level of efficiency of the vortex machine with an external peripheral channel in the range of variation Пт = 2-6 (Пт is the degree of pressure reduction in the expansion machine) for:

- the single-flow scheme: $\overline{U}=0.12...0.18$; $\overline{d}_{ph}=0.08...0.10$; $\overline{d}_s=0.28...0.38$; $\alpha_{ns}=25...50deg$.
- the dual-flow scheme: $\overline{U}=0.15...0.19$; $\overline{d}_{ph}=0.05...0.07$; $\overline{d}_s=0.28...0.38$; $\alpha_{ns}=25...50deg$.
- the three-flow scheme: $\overline{U}=0.18...0.22$; $\overline{d}_{ph}=0.02...0.04$; $\overline{d}_s=0.28...0.38$; $\alpha_{ns}=25...50deg$.

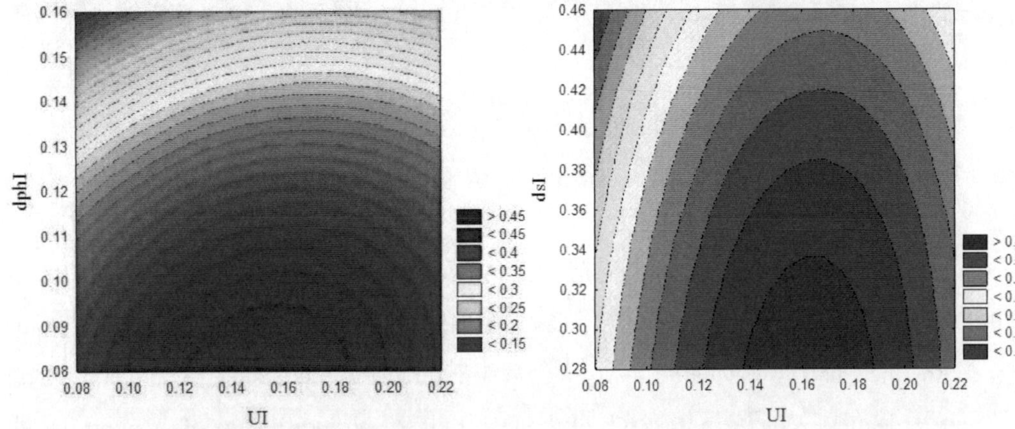

Figure 3. Maps of level lines (response surfaces) for the output parameter eff - adiabatic efficiency of the single-flow flow part.

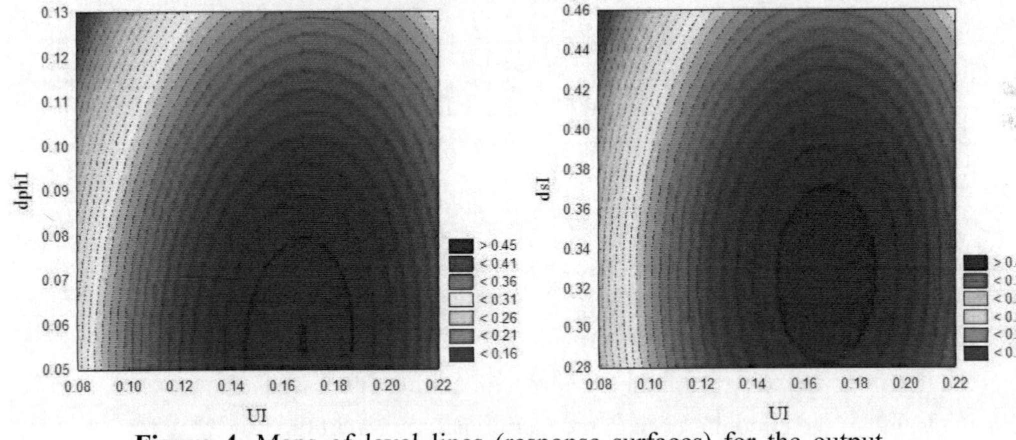

Figure 4. Maps of level lines (response surfaces) for the output parameter eff - adiabatic efficiency of the dual-flow flow part.

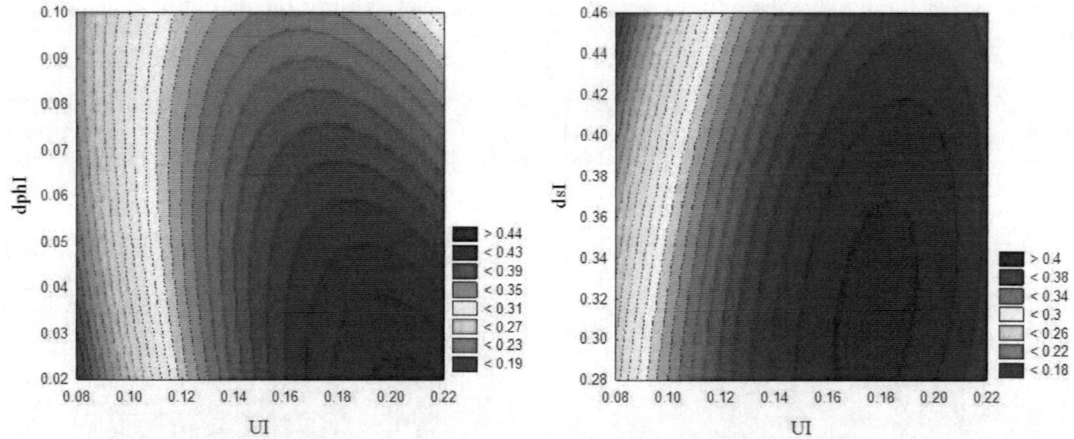

Figure 5. Maps of level lines (response surfaces) for the output parameter eff - adiabatic efficiency of the three-flow flow part.

On the basis of the obtained results, the ratios of the geometric and gas dynamic parameters generalized for the three circuits providing a relatively high level of efficiency of the vortex machine with an external peripheral channel in the range of the pressure ratio ΠT = 2-6 are determined: . $\overline{U}=0.12...0.2$; $\overline{d_s}=0.28...0.38$; $\alpha_{ns}=25...50deg$.

On the parameter $\overline{d_{ph}}$ the optimal values for the single-flow scheme are about the value $\overline{d_{ph}}=0.09$, for the dual-flow scheme - $\overline{d_{ph}}=0.06$ the three-flow scheme - $\overline{d_{ph}}=0.03$. Parameter $\overline{d_{ph}}$ for the single-flow scheme is the ratio of the length of the circumference of the meridian section of the flow part to its length in the circumferential direction:

$$\overline{d_{ph}}=\frac{\pi\cdot d_k}{\pi\cdot D}=\frac{d_k}{D} \tag{3}$$

For the multi-flow schemes it is advisable to maintain the parameter characterizing the relative length of the flow part in the circumferential direction for one flow $\overline{L_p}$, which is determined by the formula:

$$\overline{L_p}=\frac{L_p}{l_{ph}}=\frac{\pi\cdot D}{i\cdot\pi\cdot d_k}=\frac{D}{i\cdot d_k} \tag{4}$$

where

$\overline{L_p}$ is the length of the flow part in the circumferential direction for one flow, m;

l_{ph} is the length of the circumference of the meridional section of the flow part, m;

i is the number of flows.

The optimal values of the parameter $\overline{L_p}$ for the three schemes of the flow parts of a vortex machine with the external peripheral channel are about the value of $\overline{L_p}=10$ $\left(\overline{L_p}=8...11\right)$.

As one can see from Figures 3-5, the vortex stages have a relatively wide range of parameters in the optimum region. This allows the parameters $\overline{d_{ph}}$ and $\overline{d_s}$ to be adjusted by the range of their optimum to achieve the required power without significantly reducing the efficiency, without increasing the radial dimensions of the flow part. To achieve the required turbo-unit power with optimal parameters $\overline{d_{ph}}$ and $\overline{d_s}$, using the above single-channel multi-flow schemes, one can obtain the multichannel flow part of Figure 1. Such a way one can create a parametric series of vortex turbo-units for a power range up to 500 kW.

To determine the area of optimal use of the flow parts of the vortex expansion machines with the external peripheral channel, the dependences of the maximum radial load on the rotor of the vortex single-flow stage on the initial pressure at D = 0.360 m, $d_k = 0.032 \cdot m$, $d_{ph} = 0.09$ and different pressure ratios were determined (Figure 6). To build the dependencies shown in Figure 6, we used the data on pressure distribution along the length of the flow section in the single-flow vortex expansion machines. The values of the radial load on the impeller of the expansion machine were compared with the values of permissible radial loads on the rotor of standard electric generators of different power (Figure 7).

Analyzing the dependencies of Figures 6 and 7, it can be concluded that a single-flow flow part can be used without complicating the design of the turbo-generator (for example, with the impeller located on the shaft of a standard electric generator) in a limited region at input pressure (up to 1.2 ... 1.8 MPa). At higher input pressures, it is necessary to use multi-flow schemes that unload the turbo-machine rotor from the action of radial loads.

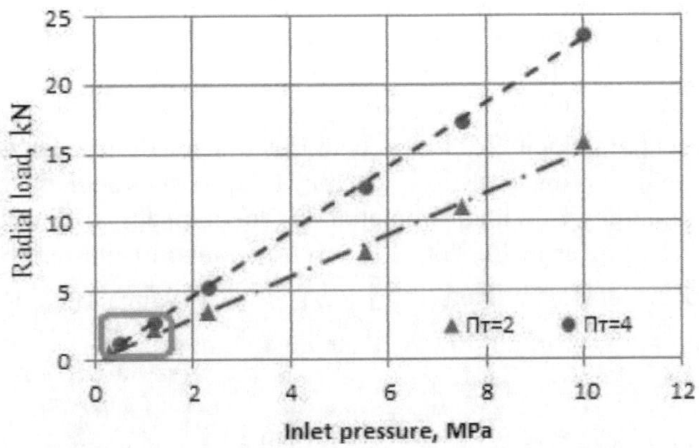

Figure 6. - Dependence of the maximum radial load on the rotor of the vortex single-flow stage on the initial pressure at (D = 0.360 m, $d_k = 0.032 \cdot m$, $d_{ph} = 0.09$).

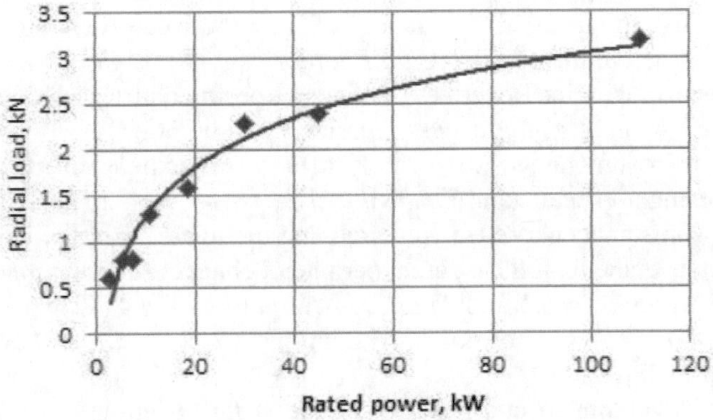

Figure 7. - Dependence of the maximum radial load on the rotor of a standard generator on its power.

4. Conclusions

A parametric model and methods for numerical study of the multi-flow vortex expansion machine in the ANSYS software complex were developed, which makes it possible to investigate the effect of geometric and gas dynamic parameters on its efficiency and characteristics. The most influential factors were identified and the ranges of their changes have been determined. Computational experiments were planned (with the application of the theory of experiment planning) and carried out, on the results of which the multi-criteria optimization was performed, which allowed to find the geometric parameters of the flow parts and their ratios providing the maximum adiabatic efficiency.

To characterize the multi-flow schemes, a dimensionless parameter $\overline{L_p}$ was added, which connects the length of the flow part in the circumferential direction for one flow with the circumference of the meridian section.

The generalized optimum values of the parameters for the three schemes of the vortex expansion machines with an external peripheral channel were obtained in the range of the ratio of pressures Пт = 2-6: $\overline{U}=0.12...0.2$; $\overline{d_s}=0.28...0.38$; $\alpha_{ns}=25...50deg$, $\overline{L_p}=8...11$

Optimal values of the efficiency of the vortex multi-flow expansion machine with an external peripheral channel were increased by more than 15% (from the level of less than 30% to the level of more than 45%).

From the point of view of radial forces, vortex single-flow stages can be used up to a gas inlet pressure of up to 1.2 ... 1.8 MPa, at higher input pressures, multi-flow schemes must be used. In comparison with the three-flow scheme, the dual-flow one has a simpler design and the optimal value of the reduced circumferential velocity $\overline{U}$ is less than in the three-flow scheme. For utilization of low-power expander units, the use of the dual-flow scheme is the most expedient, since in this scheme it is possible to balance the radial forces with a simpler and more compact design.

To achieve the required power in the entire low-power range of the turbo-unit and the maximum efficiency (with optimal design parameters of $\overline{d_{ph}}$, $\overline{d_s}$, α_{ns}) using a single-channel multithreaded scheme, a more powerful multi-channel flow part can be obtained.

References

[1] Moiseev A N, Boinov V V 2015 Relay control of the output voltage of an autonomous power supply for gas transmission networks *Sbornik nauchnykh trudov "Vestnik NTU "HPI" : Problemy udoskonalennya elektrychnykh mashyn I aparativ №13* (Vestnik NTU "HPI")

[2] Vaneev S M, Korolev S K, Rukhlov Yu A, Fedotov Yu T 1991 Compressed-air reaction motor and possibilities of its use *Chemical and Petroleum Engineering* **26** (5,6) pp 304-306

[3] S Vanyeyev, A Berezhnoi 2012 Influence of gap between driving wheel and corps on characteristics of jet-reactive turbine *Procedia Engineering* **39** pp 1-8

[4] S Vanyeyev, V Getalo 2014 Jet-Reactive Turbine: Experimental Researches and Calculations by Means of Softwares *Applied Mechanics and Materials* **630** pp 66–71

[5] S M Vaneev, D V Miroshnichenko 2015 Study of a vortex expansion machine with the external peripheral channel using a virtual stand *Zhurnal inzhenernykh nauk* **2** (2)

[6] S M Vaneev, D V Miroshnichenko 2015 Research and optimization of the design of the flow part of the vortex expansion machine with the external peripheral channel *Kompressornoye I energeticheskoye mashinistriyeniye: nauchno tekhnicheskiy I proizvodstvenniy zhurnal* **4** (42) pp 9-14

[7] Y P Adler, E V Markov, Y V Granovskii 1976 Planning an experiment when searching for optimal conditions (M: Nauka) 280 p

[8] Vaneev S M 1986 Development and research of the vortex pneumatic actuator with the external peripheral channel and a nozzle device dis. … kand. tekhn. nauk : 05.04.06 (Moskva) 183 p

[9] Staroverov K V 1990 Perfection and research of the vortex machine with the peripheral channel in a pneumatic drive mode *dis. ... kand. tekhn. nauk : 05.04.06* (Moskva) 128 p

IOP Conf. Series: Materials Science and Engineering **233** (2017) 012021 doi:10.1088/1757-899X/233/1/012021

Determination of the mass-flow coefficient of jet-reactive turbine's feed-in nozzle

A Berezhnoi[1], S Vaneev[1], S Meleychuk[1]

[1]Sumy State University, R-Korsakova str., 2, 40007, Sumy, Ukraine

[*]E-mail: o.berezhnyi@ukr.net

Abstract. The possibility of adjusting of the jet-reactive turbine's head parameters through changing of mass-flow of the working body with the help of adjustable feed-in nozzle is examined. The mass-flow coefficient of the feed-in nozzle is determined experimentally.

1. Introduction

Nowadays, the requirements to the efficiency, environmental friendliness, reliability, and safety of running engines, operating in the national economy, are becoming much more severe. It fully refers to the drives of ball valves of compressor stations (CS) and cross-country pipelines (CCP).

Following the latest demands of the customers the drives of such ball valves should use raw gas (right from the pipe) as the working body and operate the valve under the drive inlet pressure from min 2MPa to maximum pressure (16MPa). Herewith, the drive inlet gas temperature can vary within the range from -60°C till +80°C depending on the ball valve placement and environmental climatic conditions.

For ball valves with flow path diameter DN>300mm the most wide-spread are pneumohydraulic piston-type drives, which are more simple, safer and cheaper than electric motor drives. To provide drive's operation two kinds of fluids are necessary: raw gas, running right from the gas pipe-line through the control unit to pneumatic cylinder and nonfreezable oil running to the hydraulic cylinder and manual pump. The main disadvantages of such drives are complexity and clunkiness of the construction, low operating efficiency, difficult operation, as constant presence of oil in the system, control over it and its' refilling (in case of oil leakage emergency conditions appear) are required.

Production of air-powered drives of volumetric operational mode for the required settings of ball valves with big flow-path diameter is quite complicated. It is determined by the necessity of dramatic increase of dimensions of drive's power section in order to provide the workability of drive under low pressure.

The solving of this problem is possible through the production of the air-powered drive with an engine, making no excess forcing within the whole range of pressures and being safely operated. As such type of engine we offer a simple, relatively cheap, easily reversing turbine's engine based on the jet-reactive turbine (JRT) [1].

Air-driven drives operating turbine's air-driven engines are sensitive to the change of the load and are basically suitable for the devices with the stable load and settings of the working body at the inlet. However, in case of necessity, they can be quite easily mechanically adjusted to the computation parameters with the help of regulators of mass-flow, pressure, rotating frequency, etc. It should be noted that while the usage of turbine engine as the air-driven engine there appears the possibility for

IOP Conf. Series: Materials Science and Engineering **233** (2017) 012021 doi:10.1088/1757-899X/233/1/012021

the efficiency usage of the great difference of pressures, and taking into account that real pressure in the pipelines is more than 5MPa, turbine engines can compete with the piston ones even according to the coefficient of efficiency).

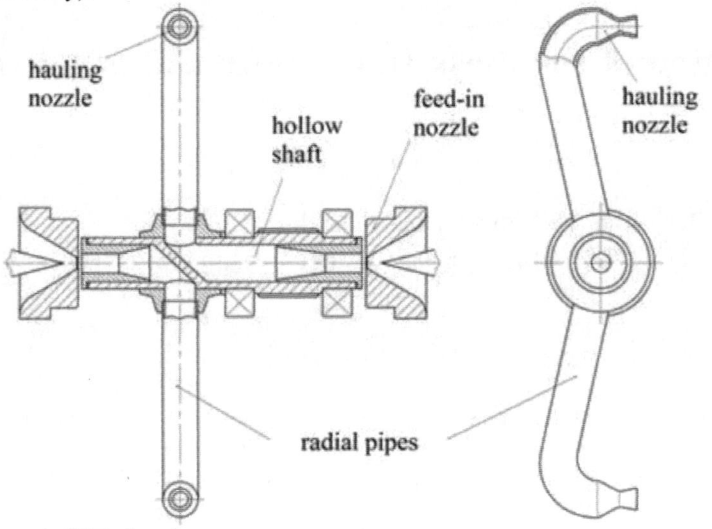

Figure 1. JRT sketch.

While the usage of turbine-driven engines for the examined series of valves it should also be taken into account that the critical sections of flow section of the turbine-driven engine should have as small dimensions as possible to provide the smallest flow, as energy losses in the supply pipeline are equal to the quadrupled flow. Taking into account all ideas expressed above (possibility for efficiency usage of the great difference of pressures, small dimensions of critical sections) jet-reactive engine and turbine-driven engine, using classical one-nozzle turbines (axial-flow and inward-flow), show themselves up among the series of turbine-driven engines. Jet-reactive engine is easily reversed, has lower flywheel effect and integral pipeline (Figure 1) that decreases the possibility of its freezing, though the coefficient of its efficiency can be lower. The main advantages of the turbine-driven engines comparing to the piston ones is excluding of the emergencies and smaller weight-and-dimensional characteristics.

2. Objectives settings

According to the information stated above it is clearly seen that jet-reactive drives should work with the maximum efficiency or close to it. However, the inlet parameters are not always stable; as a result of this the fast reaction to these changes (regulation) is required. As it has already been mentioned above, the basic methods of JRT parameters regulation is by inlet pressure, rotor spinning frequency, and mass-flow. The last method we shall study in details, as the change of mass-flow can be achieved without any additional expenditures and complication of the drive's construction.

3. Aim of research JRT

The aim of the work is to receive experimental dependences of the mass-flow and the flow coefficient of feed-in nozzle depending on the JRT inlet pressure, what, being based on these dependences, will allow in future using the feed-in nozzle even as the flow-meter.

As the first stage of researches in Sumy State University (Technical Thermophysics Department) there was made a test stand for the researching of the jet-reactive turbines and based on them pneumatic units, for the improving of the calculation methodology of JRT characteristics and receiving basic nondimensional dependences.

The model of the jet-reactive turbine as the basic element of the test stand is presented in Figure 2.

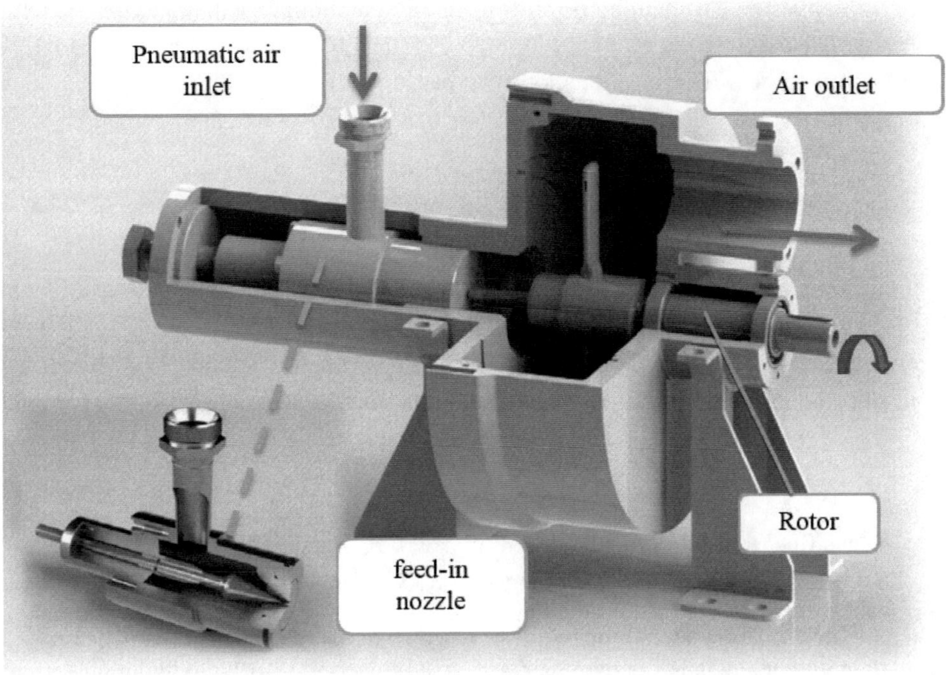

Figure 2. Model of JRT and feed-in nozzle.

Jet-reactive turbine is a machine of the dynamic action principle. The main elements of jet-reactive turbine are flow-in (feed-in) nozzle (FN) and rotor, consisting of the hollow shaft with the radial pipes cantilevered on it. At the end of the pipes there are hauling nozzles (HN) placed in tangentially.

Gas is running through the feed-in nozzle to the hollow rotor shaft and further through the gas path to the HN. In the feed-in nozzle the potential energy of compressed gas is transformed into the kinetic energy releasing together with sonic and supersonic speed of gas flow, which in its turn creates jet reaction and as a consequence rotation moment on the turbine shaft, which in its turn is calculated according to the famous dependence [2]:

$$M_T = M_U - M_{r.r} = M_U - K_{r.r}\omega_T^2 \tag{1}$$

in which: M_U – moment, which is defined according to the theorem of moment of momentum of the gas flow relatively to the spinning axis, based on the co-operation of gas flow with JRT flow part; $M_{r.r}$ – rotor resistance moment in the environment (aerodynamic moment); $K_{r.r}$ – aerodynamic resistance coefficient; depends on density of the environment, rotor outside diameter, shape, quantity and mutual position of pipes of shoulder tubes of rotor; rotation frequency of the turbine shaft; it can be defined only experimentally; ω_T - angular rate of turbine rotation.

For the no-load conditions, when $M_T = 0$, we shall receive (not taking into account mechanical losses in bearings)

$$M_U = M_{r.r} = K_{r.r}\omega_T^2 \tag{2}$$

Moment M_U is equal [3]:

$$M_U = R_W \frac{D}{2} - G_T \omega_{nl} \frac{D^2}{4} \tag{3}$$

where: R_W - tractive force of hauling nozzles in relative motion; D – diameter of the axes position of hauling nozzles relatively to the axis of spinning; G_T – gas flow through the hauling nozzles; ω_{nl} – angular rate of spinning of JRT rotor in the non-load mode.

Then

$$K_{r.r} = \frac{M_U}{\omega_{nl}^2} = \frac{R_W \frac{D}{2} - G_T \omega_{nl} \frac{D^2}{4}}{\omega_{nl}^2} \quad (4)$$

Application programs, which allow calculation of the JRT characteristics taking into account this coefficient are worked out: dependencies are the following $M_T = f(n)$; $N_T = f(n)$; $\eta_T = f(n)$ both, under $K_{r.r} = const$ and under approximate dependence $K_{r.r} = f(n)$.

Power on the turbine shaft and efficiency coefficient is calculated according to the formulas:

$$N_T = M_T \omega_T \quad \text{and} \quad \eta_T = \frac{N_T}{G_n h_s} \quad (5)$$

where: G_n – gas flow through the JRT feed-in nozzles; h_s – isoentropic work of 1 kg expansion of the working body from the stagnation parameters at the JRT inlet to the environmental pressure.

For the calculation of the head parameters and characteristics of JRT and pneumatic units, based on it, it is necessary to know actual gas flow through feed-in nozzle and that is the coefficient of this nozzle flow. Besides, the receiving of experimental dependencies of mass-flow and coefficient of feed-in nozzle flow, depending on the pressure on the JRT inlet will allow in future using the feed-in nozzle also as a flow-meter. That is why the experimental researches on the determination of the coefficient of feed-in nozzle flow in different positions of valve needle and pressures at the inlet were conducted. (see Figure 3)

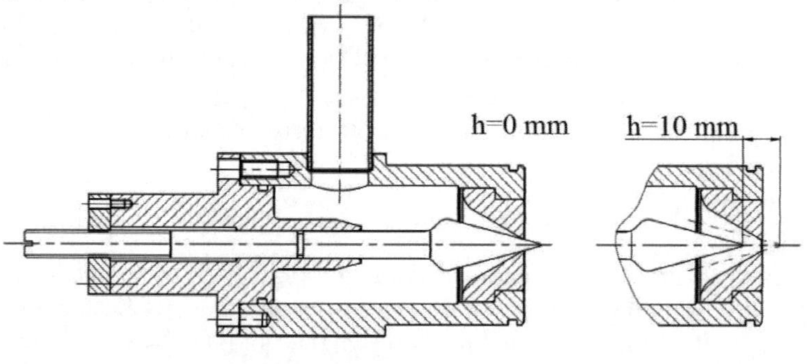

Figure 3. Feed-in nozzle structure (FN).

In its structure feed-in nozzle consists of: supply pipe with pressure gage placed on it for inlet pressure measurements; casting, in which needle seat with expansion angle 60º, constructing outer nozzle closure, is placed; regulation needle, spinning along the thread of screw in the casting and having possibility to move in the axial direction tightening to the seat. Thus, nozzle flow path is reduced. Pith of metric screw thread in the casting for the needle is equal 1mm.

Position, in which, the needle was completely tightened to the seat (nozzle fully closed) was considered to be the starting mode. For this needle position h=0 mm (see Figure 3). Then, needle was gradually moving back to h=16.

For the conduction of measurements and obtaining basic gas-dynamic dependencies there was constructed the test stand, structural drawing of which, is presented in Figure 4. The test stand allows examining starting and no-load modes of jet-reactive turbine.

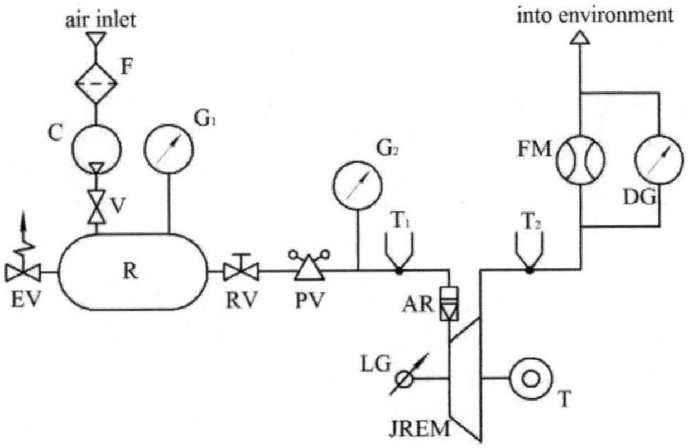

Figure 4. Structural drawing of the experimental unit.

The figure shows: F – air-intake filter; C – compressor; V – shut-off valve; EV – emergency valve; R – receiver; RV – regulation valve, G_1, G_2 – pressure gages; PV – pressure-control valve of vessel-type; T_1, T_2 – temperature meters (electronic multi-meter); AR – air-flow regulator (feed-in unit of JREM); LG – load gage of weight type; JREM – jet-reactive expansion machine; T – tachometer; FM – flow-meter; DG – differential pressure gages.

Figure 5 shows the results of the measurements. As it is seen from the graphical curves, while the increasing of the needle move h>10 mm lines $m = f(p)$ overlap each other and that is the proof of the full opening of the nozzle section, if h=10 mm. This conclusion is traced better in the diagram of dependence $m = f(h)$, presented for some values of pressure in Figure 6.

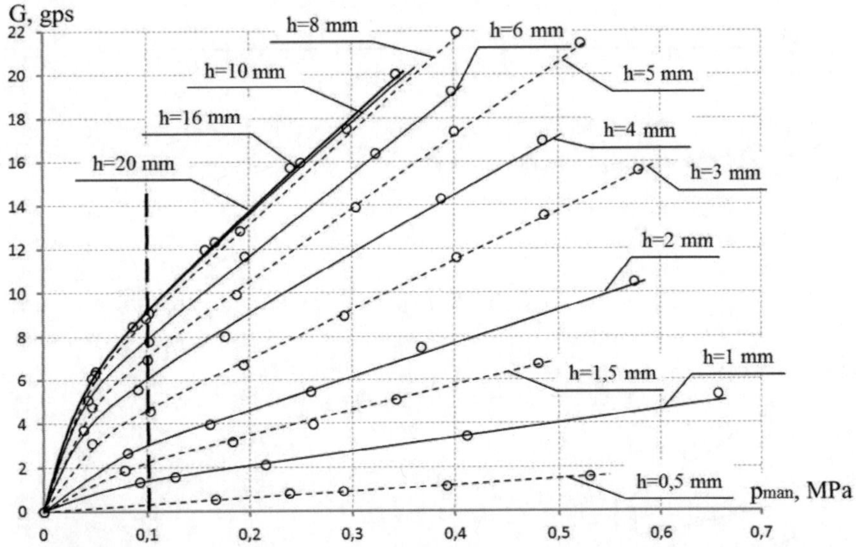

Figure 5. – Dependence of mass-flow from inlet pressure in the feed-in nozzle.

Diagrams character proves theoretical dependencies of the gas flow through the nozzle and the nozzle head [4]. Gas mass-flow while the process of isentropic flow through the short nozzle head from the hollow of unlimited volume is described by two equalizations:

- for the pre-critical mode suits the equalization:

$$G_{ns} = f_{\min} p_{in} \sqrt{\frac{2k}{RT_{in}(k-1)} \left[\left(\frac{p_{out}}{p_{in}} \right)^{\frac{2}{k}} - \left(\frac{p_{out}}{p_{in}} \right)^{\frac{k+1}{k}} \right]}, \quad \frac{p_{out}}{p_{in}} \geq \beta_{cr} \tag{6}$$

- for the post-critical mode suits the equalization:

$$G_{ns} = f_{\min} p_{in} \sqrt{\frac{k}{RT_{in}} \left(\frac{2}{k+1} \right)^{\frac{k+1}{k-1}}}, \quad \frac{p_{out}}{p_{in}} \leq \beta_{cr} \tag{7}$$

in which, $f_{\min}$ - minimal flow area for the feed-in nozzle.

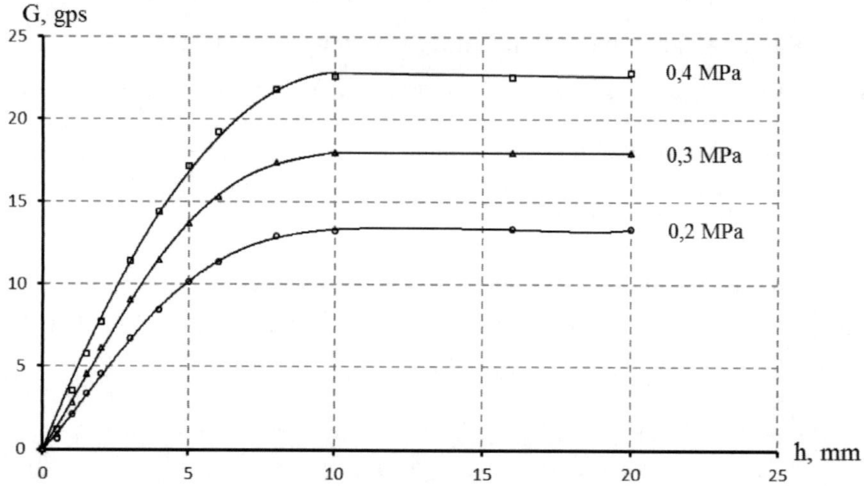

Figure 6. – Dependence of mass-flow from the needle move for different pressures in the nozzle inlet.

Because of the existence of the of the intake center body (needle) of complicated shape and cone-type surface of nozzle seat, under the low values of needle move h (needle is placed in the critical cross-section) gas flows through the axisymmetric cone-type gap. In this case, the minimal side surface area of the truncated cone must be considered to be the area of the flow-path nozzle (Figure 7 (a)). For the fully opened nozzle the area of the flow-path is determined by the area of critical size of the flow-path with the diameter $d_{cr.f}$ (Figure 7 (b)).

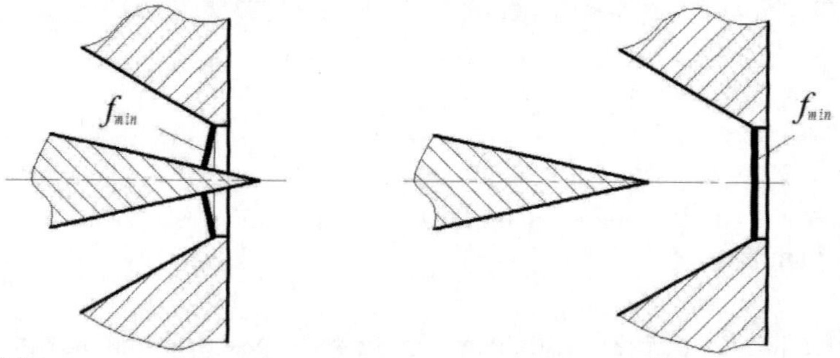

(a) (b)

Figure 7. Calculation of the area of the flow path of the feed-in nozzle: (a) for low values of needle move; (b) for fully opened nozzle.

For the correct calculation of the area of the flow path of nozzle of determined geometry there is created a computer program, which calculates the minimal value of the area $f_{\min}$ for any intake center body position (Figure 8).

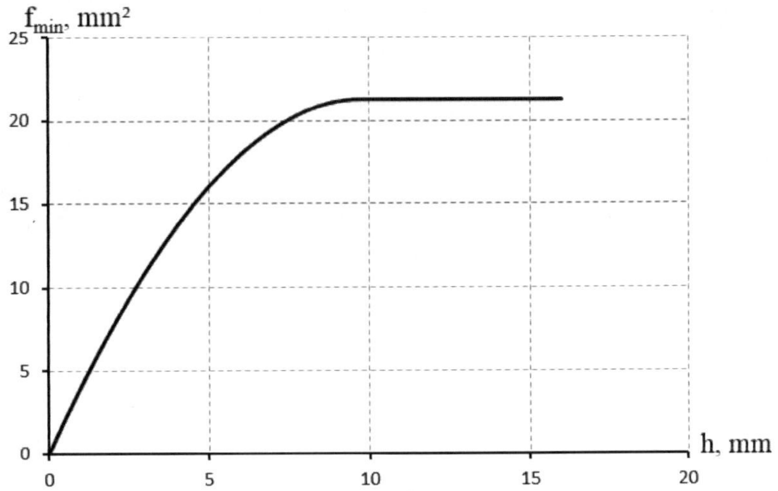

Figure 8. Dependence of the $f_{\min}$ on the needle move.

The true process of gas flow from the nozzle is calculated by the flow coefficient $\mu = \frac{G_n}{G_{ns}}$, equal to ratio of actual gas flow through the nozzle to the theoretical one.

After determination of actual air flow through the feed-in nozzle it was necessary to find the value of flow coefficient for this nozzle. For each experimentally obtained value of gas flow according to the needle move and inlet pressure in the JRT it was distinguished the correspondent theoretical value, and consequently the flow coefficient.

In the result the graphical diagram of changes $\mu = f(h)$ was received. It is presented in Figure 9.

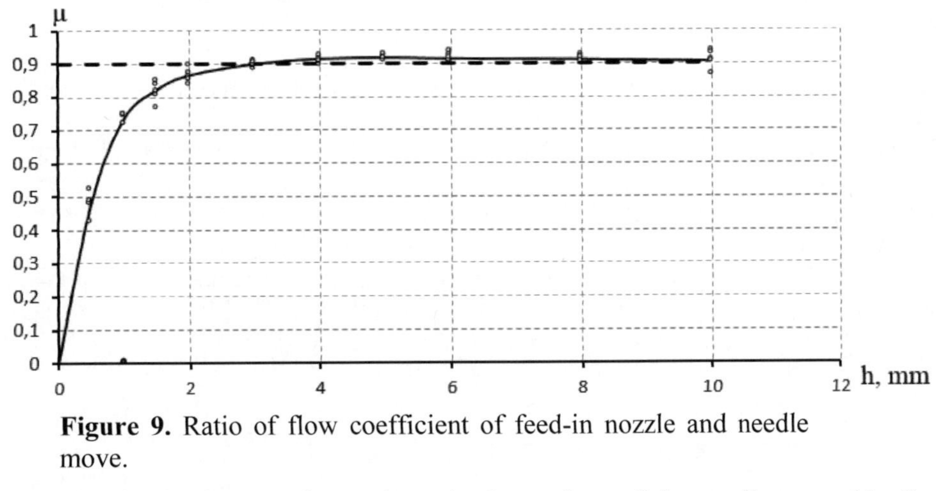

Figure 9. Ratio of flow coefficient of feed-in nozzle and needle move.

In the diagram it is clearly seen that under quite low values of the needle move ($h = 0 \div 2\,mm$), flow coefficient has relatively low value and is dependent on h, that is conditioned by the construction specifications of the feed-in nozzle (huge losses while the gas flow through the "narrow" gap between the needle and the seat exist). For other needle positions the coefficient μ becomes constant, close to

0,9 (Figure 10). Then, this value is input into the program for the calculation of head settings and characteristics of jet-reactive expansion machine and unit in general.

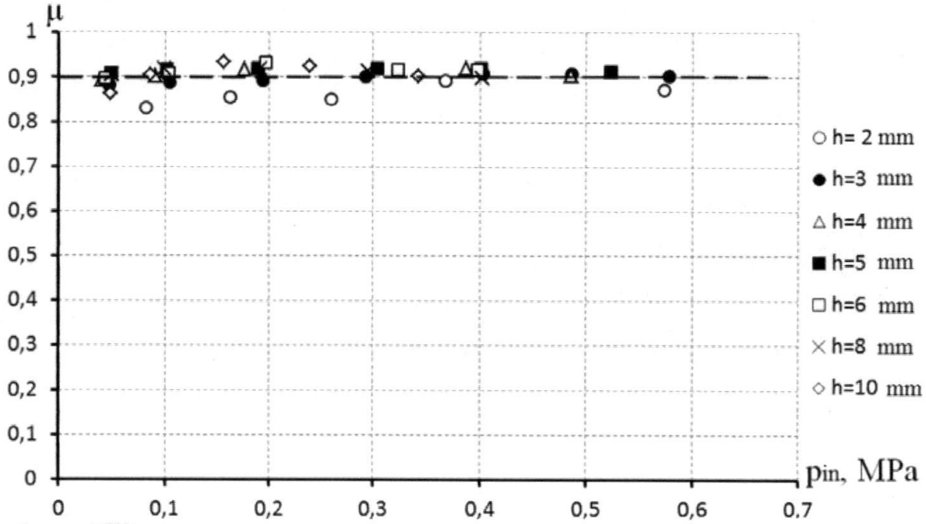

Figure 10. Dependence of flow coefficient on the inlet pressure for the needle move value $h \geq 2\,mm$.

4. Conclusions

Experimental dependences of mass flow and flow coefficient for the regulating feed-in nozzle of jet-reactive turbine in different positions of regulating needle and with different pressure values in the nozzle inlet were received, that allowed clarifying of techniques for calculation of turbine head settings and characteristics.

It was defined that while the needle move from the stop mode into the casting $h \geq 2\,mm$ flow coefficient of feed-in nozzle remains actually constant and equal to 0,9.

Obtained dependencies also allow refusing from installing flow-meters, and defining actual gas flow right by the diagram for jet-reactive turbines with similar construction of supply nozzle.

References

[1] Vanyeyev S M, Korolev S K, Rukhlov Y L, Fedotov Y T 1990 Jet-reactive engine and possibilities of its usage *Chemical and Oil Engineering* **6** pp 16-17

[2] Vanyeyev S M 1999 Calculation of jet-reactive turbine characteristics *Reporter of "KPI": Mashinostroenie* **36** pp 263-269

[3] Vanyeyev S, Berezhnoi A 2012 Influence of gap between driving wheel and corps on characteristics of jet-reactive turbine *Procedia Engineering* **39** pp 1-8

[4] Abramovich G N 1991 *Applied gas dynamics / 5-th edition, overworked and extended* (M: Nauka) p 600

The results of pre-design studies on the development of a new design of gas turbine compressor package of GPA-C-16 type

A V Smirnov[1], V M Chobenko[2], O M Shcherbakov[1], S M Ushakov[1], V P Parafiynyk[1], R M Sereda[1]

[1] Special Design Bureau, Sumy NPO, 58, Gorkogo Str., Sumy, 40004, Sumy, Ukraine
[2] Gas Turbine Research and Production Complex Zorya-Mashproekt, 42a, Zhovtnevy str., Mykolaiv, 54018, Ukraine

E-mail: ang8l@yandex.ru

Abstract. The article summarizes the results of analysis of data concerning the operation of turbocompressor packages at compressor stations for the natural gas transmission system of Ukraine. The basic requirements for gas turbine compressor packages used for modernization and reconstruction of compressor stations are considered. Using a 16 MW gas turbine package GPA-C-16S/76-1,44M1 as an example, the results of pre-design studies and some technical solutions that improve the energy efficiency of gas turbine compressor packages and their reliability, as well as its environmental performance are given. In particular, the article deals with the matching of performance characteristics of a centrifugal compressor (hereinafter compressor) and gas turbine drive to reduce fuel gas consumption; as well as application of energy efficient technologies, in particular, exhaust gas heat recovery units and gas-oil heat exchangers in turbocompressor packages oil system; as well as reducing emissions of carbon monoxide into the atmosphere using a catalytic exhaust system. Described technical solutions can be used for development of other types of gas turbine compressor packages.

1. Introduction

The Ukrainian gas transmission system (GTS) is one of the largest in Europe. Its total length is 38.6 thousand kilometers including 22.2 thousand kilometers of main gas pipelines. The transmission system includes 72 compressor stations (CS) with 702 turbocompressor packages with the total power of 5.44 million kW.

Industrial gas turbines, aero- and marine-derivative gas turbines, electric motor drives and gas reciprocating engines are used as mechanical drives for pipeline compressors at stations. Gas turbine drives are the dominant option; with power share approximately 82 % of the total power of drive units on the GTS [1].

As of today, a lot of gas turbine compressor packages (GTCP) have reached the end of their first design service life. The efficiency of the prior-art-type drive is 23 to 25 % [1]. Further operation of such gas turbines leads to unreasonable operating costs. In addition, compressors of turbocompressor packages installed at many CS require replacement and modernization due to changes in the operation conditions of gas transmission pipelines.

Therefore, modernization and reconstruction of the Ukrainian GTS is a pressing issue. We should proceed from the fact that, in the short term future, gas turbines will remain the main and most effective choice of compressor drive equipment at CS [1, 2]. The most important requirements for GTCP are independence of operation, reliability, efficiency, high level of automation, on-site maintainability, acceptable life-cycle cost, and compliance with current environmental requirements.

Sumy NPO has more than 40 years of experience in design and manufacture of equipment for natural gas transmission CS. GTCPs manufactured by Sumy NPO are widely used in oil and gas industry of Ukraine, Russia, Iran, Uzbekistan and other countries. Today, the Company has designed, manufactured and delivered over 2,350 turbocompressor packages of more than 100 modifications ranging from 4 to 25 MW, designed on the basis of 19 modifications of derivative gas turbines of different types, and 2 types of electric motor drives. Depending on the desired increase in pressure in the packages single-body or multi-body compressors can be used. Fluid or electromagnetic journal and thrust bearings, oil shaft-end or dry gas seals can be used. GTCP equipment has the following installation options: modular type, sandwich panel steel construction, and hard wall compressor shop building.

The article gives the results of GTCP pre-design studies for reconstruction of one of the pipeline CS of the Ukrainian GTS using GPA-C-16S/76-1,44M1 of modular type design as an example. Some technical solutions, which allow improving GTCP efficiency and reliability, and ensuring improved environmental performance by reducing carbon monoxide emissions into the environment, were considered.

2. General characteristics of GPA-C-16S/76-1,44M1

Gas turbine compressor package GPA-C-16S/76-1,44M1 (Figure 1) has the following technical characteristics at the design operating conditions: capacity reduced to a temperature of 293 K and pressure of 0.101 MPa V = 29.5 MCMPD; discharge pressure (absolute) p_d = 7.45 MPa; pressure ratio π=1.44; polytropic efficiency of a compressor η_p=87_{-1} %; compressor rotor speed n=5,100 rpm; absorbed power N=14.4 MW.

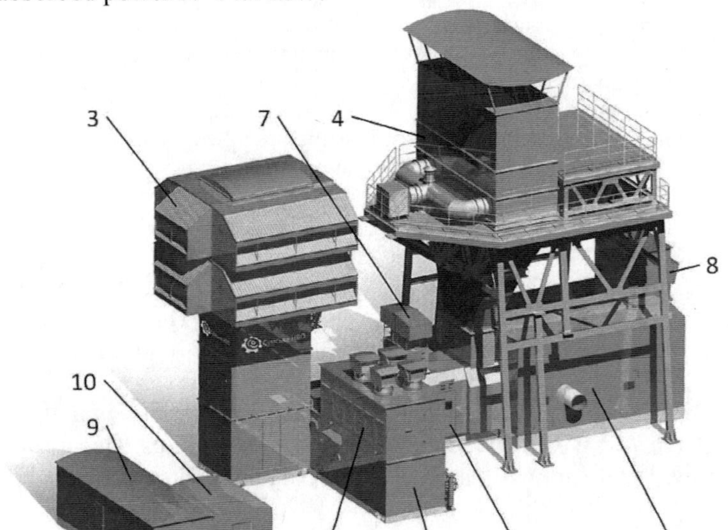

Figure 1. General view of GPA-C-16S/76-1,44M1 of modular type: 1 – power unit; 2 – compressor unit; 3 – turbine air inlet system; 4 – turbine exhaust system; 5 – gas turbine oil cooler unit; 6 – supply systems unit; 7 – power unit cooling system; 8 – compressor oil cooler unit; 9 – automation unit; 10 – fire suppression unit.

GTCP main and auxiliary equipment is supplied to the CS as complete and ready-to-operate modules. On-site installation procedure includes their installation on the foundation, interconnection and connection to external communications.

The main GTCP module is a turbine unit with a compressor and gas turbine equipped with auxiliary system elements installed in separate compartments (Figure 2).

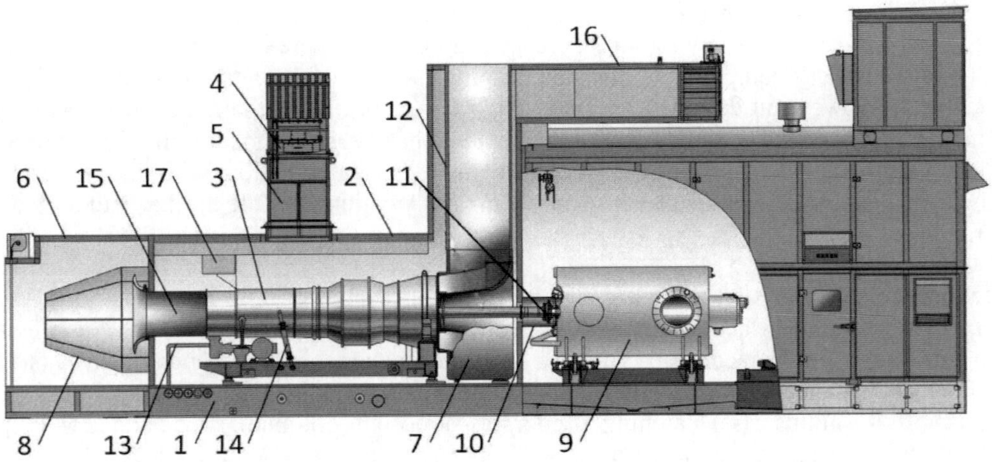

Figure 2. Turbine unit of GPA-C-16S/76-1,44M1 of modular type: 1 – frame; 2 – acoustic enclosure; 3 – gas turbine; 4 – ventilation unit; 5 – inlet ventilation silencer; 6 – turbine air inlet duct; 7 – exhaust collector; 8 – protective screen; 9 – compressor; 10 – coupling guard; 11 – coupling; 12 – exhaust duct transition; 13 – gas turbine oil system elements; 14 – fuel gas pipeline; 15 – turbine air inlet tube; 16 – exhaust ventilation duct; 17 – ventilation air distribution system.

The compressor is the main functional system that determines operating conditions of GTCP at the CS. The given GTCP uses a 3-stage barrel compressor with dry gas seals and tilting pad bearings (Figure 3).

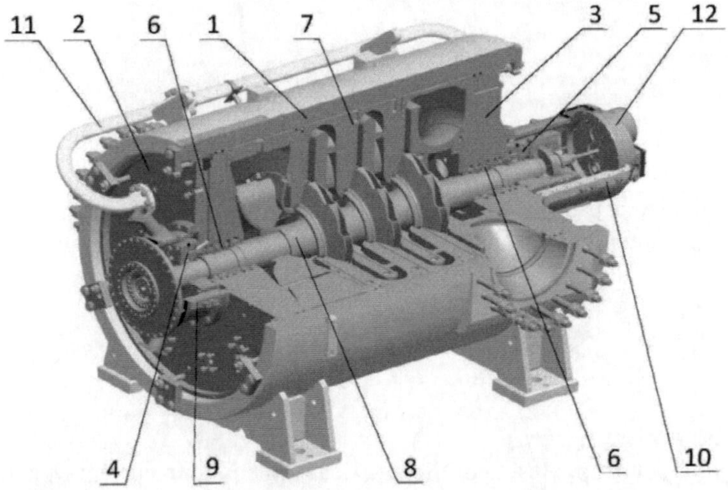

Figure 3. General view of centrifugal compressor: 1 – outer casing; 2 – front cover; 3 – rear cover; 4 – journal bearing; 5 – journal-thrust bearing; 6 – dry gas seal; 7 – inner casing; 8 – rotor; 9, 10 – enclosure; 11 – pipeline; 12 – oil pump.

The drive is designed on the base of 16,7 MW marine-derivative engine UGT 16000 with efficiency η_e=35% (ISO 2314) manufactured by Gas Turbine Research and Production Complex Zorya-Mashproekt (Mykolaiv, Ukraine). It should be noted that in this article gas turbine drive (GTD) means a drive unit consisting of a gas turbine supplied by gas turbine manufacturers and auxiliary systems (e.g. turbine air inlet and exhaust system, lubricating oil and fuel supply, etc.) designed and manufactured by Sumy NPO, which provide its effective and reliable operation as a part of the GTCP.

To reduce noise and protect the station staff and equipment against radiant heat, which occurs during gas turbine operation, the engine together with its exhaust collector, lubrication system elements, fuel gas pipelines, instrumentation and wires and cables is installed in the special acoustic enclosure that is a part of the power unit. To ensure acceptable operating temperature conditions for the engine and auxiliary equipment of the power unit, a mechanical ventilation system is used. The design feature of power unit designed by Sumy NPO is the special ventilation air distribution system inside the acoustic enclosure [3]. The ventilation system has been designed on the base of numerical simulation results [3, 4] and test-based data obtained in collaboration with the experts of the N. E. Zhukovsky National Aerospace University Kharkiv Aviation Institute during field tests of power unit thermal conditions at the CS [5].

3. Matching of characteristics of a compressor and a gas turbine

The most important issues when designing highly efficient GTCP are selection of the compressor flow path and matching of its characteristics with gas turbine characteristics when operating as a one unit.

Matching of characteristics is based on the formation and analysis of the GTCP system or integral characteristic [6-9]. In this case, matching of compressor design (optimum) mode and GTCP optimum mode should be reached. At the same time, the compressor optimum conditions mean those operating conditions of the drive unit, at which the maximum exergy or integral efficiency value of the package, determined at the stage of thermodynamic analysis of GTCP scheme [6, 7], can be reached. This mode corresponds to the minimum flow rate of fuel gas in the gas turbine.

Taking into account the nature of different forms of energy conversion during GTCP operation, as a criterion of its effective operation, it is more correct to use GTCP exergy efficiency in the form of:

$$\eta_{ex}^{GTCP} = \eta_{ex}^{CC} \cdot \eta_{ex}^{GTD} \tag{1}$$

where η_{ex}^{CC} and η_{ex}^{GTD} – compressor exergy efficiency and GTD exergy efficiency, respectively.

However, its use in the initial design stage poses a number of difficulties, so when selecting the compressor flow path, it is permissible to use as a criterion for the GTCP efficiency the integral efficiency [8] in the form of:

$$\eta_{\Sigma}^{GTCP} = \eta_e \cdot \eta_p \tag{2}$$

where

η_e – GTD efficiency;

η_n – compressor polytropic efficiency.

Fuel consumption ratio, which is calculated based on the basic parameters of the compressor and GTCP, can be used as a criterion of the package fuel efficiency [9]:

$$K_f^{GTCP} = \frac{G_f}{G_{CC} \cdot \overline{\psi}_p} \tag{3}$$

where

G_f – fuel gas mass flow rate, *kg/s*;

G_{CC} – process (compressed) gas mass flow rate, *kg/s*;

$\overline{\psi}_p = \psi_p / (\psi_p)_{des}$ – compressor relative polytropic head coefficient, determined at the current ψ_p and design $(\psi_p)_{des}$ operation conditions.

Matching of characteristics and operation conditions of the GTD and compressor reduces to the system analysis of GTCP process with the implementation of the following sequence of calculation tasks [6-11]:

- Selection of the flowing path, determination of compressor gas-dynamic and power characteristics based on customer`s technical requirements (Figure 4, $\eta_p = f(N_{PT})$), where N_{PT} – power on the shaft of engine power turbine is equal to N_{CC} including mechanical losses in the rotor system).

- Analysis of gas turbine characteristics in the required power and speed ranges and determination of gas turbine effectiveness including external losses (Figure 4, $\eta_e = f(N_{PT})$).
- Determination of GTCP integral ($\eta_\Sigma^{GTCP} = f(N_{PT})$) or system ($\eta_{ex}^{GTCP} = f(N_{PT})$) characteristics.
- Selection of the compressor/turbine drive operating modes to match the compressor operating mode with $(\eta_{ex}^{CC})_{max}$ and GTCP operating mode with $(\eta_{ex}^{GTCP})_{max}$. This mode corresponds to the minimum fuel gas flow rate.

Matching of characteristics is determined by behavior of the fuel consumption ratio as a function $K_f^{GTCP} = f(N_{PT})$ [9].

Matching of the package rated operating mode and the modes corresponding to $(\eta_{ex}^{GTCP})_{max}$ and $(K_f^{GTCP})_{min}$ confirms the accuracy in flowing path selection. Achieving the minimum value of K_f^{GTCP} proves the best efficient mode of GTCP operation.

When selecting the compressor flowing path geometry for GPA-C-16S/76-1,44M1, some kind of studies were performed to match compressor/turbine characteristics using a GTCP efficiency η_Σ^{GTCP}. Figure 4 shows that at the GTCP rated operating mode $N_{PT}=14.4$ *MW* matching of the compressor design operating mode ($\eta_p=0.87$) and GTCP optimum operating mode ($\eta_\Sigma^{GTCP}=0.283$) is provided. It corresponds to the GTCP best efficient operating mode. It is also confirmed by the corresponding minimum value K_f^{GTCP} (Figure 4 (b)).

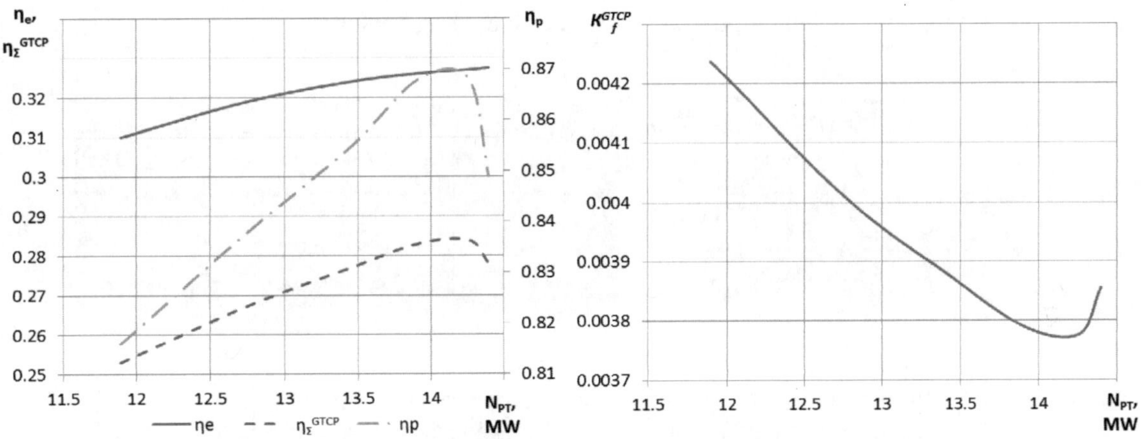

(a) (b)

Figure 4. Efficiency of compressor, GTD and GTCP (a) and fuel consumption ratio (b) as a function of power at n_{cT}=5,100 rpm.

As the study results [8] show, using low- and medium-head impellers ensures optimization of the compressor and turbine drive operating modes. In this case, consumption of UGT 15000 fuel gas can be reduced by 5.4…7.6% (3 ... 3.5 million nm^3/year) [8].

4. Improvement of GTCP energy efficiency using heat recovery systems

GTCPs are significant sources of thermal pollution to the environment. Most heat losses are caused by emission of exhaust gases into the atmosphere.

Heat recovery from the GTCP exhaust gases can be used for increasing gas turbine efficiency using regenerative or water (steam) injection cycles, as well as being used to produce heat, electric power or cold for needs of the CS or the external consumers.

Considering the high power consumption of the economy of Ukraine, development of combined power, heat and cold generation facilities on the basis of the CS is a prospective line of GTS development. This will reduce the costs of natural gas transmission and can also become the basis for development of effective regional power production and infrastructure of new agricultural and industrial enterprises [12, 13]. It should be noted that for existing simple-cycle gas turbines with the efficiencies of 30 to 40% (for 6.3 to 32 MW power respectively), complete recovery of thermal resources of exhaust gases provides the increase of thermal efficiency of the system up to 0.8 to 0.85.

However, the implementation of this approach is associated with high capital costs (especially in the context of combined cycles), the need to improve reliability of equipment, and also requires the development of the appropriate legislative and regulatory framework for the practical solution of the issues connected with sale of electric power, heat and cold produced by heat recovery facilities operating as a part of the CS [13, 14].

Therefore, currently waste heat of GTCP exhaust gases is mostly used for heating of CS, residential settlements or greenhouse facilities near the station, etc. The disadvantage of this solution is inability to use all turbine exhaust gases thermal resources due to lack of the regular consumers of generated heat at the CS.

To produce utility heat GTCP of GPA-C-16S type are equipped with heat recovery units, of between 3.5 to 9 MW which are designed and manufactured by Sumy NPO on the basis of finned tube heat exchangers (Figure 5). Their application allows achievement of thermal efficiencies of 0.36 to 0.46. The heat recovery units can contain several separate heat exchangers. Heat power control of the heat recovery units is carried out by heat exchangers on-off and by controlling the flow rate of exhaust gases through them. The design of the heat recovery units allows filling heat exchangers with water without shut-down of the gas turbine. To avoid high thermal stresses heat exchangers are cooled with atmospheric air supplied by fan of the heat recovery unit cooling system.

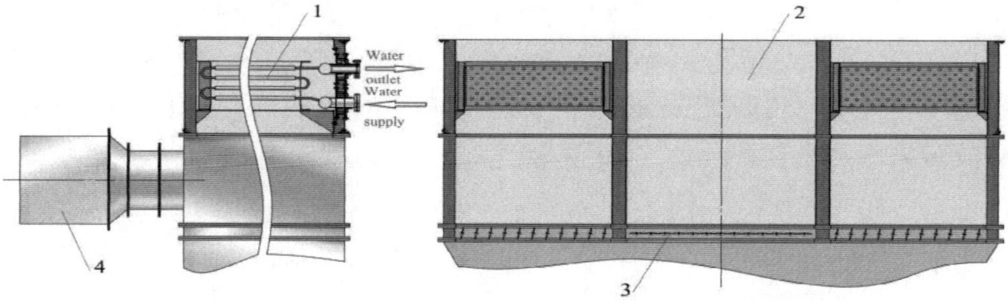

Figure 5. General view of the heat recovery unit with 3.48MW heat power used in GPA-C-16S type packages: 1 – heat exchanger; 2 – bypass; 3 – diverter valve; 4 – heat recovery unit cooling system.

One of the promising ways to improve GTCP efficiency is also the recovery of low-grade heat, which is obtained with oil from the compressor and gas turbine bearings. This heat can be used for fuel gas heating in special gas-oil heat exchangers [15-19]. The use of gas-oil heat exchangers allows reduction of fuel gas consumption and decrease of the electric power consumption to cool the oil. A general view of the gas-oil heat exchanger designed for use in the gas turbine lubrication systems for GPA-C-16S type packages is shown in Figure 6. Its main technical characteristics are given in Table 1.

According to the results of calculations [19], use of the gas-oil heat exchanger in gas turbine lubrication system of GPA-C-16S type package allows reducing the fuel gas consumption by 42,000 nm^3/year and decreasing the electric power consumption of oil coolers fans by 23,000 kWh/year.

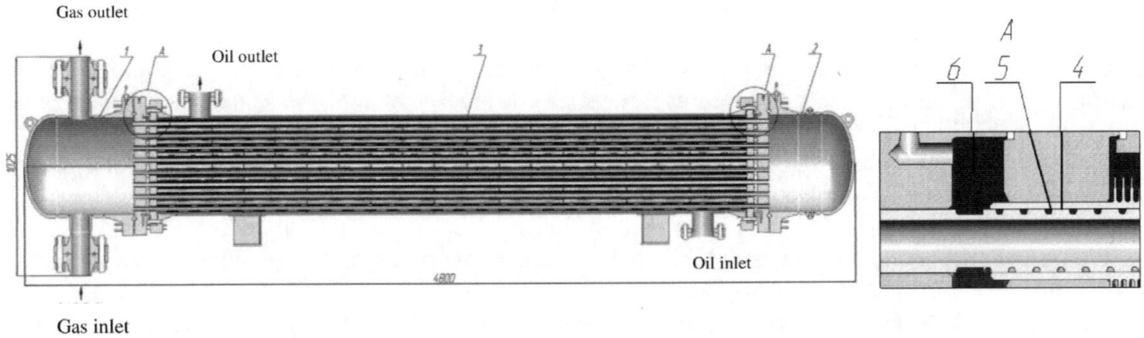

Figure 6. General view of the heat recovery gas-oil exchanger: 1 – inlet-outlet chamber; 2 – intermediate chamber; 3 – shell; 4 – bimetal finned tube; 5 – leakage path; 6 – leakage safety chamber.

Table 1. Main technical characteristics of the gas-oil heat exchanger.

Parameter	**Value**	
	Shell	**Tubes**
Medium	Oil	Natural gas
Rated flow rate, kg/h	8,700	3,500
Inlet temperature, °C	100	20…40
Outlet temperature, °C	89.5	40…60
Operating pressure at inlet, MPa	0.44	3.0

5. Reduction of carbon monoxide emissions using the catalytic exhaust system (CES)

One of the most important objectives of the CS reconstruction and modernization is reduction of emissions of pollutants, primarily nitrogen oxides (NOx) and carbon monoxide (CO) in the atmosphere.

Thanks to the implementation in Ukraine of Directive 2010/75/EU of the European Parliament and of the Council of 24 November 2010 on industrial emissions, the concentration of NOx and CO in GTCP exhaust gases (at loads more than 70% of the gas turbine rated power) should not exceed 75 and 100 mg/nm^3 respectively. Currently, Gas Turbine Research and Production Complex Zorya-Mashproekt has designed and tested a low-emission combustion chamber for UGT15000 gas turbine meeting the requirements of the Directive. However, according to the Terms of Reference for packages intended for reconstruction of the CS of Ukrtransgaz, the CO concentration in exhaust gases should not exceed 100 mg/nm^3 within the whole range of gas transmission modes. It corresponds to the load range of 45 to 90% of the gas turbine rated power. Ensuring the acceptable environmental performance of the gas turbine in such a wide range of capacities is a complex scientific and technical problem [14, 20, 21].

The main methods for reducing the pollutant emissions with products of fuel combustion in the gas turbines are as follows [14, 20, 21]:

- improving the methods of fuel combustion (so-called "dry" methods DLN, DLE; catalytic combustion chambers);
- injection of water (steam) into the gas turbine combustion chamber;
- using special systems for combustion product purification in the exhaust systems.

The first two methods require a significant change in a gas turbine design. In this regard, to meet the environmental requirements specified by the customer, the package design is provided for use of CO removal catalytic system installed in the exhaust system of GTCP. The principle of CES operation is based on the oxidation of carbon monoxide to carbon dioxide under the action of the catalyst layer at a certain temperature.

Currently, catalytic purification of combustion products is mainly used in power engineering at thermal power plants, boilers, combined cycle and gas turbine plants. At the same time, they are not widely used at CS because of relatively high capital costs [14].

There is relatively small number of CES international manufacturers for gas turbines, among which are BASF (Germany), Haldor Topsoe (Denmark), ECAT (RF).

Traditionally the CES consists of units that are installed normal to the direction of exhaust gases flow. The units are made on the basis of highly porous ceramic or metal honeycomb structures coated with catalysts.

Use of CES in GTCP exhaust systems is associated with some problems. To ensure effective chemical oxidation of carbon monoxide, it is required to meet the following conditions: the temperature of the exhaust gases should be within the certain limits (200 to 450°C depending on the type of the used catalyst); the CES inlet flow should be uniform.

As was mentioned above, to recover waste heat of gas turbine exhaust gasses, the exhaust system of GPA-C-16S/76-1,44M1 package is equipped with a heat recovery unit. Therefore, to ensure the required exhaust gas temperature at the CES inlet the catalyst should be installed before the heat recovery unit. To select the CES location, a series of numerical simulations have been performed in the exhaust system of the existing design (Figure 7 (a)). Figure 7 (b) shows the streamlines in the existing exhaust system of GPA-C-16S/76-1,44M1 package at the turbine rated operation conditions.

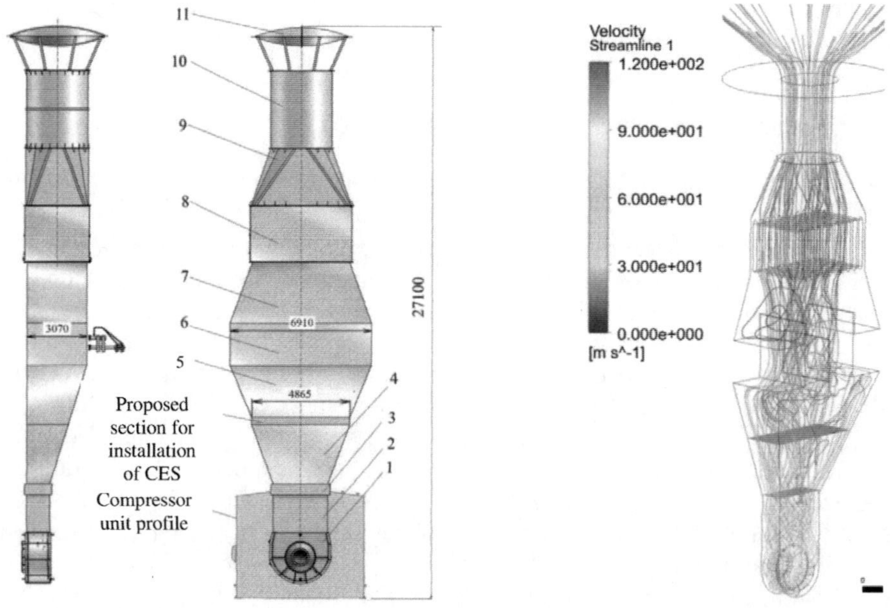

(a) (b)

Figure 7. Structural scheme (a) and streamlines (b) in the existing exhaust system of GPA-C-16S type package: 1 – exhaust collector; 2 – exhaust duct transition; 3 – expansion joint; 4 – the 1st stage diffuser; 5 – the 2nd stage diffuser; 6 – heat recovery unit; 7 – transition piece; 8 – silencer; 9 – confuser; 10 – exhaust stack; 11 – hood.

Figure 7b shows that flow in the exhaust system has a complex structure with multiple vortex zones and reverse flows. The total pressure loss in the exhaust system without catalyst is 2,550 Pa. The maximum deviation of velocity from the mass flow averaged value in the proposed cross-section of CES (between the diffusers of the first and second stages) is 400 %, RMS is 128 %. Such non-uniform flow structure in the exhaust system is associated primarily with features of the flow at the exhaust collector outlet. The flow at the exhaust collector outlet has symmetrical swirling, which occurs as a result of stream rotation flowing on the plenum of the exhaust collector after coming out the axial-radial diffuser. This swirl remains until the outlet section of the exhaust collector and extends to the silencer.

According to the CES designers at such cross-sectional area of the catalyst and the degree of non-uniformity of the inlet flow, the total pressure loss in CES at the rated operating mode of the turbine exceeds 5,000 Pa, which is unacceptable. To ensure the guaranteed parameters of the gas turbine operation, the total pressure loss in the exhaust system from the exhaust collector inlet to the outlet to the atmosphere should not exceed 4,905 Pa.

To achieve the goal, a series of numerical simulations have been done to develop a new design of the exhaust system using various types of flow straighteners. According to the results of the study, the exhaust duct design with tubular straightener (Figure 8(a)) has been selected.

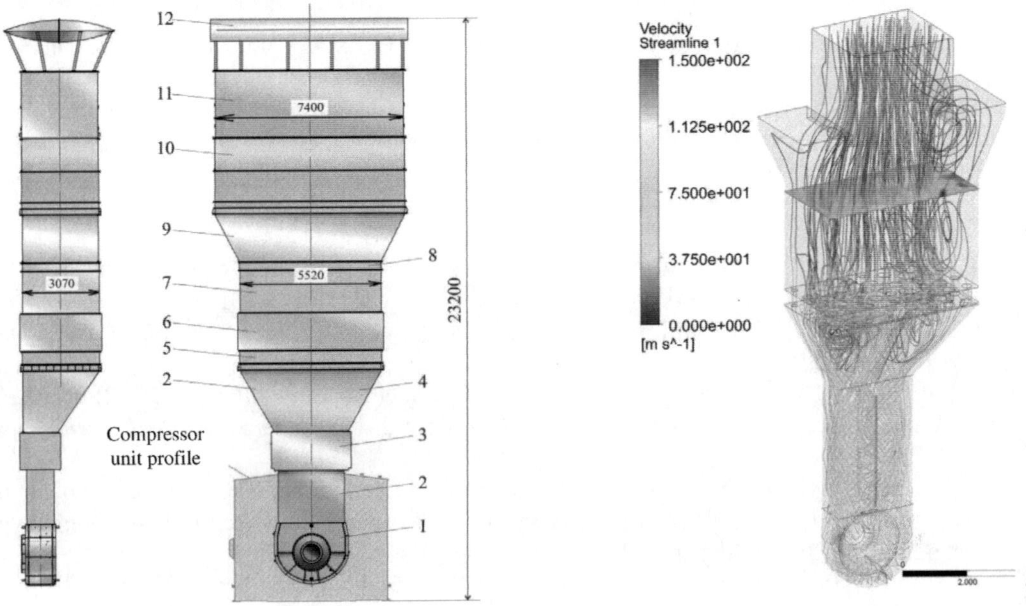

(a) (b)

Figure 8. Structural scheme (a) and streamlines (b) in the exhaust system of GPA-C-16S package with the flow straightener: 1 – exhaust collector; 2 – exhaust duct transition; 3 – expansion joint; 4 – the 1st stage diffuser; 5 – straightener; 6 – expansion joint; 7 – spacer; 8 – CES; 9 – the 2nd stage diffuser; 10 – heat recovery unit; 11 – silencer; 12 – hood.

According to the results of calculations, the total pressure loss in the exhaust system of the new design is 4,630 Pa, maximum velocity deviation from the mass flow averaged value at the CES inlet is 65 % and RMS deviation is 18 %.

6. Conclusion

Taking into account a large power of GTCPs used at CSs of the Ukrainian main gas pipelines and strategic importance of the GTS for Ukraine and other European countries, further improvement of GTCPs to increase their energy efficiency, reliability and improve environmental performance is of great importance. Tasks in the improvement process are different and knowledge-intensive. The main ones are as follows:

- Improvement of gas turbines based on the design principles of industrial gas turbines and achievements in thermo-gas dynamics, materials technology, dynamics and strength of components and systems used in aircraft and marine engines building.

- Development of new gas turbine designs based on regenerative and combined operation cycles.
- Improvement of design methods for compressor flow paths and other energy-engineering equipment.
- Application of heat recovery technologies; promising direction of further improvement of GTCPs and CSs to increase their energy efficiency is designing heat recovery facilities for power, heat and cold generation on the base of CSs.
- Improvement of auxiliary GTCP systems to provide efficient, reliable and safe operation of the package. Such systems include turbine air inlet and exhaust systems, turbine unit systems, gas turbine and compressor lubrication systems, etc.
- Improvement of GTCP environmental performance, in particular, reduction in pollutant emissions by improvement of gas turbine combustion chambers, as well as due to use of CES.

References

[1] Paton B, Khalatov A, Kostenko D, Bileka B, Pismenny O, Botsula A, Parafiynyk V and Konyakhin V 2008 *Bulletin of NAS of Ukraine* **4** pp 3-9
[2] Pirani S 2007 *Ukraine's gas sector* (Oxford: Oxford Institute for Energy Studies) p 115
[3] Smirnov A, Shcherbakov O, Tkachenko D, Parafiynyk V and Slabko Yu 2016 *Aerospace technic and technology* **7** pp 179-187
[4] Smirnov A, Kostyuk V, Tkachenko D, Kyrylash E and Slabko Yu 2013 *Herald of aeroenginebuilding* **2** pp 99-107
[5] Shcherbakov O, Tkachenko D, Parafiynyk V, Gurinenko V, Kostyuk V, Skrypka O and Kyrylash O 2015 *Eastern-European Journal of Enterprise Technologies* **6/7** (78) pp 35-42
[6] Tertyshny I, Prylypko S, Miroshnichenko E and Parafiynyk V 2015 *Journal of mechanical engineering* **18** (4/1) pp 9-17
[7] Tertyshny I, Prylypko S and Parafiynyk V 2016 *Journal of mechanical engineering* **19** (2) pp 10-18
[8] Parafiynyk V, Nefedov A, Yevdokymov and V, Tertyshny 2012 *Compressors and Pneumatics* **2** pp 10-17
[9] Tertyshny I, Parafiynyk V, Nefedov A, Rogalsky S, Kotenko N, Tymoshadchenko D and Mikhaylenko S 2014 *Proc. of XVI Int. scientific-technical Conf. on Compressor Manufacturing* vol 1 (Saint Petersburg, REP-Holding) pp 328-339
[10] Taher M and Meher-Homji C 2012 *Proc. of ASME Turbo Expo 2012: Turbine Technical Conference and Exposition* (Copenhagen) pp 51-61
[11] Morosuk T and Tsatsaronis G 2008 *Energy* **33** pp 890-907
[12] Bileka B 2006 *Industrial heat engineering* **28** (2) pp 132-149
[13] Smirnov A, Parafiynyk V, Shcherbakov O, Epifanov S, Kostiuk V, Chobenko V and Shevchuk V 2016 *Bulletin of National Technical University "KhPI": coll. of sci. papers. Ser: Power and Heat Engineering Processes and Equipment* **9 (1181)** pp 13-25
[14] Galiullin Z, Salnikov S and Shchurovsky V 2014 *Modern Gas Transmission Systems and Technologies* ed Shchurovsky V (Moscow: Gazprom VNIIGAZ) p 346
[15] Bodunov D. 2013 *Gas Turbo Technology* **6** pp 18-19
[16] Yokell S 1973 *Chemical Engineering* **14** pp 133–136
[17] Kuznetsov L, Kuznetsov Yu, Efremov A, Burakov A and Abramov A 2009 Fuel gas treatment unit *Patent RU 92934*
[18] Belousov Yu, Vereshchagin N, Pchelintsev V and Golubtsov S 2012 Tubular bundle with buffer chambers of gaseous oil shell-and-tube heat exchanger *Patent RU117599*
[19] Smirnov A, Sereda R and Borisov M 2016 *Bulletin of National Technical University "KhPI" coll. of sci. papers. Ser. : Power and Heat Engineering Processes and Equipment* **10** (1182) pp 93-98
[20] Shchurovsky V, Sinitsyn Yu and Cheremin A 2004 *Reduction in Pollutant Emissions of Gas Turbines at GASPROM Compressor Stations: Report* (Moscow)
[21] Varlamov G, Khalatov A, Poznyakov P and Yurashev D 2012 *Eastern-European Journal of Enterprise Technologies* **10** (57) pp 9-14

IOP Conf. Series: Materials Science and Engineering **233** (2017) 012023 doi:10.1088/1757-899X/233/1/012023

Maisotsenko cycle applications for multistage compressors cooling

D Levchenko[1], I Yurko[2], A Artyukhov[3], V Baga[4]

[1] Department of Technical Thermal Physics, Sumy State University, Laboratory Building, Room B-104, 2 Rymskogo-Korsakova St, Sumy, 40007, Ukraine
[2] TRIZ Ltd., Mashinostroiteley 1 St, Sumy, Ukraine
[3] Processes and Equipment of Chemical and Petroleum-Refineries Department, Sumy State University, 2 Rymskogo-Korsakova St, Sumy, 40007, Ukraine
[4] Department of Technical Thermal Physics, Sumy State University, Laboratory Building, Room B-104, 2 Rymskogo-Korsakova St, Sumy, 40007, Ukraine

E-mail: levchenkodmitriyalex@gmail.com

Abstract. The present study provides the overview of Maisotsenko Cycle (M-Cycle) applications for gas cooling in compressor systems. Various schemes of gas cooling systems are considered regarding to their thermal efficiency and cooling capacity. Preliminary calculation of M-cycle HMX has been conducted. It is found that M-cycle HMX scheme allows to brake the limit of the ambient wet bulb temperature for evaporative cooling. It has demonstrated that a compact integrated heat and moisture exchange process can cool product fluid to the level below the ambient wet bulb temperature, even to the level of dew point temperature of the incoming air with substantially lower water and energy consumption requirements.

1. Introduction

Gas compression and transportation is one of the most energy consuming technologies in the world. Compressors are a vital link in the conversion of raw materials into refined products. Compressors also handle economical use and transformation of energy from one form into another. They are used for the extraction of metals and minerals in mining operations, for the conservation of energy in natural gas reinjection plants, for secondary recovery processes in oil fields, for the utilization of new energy sources such as shale oil and tar sands, for furnishing utility or reaction air, for oxygen and reaction gases in almost any process, for process chemical and petrochemical plants, and for the separation and liquefaction of gases in air separation plants and in LPG and LNG plants. The economy and feasibility of all these applications depend on the reliability of compressors and the capability of the compressors selected to handle a given gas at the desired capacity [1]. Since, value of specific work for all compressors is determined by several factors as: specific heat ratio k, specific gas constant R, pressure ratio π and inlet temperature T_{in}. As a rule, all the enumerated factors except the inlet temperature are defined by the initial conditions and/or performance specification for the designed compressor. That is why inlet temperature reduction remains the most feasible way to reduce compressors' specific work.

There are a lot of approaches to reduce T_{in} value (depending on the place of cooling):

IOP Conf. Series: Materials Science and Engineering **233** (2017) 012023 doi:10.1088/1757-899X/233/1/012023

- Pre-cooling (cooling at suction pipe).
- Cooling in compression process (cooling of cylinder jacket, envelope of compression channels).
- Interstage cooling (cooling after each compressor stage or section).
- Aftercooling (cooling at discharge pipe).

However, these approaches could be implemented by different means:

- Vapor compression refrigeration (VCR)
- Adsorption refrigeration (AR)
- Injection cooler
- Air/water-cooled heat exchanger
- Indirect evaporative cooling (IEC) heat and mass exchanger (HMX)
- Maisotsenko cycle (M-cycle) heat and mass exchanger (HMX)

Considering first two methods of refrigeration, one should say that VCR and AR systems could provide the highest level of cooling due to using a low-boiling refrigerant. Though, vapor compressor refrigerator has lower efficiency than absorption refrigerator , but has smaller size and simple construction. However, VCR and AR have common disadvantages, which restrict or turn off their benefits: 1) consume too much energy; 2) refrigerant leakage risk; 3) high price of the refrigeration units, their installation and maintenance.

Injection cooler is applicable restrictively as to excessive moistening of the gas. Moreover, it could lead to compressors parts corrosion in case of off-designed cooling regime.

Air/Water-cooled heat exchanger has big specific quantity of metal and installation costs. It is necessary to provide a large amount of water and its cooling in case of closed circuit of water supplying system. Air- cooled heat exchanger has poor low cooling capacity due to low specific heat of cooling air. Therefore these type of heat exchangers requires large area of cooling surfaces.

IEC has acceptable cooling capacity, but wet bulb temperature is the limit of gas cooling for one-stage design apparatus.

The most promising technology of gas cooling is M-cycle process. This process allows to reach dew-point temperature in one HMX stage.

2. Fundamentals of M-cycle HMX

The M-Cycle is a thermodynamic process which captures energy from the air by utilizing the psychrometric renewable energy available from the latent heat of water evaporating into the air. It combines thermodynamic processes of heat transfer and evaporative cooling to facilitate product temperature to reach the dew-point temperature of the ambient air [2]. The basic principle and features of the M-Cycle can be explained from Figure 1 (a) and Figure 1 (b) representing the partial extraction of air and product air cooling M-Cycle, respectively [3]. The psychrometric representation of both variations of M-cycle is shown in Figure 1 (c). It consists of two kinds of primary channels named as wet and dry channels.

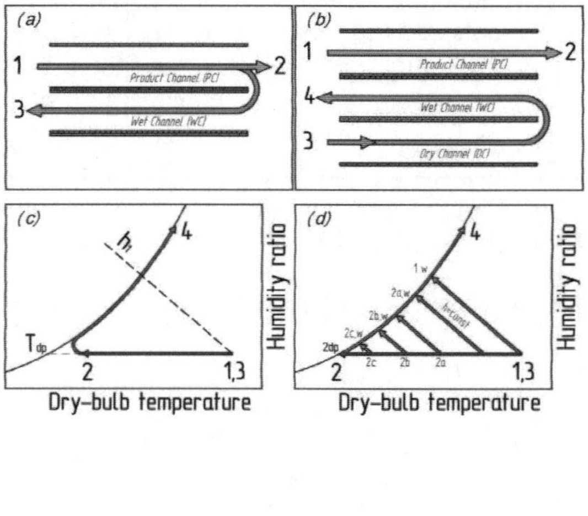

Figure 1. Schematic diagram of Maisotsenko Cycle for: (a) old M-Cycle, (b) modified M-Cycle, (c) psychrometric representation, and (d) sequential temperature decrement in wet-channel.

Product Air Cooling arrangement keeps working and product dry channels separated. Partial Extraction of Air arrangement fractions a portion of the dry channel as usable product. Both arrangements are psychrometrically identical while arrangement in Figure 1 (b) with product channel has essential advantages as compared to arrangement in Figure 1 (a). Thus, any gas or liquid matter (for example hot waste product) can be used as a product flow in arrangement in Figure 1 (b) while arrangement in Figure 1 (a) is limited to the same gas for both channels. Moreover using the product channel allows reducing pressure drop in the system. Another advantage of arrangement in Figure 1 (b) compared with Figure 1a would be a possibility to use the product channel as a condenser for gas dehumidification [4].

For example, ambient air (1) is flowed into the dry-channel where it is sensibly cooled at constant humidity to cycle point (2) by transferring the heat to the wet-channel. The operational principle of M-Cycle is based on diverting the cooled air (2) to the wet-channel in order to use as working air. It results in subsequently decrement of effective dry-bulb (1 - 2a; 2b; 2c; 2) and wet-bulb (1w - 2a,w; 2b,w; 2c,w; 2dp) temperatures of the working air in the wet-channel as shown in Figure 1d. Sequential decrement of dry-bulb temperature in the wet-channel brings the effective wet-bulb temperature to be ideally equal to the dew-point temperature. Hence for an ideal heat transfer surface, the product air can be sensible cooled to the dew-point temperature of the ambient air. Moreover, saturated hot air (3) is rejected from the wet-channel equivalent to the evaporated water and recovered heat.

Thereby, M-Cycle HMX is suitable for all kind of air cooling (pre-cooling, inter-cooling and aftercooling). The basic scheme of transported gas cooling is represented in Figure 2. Gas is cooled in shell and tube heat exchanger. Water absorbs sensible heat which is transferred from the gas to the water through the tube walls. Hot water then goes to cooling tower, where its temperature decreases by direct evaporating process.

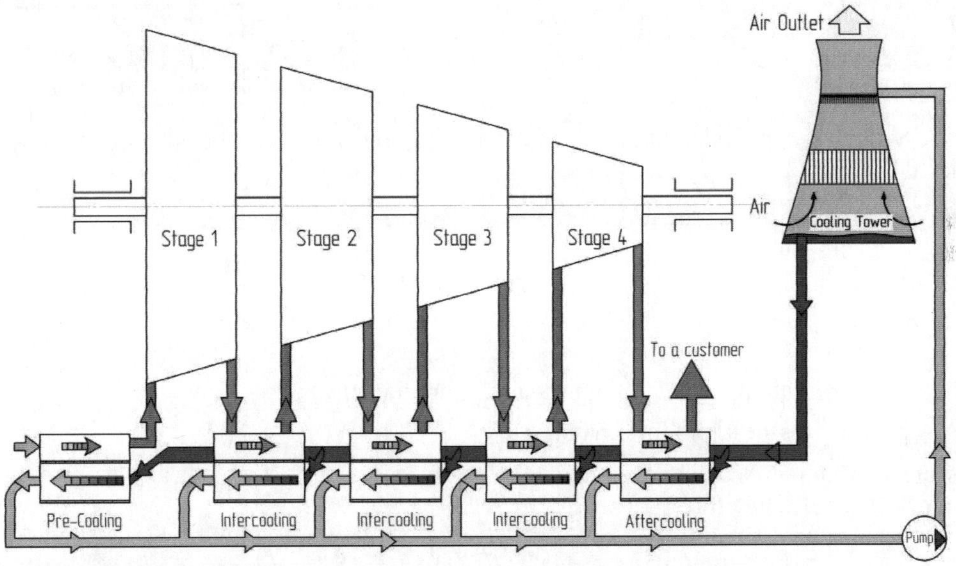

Figure 2. Conventional scheme of a compressed gas cooling.

Proposed cooling scheme of gas cooling via M-cycle is represented in Figure 3. The M-cycle utilizes the air potential energy available from latent heat of water evaporating into the air. The main advantage of proposed M-cycle cooling is possibility to get rid off cooling tower and closed cycle water supplying system. It is connected with the fact that M-cycle HMX required significantly less amount of water for cooling process than conventional cooling scheme shown in Figure 2.

Conceptual design of M-cycle HMX is represented in Figure 4. The predicted heat and mass transfer of the IEC could only be achieved on the condition of obtaining complete wetting of the wet channels surfaces. Poor water distribution on the surfaces would reduce the heat/mass transfer rate

IOP Conf. Series: Materials Science and Engineering **233** (2017) 012023 doi:10.1088/1757-899X/233/1/012023

between the primary air and the water film, and between the water film and secondary air, which would eventually undermine the effectiveness of the indirect evaporative cooler [5]. The most preferable way of water distribution system is applying superhydrophilic capillary porous medium with high moisture wicking and evaporation ability. A wide range of materials have been commonly used as the heat/mass exchanging medium. However, it is almost impossible to choose one all-mode material for various M-cycle heat exchanger applications. That is why research activities aimed to improve ability of wet channel surface to retain a water film become very important.

Posessing already installed cooling tower and sufficient amount of circulating water at the place where compressor works leads to considerable increase of M-Cycle HMXs operating effectiveness. This is connected with the phase change process in the wet channel that intensify heat transfer from hot compressed gas to working medium. For preliminary M-cycle HMX calculation it is proposed the following algorithm. It allows to estimate material and heat values in relevant (coresponding) points of M-cycle.

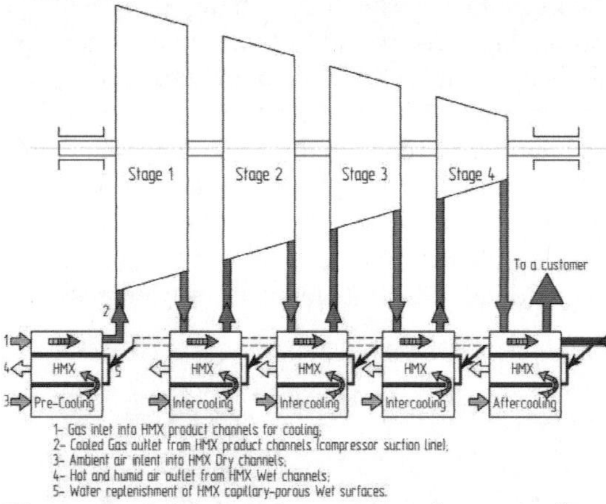

Figure 3. Scheme of compressed gas cooling via M-cycle.

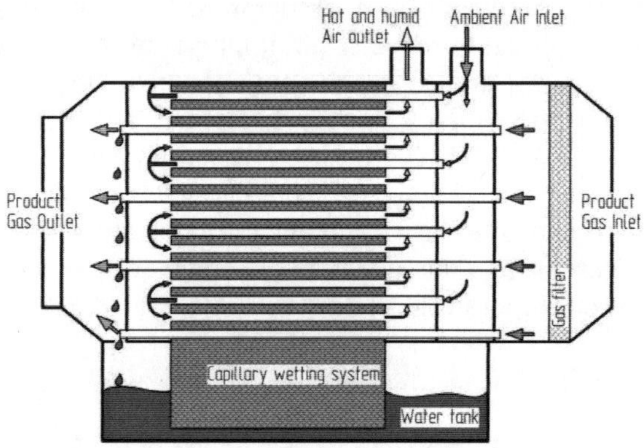

Figure 4. Proposed design of M-Cycle HMX.

There is pressure drop in product channel. Therefore:

$$P_1 = P_2 + \Delta P_p \tag{1}$$

where

P_1 and P_2 – absolute inlet and outlet pressure in product channel;

ΔP_p – pressure losses in product channel.

Pressure of water in wet working channels should be equal to ambient air(gas) pressure at inlet working dry channel. Therefore:

$$P_5 = P_3 \tag{2}$$

where P_5 is water pressure and P_3 is absolute inlet pressure in dry working channel.

Absolute outlet pressure in working wet channel will be lower then inlet pressure due to pressure drop:

$$P_4 = P_3 - \Delta P_w \tag{3}$$

where

P_4 – absolute outlet pressure in working wet channel;

P_w – pressure losses in working channel.

IOP Conf. Series: Materials Science and Engineering **233** (2017) 012023 doi:10.1088/1757-899X/233/1/012023

The energy balance then would be:

$$h_1 \cdot G_1 + h_3 \cdot G_3 + h_5 \cdot G_5 = h_2 \cdot G_2 + h_4 \cdot G_4 \tag{4}$$

where

h_1, G_1 and h_2, G_2 – gas enthalpy and massflow rate at the product channel inlet and outlet respectively;

h_3, G_3 and h_4, G_4 – gas enthalpy and mass flow rate at the working channel inlet and outlet respectively;

h_5, G_5 – water enthalpy and mass flow rate in M-cycle.

Material balance in apparatus can be written as:

$$G_1 + G_3 + G_5 = G_2 + G_4 \tag{5}$$

Material balance in product channel:

$$G_1 = G_2 \tag{6}$$

Material balance in working channel:

$$G_4 = G_3 + G_5 \tag{7}$$

Gas temperature in product channel outlet will be slightly larger then dew point of ambient working air:

$$t_2 = t_{dp} + \Delta t_{dp} \tag{8}$$

where

t_2 – outlet product gas temperature;

Δt_{dp} – temperature drop regarding the heat exchange imperfection (dew-point temperature unattainability);

t_{dp} – dew-point temperature of ambient working air which is defined for pressure equal to partial pressure of water vapour P_{H3}. This pressure is defined by:

$$P_{H3} = P_3 \cdot r_{H2O\_3} \tag{9}$$

where r_{H2O_3} – mole fraction of water vapour at working channel inlet.

Outlet temperature in working channel t_4 is defined depending on the user-specified heat transfer at the ends of HMX:

$$t_{dp} < t_4 = \max(t_1,\ t_3) - \Delta t_{in-out} < \max(t_1,\ t_3) \tag{10}$$

where Δt_{in-out} – the temperature difference at the ends of the HMX, set by the user as a parameter.

Mole fraction of water vapour at working channel outlet:

$$r_{H2O\_4} = \frac{P_H(t_4)}{P_4} \cdot \frac{\varphi_4}{100\%} \tag{11}$$

where

$P_H(t_4)$ – water vapour partial pressure at working channel outlet;

φ_4 – relative humidity of air at working channel outlet (set by user as a parameter in the range of 90%-100%).

Water mass flow rate G_5 is proportional to difference between absolute humidity ratio at working inlet (dry) and outlet(wet) channels. Therefore it can be expressed as:

$$G_5 = G_{W4} + G_{W3} \tag{12}$$

where $G_{W4} + G_{W3}$ – working mass flow rate increment caused by water evaporation.

$$G_{W3} = G_3 \cdot m_{H_2O\_3} = G_3 \cdot \frac{r_{H_2O\_3} \cdot \mu_{H_2O}}{\mu_{mixture\ 3}} \tag{13}$$

$$G_{W4} = G_4 \cdot m_{H_2O\_4} = G_4 \cdot \frac{r_{H_2O\_4} \cdot \mu_{H_2O}}{\mu_{mixture\ 4}} \tag{14}$$

$$G_5 = \left(G_3 + G_5\right) \cdot \frac{r_{H_2O\_4} \cdot \mu_{H_2O}}{\mu_{mixture\ 4}} - G_3 \cdot \frac{r_{H_2O\_3} \cdot \mu_{H_2O}}{\mu_{mixture\ 3}} \tag{15}$$

$$\frac{G_5}{G_3} = \frac{\frac{r_{H_2O\_4} \cdot \mu_{H_2O}}{\mu_{mixture\ 4}} - \frac{r_{H_2O\_3} \cdot \mu_{H_2O}}{\mu_{mixture\ 3}}}{1 - \frac{r_{H_2O\_3} \cdot \mu_{H_2O}}{\mu_{mixture\ 3}}} = X \tag{16}$$

Equation (16) establishes relation between G_5 and G_3. Required amount of working air could be defined as:

$$G_3 = \frac{G_1 \cdot \left(h_1 - h_2\right)}{h_4 - h_3 + X \cdot \left(h_4 - h_5\right)} \tag{17}$$

3. M-cycle enhanced cooling tower concept

The total effectivenes of cooling system could be increased significantly by using advanced fill for cooling tower based on M-cycle [6]. Figure 4 shows comparison between conventional and advanced fills for cooling tower.

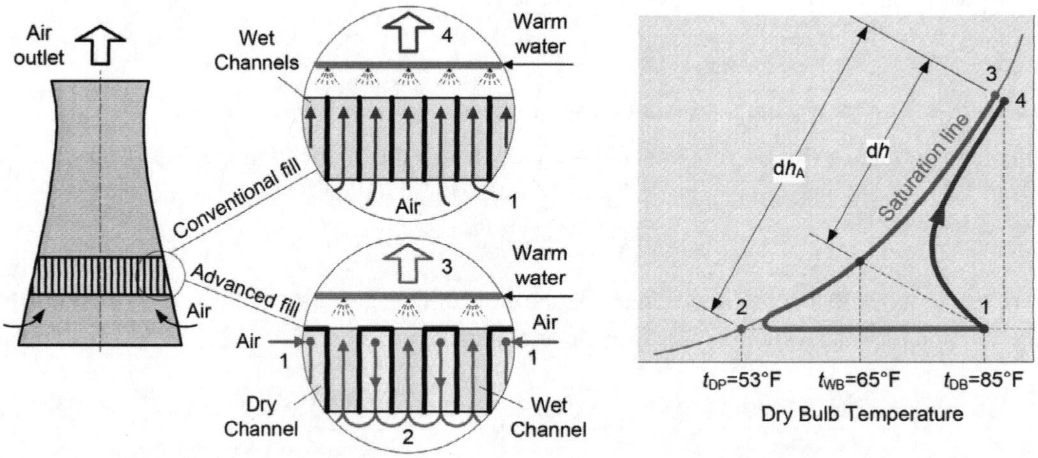

Figure 4. Cooling tower: conventional design and advanced design.

The lowest attainable water temperature using conventional fill is ambient air wet bulb temperature. Through an integrated design whereby the ambient air flow is in effect pre-cooled by itself, the "dew point" (M-cycle) approach reaches cooled water temperature level nearing the ambient dew point temperature. This "dew point" approach, using advanced fill, provides higher performance via the psychrometric principle - when a parcel of air is sensibly cooled, the saturated water vapor pressure decreases, therefore, reducing the wet bulb temperature and increasing the evaporative cooling potential down to the ambient air dew point temperature.

In additional, solid or liquid desiccants can be used in dry channels to dehumidify the air to increase its latent heat potential before it is sent into the wet channels. This process will lower the dew point attainable in the second stage of cooling. Also, continuous cooling of the desiccant by air increases its absorption capacity and rate, thus drying the air faster. In humid environments, the incoming ambient air should be heated so as to increase its saturation vapor pressure and latent heat capacity [7].

A truly integrated precooling was theorized for air cooling [8] and then developed for water cooling [9] by Maisotsenko et al., achieving dew point cooling without an initial external indirect-direct evaporative precooling stage and with an integrated heat and mass exchanger (fill) optimized for sensible precooling and pressure drop with a hybrid cross/counter flow arrangement.

Dew point cooling tower concept which is proposed by Maisotsenko [9] has numerous benefits, comparing to traditional cooling techniques:

- Reduced capital and installation costs due to uselessness of additional external heat exchangers and/or cooling equipment (e.g. chillers).
- Extended performance in certain climate zones.
- Decreasing of cooling tower size and weight.
- Decreased evaporation losses, requiring less makeup water.
- Decreased air side pressure drop, due to absence of an external heat transfer equipment.

Critical to achieving these claims is the advanced fill, which integrates the air precooling and direct evaporative cooling tower segments into a single heat and mass exchanger, bringing innovative advancements in evaporative cooling to cooling towers. Integrating these stages into one system realizes a positive feedback mechanism that cannot be realized with separate (external) air precooling. With the humid air in the direct evaporative section as the air precooling heat sink, sensible heat pickup on the humid air side drives further evaporation, thus cooling, already enhanced by the reduced wet bulb temperature (higher evaporative potential) of the inlet air.

It is feasible to achieve the benefits of integrated precooling-cooling tower operation in a truly integrated and compact heat and mass exchanger, capable of effective heat and mass transfer. Unlike direct evaporative or staged precooling technologies employing an initial direct evaporative precooling step (e.g. [10, 11] concept), heat transfer within this advanced dew point cooling tower is driven by the dew point temperature gradient, not wet bulb temperature. As such, only the absolute outside ambient humidity is important, not the temperature of incoming air or water.

GTI and Idalex Technologies, Inc. are currently developing and optimizing the advanced fill for the dew point cooling tower. The concept is shown in Figure 5, whereby the diagrams show the integration of vertical dry precooling and wet direct evaporative channels in a unique flow arrangement. This integrated method of achieving dew point cooling with a single heat and mass exchanger, is referred to as the thermodynamic M-Cycle [12, 13]. In comparison to indirect-direct technologies, researchers have demonstrated water and process air (fan power) reductions of 45% and 74% respectively are feasible with this method [14]. In this case, it is not absolutely necessary to precool the incoming air outside of the fill, or use any multistage cooling [10, 11] in order to get the dew point temperature level achieved with supplementary mechanical refrigeration. Instead, the air flow can be specially arranged in the fill to be precooled by evaporating water, which in principle is what drives this novel cycle.

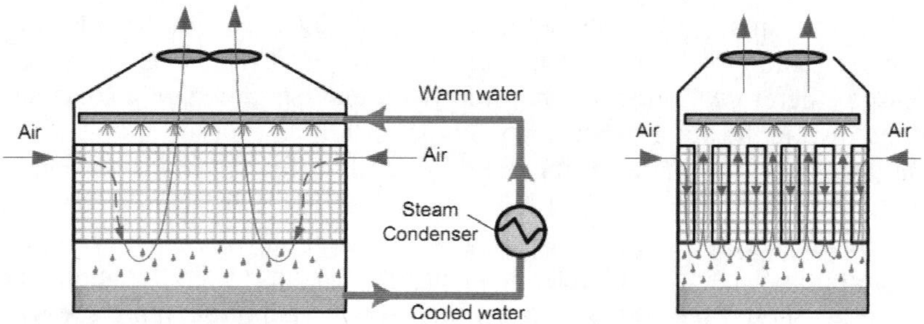

Figure 5. Dew-point cooling tower concept.

In a regenerative evaporative cooler, dry and wet channels are placed close together in contact with each other. A fan draws outside ambient air into the dry channels. A second fan draws a part of the primary air coming out from the wet channels into the dry channels. The directions of air flow are opposite to each other in dry and wet channels. Fins are provided in order to enhance heat transfer between air and water [15]. The fluid is evaporatively cooled by outside air. The cooled fluid is then divided into two parts. The first part is circulated into the section to be cooled using a heat exchanger. The second part is again sensibly cooled using outside air to a lower wet bulb temperature without changing its humidity. The second stream is then mixed with the first air stream and recirculated into the section to be cooled. In this process, the fluid can be cooled to even lower than the wet bulb temperature and ideally to the atmospheric dew point temperature [16-19].

4.Conclusions

Compressor energy consumption can be reduced dramatically using M-cycle HMX for cooling.
This HMX is not directly comparable to any existing heat-rejection/recovery cycle but it allows to cool any fluid, below the wet bulb temperature and approach the dew point temperature of the working air, without a compressor or refrigerant. Unlike conventional cooling towers, the COP of the Maisotsenko cooling tower increases with the increase in ambient air temperature which distinguishes its applicability in hot climate. So, the advanced heat rejection characteristics of M-cycle HMX can revolutionize the industrial gas cooling and heat rejection processes. Therefore, a lot of research work still needs to be done in order to find the optimum performances of M-Cycle for each application. Perspective application of the M-cycle is heat utilization during the dehydration of materials in multi-stage dryers [20] as well as granulation plants [21-24].

References

[1] Bloch H P 2006 *A practical guide to compressor technology* John Wiley & Sons, 2nd ed. p 555
[2] Miyazaki T, Akisawa A, Nikai I. 2011 The cooling performance of a building integrated evaporative cooling system driven by solar energy *Energy and Buildings* **43** pp 2211–18
[3] Mahmood M, Sultan M, Miyazaki T, Koyama S, Maisotsenko V 2016 Overview of the Maisotsenko cycle – A way towards dew point evaporative cooling *Renewable and sustainable energy reviews* **66** pp 537-555
[4] Pandelidis D, Anisimov S, Drag P Performance Comparison between Selected Evaporative Air Coolers *Energies* **10** (4) pp 577
[5] Duan Z, Zhan C, Zhang X, Mustafa M, Zhao X, Behrang A, Hasan A 2012 Indirect evaporative cooling: Past, present and future potentials *Renewable and sustainable energy reviews* **16** (9) pp 6823-50
[6] Anisimov S, Kozlov A, Glanville P, Khinkis M, Maisotsenko V, Shi J 2013 Advanced cooling tower concept for commercial and industrial applications *Proc., of the ASME Power Conf.* (Boston, Massachusetts, USA POWER 2013-98230)

[7] Maisotsenko V et al. 2002 US Patent 6,497,107, Method and Apparatus of Indirect-Evaporation Cooling
[8] Maisotsenko V et al. 1978 SU Patent 620,745, Device for Indirect Evaporative Cooling of Air
[9] Maisotsenko V et al. 2005 US Patent 6,854,278, Method of Evaporative Cooling of A Fluid and Apparatus Therefore
[10] Schlom L et al. 1977 US Patent 4,023,949, Evaporative Refrigeration System
[11] Schlom L et al. 1979 US Patent 4,156,351, Depressed Wet Bulb Water Cooler Refrigeration System
[12] Gillan L 2008 Maisotsenko Cycle for cooling processes *International journal of energy for a clean environment* **9** pp 47-64
[13] Gillan L et al. 2011 Maisotsenko-Cycle Enhanced Cooling Towers *Proc., Cooling Technology Institute Annual Conf.*, San Antonio, Texas TP11-13
[14] Glanville P, Kozlov A, and Maisotsenko V 2011 Dew Point Evaporative Cooling: Technology Review and Fundamentals. *Proc., of the ASHRAE Winter Meeting*, Las Vegas, NV
[15] Lee D et al. 2002 US Patent 6,338,259, Regenerative Evaporative Cooler
[16] Lee D et al. 2007 US Patent 7,234,309 B2, Method and Apparatus for Evaporative Cooling of a Cooling Fluid
[17] Maisotsenko V et al. 2004 US Patent 6,776,001, Method and apparatus for dew point evaporative product cooling
[18] Maisotsenko V. et al. 2002 US Patent 6,497,107, Method and Apparatus of Indirect-Evaporation Cooling
[19] Maisotsenko V. et al. 2007 US Patent 7,197,887, Method and plate apparatus for dew point evaporative cooler
[20] Artyukhova N A 2014 Energy efficiency assessment of multi-stage convective drying of concentrates and mineral raw materials *Naukovyi Visnyk Natsionalnoho Hirnychoho Universytetu* **1** pp 92-98
[21] Artyukhov A E 2016 3D Nanostructured Porous Layer of Ammonium Nitrate: Influence of the Moisturizing Method on the Layer's Structure *Journal of Nano- and Electronic Physics.* **8** (4) 04051-1-04051-5
[22] Artyukhov A E 2016 Thermodynamic conditions for obtaining 3D nanostructured porous surface layer on the granules of ammonium nitrate *Journal of Nano- and Electronic Physics* **8** (4) 04083-1-04083-5
[23] Artyukhov A E, Voznyi A A 2016 Thermodynamics of the vortex granulator's workspace: the impact on the structure of porous ammonium nitrate *Proc. Int. Conf. NAP-2016* **5** (2) 02NEA01
[24] Artyukhov A E 2016 Kinetics of heating and drying of porous ammonium nitrate granules in the vortex granulator. *Proc. Int. Conf. NAP-2016* **5** (2) 02NEA02

IOP Conf. Series: Materials Science and Engineering **233** (2017) 012024 doi:10.1088/1757-899X/233/1/012024

Maisotsenko cycle applications in multi-stage ejector recycling module for chemical production

D O Levchenko[1], A E Artyukhov[2], I V Yurko[3]

[1]Department of Technical Thermal Physics, Sumy State University, 2 Rymskogo-Korsakova st., 40007 Sumy, Ukraine Sumy Ukraine
[2]Processes and Equipment of Chemical and Petroleum-Refineries Department, Sumy State University, 2 Rymskogo-Korsakova st., 40007 Sumy, Ukraine Sumy Ukraine
[3]TRIZ Ltd., Mashinostroiteley 1 St, Sumy, Ukraine

E-mail: d.levchenko@kttf.sumdu.edu.ua

Abstract. The article is devoted to the theoretical bases of multistage (multi-level) utilization modules as part of chemical plants (on the example of the technological line for obtaining nitrogen fertilizers). The possibility of recycling production waste (ammonia vapors, dust and substandard nitrogen fertilizers) using ejection devices and waste heat using Maisotsenko cycle technology (Maisotsenko heat and mass exchanger (HMX), Maisotsenko power cycles and recuperators, etc.) is substantiated. The principle of operation of studied recycling module and prospects for its implementation are presented. An improved technological scheme for obtaining granular fertilizers and granules with porous structure with multistage (multi-level) recycling module is proposed.

1. Introduction

The problem of the ecological safety of chemical production is becoming more urgent. The main environmental problems are: pollution of water, atmospheric air, production waste increasing and consumption, etc.

The production of granular products (including granules of nitrogen fertilizers - ammonium nitrate, carbamide and granules of porous ammonium nitrate for the needs of mining industry) is a major sector of the chemical industry.

Improving the quality of nitrogen fertilizers and indicators for energy consumption and environmental safety of their production to the level of world requirements is an important task for manufacturing enterprises. Increasing the efficiency indicators of production and improving the quality of products is proceeding in two directions: change in technology and processes; increasing of energy and environmental efficiency of equipment in existing production facilities.

In the process of production of granules of ammonium nitrate in vortex granulators [1-3] there is an intensive separation of gaseous nitrous oxide (concentration in the exhaust gases in the range 0.005-0.01 kg / m^3), small unconditioned granules and dust of ammonium nitrate (approximately 1% of the capacity of the vortex granulator on the finished product, the size is less than 0.5 mm), ammonia (concentration in the exhaust gases within 0.003-0.005 kg / m^3). In this context the issue of transfer of nitrous oxide (I) and thermal decomposition of substandard ammonium nitrate with the formation of higher oxides and their return to the technological process is of great practical interest. This innovation will increase the efficiency of the nitrogen conversion cycle in the enterprise and reduce the emission of nitrogen-containing compounds into the atmosphere.

2. Main methods of utilization of wastes from the ammonium nitrate production

The basic technological schemes for ammonium nitrate production function in such way (Figure 1) [4,5]. Process air is fed by gas blowing to the air heater, where it is heated for further use as a coolant and liquefying agent to create a vortex weighted layer. The heated air is fed to the vortex granulator I. After contact with the melt of ammonium nitrate, the gas leaves the volume of the granulator. Granules of finished products are discharged from the lower part of the granulator and enter the cooler of fluidized bed, they are cooled there by means of air supplied to the apparatus by gas blowers; then the granules are removed from the cooler and sent to the packaging. Contaminated air leaving the granulator as well as gas from the fluidized bed cooler contain dust, small granules of ammonium nitrate, nitrogen oxides and ammonia. It enters the lower part of the absorber II, where its absorption cleaning takes place. The variant of circuit, shown in Figure 1 (b), provides for an additional cyclonic purification of the off-gases II [6,7].

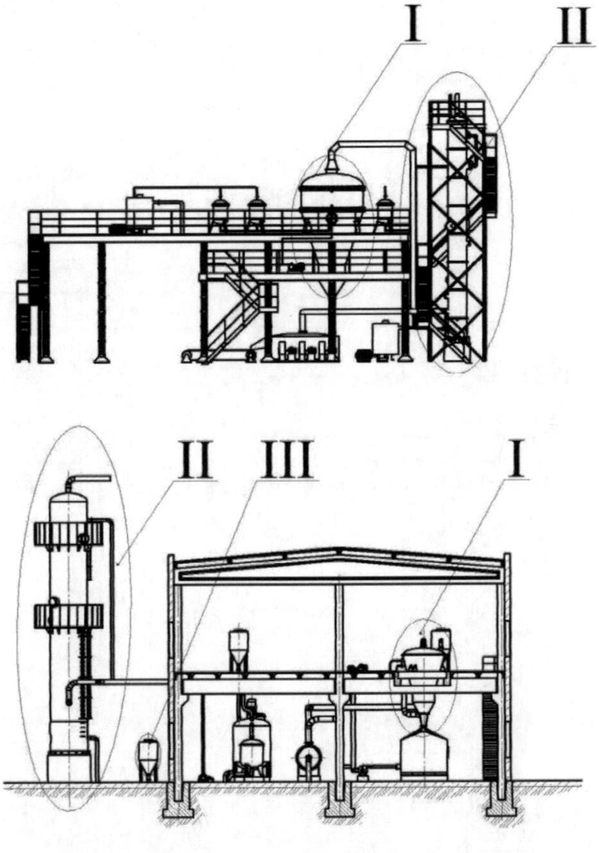

(a)

(b)

Figure 1. Basic schemes for the granular ammonium nitrate production in vortex-type granulators: (a) - with the unit of absorption purification of waste gases; (b) - with the unit of cyclone and absorption purification of exhaust gases; I - vortex granulator; II - absorber; III – cyclone.

The schemes described above do not imply the conversion of nitrous oxide to higher oxides. To solve this problem, the use of ejectors (Figure 2) was proposed in [8]. The essence of the method is as follows: a mixture of waste gases from the granulator and decomposition products of substandard ammonium nitrate are sent to a bubbling chamber where they are cooled and purified from ammonia and mechanical inclusions; Then the purified cold gas is supplied by a passive flow to the ejector, where the active (ejecting) stream is oxygen. A mixture of nitrous oxide and oxygen passes through the ejector diffuser path through a heated spiral coated with a catalyst bed and is oxidized to higher nitrogen oxides. The resulting gas passes through a coiled heat exchanger and is cooled to the liquid transition temperature. The resulting product is sent to a storage tank or closes the cycle for the production of nitric acid. Such wasteless technology allows to increase the specific energy efficiency of the plant and to reduce the emission of nitrogen-containing compounds into the atmosphere.

Explication of pipelines

Designation	Name of process streams
1 - 1	Ammoniun nitrate
2 - 2	Humidified air
3 - 3	Dry air
4 - 4	Humidified ammonium nitrate
5 - 5	Cold air
6 - 6	Steam
7 - 7	Water
8 - 8	Hot air
9 - 9	Porous ammonium nitrate
10 - 10	Dusty gas-air mixture
11 - 11	Gas-air mixture
12 - 12	Dust and small fraction of ammonium nitrate
13 - 13	Nitrous oxide (I)
14 - 14	Oxygen
15 - 15	Nitrous oxide (II)
16 - 16	Cold air
17 - 17	Porous ammonium nitrate per package

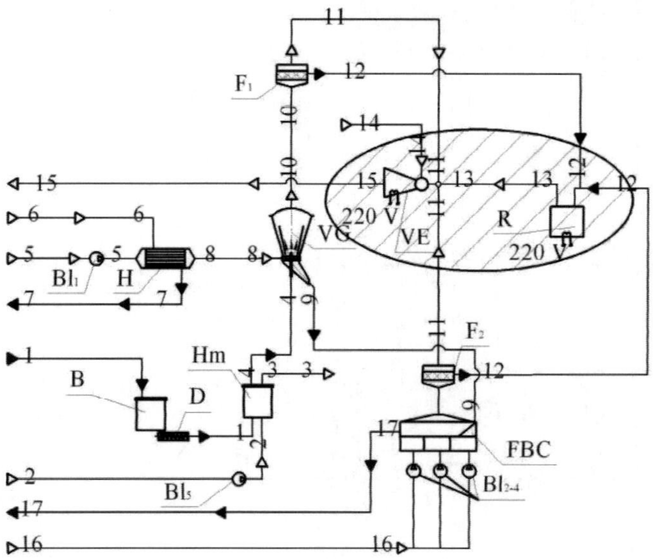

Explication of equipment

Designation	Name of equipment
Bl_{1-5}	Blower
B	Bunker
D	Dosimeter
Hm	Humidifier
VG	Vortex granulator
VE	Vortex ejector
R	Reactor
H	Heater
F_{1-2}	Filter
FBC	Fluidized bed cooler

Figure 2. The scheme for obtaining granular ammonium nitrate in vortex-type granulators using an ejection module [9].

3. Theoretical basics and results

The considered methods of ammonium nitrate production wastes utilization do not provide for the presence of heat recovery units for the drying and cooling agent. The receipt of additional energy by utilizing the heat of waste gases will reduce the amount of external heat, for example, at the stage of reactive decomposition of substandard ammonium nitrate. The authors of this article propose a comprehensive scheme for waste products processing from the ammonium nitrate production in the granulation stage with simultaneous utilization of the heat of the fluidizing agent (drying agent in the granulator and cooling agent in the fluidized bed cooler).

Since the ammonium nitrate production line contains many processes fulfilled with wet and dry flows of hot/warm air, the question of heat recovery arises dramatically. Considering that some units of the line consume considerable amount of energy in form of electricity and/or burned fuel (natural gas / methane, propane etc.), application the heat and mass exchangers (HMX) that utilize Maisotsenko cycle (M-cycle) become very beneficial. Moreover, integration Maisotsenko power cycles and M-cycle HMX recuperators into technological line of ammonium fertilizer production could provide by energy most of the technological processes of the line, reduce cost per tonne of the product and produce additional products like heat water and/or heating, electricity for inner or exterior consumers.

Describing M-cycle main principles and perspective one should refer to the profound paper as "Overview of the Maisotsenko cycle – A way towards dew point evaporative cooling" by Muhammad H. Mahmood, Muhammad Sultan, Takahiko Miyazaki, Shigeru Koyama and Valeriy S. Maisotsenko [10]. So, the M-Cycle is a thermodynamic process which captures energy from the air by utilizing the psychrometric renewable energy available from the latent heat of water evaporating into the air. It combines thermodynamic processes of heat transfer and evaporative cooling to facilitate product temperature to reach the dew-point temperature of the ambient air [11]. The basic principle and features of the M-Cycle can be explained from Figure 3 (a) and Figure 3 (b) representing the partial extraction of air and product air cooling M-Cycle, respectively [10]. The psychrometric representation of both variations of M-cycle is shown in Figure 3 (c). It consists of two kinds of primary channels named as wet and dry channels.

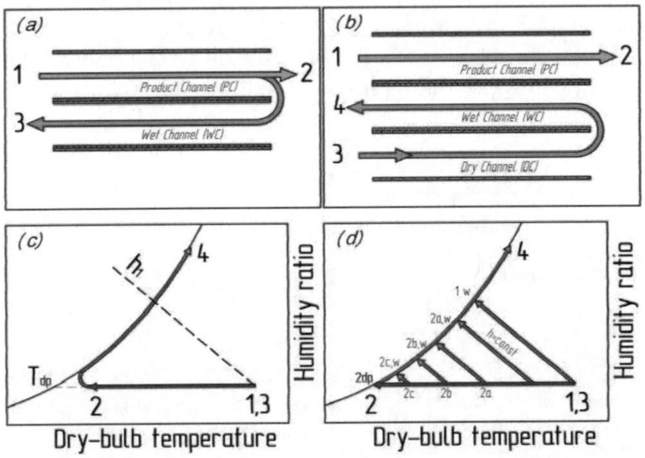

Figure 3. Schematic diagram of Maisotsenko Cycle for: (a) old M-Cycle, (b) modified M-Cycle, (c) psychrometric representation, and (d) sequential temperature decrement in wet-channel.

Product Air Cooling arrangement keeps working and product dry channels separated. Partial Extraction of Air arrangement fractions a portion of the dry channel as usable product. Both arrangements are psychrometrically identical while arrangement in Figure 1 (b) with product channel has essential advantages as compared to arrangement in Figure 1 (a). Thus, any gas or liquid matter (for example hot waste product) can be used as a product flow in arrangement in Figure 1 (b) while arrangement in Figure 1 (a) is limited to the same gas for both channels. Moreover using the product channel allows reducing pressure drop in the system. Another advantage of arrangement in

IOP Conf. Series: Materials Science and Engineering **233** (2017) 012024 doi:10.1088/1757-899X/233/1/012024

Figure 1 (b) compared with Figure 1a would be a possibility to use the product channel as a condenser for gas dehumidification [12].

For example, ambient air (1) is flowed into the dry-channel where it is sensibly cooled at constant humidity to cycle point (2) by transferring the heat to the wet-channel. The operational principle of M-Cycle is based on diverting the cooled air (2) to the wet-channel in order to use as working air. It results in subsequently decrement of effective dry-bulb (1 - 2a; 2b; 2c; 2) and wet-bulb (1w - 2a,w; 2b,w; 2c,w; 2dp) temperatures of the working air in the wet-channel as shown in Figure 1d. Sequential decrement of dry-bulb temperature in the wet-channel brings the effective wet-bulb temperature to be ideally equal to the dew-point temperature. Hence for an ideal heat transfer surface, the product air can be sensible cooled to the dew-point temperature of the ambient air. Moreover, saturated hot air (3) is rejected from the wet-channel equivalent to the evaporated water and recovered heat. This hot and wet air could be used in Maisotsenko power cycles hence increasing the overall energy recovery index.

Utilization of hot and humid air after HMX using Maisotsenko power cycles will increase thermal efficiency of the fertilizer production line. Using the modern gas turbine engines, based on well-known open Brayton cycle [13, 14], has overall disadvantage of the cycle as the significant amount of waste heat discharged into the atmosphere which results in poor thermal efficiency [15]. In this regard, inverse Brayton cycle [16] have been investigated with different configuration in order to increase the overall cycle performance [17-22]. In this cycle heated working medium at atmospheric pressure is initially expanded in the gas turbine. After that the working medium heat is recovered by the heat exchanger, and finally the cooled gas is sucked by the compressor to the atmospheric pressure. The cycle works below the atmospheric pressure and referred as sub-atmospheric cycle [15]. It is believed that the reverse Brayton cycle is not commercially feasibly because of the greater compressor size and employed higher operational energy [15]. On the other hand, the M-Cycle as an innovative humidifying recuperator can significantly improve the cycle performance by providing extremely saturated hot air to the combustion chamber (before turbine) and cooled air to the compressor (after turbine) simultaneously [15, 23-25]. Consequently, it will improve the fuel combustion efficiency as well as compressor efficiency at the same time. Furthermore, simple designs of atmospheric combustion chamber and cheaper materials could be employed in the turbine industry [15]. The Maisotsenko sub-atmospheric Brayton (M-SAB) cycle conception was realized recently by Maisotsenko et al. [24, 25] in which the authors proposed various possible configurations of M-SAB cycle. On the bas is of available literature, the present study discuses two kinds of M-SAB cycle which are based on: (1) compressor [15], and (2) ejector [23]. The compressor based M-SAB cycle is similar to the conventional reverse Brayton cycle.

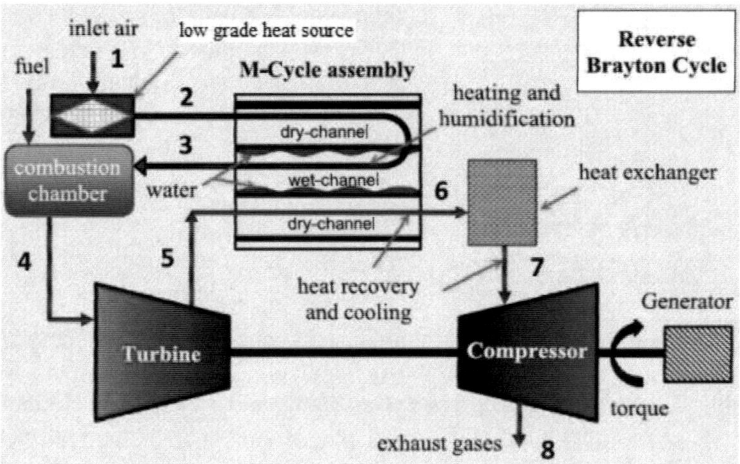

Figure 4. Schematic diagram of the compressor based M-SAB cycle, reproduced from [10, 16, 24, 25].

Khalatov et al. [15] analyzed the compressor based M-SAB cycle while recovering the turbine waste heat. The schematic diagram of the cycle is shown in Figure 4 [15, 24, 25]. The cycle configuration is similar to the [24, 25], however an additional low grade heat source unit is proposed for pre-heating (process 1-2) in order to improve the cycle efficiency. The air is heated and humidified simultaneously by the M-Cycle assembly (process 2-3) while recovering the turbine waste heat (process 5-6). The saturated hot air improves the combustion efficiency as well as reduces the NOx emission when used in combustion chamber (process 3-4). The combust hot gases at atmospheric conditions are expanded in the gas turbine (process 4-5). The energy from the hot gases at state (5) is recovered before it goes to the compressor by means of M-Cycle assembly (process 5-6) and an additional heat exchanger (process 6-7), which consequently improves the compressor efficiency. Analysis showed that the M-SAB cycle can achieve the thermal efficiency of 0.45-0.82 at preheating (T2) and combustion temperature (T4) of 40–90 °C and 160–340 °C, respectively. It is worth mentioning that the preheating shows significant improvement in thermal efficiency by the M-SAB cycle because of the versatile features of M-Cycle at higher temperature. Unlike open Brayton cycle the higher regeneration rate promotes the thermal efficiency of the M-SAB cycle. The study concluded that the M-SAB cycle possesses higher efficiency as compared to conventional open Brayton cycle at certain conditions.

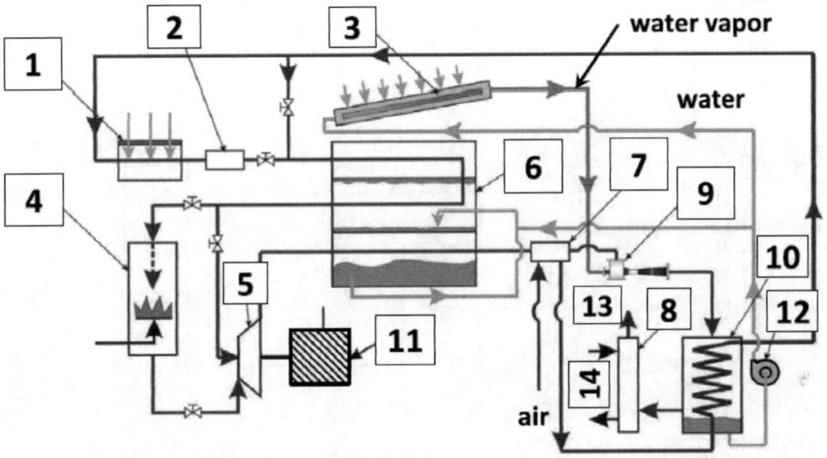

1	low grade heat source	8	exhaust air heat recuperator
2	heat storage	9	steam-air ejector
3	solar thermal collector	10	condenser/air recuperator
4	combustion chamber	11	electric generator
5	turbine	12	water pump
6	M-Cycle assembly	13	air exhaust
7	heat recuperator	14	heated water

Figure 5. Schematic diagram of the ejector based M-SAB cycle, reproduced from [10, 23]

Buyadgie et al. [23] proposed the ejector based M-SAB cycle and investigated its performance for various applications. The schematic diagram of the turbo-ejector based M-SAB cycle is shown in Figure 5 [23]. The principle operation of the cycle is similar the one based on compressor as explained in Figure 4 [10, 15, 24, 25], though the compressor is replace by the steam-air ejector. Each process of the cycle is labelled on the Figure 5 which gives the detail insight of the cycle. However, solar thermal collector 3 could be replaced by high grade heat source or used as recuperator in technological processes of ammonium nitrate production. According to the results the replacement of mechanical

IOP Conf. Series: Materials Science and Engineering **233** (2017) 012024 doi:10.1088/1757-899X/233/1/012024

compressor with the steam-air ejector results in 2-4 times higher power generation, and yields 15–20% capital cost reduction of the system. In addition, the electricity used to operate the fans for the M-Cycle assembly decreases two times per power unit. The authors concluded that the turbo-ejector M-SAB cycle design is the optimum choice when the electricity price is high and heat price is low. Furthermore, it is more beneficial when the power generation and low temperature cooling is required simultaneously despite of the available heat cost. From the above prospective the present study concludes that the M-Cycle possesses huge energy recovery potential in various power producing gas turbines. It addition to provide hot and humidified air for combustion, the M-Cycle recovers the turbine waste heat efficiently as compared to conventional heat exchangers. Furthermore, the nature of the M-Cycle helps to provide the cooled air to the compressor simultaneously, which increases the compressor efficiency. Another silent feature of the M-Cycle is the pollution control by reducing NOx formation during combustion which can lead towards an environment friendly gas turbine power cycle.

Figure 6 represents the proposed technological scheme of fertilizer production line which contains ejector recovery module UM1 (for N_2O acidification), number of HMX recuperators (for heat recovery, hot-water and heating supplying) and recovery module UM2 which includes M-SAB cycle plant (for electricity and/or cold production).

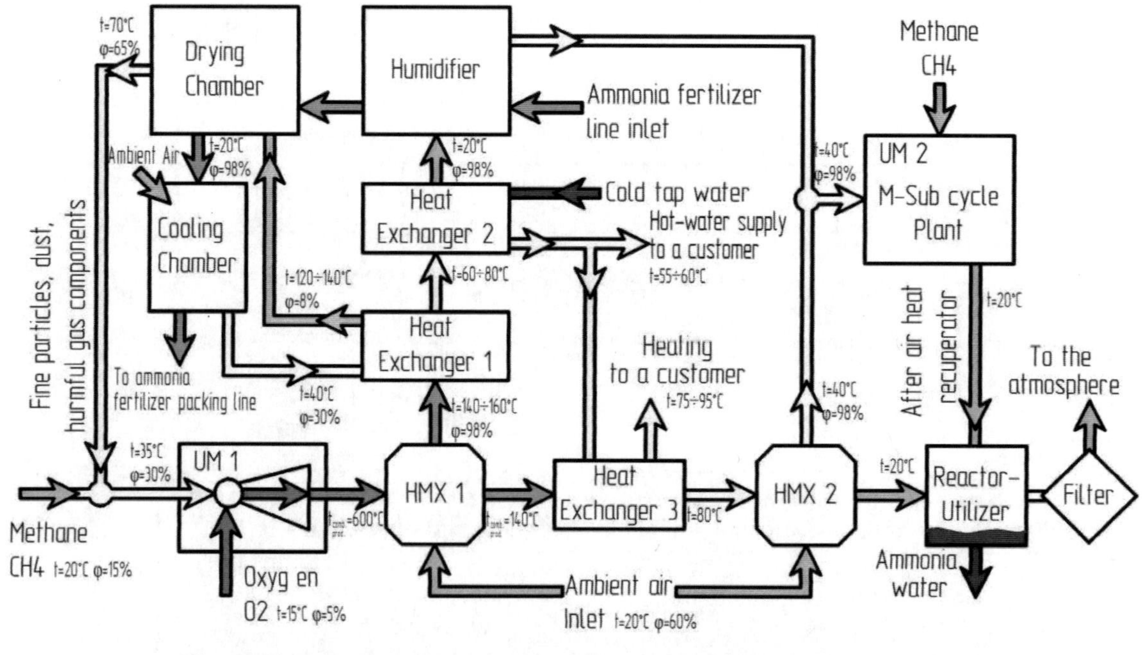

Figure 6. Schematic diagram of technological scheme of fertilizer production line.

The essential role in all M-cycle applications plays M-cycle HMX which conceptual design represented in Figure 7. There is a direct relation between M-cycle heat exchanger's performance and evaporating cooling efficiency which is directly affected by factors such as the type of wet surface material, surface area, mass flowrates of air, moisture evaporation rate from the wet surface, pad efficiency, dryness of the air and relative humidity of air passing through the wet channels and volume of water used. Hence selection of wet surface media requires careful attention.

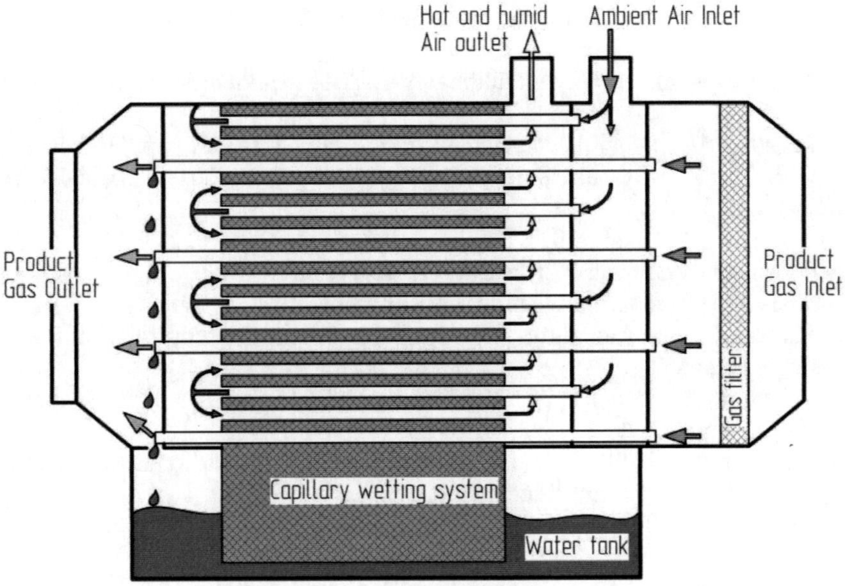

Figure 7. Proposed design of M-Cycle HMX.

A wide range of materials have been commonly used as the heat/mass exchanging medium. However, it is almost impossible to choose one all-mode material for various M-cycle heat exchanger applications. Additionally, one should notice that porous medium for HMX within fertilizer production line has much more strict requirements, comparing to porous materials for other field of M-cycle application, as the flows possess high corrosiveness. That is why research activities aimed to improve ability of wet channel surface to retain a water film become very important. Such improvement will play a key role in development new high-performance recovery modules.

4. Conclusion

In the article the advanced scheme of recovery-recycling module for chemical productions is represented on the basis Maisotsenko cycle and ejector technologies. The conceptual scheme of fertilizer/porous ammonium nitrate production and new design of M-Cycle HMX are developed. Using of this technical solution will increase thermal efficiency of the granulated materials production, reduce cost per tonne of the product and produce additional products like heat water and/or heating, electricity for inner or exterior consumers.

However, future spreading of the proposed recovery technology connected with its reliability and performance improvement. On its turn, HMX design, using appropriate material in wet channels, plays the key role in the improvement of overall M-cycle technology. Thus, the forthcoming research efforts will be directed to careful investigation of HMX optimal design and wet medium properties.

References

[1] Artyukhov A E, Sklabinskyi V I 2015 *Chemistry & chemical technology* **9(2)** pp 175-180
[2] Artyukhov A E, Sklabinskyi V I 2015 *Chemistry & chemical technology* **9(3)** pp 337-342
[3] Artyukhova N A, Shandyba A B, Artyukhov A E 2014 *Naukovyi Visnyk Natsionalnoho Hirnychoho Universytetu* **1** pp 92-98
[4] Artyukhov A E, Sklabinskyi V I 2013 *Naukovyi Visnyk Natsionalnoho Hirnychoho Universytetu* **6** pp 42-48
[5] Artyukhov A E, Fursa A S, Moskalenko K V 2015 *Chemical and Petroleum Engineering* **51**(5-6) pp 311-318
[6] Artyukhov A 2016 *CEUR Workshop Proceedings* **1761** pp 363-373

[7] Artyukhov A, Sklabinskiy V , Ivaniia A, Moskalenko K 2016 *CEUR Workshop Proceedings* **1761** pp 374-385
[8] Prokopov M G, Levchenko D A, Artyukhov A E 2014 *Applied Mechanics and Materials* **630** pp 109-116
[9] Artyukhov A E 2014 *Chemical and Petroleum Engineering* **49(11-12)** pp 736-740
[10] Mahmood M, Sultan M, Miyazaki T, Koyama S, Maisotsenko V 2016 *Renewable and sustainable energy reviews* **66** pp 537-555
[11] Miyazaki T, Akisawa A, Nikai I 2011 *Energy and Buildings* **43** pp 2211-18
[12] Pandelidis D, Anisimov S, Drag P 2017 *Energies* **10(4)** pp 577
[13] Cengel Y, Boles M 1994 Thermodynamics: An Engineering Approach. New York: McGraw-Hill
[14] Cohen H, Rogers G, Saravanamuttoo H 1996 Gas Turbine Theory-4th Edition. Reading MA. Addison Wesley Longman
[15] Khalatov A, Severin S, Brodetsky P, MaisotsenkoV. 2015 *Reports of the National Academy of Sciences of Ukraine* **1** pp 72-79
[16] Wilson D, Korakianitis T 1984 The design of high-efficiency turbomachinery and gas turbines. London, England: MIT Press Cambridge
[17] Besarati S, Atashkari K, Jamali A, Hajiloo A, Nariman-zadeh N 2010 *Energy Convers Manag* **51** pp 212-217
[18] Zhang W, Chen L, Sun F 2009 *Appl Therm Eng* **29** pp 2885-2894
[19] Alabdoadaim M, Agnew B, Potts I. 2006 *Appl Therm Eng* **26** pp 1448-1454
[20] Agnew B, Anderson A, Potts I, Frost T, Alabdoadaim M. 2003 *Appl Therm Eng* **23** pp 953-963.
[21] Zhang W, Chen L, Sun F. 2012 *Sci Iran* **19** pp 1638-1652
[22] Zhang Z, Chen L, Sun F. 2012 *Sci Iran* **19** pp 1279-1287
[23] Buyadgie D, Buyadgie O, Drakhnia O, Brodetsky P, MaisotsenkoV 2015 *Int J Low-Carbon Technol* **10** pp 157-164
[24] Maisotsenko V, Gillan L, Heaton T, Gillan A 2006 US Patent No. US7007453B2
[25] Maisotsenko V, Gillan L, Heaton T, Gillan A 2004 US PatentNo.US20040103637A1

Acknowledgments
This work was carried out under the project «Improving the efficiency of granulators and dryers with active hydrodynamic regimes for obtaining, modification and encapsulation of fertilizers», state registration No. 0116U006812. The authors thank researchers of department "Processes and Equipment of Chemical and Refining Industries", and department of Technical Thermal Physics, Sumy State University, for their valuable comments during the article preparation.

Appliance of Inertial Gas-Dynamic Separation of Gas-Dispersion Flows in the Curvilinear Convergent-Divergent Channels for Compressor Equipment Reliability Improvement

O O Liaposhchenko[1], V I Sklabinskyi[1], V L Zavialov[2], I V Pavlenko[3], O V Nastenko[1], M M Demianenko[3]

[1]Department of Processes and Equipment of Chemical and Petroleum-Refineries, Sumy State University, 2, Rymskogo-Korsakova St., Sumy, 40007, Ukraine
[2]Department of Process and Equipment in Food Production, National University of Food Technologies, 68, Volodymyrska St., Kyiv, 01601, Ukraine
[3]Department of General Mechanics and Machine Dynamics, Sumy State University, 2, Rymskogo-Korsakova St., Sumy, 40007, Ukraine

E-mail: lyaposchenko@pohnp.sumdu.edu.ua

Abstract. The new methods of vibration and inertial gas-dynamic separation of gas-condensate and dusty flows and the corresponding separation devices are proposed in order to avoid emergencies and premature wear of parts and components of the compressor equipment. The formation of the gas flow and disperse particles in the curvilinear convergent-divergent channels are investigated. The optimizing hydrodynamic profiling of a geometrical configuration of curvilinear separation channels with rigid and flexible walls of baffles is carried out.

1. Introduction

Application of the highly effective gas-separation equipment for cleaning gas-condensate and dusty flows is one of the important and necessary techniquesfor increasingthe reliability of compressor equipment for compression and pumping of gases. The occurrence of dropped liquid and mechanical impurities in gas flow leads to emergencies and premature wear of rotor parts, impellers of compressors and superchargers. In terms ofevaluation of specific energy consumption and separation efficiency, ways of inertial and inertial-filtering separation are optimal [1,2]. Herewith, it is necessary to consider that the secondary processes accompanying the main process of separation have the significant impact on efficiency and hydraulic resistance of separation devices [2-4].

Due to complexity of secondary processes and their diversity, they remain insufficiently studied despite of their decisive practical importance. Most often secondary processes have a negative impact on the efficiency and intensity of the main separation process, and decrease hydraulic and separation characteristics of the compressor equipment. However, the secondary processes in inertial-filtering separation deviceschange the separation degree of gas-liquid flow due to various factors, such as co-deposition caused by manifestation of adhesion effects, when drops are settled mainly on already deposited ones, which form an irregular liquid film on the surface of channels; coagulation of

IOP Conf. Series: Materials Science and Engineering **233** (2017) 012025 doi:10.1088/1757-899X/233/1/012025

dispersed particles, when highly dispersed particles move to further coalesce into larger drops with following accumulation of fluid in cavities of blinds and crossing fiber filter; mechanical vibrations and deformations, which impacton the effectiveness of the inertial collision and hydraulic resistance; capillary phenomena and mechanical destruction of fiber filters due to solid particles; secondary entainment of particles from the curved inertia-filtering separation channels by turbulent gas flow [3,5,6].

Notwithstanding the foregoing, along with the negative impact of secondary processes, positive effects can be obtained. For example, in the inertial gas-dynamic separation devices with curvilinear convergent-divergent separation channels, the separation degree can be increased due to coagulation and flocculation of the weighed liquid drops and solid particles in a turbulent gas flow [8,13].

Consequently, implementation of processes of inertial gas-dynamic separation is accompanied by setting as the aimmutual avoidanceof negative impact and using positive impact of secondary processes for an intensification and increasing efficiency of the main separation process, as well as improvement of hydraulic and separation parameters of the equipment.

To achieve the abovementioned aim, two approaches are proposed:

1) implementation ofthe method of imposing acoustic vibrations on the gas-disperse flow, which causes an intensive mechanical vibration of the high-disperse weighed liquid drops and solid particles in gas-disperse flow for sharp increasing the number of collisions and coagulation / flocculation of droplets / particles due to acoustic coagulation into drops of significantly larger sizes, which can be captured more efficientlydue to mechanisms of inertial trapping, which allows increasing the efficiency and intensity of the separation process [7,9,14];

2) application the curvilinear separation channels with elastic and mobile elements, such as channel borders or directing bafflestothe design of inertial gas-dynamic separation devices [10-12] for dynamic regulation of the flow cross-section in curvilinear convergent-divergent channel, the velocity of a gas-disperse flow, as well as for deflection of flow trajectories or change the attack angle of separation elements for increasingthe efficiency of separation in a wide range of loadingof a gas phase.

Combining the abovementioned approaches in order to create an effect, similar to the acoustic coagulation as a source of vibration is proposed for application of mechanical vibrations directly to the borders of convergent-divergent channels [15].

Furthermore, the objective of this work is the mathematical formulation and solving the problem of modeling the gas-dispersion flow motion in curvilinear separation channels with curved borders, which oscillate in the transverse direction.

2. General Problem Description and Mathematical Model

The system of Navier-Stokes and continuity differential equations can be applied for describing the motion of a continuous phase in curvilinear channels (Figure 1) withoscillating deformable borders. Analytical solving this system is realized under the following assumptions and simplifications:

a) two-dimensional isothermal viscous incompressible liquid flowin the curvilinear channel of infinite length;

b) small oscillations ofdeformable borders;

c) neglecting the convective terms of the Navier-Stokes equation.

IOP Conf. Series: Materials Science and Engineering **233** (2017) 012025 doi:10.1088/1757-899X/233/1/012025

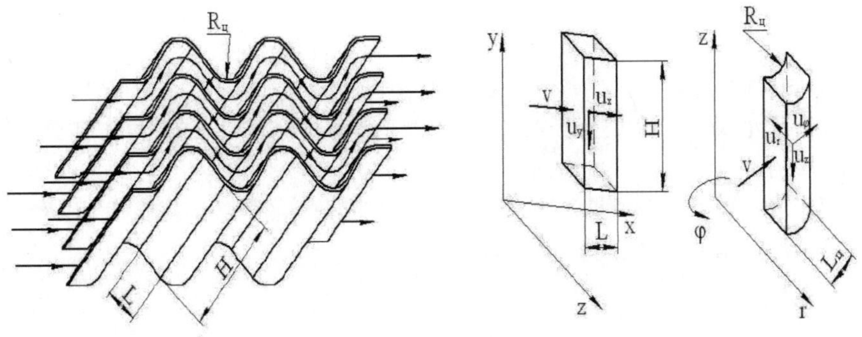

(a) (b)

Figure 1. The design model of louver unit and the separate sections of curvilinear separation channels: (a) – curvilinear inertial separation channels; (b) – considered section.

Considering the abovementioned simplifications and assumptions allows rewriting these equations:
–the law of massconservation as the continuity equation:

$$\frac{\partial u}{\partial x}+\frac{\partial v}{\partial y}=0 \tag{1}$$

–the system of Navier-Stokes equations:

$$\begin{cases} \rho\dfrac{\partial u}{\partial t}=-\dfrac{\partial p}{\partial x}+\mu\left(\dfrac{\partial^2 u}{\partial x^2}+\dfrac{\partial^2 u}{\partial y^2}\right) \\ \rho\dfrac{\partial v}{\partial t}=-\dfrac{\partial p}{\partial x}+\mu\left(\dfrac{\partial^2 v}{\partial x^2}+\dfrac{\partial^2 v}{\partial y^2}\right) \end{cases} \tag{2}$$

where
u, v – components of flow velocity in the directions of axes x, y respectively, m/s;
p – pressure at the considering point, N/m^2.
It should be noted that in such problem shaping, the components u and v and pressure p are function of coordinates x, y and time t.
To solve differential equations (1), (2), the initial and boundary conditions are taken into account:

$$\begin{cases} u(0,y,t)=u_0(y) \quad u(x,0,t)=0 \quad u(x,h,t)=0 \\ v(0,y,t)=0 \quad v(x,0,t)=f(x)\phi(t) \quad v(x,h,t)=0 \\ p(0,y,t)=p_0 \end{cases} \tag{3}$$

where
$u_0(y)$ – flow velocity profile in the inlet section of the channel, m/s;
h – characteristic linear dimension (height, width) of the channel, m;
$f(x)$ – function of wall oscillation form;
$\varphi(t)$ – timefunction for oscillating border;
p_0 – inlet pressure, Pa.

3. Solving the Equations

For solving the equations (1), (2) under the boundary conditions (3) the Laplace transform is used with respect to coordinate x:

$$U_1(y,t)=L_x\left[u(x,y,t)\right]=\int_0^\infty u(x,y,t)e^{-sx}dx,\quad U(y)=L_t\left[U_1(y,t)\right]=\int_0^\infty U_1(y,t)e^{-\tau t}dt$$
$$V_1(y,t)=L_x\left[v(x,y,t)\right]=\int_0^\infty v(x,y,t)e^{-sx}dx,\quad V(y)=L_t\left[V_1(y,t)\right]=\int_0^\infty V_1(y,t)e^{-\tau t}dt \tag{4}$$
$$P_1(y,t)=L_x\left[p(x,y,t)\right]=\int_0^\infty u(x,y,t)e^{-sx}dx,\quad P(y)=L_t\left[P_1(y,t)\right]=\int_0^\infty P_1(y,t)e^{-\tau t}dt$$

where
U_1, V_1, P_1, U, V, P – Laplace integrals;
s, τ – parameters of transformation.

Consequently, the continuity equation (1), the system of Navier-Stokes equations (2) and boundary conditions (3) take the following form:

$$\begin{cases} sU-\dfrac{u_0}{\tau}+V'=0 \\ sP-\dfrac{p_0}{\tau}+\rho\tau U=\mu\left(s^2U-\dfrac{su_0}{\tau}+U''\right) \\ P'+\rho\tau U=\mu(s^2V+V'') \\ U(0)=0\quad U(h)=0\quad V(0)=L_xL_t\quad V(h)=0 \end{cases} \tag{5}$$

where
«$'$» – sign for the differentiation with respect to coordinate y;
L_x, L_t – Laplace integrals of functions $f(x)$ and $\varphi(t)$ in terms of appropriate parameters.

The first equationof the system (5) allows obtaining Laplace integral U:

$$U=\frac{u_0}{s\tau}-\frac{V'}{s} \tag{6}$$

and Laplace integral P can be obtained from the second equation (5):

$$P=\frac{1}{s}\left[\frac{p_0}{\tau}+\mu\left(-sV'+\frac{u_0''}{s\tau}-\frac{V'''}{s}\right)-\rho\tau\left(\frac{u_0}{s\tau}-\frac{V'}{s}\right)\right] \tag{7}$$

as well as the third equation of the system (5) due to formulas (6) and (7) after equivalent transformations takes the form:

$$V^{IV}+2s^2(1-k)V''+s^4(1-2k)V=\frac{u_0'''}{\tau}-\frac{u_0'}{\nu} \tag{8}$$

where $k=\tau/(2\nu s^2)$ – dimensionless parameter, which takes into account the influence of local inertia forces.

It should be noted that that in case of κ>>1,solution of the general equation (8) satisfied the boundary conditions is proportional to $sin(s(h-y))/sin(sh)$, and the original function $v=0$ is contrary to the physical meaning of the problem. Further consideration of equation (8) for the case of κ<< 1 corresponds to the quasi-stationary problem statementdue to the final value theorem.

In case of parabolic flow velocity profile in the channel

$$u_0=\frac{6q_0}{h^3}y\left(h-y\right) \tag{9}$$

where q_0 - mass flowrate of a fluid, allows determining the flow rate:

$$q_0 = \int_0^h u_0(y)dy \tag{10}$$

Due to taking into consideration the velocity profile, equation (8) can be given in the form of inhomogeneous equation biquadratic equation:

$$V^{IV} + 2s^2V'' + s^4V = -\frac{u_0'}{\nu} \tag{11}$$

the general solution of which

$$V(y) = C_1 \sin sy + C_2 \cos sy + sy(C_3 \sin sy + C_4 \cos sy) - \frac{u_0'}{\nu s^4} \tag{12}$$

Substitution formula (12) to equation (6) allows obtaining the resulting expression in the form:

$$U(y) = \frac{u_0}{s\tau} - C_1 \cos sy + C_2 \sin sy - C_3(\sin sy + sy \cos sy) - C_4(\cos sy - sy \sin sy) \tag{13}$$

The constants of integration $C_{1,2,3,4}$ are determined from the boundary conditions $u_0(0) = u_0(h) = 0$ as the result of algebraic linear equations solutions:

$$C_1 = -C_4 = \frac{sh + \frac{1}{2}\sin 2sh}{s^2h^2 - \sin^2 sh} L_x L_t + \frac{2\sin^2 sh}{sh + \sin sh}\frac{u_0'}{\nu s^4}$$
$$C_2 = L_x L_t + \frac{u_0'}{\nu s^4} \qquad C_3 = -\frac{\sin^2 2sh}{s^2h^2 - \sin^2 sh} L_x L_t + \frac{\sin sh}{sh + \sin sh}\frac{u_0'}{\nu s^4} \tag{14}$$

and correspondingly can be written as

$$U(y) = \frac{u_0(y)}{s\tau} + L_x L_t F_{U10}(y) + \frac{u_0'}{\nu} F_{U2}(y)$$
$$V(y) = L_x L_t F_{V1}(y) + \frac{u_0'}{\nu} F_{V2}(y) \tag{15}$$
$$P(y) = \frac{p_0}{s\tau} + \frac{\mu}{s^2}\left(\frac{u_0''}{\tau} - \frac{u_0}{\nu} \right) + \mu L_x L_t s\left[F_{p1}(y) - \frac{\tau}{\nu s^2} F_{U1}(y) \right] + \rho s\left[F_{P2}(y) - \frac{\tau u_0'}{\nu s^2} F_{U2}(y) \right]$$

where Laplace integrals are introduced:

$$F_{U1}(y) = -\frac{sh + \frac{1}{2}\sin 2sh}{s^2h^2 - \sin^2 sh} sy \sin sy + \frac{\sin^2 sh}{s^2h^2 - \sin^2 sh}(\sin sy + sy \cos sy) + \sin sy$$
$$F_{U2}(y) = \frac{1}{s^4}\left[\frac{\sin^2 sh}{sh + \sin sh} sy \sin sy - \frac{\sin sh}{sh + \sin sh}(\sin sy + sy \cos sy) + \sin sy \right]$$
$$F_{V1}(y) = \frac{sh + \frac{1}{2}\sin 2sh}{s^2h^2 - \sin^2 sh}(\sin sy + sy \cos sy) - \frac{\sin^2 sh}{s^2h^2 - \sin^2 sh} sy \sin sy + \cos sy \tag{16}$$
$$F_{V2}(y) = \frac{1}{s^4}\left[\frac{2\sin^2 sh}{sh + \sin sh}(\sin sy + sy \cos sy) + \frac{\sin sh}{sh + \sin sh} sy \sin sy + \cos sy - 1 \right]$$
$$F_{P1}(y) = F_{U1}(y) + \frac{F_{U1}''(y)}{s^2} \qquad F_{P2}(y) = u_0' F_{U2}(y) + \frac{\left[u_0' F_{U2}(y) \right]''}{s^2}$$

Analytical definition of original functions by Riemann-Mellin's formulas is not possible. In this case due to the conditions of absolute convergence of $F_{U1,2}$, $F_{V1,2}$, the original functions can be determined as auxiliary images by the residue theorem:

$$\begin{aligned} f_{u1,2}(x,y) &= \underset{s=s_0}{Res}(e^x F_{U1,2}) = \frac{1}{(n-1)!}\lim_{S\to S_0}\frac{d^{n-1}}{ds^{n-1}}\left[(s-s_0)^n e^{sx} F_{U1,2}\right] \\ f_{v1,2}(x,y) &= \underset{s=s_0}{Res}(e^x F_{V1,2}) = \frac{1}{(n-1)!}\lim_{S\to S_0}\frac{d^{n-1}}{ds^{n-1}}\left[(s-s_0)^n e^{sx} F_{V1,2}\right] \\ f_{p1,2}(x,y) &= \underset{s=s_0}{Res}(e^x F_{P1,2}) = \frac{1}{(n-1)!}\lim_{S\to S_0}\frac{d^{n-1}}{ds^{n-1}}\left[(s-s_0)^n e^{sx} F_{P1,2}\right] \end{aligned} \tag{17}$$

where

$s_0 = 0$ – the only root of equation $s^2h^2 - sin^2 sh = 0$ and $s^4(sh + \sin sh)$of the order $n = 4$.

Application of the L'Hopital's rule for regular functions of a complex variable s, which turn into uncertainty {0/0} allows receiving dependences (17) in an explicit form. Taking into account the abovementioned dependences for auxiliary images, it can be obtained:

$$\begin{gathered} f_{u1}(x,y) = \frac{6y(h-y)}{h^3} \quad f_{v1}(x,y) = \left(1-\frac{y}{h}\right)^2\left(1+\frac{2y}{h}\right)\delta(x) \quad f_{p1}(x,y) = \frac{3x^2}{h^3} \\ f_{u1}(x,y) = f_{v2}(x,y) = f_{p2}(x,y) = 0 \end{gathered} \tag{18}$$

which corresponds to Laplace integrals:

$$\begin{gathered} F^*_{U1}(y) = \frac{6y(h-y)}{h^3 s} \quad F^*_{V1}(y) = \left(1-\frac{y}{h}\right)^2\left(1+\frac{2y}{h}\right) \quad F^*_{P1}(y) = \frac{6}{h^3 s^3} \\ F^*_{U1}(y) = F^*_{V1}(y) = F^*_{P1}(y) = 0 \end{gathered} \tag{19}$$

where $\delta(x)$ – Dirac delta function. The sign "*" shows the alternative variant of Laplace integral notation.

Application of the inverse Laplace transformationto theformulas (16) with respect to parameter τ, and taking into account the dependencies (19) for$t$>>0 allows obtaining:

$$\begin{gathered} U_1(y,t) = \frac{u_0(y)}{s} + L_x\phi(t)F_{U1}(y) \\ V_1(y,t) = L_x\phi(t)F_{V1}(y) \\ P_1(y,t) = \frac{p_0}{s} + \mu\left[\frac{u_0''}{s^2} + sL_x\phi(t)F_{P1}(y)\right] \end{gathered} \tag{20}$$

Using the Duhamel's integral and theorem of convolution for Laplace integrals

$$\begin{gathered} f(x)\circ\delta(x) = \int_0^\infty f(x-\xi)\delta(x)d\xi = f(x) \\ L_x^{-1}\left(\frac{L_x}{s}\right) = \int_0^x f(\xi)d\xi = F_1(x) \\ L_x^{-1}\left(\frac{L_x}{s^2}\right) = \int_0^x\int_0^\eta f(\xi)d\xi d\eta = F_2(x) \end{gathered} \tag{21}$$

allows determining the initial original function

$$u(x,y,t)=u_0(y)+\frac{6y(h-y)}{h^3}F_1(x)\phi(t)$$
$$v(x,y,t)=\left(1-\frac{y}{h}\right)^2\left(1+\frac{2y}{h}\right)f(x)\phi(t) \quad (22)$$
$$p(x,y,t)=p_0+\mu\left[u_0''(y)x-\frac{12}{h^3}F_2(x)\phi(t)\right]+\frac{6\mu y(h-y)}{h^3}f(x)\phi(t)$$

Taking into account the sinusoidal form of borders of curvilinear channel with the wavelength L_0, and under assumption of mono-harmonic oscillations with amplitude a and vibration frequency ω_0

$$f(x)=sin(\lambda x)$$
$$\phi(t)=a\,sin(\omega_0 t) \quad (23)$$

where $\lambda = 2\pi/L_0$ – introduced parameter,the following expressions for the pressure field and components of flow velocity can be obtained:

$$u(x,y,t)=\frac{6y(h-y)}{h^3}\left[q_0+\frac{a}{\lambda}(1-cos\,\lambda x)sin\,\omega_0 t\right]$$
$$v(x,y,t)=\left(1-\frac{y}{h}\right)^2\left(1+\frac{2y}{h}\right)a\,sin(\lambda x)sin(\omega_0 t) \quad (24)$$
$$p(x,y,t)=p_0+\frac{12\mu}{h^3}\left[q_0 x+\frac{\lambda x-sin(\lambda x)}{\lambda^2}\right]+\frac{6\mu y(h-y)}{h^3}a\,sin(\lambda x)sin(\omega_0 t)$$

Due to the fact, that the elastic borders of the channel are used as a source of vibrations, the spatial position of which is changed by the flow, it is necessary to investigate not only the flow motion in these channels, but wall deformations.

Two-dimensional problem of deformation of a curvilinear elastic element under the influence of pressure in stationary formulation is considered in the work [10]. The fields of velocities is defined, and analytical dependences for determination of the maximum deflection of elastic borders is obtained, which allows improving the geometry of borders and sections of curvilinear convergent-divergent channels in the inertial gas-dynamic separation devices as a result of their deformation.

4. Conclusions

A possibility of mutual avoiding of negative impact and using of positive effects of the secondary processes on the main process of inertial gas-dynamic separation is obtained for improvement of hydraulic and separation parameters.Ways for implementation of vibration coagulation and gas-dynamic separation are proposed.

Expressions for the flow velocity and pressure fields in case of isothermal incompressible flow on the channel of a sinusoidal form with small oscillating borders have been definedanalytically.

For achieving the main aim of the research, unificationof abovementioned approaches and usingthe elastic walls of curvilinear channels as the source of vibration is offered. Theseparation of gas-disperse flows was considered as a way to make possible carrying out coagulation not only of high-disperse fogs, but also effective cleaning of dusty flows.

References

[1] Sklabinskyi V, Liaposchenko O, Pavlenko I, Demianenko M 2015 Solution of the Navier–Stokes equations for the processes of inertial gas-dynamic separation in the curvilinear channels *Proceedings of the International Symposium "Discrete Singularities Methods in Mathematical Physics", Kharkiv, V. N. Karazin Kharkiv National University* pp 232–235

[2] Sklabinskyi V I, Liaposhchenko O O, Nastenko O V, Al-Rammahi M M 2014 Modeling and designing of the inertial-filtering gas separators-sondensers for compressor units of oil and gas industry *Applied Mechanics and Materials, Problems of Mechanics in Pump and Compressor Engineering* **630** pp 117–123
[3] Liaposchenko O 2016 The theoretical basics of inertial-filtering separation *DSc. thesis, 05.17.08, (*Lviv, Lviv Polytechnic National University)
[4] Motin A, Tarabara V V, Petty C A, Benard A 2017 Hydrodynamics within flooded hydrocyclones during excursion in the feed rate *Understanding of turndown ratio. Separation and Purification Technology* **185** pp 41–53
[5] Liaposhchenko O, Sklabinskyi V, Logvin A 2009 Inertial-filtering separation equipment for thermochemical processing of oil sludge of oil production *Proceedings of the N Lviv Polytechnic National University, Heat and power. Environmental Engineering. Automatization* **659** pp 148–150
[6] Setnickova K, Sima V, Petrychkovych R, Reznickova J, Uchytil P 2016 Separation of gas mixtures by new type of membranes *Dynamic liquid membranes, Separation and Purification Technology* **160** pp 132–135
[7] Sklabinskyy V, Liaposchenko O, Logvyn A, Al-Rammahi M 2014 Hydrodynamics modeling of the gas separator's inertial and filtering elements for natural gas fine cleaning *Chemistry & Chemical Technology* **8** (4) pp 479–485
[8] Nastenko O, Broniarz-Press L, Liaposchenko O 2016 Mathematical modelling of separation process by coupled heat transfer in the inertial-filtering gas separator-condenser *Engineering and Chemical Equipment* **2** pp 62–63
[9] Liaposhchenko O, Pavlenko I, Demianenko M 2015 Investigation of forced oscillations of gas-dynamic separator baffles *Scientific papers of the International Scientific and Technical Conference "Young Engineer – the Basis of Scientific and Technical Progress", Kursk, South-West State University* pp 262–265
[10] Liaposchenko O, Pavlenko I, Nastenko O, Usik R, Demianenko M 2015 The method of capturing highly dispersed dropped of gas-liquid flow *Patent 102445, Ukraine, B01D 45/04 (2006.01), u201505124* **20**
[11] Sklabinskyi V I, Liaposhchenko O O, Lohvyn A V, Skydanenko M S 2011 Device for capturing highly dispersed liquid drops from the gas-liquid flow *Patent 57386, Ukraine, 6 B 01 D 45/04. U201009488* **4**
[12] Liaposhchenko O, Demianenko M, Pavlenko I, Sklabinskyi V 2015 Solution of the Navier–Stokes equations for the processes of inertial gas-dynamic separation in the curvilinear channels *Bulletin of V. N. Karazin Kharkiv National University, Series "Mathematical Modeling. Information Technology. Automated Control Systems"* **27** pp 53–64
[13] Liaposhchenko O, Nastenko O, Pavlenko I 2017 The model of crossed movement and gas-liquid flow interaction with captured liquid film in the inertial-filtering separation channels *Separation and Purification Technology* **173** pp 240–243
[14] Liaposhchenko O, Nastenko O 2015 Analysis of the conditions of phase equilibrium and influence of the united heat and mass transfer on the effectiveness of separation in the inertial-filtering separator *Chemistry & Chemical Technology* **9** (1) pp 125–130
[15] Jia W, Murad S 2005 Separation of gas mixtures using a range of zeolite membranes: A molecular-dynamics study *The Journal of Chemical Physics* **122**

Estimation of gas flow dustiness in the main pipelines of booster compressor stations

M Yukhymenko[1], R. Ostroha[1], A Litvinenko[1] and J Bocko[2]

[1]Sumy State University, Department of Processes and Equipment of Chemical and Petroleum-Refineries, 2, Rimsky-Korsakov Street, Sumy 40007, Ukraine.
[2]Technical University of Košice, Department of Applied Mechanics and Mechanical Engineering, 9, Letná, Košice 04200, Slovak Republic.

E-mail: r.ostroga@pohnp.sumdu.edu.ua

Abstract. The article provides groundings for cleaning the gas flows of major pipelines from soild particles in order to improve the compressor operation reliability. One obtained formulas for determination of the dust level in the vertical and horizontal sections of the pipelines supplying gas to the booster compressor stations. Ways of structural modernization of vertical pipeline sections using the inbuilt separation devices for more efficient removal of dust particles from the gas flow are described.

1. Introduction

Gases and their mixtures are treated in the chemical and gas processing industries. Compressor machines transfer gas flows. Gases conveyed by process and main pipelines often contain various solid impurities like sand particles, dust, welded slag, corrosion scale. In pneumatic transport installations gas recirculation flows, feeding to the suction line of the booster compressors, contain a small fraction of transported materials.

Solid particles in a gas flow, entering piston compressors, cause increased wear of piston rings, valves and cylinders. They also impair impellers and inner walls of centrifugal blowers. Therefore, they decrease performance of gas compressor stations and units, namely, reducing reliability and economical operation.

At the principal facilities of the main gas pipelines and technological pneumatic transport units there are used devices for cleaning gas from mechanical impurities: oil and dry dust collectors, filters, scrubbers, adsorbers, separators of various designs. At the same time, an important issue is a stable conveyance of solid particles and impurities by a gas flow through pipelines in order to prevent them from lying on the walls of pipeline horizontal sections and dropping them out of the flow in vertical sections.

It is rational to install separating devices for preliminary separation of a gas flow from solid impurities on the vertical sections of the pipeline. It enables to maintain pipeline performance at the target level and purify gases in the main separators more efficiently.

2. Analysis and statement of a problem. Purpose and objectives of the research

For stable conveyance of gas together with solid impurities through the pipeline, it is necessary to maintain a certain value of critical velocity of the gas flow. A number of formulas have been proposed for determining the conveyance velocity of the gas flow [1].

Empirical formulas are proposed for calculating critical velocity of the gas flow, wherein solid particles do not sediment at the bottom of a horizontal pipeline [1]:

$$W = 5,6 \cdot D^{0,34} \cdot d^{0,36} \cdot \left(\frac{\rho_T}{\rho} \right) \cdot m^{0,25} \tag{1}$$

$$W = A \cdot \sqrt{\rho_T} + B \cdot L^2 \tag{2}$$

where

D – pipeline diameter, m; d – average particle diameter, m; ρ_T – particle density, kg / m^3 (in the formula (2) – t / m^3); ρ – is the gas flow density, kg / m^3; m – mass coefficient of the suspension, kg / kg; L – length of the pipeline, m; A, B – empirical coefficients.

As it follows from the presented equations, the speed of horizontal conveyance is mainly influenced by the pipeline diameter and suspension coefficient, and pipeline length affect is insignificant.

Pipeline length affects the flow state and the conveyance nature of the dusty gas flow if the flow properties vary along the length of the pipe. It is possible if temperature of the flow conveyance varies along the length of the pipeline. Then it is necessary to determine the velocity, based on the physical properties of the flow in the middle part of the pipe, or to calculate it separately for individual pipeline sections.

In case of vertical conveyance of the dusty gas flow, a blockage phenomenon can occur. If a gas flow velocity is below the critical value, the distance between particles decreases, their concentration in the flow increases and particles get into the hydrodynamic trajectory of the particles moving ahead, their front resistance falls and therefore their terminal velocity rises causing particle dropping out i.e. a phenomenon of blockage takes place.

To determine the ascending gas flow velocity at which the phenomenon of suspended material blockage arises, there are proposed [2] the following formulas:

$$W = 10,5 + 0,57 \cdot U_B \tag{3}$$

$$W = 1,4 \cdot m^{0,28} \cdot U_B \tag{4}$$

where

U_B – velocity of particles of a given size, m / s.

For steady conveyance of a dusty gas flow in a vertical pipeline, it is proposed [2] to determine gas flow velocity using the following formula:

$$W = k \cdot U_B \tag{5}$$

where

k – is an experimental coefficient, the choice of which depends on the size of the conveyed particles.

Formulas (3) – (5) have significant disadvantages. There is no specific way of choosing the experimental coefficient "k" and "m" parameter, as well as there is no record of conveyed particles concentration influence on the operating velocity of the gas flow. Pipeline diameter and some other parameters, complicating particle motion in the conveyed gas flow are also not considered.

Purpose of the paper is to study the influence of solid phase concentration in the flow, the effect of solid particles sizes and pipeline diameter on the velocity of sustainable conveyance of the dusty gas flow along the vertical and horizontal sections of pipelines. Objective of the paper is to ground a general analytical expression for calculating gas flow velocity and deriving equations wherein gas flow velocity would be a function of such parameters like concentration of particles in the flow, particle terminal velocity and pipeline diameter.

IOP Conf. Series: Materials Science and Engineering **233** (2017) 012026 doi:10.1088/1757-899X/233/1/012026

3. Regularities of weighing and conveyance of solid particles by a gas flow

3.1 Theoretical grounding

In stationary conditions behind the accelerating section or at the outlet of the gas pipeline, concentration of particles of a given size (dy), weighted by the gas flow over the elementary time ($d\tau$), is proportional to the limiting concentration of particles of this size (y_{PR}), that can arise in a gas flow at its complete saturation with suspended particles, that is, presented in the form of a dependence:

$$\frac{dy}{d\tau}=-k\cdot y_{PR} \tag{6}$$

In this case, the value of limiting concentration of particles is considered as a parameter that takes into account the influence of deterministic factors (particle mass, averaged velocity of the gas flow) and parameters complicating the particle motion – turbulence and flow pulsation, rotation and joint impacts of particles, their impacts against the pipeline walls.

Having divided the variables and integrating equation (6) replacing the logarithm according to the initial and boundary conditions, we obtain

$$y=y_{PR}\cdot 10^{-k_1\cdot\tau} \tag{7}$$

Particle residence time in the pipeline τ increases with growing the conveyance height and decreases with growth of gas flow velocity. We take the first approximation of:

$$\tau=k_2\cdot\frac{h}{u_T}=k_2\cdot\frac{h}{W} \tag{8}$$

For a stationary section of a gas-conveying pipeline, the particle velocity (u_T) is close to the flow velocity (w), therefore in equation one takes (8) $u_T=W$. Having substituted equation (8) in (7), we obtain:

$$y=y_{PR}\cdot 10^{-\frac{k_1\cdot k_2\cdot h}{W}}=y_{PR}\cdot 10^{-k_0\cdot\frac{h}{W}} \tag{9}$$

For gas transmission pipelines with a stationary regime of particle motion $k_0\cdot h=b$ – then

$$y=y_{PR}\cdot 10^{-\frac{b}{W}} \tag{10}$$

or

$$\lg y=\lg y_{PR}-\frac{b}{W} \tag{11}$$

We replace $a=\lg y_{PR}$ in equation (11). Further from the formula (11), one obtains an equation to determine the working velocity of the gas flow for sustainable conveyance of solid particles of the type:

$$W=\frac{b}{a-\lg y} \tag{12}$$

where

y – current concentration of solid particles in the gas flow, g / m^3; *a, b* – experimental constants.

The constant *"a"* is a logarithm of the limiting concentration of particles in the gas flow to which this value asymptotically approaches with increasing of gas flow velocity.

The constant *"b"* characterizes particles' ability to pass into suspended state in a pipeline of a given diameter. It represents the energy, expended by the flow on the distribution of particles along the channel section. For fine particles smaller than 100 μm, this constant has lower value. In this case, the particles are distributed more evenly over the cross section of the pipeline, and their concentration approaches the limiting value. As the constant value increases, the gradient of particles concentration in the gas flow grows and the current value of the particle concentration differs more and more from the limiting value.

3.2 Experimental research

The experimental setup consists of vertical and horizontal pipelines of various diameters (0.02 – 0.2 m) and a length up to 3 m, that are connected to a cyclone and a bag filter. Model materials, quartz, aluminum and copper powders, were fed uniformly by a belt feeder into the initial section of the pipeline. The air flow, upon which the averaged gas flow velocities were calculated, were measured at the inlet of the initial section of the pipeline by a calibrated collector. Pressure drops were measured by a manometer with an inclined moving scale connected to several (up to ten) piezometric rings behind the accelerating section. A turbogas blower acted like agitator. With the reduction of flow velocity, its value (W) was fixed at a constant flow rate of material (G) to the section of the pipeline (F). When specific pressure drops reached the minimum value the formula $y = G/FW$ was used for each material to determine the particle concentration y. Up to reaching the maximum concentration of particles, pressure pulsations in the stationary section of the pipeline were relatively small and were not detected, if distances between piezometric rings for measuring the pressure exceeded 0.5 m. Pressure fluctuations in the accelerating section of the pipeline were somewhat larger, but in all cases they were in range of 5 to 10% of the measured value.

They increased sharply after concentration of particles in the gas flow exceeded its limiting value, at which elements of a dense phase in the form of aggregates of particles began to separate from the flow. In accordance with the work objectives, only the data obtained at the stationary section of the pipeline with the concentration of particles not exceeding the limiting values were recorded.

The graphs (Figure 1 (a), (b)) in the coordinates $\lg y = f(1/W)$ show the dependence of the concentration of mono-fraction particles on the gas velocity for the vertical and horizontal pipelines of various diameters, respectively.

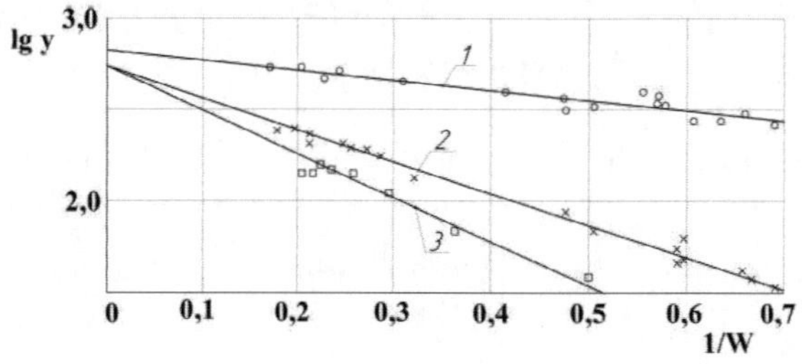

(a)

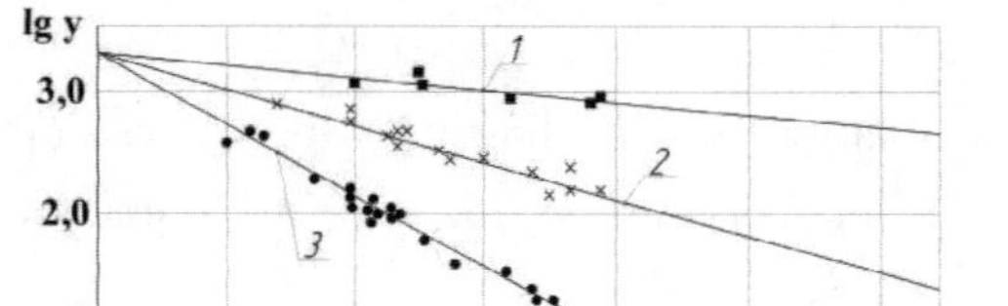

(b)

Figure 1. Dependence of the particle concentration in the vertical (a) and horizontal (b) pipelines on the gas flow velocity: 1 – 3 – diameter of the pipeline, 0.03; 0.04 and 0.08 m. respectively.

As it can be seen from Figure 1 (a), (b) the experimental data for stationary sections are expressed by straight lines, which correspond to equation (11). The ordinate of intersection point of each straight line with the ordinate axis provides a value of the constant *"a"*, which is a logarithm of that value of the particle concentration in the gas flow to which it tends with increasing flow velocity.

Constant *"a"* value increases with growing of the conveyed particle size and is practically independent on the diameter of the pipeline, since straight lines within the monofraction for pipelines of different diameters intersect on the ordinate axis at one point. Constant *"b"* is a tangent of lines inclination to the abscissa axis (Figure 1 a, b). Values of these constants, obtained experimentally, are given in Table 1.

Table 1. The main properties of particles and values of constants *"a"* and *"b"*.

Monofraction, mm	Average particle size, mm	Density, kg / m^3	Terminal velocity U_B, m / s	Pipe-line diameter D, m	Constants			
					vertical		horizontal	
					a	*b*	*a*	*b*
–0,105 +0,075	0,09	2600	0,67	0,02	3,33	0,71	4,25	6,0
				0,03	3,26	1,81	4,20	11,5
				0,04	3,26	3,1	4,18	16,1
–0,105 +0,075	0,09	2700	0,45	0,03	2,82	0,68	3,31	2,58
				0,04	2,80	1,76	3,30	6,02
				0,08	2,78	2,48	3,30	11,4
–0,105 +0,075	0,09	8800	0,92	0,02	3,68	3,0	–	–
				0,04	–	–	4,31	16,7
				0,08	3,68	4,78	–	–

3.3 Analysis and summarizing of the research results

Analysis of the experimental data showed that the constants "a" and "b" are single-valued non-linear functions of easily determined characteristics, such as a particle velocity (U_B) and a diameter of the pipeline (D). i.e

$$a = f\left(U_B\right) \tag{13}$$

$$b = f\left(U_B \cdot D\right) \tag{14}$$

which can not be expressed by a single equation for particles with different terminal velocities. It is due to a complex nature of the phenomena occurring under the interaction of particles of different diameters with the turbulent flow in the pipeline. When a particle size and a pipeline diameter change, flow regime of the particles, flow turbulence, boundary layer on the particle surface, and other parameters also change.

Therefore, non-linear connections of the form (13) and (14) will be presented in a wide range of terminal velocity changes for solid particles by several equations of the form:

$$a = m_1 \cdot 10^{\frac{-n_1}{U_B}} \tag{15}$$

$$b = m_2 \cdot (U_B \cdot D)^{n_2} \tag{16}$$

Formulas for determining the constants *"a"* and *"b"* under the conditions for conveying of particles in a vertical pipeline are described in Table 2.

Table 2. Formulas for determining the experimental constants *"a"* and *"b"*.

Terminal velocity of the conveying particles U_B, m / s	Constant	
	a	*b*
0,1 – 0,4	$2,9 \cdot 10^{-0,024/U_B}$	0,25
0,4 – 1,0	$4,3 \cdot 10^{-0,08/U_B}$	$10^{5,2} \cdot (U_B \cdot D)^{2,9}$
1,0 – 10	$5,8 \cdot 10^{-0,21/U_B}$	$180 \cdot (U_B \cdot D)^{1,26}$
10 – 20	$7,0 \cdot 10^{-1,3/U_B}$	$47 \cdot (U_B \cdot D)^{0,33}$

Formulas for determining the constants *"a"* and *"b"* under conditions of conveying the particles (U_B = 0.4 – 12 m / s) in the horizontal pipeline are:

$$a = 2,9 \cdot 10^{-0,024/U_B} \quad (17)$$

$$b = 10^{4,5/(0,2/U_B) - 0,3/D} \quad (18)$$

The discrepancy between calculated and experimental data is in the range of 10 – 25%.

4. Hardware design of the separator on the vertical section of the pipeline

It is advisable to install a conical-shaped separator with the rectangular cross-section (Figure 2) on vertical pipeline sections for conveying a dusty gas flow. The opening angle of the conical-shaped housing is in the range 5 – 7 °.

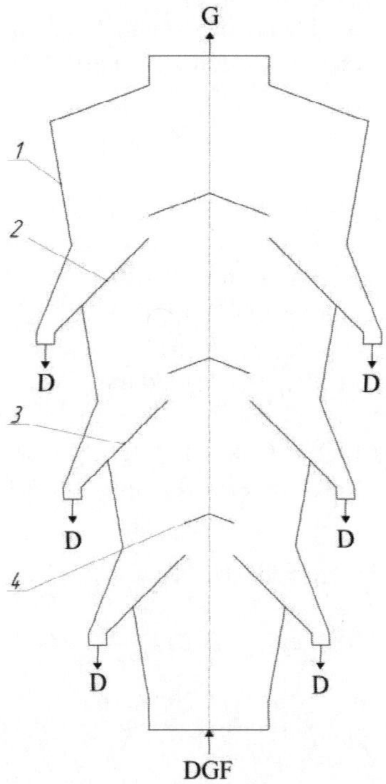

Figure 2. Separator scheme:
1 – housing; 2 – inclined shelf; 3 – discharge sleeve; 4 – reflector;
DGF – dusty gas flow; D – dust; G – purified gas.

On the height of the housing 1, there are installed inclined shelves 2, forming channels for the formation of an ascending jet. Under each shelf there are discharge sleeves 3 for removing separate fractions of captured dust particles. The gaseous flow is introduced into the separator and moves in its housing from the bottom up. A suspended layer with a relatively small concentration of particles takes place on the lower shelves. Gas velocity in this section is approximately equal to the velocity of the largest particles. That is, the principle of equilibrium classification takes place. Particles with a velocity greater than the velocity of the gas flow fall onto the surface of the shelves and are directed to the discharge sleeve. Particles with a terminal velocity less than the gas flow rate are carried to the middle shelves, where they are removed from the gas flow. On the upper shelves, the smallest fractions are removed from the gas flow. The results of particle separation from a dusty gas flow, using coal powder as an admixture (particle density 1500 kg / m^3), are shown in Table 3.

Table 3. Sieve analysis of the initial mixture and fractions captured in the separator.

Fraction type	Fraction content, %					
	+3 mm	−3+2 mm	−2+1 mm	−1+0,5 mm	−0,5+0,1 mm	−0,1 mm
Initial	28,2	21,2	16,4	5,2	23,0	6,0
Undersize	56,6	26,8	12,9	1,1	2,3	0,3
Middle-size	29,7	25,0	28,5	8,9	7,5	0,4
Extra-size	–	–	7,0	8,7	76,6	7,7

5. Conclusion

Results of the sieve analysis of separation products obtained in the apparatus show that two lower shelves provide effective removal of particles larger than 1 mm, and the upper shelf withdraw mainly fine particles smaller than 0.5 mm.

The proposed separator design operates effectively to clean dusty gas flows with particle sizes greater than 50 – 70 µm. Gravitational forces do not provide so efficient separation of particles of the smaller size. In addition, dusty gas flows containing highly disperse dust must be cleaned in centrifugal separators [4 – 9]. In this case the proposed separator serves as the first stage of purification to separate fractions higher than 70 – 75 µm from the dusty gas flow. Therefore, the efficiency of separation and productivity of the centrifugal separator of the second stage of purification will be much higher. The advantage of the separator built up in the vertical section of the pipeline is a sufficient depuration efficiency (about 70 – 75%) at a relatively small hydraulic resistance (500 – 700 Pa).

References

[1] Razumov I M 1972 Fluidization and pneumatic transport of bulk materials (Moscow: Chemistry)
[2] Yukhimenko N P, Mykhailovskiy Y E 2011 Analytical determination of flow velocity in pneumatic conveying systems *Bulletin of the Sumy National Agrarian University* **8** (23) pp 48–50
[3]Yukhimenko N P, Vakal S V, Kononenko N P, Filonov A P 2003 Apparatus of the suspended layer (Sumy: Sobor)
[4] Burov A A, Burov A I, Karamushko A V, Maksimovskaya, Tsabiev O N 2013 Separation and filtration of gas emissions into the atmosphere *Modern science: research, ideas, results, technology* **2** (13) pp 12-17
[5] Khazbulatov A I 2012 Development and investigation of a direct-flow centrifugal separator for purification from mechanical impurities *Scientific and Technical Bulletin of the Volga Region* **2** pp 35-38
[6] Serebryanskiy D A, Zakharov A A, Plashikhin S V 2013 Cyclonic dust collectors. Low-cost modernization *Chemical Industry of Ukraine* **3** pp 70–74
[7] W Peng, A C Hoffmann, P J A J Boot, A Udding, H W A Dries, A Ekker, J Kater 2002 Flow pattern in reverse-flow centrifugal separators *Powder Technology* **127** (3) pp 212–222

[8] A B Nesbitt, S Wanliss, F W Petersen 2002 The Feasibility of Ferro-alloy fines recovery by using *Centrifugal Separator Technology* (Kona: Powder and Particle) **20** pp 98–103
[9] G Gronald, J J Derksen 2011 Simulating turbulent swirling flow in a gas cyclone: A comparison of various modeling approaches *Powder Technology* **205** pp 160–171

IOP Conf. Series: Materials Science and Engineering **233** (2017) 012027 doi:10.1088/1757-899X/233/1/012027

The applicability of existing renovation methods in technology of screw compressors repairing

P V Kosenko[1]

[1] Sumy National Agrarian University, 160 G. Kondrateva Str., 40021, Sumy, Ukraine

E-mail: kosenkoPV@hotmail.com

Abstract. The article deals with the basic methods of machine parts recovery. Advantages and disadvantages of these methods had been widely spoken. The effect of these methods on the surface properties had been analysed. The main reasons of the wear of screw compressor parts are described in short. The authors paid attention to the evaluation of the applicability of existing renovation methods in technology of screw compressors repairing. In choosing the optimal method of renovation the surfaces of screw compressor parts are guided by the criteria of RAMS: reliability, availability, maintainability and safety. Also, it is specially noted about the evaluation of the effect of renovation methods on the screw surface layer by the following criteria: adhesion, ecological safety, cost, wear resistance, hardness, roughness, layer thickness, capacity, hogging, porousness, surface preparation, changes in the structure and physical properties of the base metal and ease of use. The most appropriate method for the renovation of parts of screw compressors had been proposed.

1. Introduction

A screw compressor is a type of compressor that uses a rotary-type positive-displacement mechanism. They are commonly used to replace piston compressors where large volumes of high-pressure air are needed, either for large industrial applications or to operate high-power air tools such as jackhammers.

Screw compressors don't have any assembly units that perform reciprocating movements. Therefore, there are no alternating loads on parts and the increased wear of friction responsible pairs. That's why the life cycle of screw compressors is long. Fulfilling the requirements of exploitation can increase their life cycle up to 40 000 hours. The units, limiting resource of screw compressors, are bearings, installed on drive and driven shafts. In addition, the abuse could lead to mechanical damage of the screw's guide.

2. Formulation of the Problem

If there is continued operation of compressor with the hot oil, if the frequent starting and stopping of the compressor is made, bearings can become damaged prematurely. This situation can lead to the failure of the compressor, related to the jamming of the screws, prehension of screws and the compressor's housing. The repair of such compressor can be pursuance only in plant conditions, but in a number of cases the compressor unit must be changed. Basically, there are following types of wear during the screw compressors operation: wear of bearings' seats and wear of the working surfaces of screws (Figure 1).

Generally, the wear rate is not over than 0.3 mm in diameter, but it can lead to reduce the capacity of the compressor, because the initial clearance between screws is 0.05 mm. To prevent such situations, it needs to make a diagnostic review of bearings with special handheld devices. This reliable equipment is readily available. It should be included to the list of necessary repairing tools for maintenance services of the plants, where the screw compressors are being used. This notwithstanding the damaged screws usually must be replaced with new ones.

Figure 1. Places of screw's wear: 1 - working surfaces of screws; 2 - bearings' seats.

Therefore, it is considerably expensive from an economic standpoint. Hence, the main aim of the study is to find the most effective method of renovation of worn screw surfaces and bearings' seats.

3. Findings of Investigation

Most methods of recovery should be considered as an alternative. The same coating material can be applied in several ways. In this case, the properties of the coating layer and cost of its application can vary significantly. The application conditions can change the mechanical properties of the base material. In this way, the properties of the coated part depend on the method of application. To renovate the worn part means to restore the primary (or close to them) geometric, physical, mechanical, chemical and other properties, to eliminate operational defects, to restore dimensions and geometric shape according to technical requirements. According to [1], the operability and resource of the renovated parts averages by 60...80% of these indicators for new parts. However, at present there are some technological methods which can completely restore the original resource of parts or even increase it. Restoring of the parts allows saving a significant amount of materials. Also, it leads to noticeably extend the life cycle of the parts and reduce the cost of equipment repair. Currently, there are many different technological methods to compensate the worn-out layer of parts [1 - 4]. The most common methods, their advantages and disadvantages are shown in Table 1.

Analysing the data of Table 1, we can conclude that each of these methods has its advantages and disadvantages. But to select the recovery method for a narrower application (namely, for renovation of the surfaces of the screws and seats for the bearings of screw compressor units), several criteria must be taken into account. There are the following criteria to selecting the optimal method of restoring parts: applicability, durability and economic [5].

The criterion of applicability determines the fundamental possibility of applying various recovery methods for specific details. This criterion is described by the function:

$$K_З = \varphi(M_Д ; \Phi_Д ; Д_Д ; C_Д ; H_Д ; \sum_{i=1}^{m} T_i) \quad (1)$$

where:
$M_Д$ – the material of the part;
$\Phi_Д$ – the shape of the surface to be restored;
$Д_Д$ – the diameter of the surface to be restored;
$C_Д$ – the wear of the part;
$H_Д$ – the load acting on the part;
$\sum_{i=1}^{m} T_i$ – the sum of technological features of the method that determine the area of its rational application.

Table 1. Methods for repairing worn surfaces.

Method	Advantages	Disadvantages
Surfacing	Increasing hardness and wear resistance, the possibility of unlimitedly increasing the worn surface	The formation of cracks, high porosity, the presence of slag inclusions, reduced fatigue strength, warpage, increased environmental hazard
Electroplating	Preserves the structure of the part, high wear resistance and hardness of the surface	Low run-in, reduced fatigue strength, low adhesion, increased environmental hazard
Metallization	The mechanical properties of the material do not change, the part is not subjected to buckling, high wear resistance	High porosity (up to 10%), reduced fatigue strength, low adhesion, increased environmental hazard
Plastic deformation	Increases hardness, reduces roughness, increases wear resistance	Low productivity, it is possible to deform the surface by 5-10 microns or more
Electroerosion alloying	Local surface treatment; a strong compound of the transferred and base metal; absence of general heating of the part during processing, possibility of using as processed materials: pure metals, alloys, metal-ceramic compositions, refractory compounds; increase hardness, heat resistance, wear resistance and corrosion resistance; no need for surface preparation	Appearance of tensile residual stresses in the surface layer, reduction of fatigue strength, roughness increase
Application of polymer materials	The ability to unlimitedly increase the worn surface, close to metal deformation characteristics, high adhesion	The need for special surface preparation, including the formation of surface roughness. Relatively low hardness

By this criterion competitive methods are selected for further evaluation by other criteria.

The criterion of durability determines the operability of the parts being restored. It is expressed by the coefficient of durability, which is meant the ratio of the durability of the restored part to the durability of the new one. This criterion is described by the function:

$$K_Д = f_1(k_C ; k_B ; K_{зч}) \quad (2)$$

IOP Conf. Series: Materials Science and Engineering **233** (2017) 012027 doi:10.1088/1757-899X/233/1/012027

where:
k_C – coefficient of stability against wear;
k_B – coefficient of endurance;
$K_{зч}$ – adhesion coefficient.
The economic criterion is the function of two arguments:

$$K_{m.e} = f_2(k_{np}; e) \tag{3}$$

where:
k_{np} – coefficient of productivity of the method;
e – an indicator of the economy of the method.
The economic effect of the introduction of the developed technological process of the restoration is determined by the formula:

$$e = \left[C^B_{n(б)} - C^B_{n(i)} - E_H(\kappa_i - \kappa_б)\right] N_B \tag{4}$$

where:
$C^B_{n(б)}$ – the full cost price of restoration with a base variant of technological process;
$C^B_{n(i)}$ – the full cost price of restoration with i-th (introduced) technological process;
$E_н$ – normative coefficient of investment efficiency ($E_н$=0,15);
κ_i, $\kappa_б$ – capital investments in accordance with the introduced and basic technological processes (equipment, instrument, design, manufacturing, on-site installation, etc.);
N_B – the program of restoration of details.
The payback period after the introduced technological process is determined by the formula:

$$\tau_{ок} = \frac{\Delta\kappa}{(C^B_{П(б)} - C^B_{П(i)})N_B} \tag{5}$$

where:
$\Delta\kappa = \kappa_i - \kappa_б$ – additional investment.
Determining the labour productivity, the process of restoring parts involves preparing the part, applying the material to the prepared part and processing it had been taken into account. Therefore, the recovery method and the process of applying the material by this method are evaluated separately by the productivity.

To visualize the objective evaluation of recovery methods for a particular application, consider Table 2.

4. Conclusion

- For the renovation of screw compressor parts, it is suggested to use combined methods.
- To use combined method consisting of ion nitriding with subsequent cementation by electroerosion alloying for the renovation of the bearings' seats is insistently recommended.
- For the renovation of worn-out surfaces of screws the authors also insisted on using a combined method consisting of electroerosive alloying of the surface with an electrode made of corrosion-resistant steel with subsequent application of polymeric materials.

Table 2. Evaluation of the criteria of recovery methods for renovation of the surfaces of screws and bearings' seats for the of screw compressor units.

Method	Criteria												
	Adhesion	Environmental safety	Cost price	Wear resistance	Hardness	Roughness	Layer thickness	Productivity	Warping	Porosity	Surface preparation	Changes in the structure and physical properties of the base metal	Ease of application	**Average score**
Surfacing	4	1	2	4	5	2	3	4	2	2	4	2	3	**2,9**
Electroplating	2	1	1	4	4	3	2	2	3	4	1	4	2	**2,5**
Metallization	2	1	2	4	4	2	2	3	4	1	2	4	2	**2,5**
Electroerosion alloying	5	3	4	4	4	2	4	4	3	4	5	4	5	**3,9**
Application of polymer materials	4	4	3	3	1	3	5	5	5	5	1	5	5	**3,8**
Combined method 1[a]	5	4	4	5	5	4	5	5	4	4	4	4	5	**4,5**
Combined method 2[b]	5	4	4	5	5	5	4	4	5	4	5	5	5	**4,6**

[a] Method for the restoration of bearings' seats, consisting of ion nitriding with subsequent cementation by electroerosion alloying.
[b] Method for the restoration of worn out surfaces of screws, consisting of electroerosive alloying of the surface with an electrode made of corrosion-resistant steel with subsequent application of polymeric materials.

References

[1] Chabanniy V Y 2007 *Repair of automobiles: Manual* (Kirovograd: Kirovograd regional printing house)
[2] Ivanov V P 2007 *Technology and equipment for the restoration of machine parts* (Minsk: Tekhnoperspektiva)
[3] Ishchenko A A 2007 *Technological bases of restoration of the industrial equipment by modern polymeric materials* (Mariupil: PHTU)
[4] Panteleyenko V P *et al* 2003 *Restoration of machine parts: Handbook* (Moscow: Mashinostroyeniye)
[5] Savulyak V I *et al* 2004 *Restoration of machine parts* (Vinnytsia: VNTU)
[6] Tarelnyk V B *et al* 2005 *Modernization and repair of rotary machines* (Sumy: Kozatskyy val)

IOP Conf. Series: Materials Science and Engineering **233** (2017) 012028 doi:10.1088/1757-899X/233/1/012028

Application of jet thermal compression for increasing the efficiency of vacuum systems

S O Sharapov[1], V M Arsenyev[1], V M Kozin[1]

[1]Sumy State University, 2, Rimsky-Korsakov, Sumy, 40007, Ukraine

E-mail: s.sharapov@kttf.sumdu.edu.ua

Abstract. The article describes the principle of jet thermal compression with reference to vacuum systems. We consider liquid-vapor ejectors, the working process of which is based on this principle, as an alternative to existing steam jet ejectors. The results of a theoretical and experimental investigations of the liquid-vapor ejector of a vacuum unit are presented. With the help of thermodynamic analysis of exergetic efficiency, the range of operation of the liquid-vapor ejector and the vacuum unit on its base is determined, at which its performance is at the maximum level. The efficiency of vacuum creation by such units is estimated by the example of various pumped media, such as saturated, superheated steam and steam-air mixture.

1. Introduction

Due to the high rates of development of modern industry, technological the vacuum processes are becoming increasingly widespread, which makes it possible to improve the quality of final products significantly, obtained by reducing the content of harmful impurities as a result of protecting technological systems from interaction with the atmosphere and increasing the degree of completeness of processes and opens up broad prospects for the development of new, more technological processes, unrealizable in the conditions of an atmospheric pressure.

One of such methods is the using of jet devices in which the creation of vacuum occurs due to the active energy propulsion jet stream. Such apparatuses include aggregates that comprise steam jet ejectors which, as a rule, are multistage at the relation of pressure 10-20 and their total efficiency is 2-10% [1].

This low efficiency of existing vacuum systems based on steam-jet ejectors leads to the need to develop new, more sophisticated devices. These liquid vapor ejectors are operating on the principle of the jet thermal compression, in which the injection of the passive flow is carried out by a working steam jet, which is formed by the boiling of liquid supplied into the active nozzle [2]. When implementing such active flow expansion cycle of the working stream occurs on the left of the lower boundary curve. In this case, the emergence of limiting critical flow regimes at the entrance to the mixing chamber, which significantly reduces the efficiency of steam jet ejectors, is virtually eliminated.

2. Principle of jet thermal compression and working process of liquid-vapor ejector

Vacuum units based on steam jet ejectors (Figure 1), used in a modern industry, have a number of disadvantages, the main one of which are the "impact" losses that occur when supercritical active and subcritical passive flows are mixed. The low efficiency of such devices is also a consequence of the fact that it is possible to increase the pressure in one steam-jet stage by no more than 2-3 times under the condition of a high level of energy conversion.

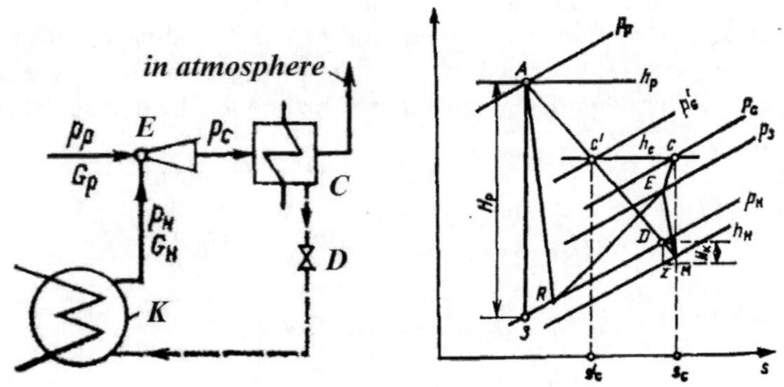

Figure 1. Scheme (left) and cycle (right) in h,s-coordinates of the classical unit based on the steam jet ejector: E – steam jet ejector, C – cooler, K – condenser, D – drainage condensate from the refrigerator.

In consequence of this situation, the using of a liquid-vapor ejector operates on the principle of jet thermal compression, which is that at the inlet to the nozzle of the active flow, the working flow is in a state of metastable superheated liquid becomes very topical. Passage of the processing medium of the active flow through the Laval nozzle is accompanied by a process of relaxation vaporization in the expanding part of it. The flow of the working medium of the active stream through the Laval nozzle corresponds to process 1_a-2_a (Figure 2).

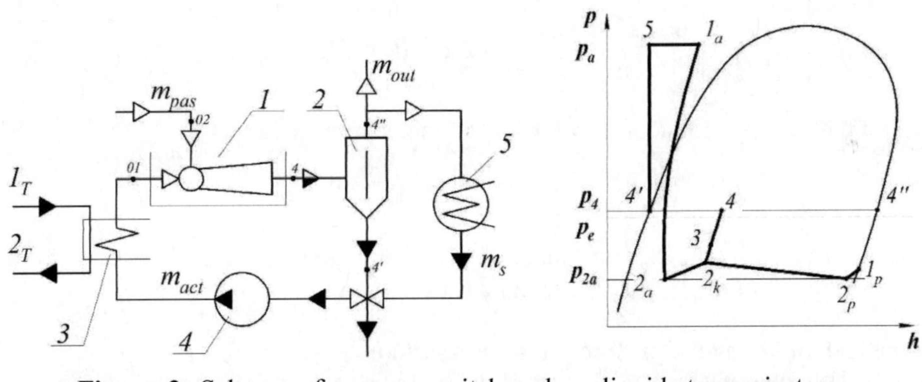

Figure 2. Scheme of vacuum unit based on liquid-steam ejector (left) and the representation of its workflow in p, h-coordinates (right): 1 – liquid-vapor ejector, 2 – separator, 3 – heat exchanger, 4 – circulating pump, 5 – condenser.

A supersonic jet of a finely dispersed vapor-drop structure with a high volumetric vapor content, that pressure is less than the pressure of the surrounding medium ($p_a < p_o$), is formed in the exit section of the nozzle of the active stream of the liquid-vapor ejector. Further, it injects the working medium of the passive stream, that enters the receiving chamber with the pressure p_{02}. At the entrance to the mixing chamber, the pressure of the active and passive stream media is equalized (processes 2_a-2_k and 2_n-2_k, respectively). In the mixing chamber, the working media of the active and passive streams are mixed into one with the reaching of the pressure p_3 (process 2_k-3). In the diffuser there is a subsequent compression of the mixed flow and a pressure p4 equal to the pressure at the exit from the ejector (process 3-4).

Produced in the liquid-steam ejector 1, a vapor is separated in the separator 2 from that the saturated liquid is taken by the pump 4 into the circulation circuit and after heating in the heat exchanger 3 is fed into the active nozzle of the ejector (Figure 2).

The mathematical model of the working process of the liquid-vapor ejector, operated in the evacuation mode, is numerically described as a system of equations for one-dimensional adiabatic motion in the quasi-equilibrium thermodynamic approximation for the distinguished boundaries of the current section of the flow:

- equation of state of a thermally metastable vapor-drop environment

$$\upsilon = \upsilon_{liq}\left(t_{liq}\right) + x \cdot \left[\upsilon_{vap}\left(p\right) - \upsilon_{liq}\left(t_{liq}\right)\right]$$

$$d\left[\frac{w(z)\cdot F(z)}{\upsilon}\right] = 0$$

- equation of conservation of mass (with allowance for the phase transition)

$$dx = \chi(z)\cdot\left[\frac{\upsilon}{w(z)}\right]dz$$

- equation of total enthalpy (I law of thermodynamics)

$$d\left[h_{liq}\left(t_{liq}\right) + x\left(h_{vap}\left(p\right) - h_{liq}\left(t_{liq}\right)\right) + \frac{w^2(z)}{2}\right] = 0$$

- momentum equation (pulse)

$$d\left[\frac{w^2(z)\cdot F(z)}{\upsilon}\right] = -F(z)dp - \tau_w \cdot \Pi(z)dz$$

- equation of the entropy production (II law of thermodynamics)

$$d\left\{s_{liq}\left(t_{liq}\right) + x\left[s_{vap}\left(p\right) - s_{liq}\left(t_{liq}\right)\right]\right\} = \delta s_{diss} > 0$$

and is also supplemented by the kinetic functions of vaporization, the characteristics of crushing and polydisperse distribution of the liquid phase, and critical regimes [3].

3. The theoretical analysis of experimental investigations

The purpose of the experimental studies was to determine the range of the initial parameters of the working fluid of the active stream, under that the maximum efflux efficiency of the boiling, underheated liquid saturation of the active flow from the expanding channels is observed. Such a range is observed at the values of the relative initial underheating ($1 - \varepsilon_{s0}$) at the level of 0,2-0,4 (Figure 3).

The relative initial underheating is the determining factor in the efficiency of the efflux of the working jet from the expanding channels and is determined by the formula:

$$\left(1-\varepsilon_{s0}\right) = \frac{P_0 - P_{s0}}{P_0} = 1 - \frac{P_{s0}}{P_0}$$

where P_{s0} is the saturation pressure at temperature t_0.

IOP Conf. Series: Materials Science and Engineering **233** (2017) 012028 doi:10.1088/1757-899X/233/1/012028

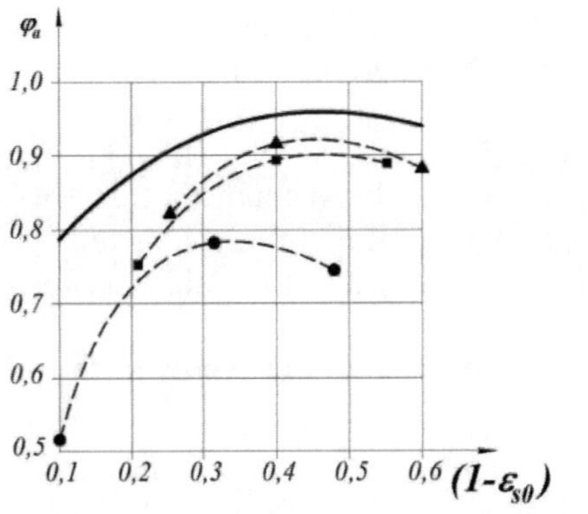

Figure 3. The experimental dependence of the speed coefficient of the active flow nozzle from the value of the relative initial underheating of the working fluid (P_{01} = 3÷10 *bar*, t_{01} = 130÷175^0*C*, P_a = 0,5÷1,0 *bar*).

The character of the process of mixing active and passive streams was studied in mixing chambers of different geometric shapes [4-5]. They were cylindrical and conical with a subsequent cylindrical section. The angle of convergence of conical mixing chambers was within 5-12^0.

On the basis of an experimental study of a liquid-vapor ejector with mixing chambers of various geometric shapes, the dependence of achievable efficiency indicators (vapor content x_4, overproduction of vapor ψ_4 and injection coefficient u) on the depth of the vacuum P_c/P_{02} created by it (Figure 4).

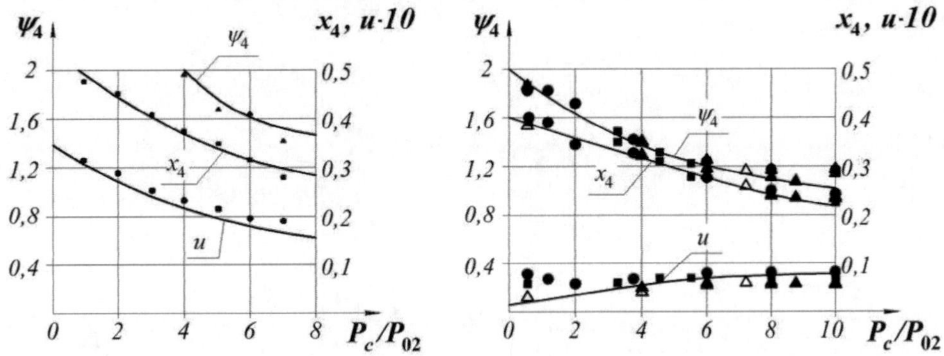

Figure 4. Experimental dependence of vapor content x_4, overproduction of vapor ψ_4 and injection coefficient u from the degree of increase in the pressure of the passive flow in cylindrical chamber (left): • – P_{01} = 4 *bar*, ■ – P_{01} = 6 *bar*, ▲ – P_{01} = 8 *bar*, in conical chamber (right) with different convergent angles: - - - - 2^0, - · - · - 4^0, -□-□-□- 6^0, -△-△-△- 8^0.

The range of effective operation of the liquid-vapor ejector with a cylindrical mixing chamber, and the initial parameters of the working fluid of the active stream specified above, is in the interval P_c/P_{02} = 4-6.

The transition from the cylindrical mixing chambers into conical chambers with the subsequent cylindrical section allows to achieve more significant depth of the vacuum. Using it in a liquid-vapor ejector, that working process is based on the principle of jet thermal compression, based on the mixing of two subcritical flows, it is possible to avoid limiting operation modes of the ejector and "locking" the mixing chamber. Having investigated the conical mixing chambers with different confusion

IOP Conf. Series: Materials Science and Engineering **233** (2017) 012028 doi:10.1088/1757-899X/233/1/012028

angles, it can be concluded that the mixing process is most effective in chambers with confusion angles of 4-8^0 in the range of working fluid pressures P_{02} = 4-9 bar.

4. Vacuum systems of exergetic efficiency

For an assessment of power efficiency of the vacuum unit on the basis of liquid-vapor ejector, it is most correct to involve the exergy method of thermodynamic analysis, the main provisions of that are presented in [6-8]. The use of this method makes it possible to express and rank different energy flows unequivocally in thermomechanical systems.

Exergetic efficiency is defined as the ratio of the exergy of the product flow E_P to the exergy of the fuel stream E_F of the system:

$$\varepsilon_{ex} = \frac{E_P}{E_F}$$

For the liquid-vapor ejector, as a product exergy, the difference between the exergy of the saturated vapor at the outlet from the separator and the exergy of the passive stream at the inlet to the liquid-vapor ejector is considered. As an exergy of the fuel flow, the difference between the exergy of the active stream at the inlet to the liquid-vapor ejector and the exergy of the liquid at the outlet from the separator is considered.

For the vacuum unit (Figure 2), as the exergy of the product flow, as well as for the liquid-vapor ejector, the difference between the exergy of the saturated vapor at the outlet from the separator and the exergy of the passive flow at the inlet to the liquid-vapor ejector is considered. The fuel flow exergy is the sum of the power consumption of the circulation pump and the exergy of the coolant flow in the heat exchanger-preheater.

As a result of exergetic analysis, theoretical and experimental values of achievable performance indicators was received (Figure 5).

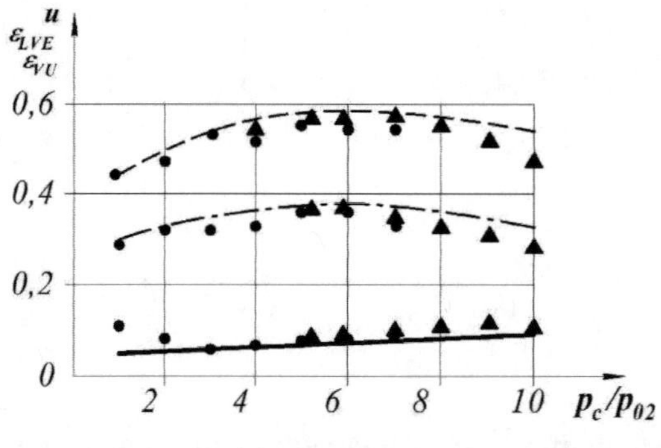

Figure 5. Dependence of achievable performance indicators of liquid-vapor ejector from the degree of increase in the pressure of the passive flow (P_{01} = 8 bar, P_a = P_{02} = 0,5 bar): results of theoretical investigations: —— – injection coefficient, - - - - – exergetic efficiency of liquid-vapor ejector, - · - · - – exergetic efficiency of vacuum unit, experimental data: • – liquid-vapor ejector with cylindrical mixing chamber, ▲ – liquid-vapor ejector with conical mixing chamber.

To estimate the efficiency of using a liquid-steam ejector in vacuum systems, the predicted efficiency of vacuum installations based on it is numerically calculated [9-10]. The considered schemes differ in pumped out media - saturated and superheated water vapor and steam-air mixture. Analyzing the obtained results of comparison of the basic schemes and the proposed scheme on the basis of a liquid-vapor ejector operating on the principle of jet thermal compression, it can be concluded that the introduction is expedient, since it allows reducing the consumption of boiler steam used in the basic version for active Flow steam jet ejector, reduce the initial parameters of the working medium of the active flow and increase the efficiency by 1,5-3 times.

5. Conclusions

Comparing the working processes of the classical steam jet ejector and liquid-vapor ejector operating on the principle of jet thermal compression, it can be asserted that the liquid-steam ejector has a more efficient workflow, which makes it possible to convert the working jet energy of the active flow with less cost and more efficiency.

Analyzing the results of numerical studies of FPE and evaluating its effectiveness, it can be concluded that it is expedient to use FET to create a vacuum in a wide range of passive flow pressures and injecting various media with a sufficiently high degree of perfection of the working process for jet devices.

In the experimental investigation of the active stream nozzle in the range of the initial parameters of the working fluid of the active flow P_{01} = 3÷10 bar, t_{01} = 130÷175^{0}C, (1 – ε_{s0}) = 0,2-0,4, the character of the change in the working fluid parameters of the active flow was confirmed. Its expiration from the expanding channels into the pressure region below the atmospheric pressure (P_{02} = 0,45-0,85 bar), with the possibility of achieving maximum efficiency (φ_a = 0,768-0,917). The character of the effect of structural and regime parameters on the level of vacuum achieved in mixing chambers of various geometric shapes has been experimentally investigated and the possibility of achieving maximum efficiency indicators by optimizing the flow part ($\alpha_к$ = 4-8^0, P_4/P_{02} = 2-3,5 – for cylindrical mixing chamber, P_4/P_{02} = 3-5 – for conical mixing chamber). By correlating the mathematical model as a result of experimental research and introducing the appropriate coefficients, a range of parameters for the active and passive flow media (P_4/P_{02} = 4-9) is established, ensuring maximum efficiency of the liquid-vapor ejector (u = 0,03-0,1, ε_{LVE} = 0,45-0,55, ε_{VU} = 0,3-0,38).

Based on the results of the exergy analysis of vacuum installations based on liquid-steam ejectors that eject saturated, superheated water vapor and steam-air mixture and their comparisons with existing installations based on steam jet ejectors with the determination of efficiency indicators for the introduction of new technology, an increase in their efficiency of 1.5-3 times.

References

[1] Arsenyev V M, Sharapov S O, Prokopov M G 2011 Estimation of the energy efficiency of a liquid-vapor ejector used to create a vacuum in metallurgy *Refrigeration engineering and technology* **5** (133) pp 62-66

[2] Marchenko V N, Osipov V A, Prokopov M G, Sharapov S O 2009 Principle of stream thermocompression: conception of energetic efficiency and prospect of realization is in small heat energetic *Motoryzacja i energetyca rolnictwa* **11A** (Simferopol-Lublin) pp 70-76

[3] Sharapov S, Arsenyev V, Protsenko M 2013 The use of liquid-vapor ejector in vacuum systems *Refrigeration engineering and technology* **4**

[4] Sharapov S O, Arsenyev V M 2016 Experimental study of a liquid-steam ejector with a cylindrical mixing chamber *Refrigeration engineering and technology* **52** (2) pp 87-92

[5] Sharapov S, Arsenyev V, Kozin V 2016 Experimental investigation of liquid-vapor ejector with conical mixing chamber *Technology audit and production reserves* **4/1** (30) pp 50-55

[6] Tsatsaronis J. 2002 Interaction of thermodynamics and economics for minimizing the cost of an energy-transforming system (Odessa: Studio Negotsiant Ltd.) p 152

[7] Morozyuk T V 2006 Theory of refrigerating machines and heat pumps (Odessa: Studio Negotsiant Ltd.) p 712

[8] Arsenyev V M 2011 Heat pump technology of energy saving (Sumy: SumDU) p 283

[9] Arsenyev V M, Sharapov S O, Borisov N A 2012 Energy efficiency evaluation of a vacuum installation using a liquid-steam jet ejector *Sustainable development and artificial cold* pp 355-359

[10] Arsenyev V M, Sharapov O O, Miroshnichenko V V 2013 Application of a liquid-steam ejector for the recompression of a secondary vapor of vacuum installations *Energy saving. Power engineering. Energoaudit* (Kharkiv) pp 57-64

IOP Conf. Series: Materials Science and Engineering **233** (2017) 012029 doi:10.1088/1757-899X/233/1/012029

Design of pneumatic proportional flow valve type 5/3

P A Laski[1], D S Pietrala[1], J Zwierzchowski[1], K Czarnogorski[1]

[1]Department of Automation and Robotics, Faculty of Mechatronics and Machine Design, Aleja Tysiaclecia Panstwa Polskiego 7, 25-314 Kielce, Poland

E-mail: pawell@tu.kielce.pl

Abstract: In this paper the 5/3-way pneumatic, proportional flow valve was designed and made. Stepper linear actuator was used to move the spool. The valve is controlled by the controlled based on a AVR microcontroller. Virtual model of the valve was created in CAD. The real element was made based on a standard 5/3-way manually actuated valve with hand lever, which was dismounted and replaced by linear stepper motor. All the elements was mounted in a specially made housing. The controller consists of microcontroller Atmega16, integrated circuit L293D, display, two potentiometers, three LEDs and six buttons. Series of research was also conducted. Simulation research were performed using CFD by the Flow Simulation addition to SolidWorks. During the experiments the valve characteristics of flow and pressure was determined.

Nowadays, proportional valves are becoming increasingly popular in industrial applications. They perfectly fill the application area between conventional control technology and servo-valve technology [1, 2] [3]. Proportional elements are very well suited for controlling the direction, flow rate and operating pressure [4][5]. This project will present a design of a pneumatic proportional flow valve type 5/3. A stepper motor integrated in the screw-nut mechanism was used to move the valve slider. The valve is controlled by a controller built on an AVR microcontroller. The mechanical parts design was made in 3D-CAD software, which created three-dimensional models that were subsequently analyzed for CFD (Computional Fluid Dynamics), as well as the preparation of technical documentation. Based on solid and commercial models, the Prema ™ 5/3 type lever-controlled mechanical valve has a prototype that is subjected to physical testing. During the experimental study, flow and pressure characteristics of the valve were determined. The valve driver consists of an Atmega16 microcontroller, an L293D integrated circuit, a display. At the design stage, a number of simulation studies were conducted using the SolidWorks Flow Simulation environment.

1. Construction assumptions

The proportional valve design is based on the prema 5/3 type divider with manual control, from which the lever to change the position of the slider was removed in order to use another type of control.

It was assumed that the proportional valve piston would control the stepper motor with an integrated screw-nut mechanism. The engine connection with the valve slider was done by cutting the thread on the plunger and screwing in the actuator's clamping handle. An adapter in 3D printing technology was used to connect the motor to the valve body. Figure 1 shows the solid model of the analyzed valve.

IOP Conf. Series: Materials Science and Engineering **233** (2017) 012029 doi:10.1088/1757-899X/233/1/012029

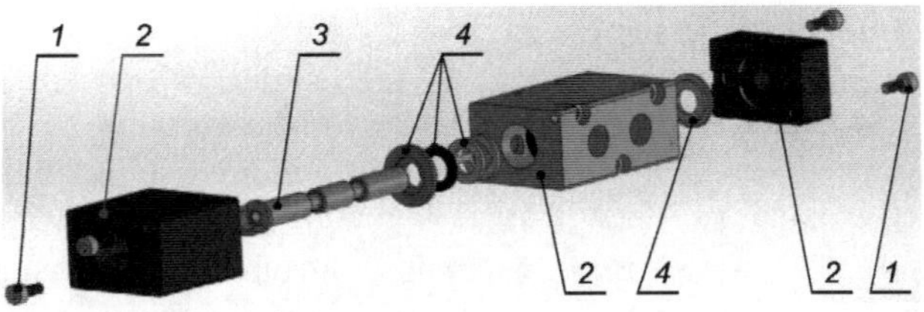

Figure 1. Components of 5/3 valve, 1 –M3 screws, 2 – valve parts, 3 – slider, 4 – sealing.

2. Valve slider drive

As already mentioned, the miniature stepper motor L35BYZ-B04601B was used to move the valve slider, which provides precise linear motion. Inside the engine, the rotary motion is converted into linear, which has a range of 0-75 mm. The feed rate of the slider is 0.025 mm per step for full-speed control and 0.0125 mm for semi-speed control. The maximum power that the drive generates is 55 N at 100 pulses per second. The actuator is powered by a rated voltage of 5 V and the phase current is 0.46 A, the resistance is 11 Ω, and the inductance of the winding is 8 mH. Such parameters are sufficient and allow dynamic control of the valve slider. In idle state, the windings supply the full torque maintenance and hence the constant valve slide force, even after the pneumatic supply is switched on. The most important advantage of the L35BYZ-B04601B engine used in the context of a proportional valve design is that it relies on the rotation of the device from the input pulses and control is possible in the open loop.

Figure 2 shows the solid models of the parts that comprise of the 5/3 divider, the linear stepper actuator L35BYZ-B04601B, the housing made on a 3D printer, and the M3 screws..

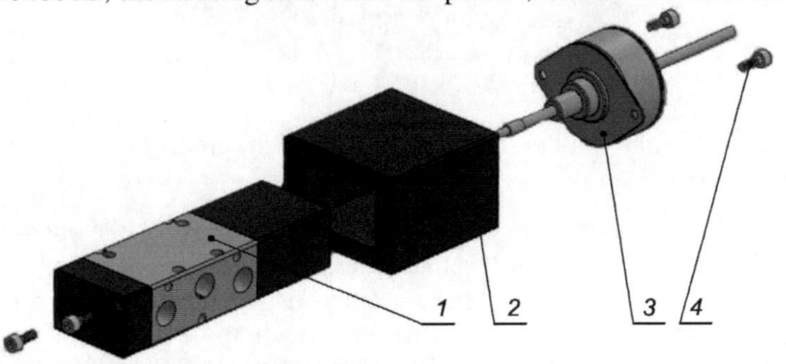

Figure 2. Components of the proportional valve design, 1 - 5/3 valve, 2 – casing, 3 –L35BYZ-B04601B linear actuator, 4 –M3 screws.

On the other hand, the actual prototype of the designed 5/3 type valve is shown on Figure 5.

Figure 3. Proportional 5/3 valve prototype.

3. Design assumptions of proportional valve control

A linear stepper actuator controller, moving the divider slider, was made, based on the Atmega microcontroller. L293D integrated circuits are used to power the motor coils. The current status of the device is displayed on the two-line display. Two operating modes are implemented in the controller - full-speed and semi-speed.

The applied drive (L35BYZ-B04601B actuator) is controlled like a typical stepper motor with bipolar winding in which the current pulses are converted into the correct rotation and then the linear motion. The stator of the stepper motor has windings that are split into two phases. Their cyclic power causes rotor rotation. In full-speed control, the motor moves at a basic angle per step. The second type of work is semi-speed control. It achieves steps two times lower than full-speed, but together with a slight increase in resolution, the torque also slightly drops.

4. Model research by cfd method

Within model research, air flow was simulated through the designed valve and simulated flow characteristics was determined. For the analysis of workflow flow, the geometry of models of parts prepared in 3D-CAD software was directly used. Figure 4 shows a model designed for use in the airflow test along with the selected work area.

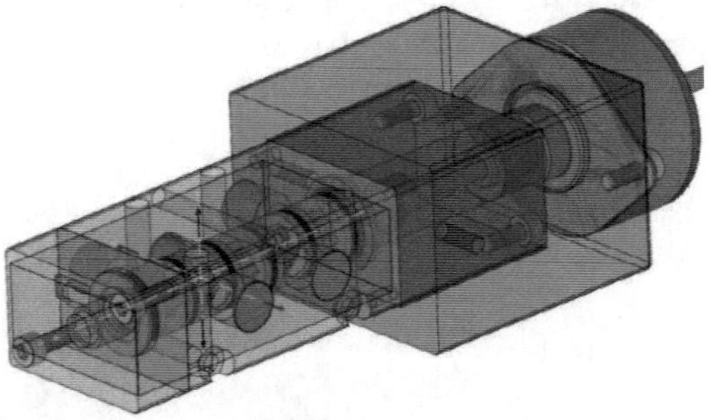

Figure 4. Solid model for flow analysis in proportional valve type 5/3.

The simulation of the air flow in the valve consisted of moving the model valve slider by a specified value equal to ten full steps of the linear actuator and then performing numerical analysis using the CFD method [6]. The input is set at a constant pressure of 0.63 MPa and at the outlet pressure of 0.1 MPa. Flow tests were performed for each of the roads. For example, the flow trajectories of the air through the valve for sliding the slider by 5.75 mm, or 230 full steps of the engine, on the road 1-2 are depicted in Figure 5. In contrast, the velocity distribution of air on the same road is shown n Figure 6.

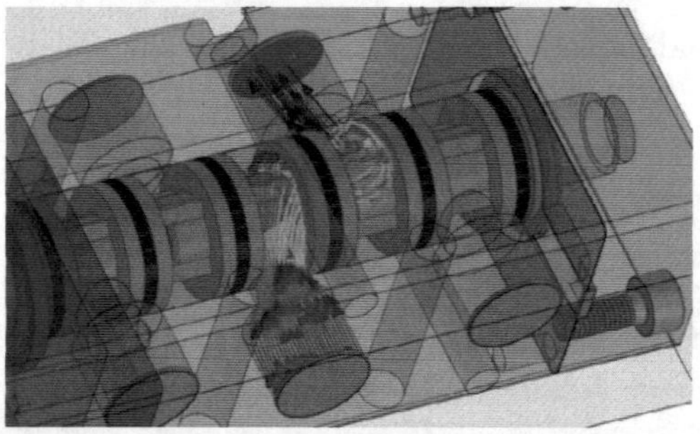

Figure 5. Exemplary air flow trajectories through the proportional valve on road 1-2, with the slider extending 5.75 mm.

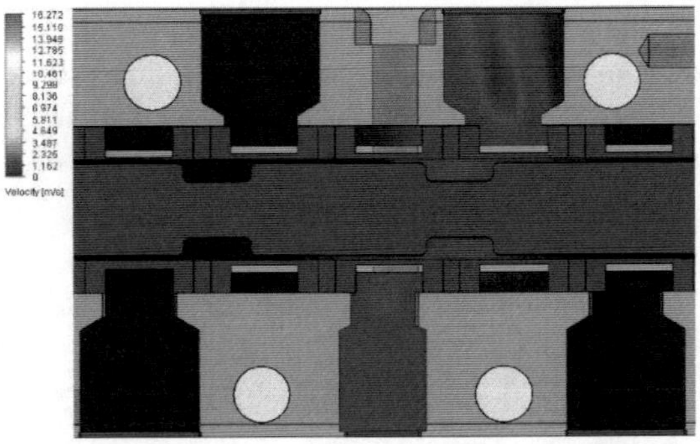

Figure 6. Distribution of air velocity through the proportional valve on the road 1-2, with the slider extending 5.75 mm.

5. Determination of experiment flow characteristics
Basic flow characteristics determines the relationship between the fluid flow and the degree of valve opening, with constant pressure drop in the setting element. The diagram of the station used to determine the flow characteristics is shown in Figure 7.

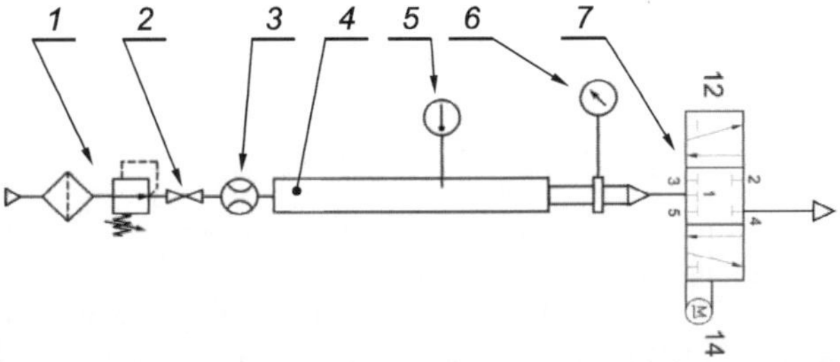

Figure 7. Diagram of the station for determining flow characteristics of the valve: 1 - air preparation unit, 2 - shutoff valve, 3 - flow meter, 4 - laminate tube, 5 - thermometer, 6 - pressure gauge, 7 – tested proportional valve.

The compressed air flows through the air preparation unit, the shutoff valve and the flow meter to the laminating tube. At this point the transient (turbulent) flow is brought to a steady state (laminar). Then the air flows through the specified valve paths, depending on the connection of the element. During this time, the flow meter measures the volume of the air stream. The actual appearance of the station is shown in Figure 8.

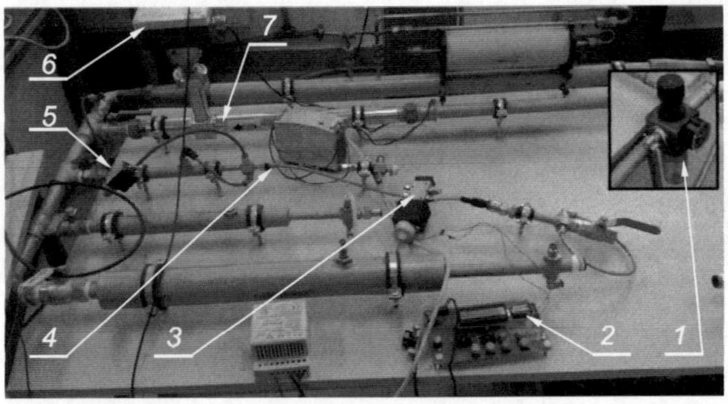

Figure 8. Appearance of the research station for testing flow characteristics: 1 – Air preparation unit, 2 - Proportional valve controller, 3 - Proportional valve tested, 4 - 1/8 "connection, 5 - Shutoff valve, 6 - Flow meter, 7 – Laminate tube.

The test involved moving the valve slider to the characteristic points where there was a distinct change in air flow through the valve and then recording the flow value with the *Sensyflow VT-S* program integrated with the ABB flowmeter. Working pressure was 0.63MPa. The full movement of the slider from one extreme position to the other required 350 full-speed linear actuators of 0.025 mm each. Determination of flow characteristics was made for each of the proportional valve operating paths. The obtained results from the simulation and experimental studies are presented in the diagram on Figure 9.

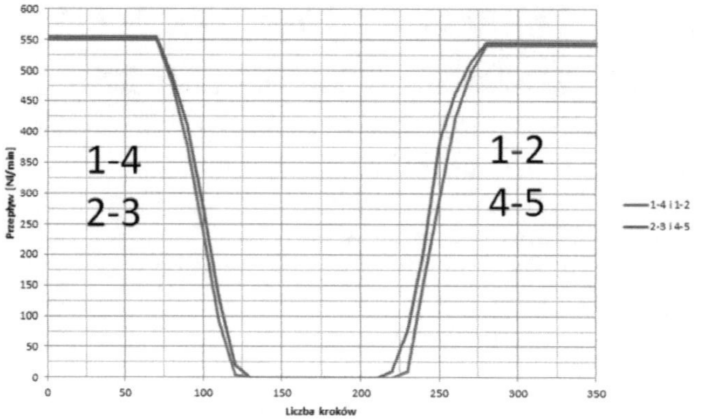

Figure 9. Flow characteristics of the designed proportional valve, blue color simulative, red color experimental.

By analyzing both the simulative and experimental results obtained, it can be seen that the valve has hysteresis in the middle and around the extreme positions of the slider. Whereas, at the opening and closing points of the control path, the curve may take shape similar to the linear characteristic, which will have a positive effect on the operation of the element. As you can see from the experimental study, the proportional valve does not work fully symmetrically. It has hysteresis in the middle part of it and around the extreme position of the slider, which has already been shown in the simulation studies. In contrast, in the opening and closing of the control path, the curve adopts a shape similar to the linear characteristic.

6. Determination of experimental pressure characteristics

The pressure characteristic determines the relationship between the pressure of the flowing fluid and the degree of opening of the valve. The test involves moving the valve slider from one extreme position to the other and continuously recording the results. The diagram of the position used to determine this characteristic is shown in Figure 10.

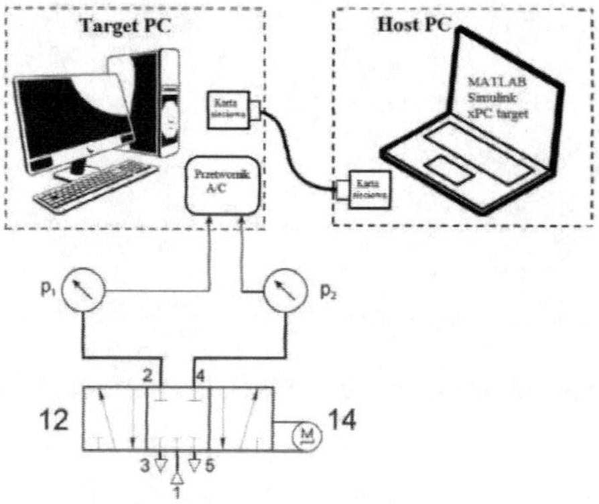

Figure 10. Diagram of the station for determining valve pressure characteristics.

Pressure gauges integrated with the measuring card are connected to the valve outlet. It transmits current measurement results to the real time machine *Speedgoat*, where they are displayed on the screen as diagrams. *Host PC* communicates with *Target PC* using *Matlab Simulink*, downloading data and displaying it as pressure characteristics. The obtained pressure characteristic of the proportional valve is shown in Figure 11.

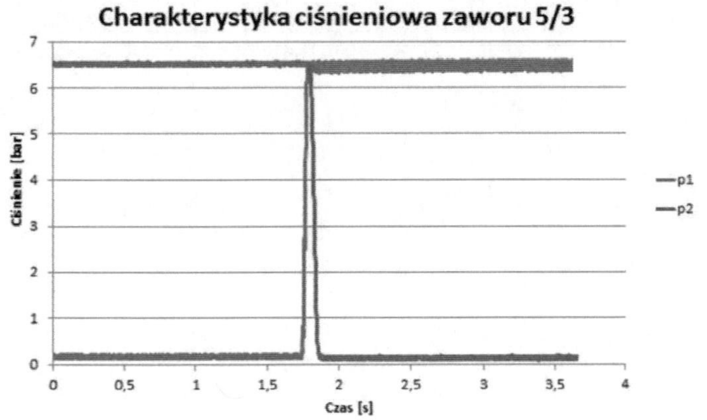

Figure 11. Pressure characteristic of proportional valve.

7. Conclusions

In this article one attempts to design and execute a pneumatic proportional flow valve type 5/3. Instead of the standard control of the electromagnet in this type of element, a linear stepper actuator is used to move the valve slider. The valve is controlled by a driver based on the AVR microcontroller. The virtual valve model was created in a 3D CAD environment. The actual valve model was based on a standard 5/3 type valve mechanically controlled by a lever that was removed from it and replaced with a stepped linear actuator.

The results of simulations and experiments for the valve are presented. Simulation testing was done using the SolidWorks add-on called Flow Simulation. During the experimental study flow and pressure characteristics of the valve were determined. By analyzing the results of the simulation and experimental studies, it can be concluded that the valve has a hysteresis in the middle and around the extreme positions of the slider. In contrast, in the opening and closing of the control path, the curve adopts a shape similar to the linear characteristic. As part of the further development of the valve, it will be necessary to modify the shape of the valve piston edge and thus the size of the flow windows. It is also opportune to introduce a closed-loop feedback loop control system by using a miniature distance sensor. It would determine the degree of ejection of the valve slider, the data received would be analyzed by the microcontroller in the controller.

The developed prototype of the 5/3 type proportional valve could be used in the control systems of slowly changing technological processes to become an alternative to the valves currently in use.

References

[1] Laski P A 2016 The design of a proportional slit valve with a piezoelectric actuator *Engineering Mechanics* ed. Zolotarev I and Radolf V (Acad Sci Czech Republic, Inst Thermomechanics, Czech Republic) pp 350–353

[2] Laski P A 2016 Proportional valve with a piezoelectric actuator *Proceedings of the International Conference Experimental Fluid Mechanics* (Techn. Univ. Liberec) pp 407–410

[3] Takosoglu J E 2016 Experimental research of flow servo-valve *Proceedings of the International Conference Experimental Fluid Mechanics* (Techn. Univ. Liberec) pp 819–823

[4] Laski Pawel A, Takosoglu Jakub E, Blasiak S 2015 Design of a 3-DOF tripod electro-pneumatic parallel manipulator *Robotics and Autonomous Systems* **72** pp 59–70

[5] Zwierzchowski J 2016 Indsustrial robot vison system for moving spherical elements *Engineering*

IOP Conf. Series: Materials Science and Engineering **233** (2017) 012029 doi:10.1088/1757-899X/233/1/012029

Mechanics ed. Zolotarev I and Radolf V (Acad Sci Czech Republic, Inst Thermomechanics, Czech Republic) pp 626–629
6. Blasiak S, Takosoglu J E, Laski P A 2014 Heat transfer and thermal deformations in non-contacting face seals *Journal of Thermal Science and Technology* **9** pp 1–8

Automatic device for indirect measurement of leakage flow rate in compressed air pipeline

R Dindorf[1], P Wos[1], K Pawelec[1]

[1]Faculty of Mechatronics and Machine Design, Kielce University of Technology, al. Tysiąclecia Państwa Polskiego 7, PL-25 314 Kielce

E-mail: dindorf@tu.kielce.pl

Abstract. The new measurement method of compressed air leakage flow rate in compressed air pipeline is proposed. In this method, the automatic measuring device is connected to a branch of the pipeline. The measuring device can be used to measure compressed air leakage in any place of compressed air pipeline: in main line, distribution line and connection line. The proposed measurement methods of compressed air leakage in pipeline are independent of receiver and compressor parameters, which is not the case with traditional method measuring leaks by emptying the receiver.

1. Introduction

The purpose of the compressed air piping system is to deliver compressed air to the points of usage. The compressed air needs to be delivered with enough volume, appropriate quality, and pressure to properly power the components that use the compressed air. Compressed air is costly to manufacture. A poorly designed compressed air system can increase energy costs, promote equipment failure, reduce production efficiencies, and increase maintenance requirements. Compressed air is one of the most expensive utilities in an industrial facility. The main causes of energy waste are leaks, pressure drops, over-pressurization, misuse of jets and poor compressor management [7], [8]. Also the leakage is usually the largest source of energy waste associated with compressed air usage. Table 1 provides an indication of the cost of leaks.

Table 1. Cost of compressed air leaks [1].

Equivalent hole diameter	Leakage flow air lost per leak (l/s)		Loss of energy (kWh)		Cost of air leak (€)	
mm	6 bar	12 bar	6 bar	12 bar	6 bar	12 bar
1	1.2	1.8	0.3	1.0	144	480
3	11.1	20.8	3.1	12.7	1,488	6,096
5	30.9	58.5	8.3	33.7	3,984	16,176
10	123.8	235.2	33.0	132.0	15,840	63,360

While designing energy saving compressed air systems various methods are applied to reduce energy losses and minimize energy consumption. Compressed air system maintenance is particularly important to avoid excessive loss through leaks:

- Leaks not only waste energy but also cause pressure drops that adversely affect the operation of air-using equipment and tools.
- Leaks are responsible for considerable waste, frequently up to 40 to 50 % of consumption.
- 15% leakage is considered to be an acceptable rate.

Leakage can occur at a number of points in a compressed air pneumatic pipeline system. Some of these are listed below: branch line connection, rubber hose, automatic drain trap, quick coupler, desiccant filter, isolating valve, filter/regulator/lubricator assembly, control valve, filter/regulator/coalescent filter assembly, coil hose, pressure regulator, pneumatic cylinder.

There are many various ways of determining or measuring the quantity of compressed air leaking out [4-6]. For compressors that have on/off controls or load/unload controls, there is an easy way to estimate the amount of leakage in the system. This method involves starting the compressor when there are no demands on the system (when all the air-operated end-use equipment is turned off). A number of measurements are taken to determine the average time it takes to load and unload the compressor. The compressor will load and unload because the air leaks will cause the compressor to cycle on and off as the pressure drops due to air escaping through the leaks. Leakage can be estimated in compressed air pipelines by emptying the compressed air receiver. This method requires an estimate of total system volume, including any downstream secondary air receivers, air mains and piping. The system is started and brought to the normal operating pressure (p_1). Measurements should then be taken of the time (t) it takes for the system to drop to a lower pressure (p_2), which should be a point equal to about one-half the operating pressure.

2. Determine the of leakage flow rate by indirect method

The new measurement method of compressed air leakage flow rate in compressed air pipeline based on the controlled flow in branch line is proposed. In this method, the automatic measuring device is connected to a branch of the pipeline. The measurement method consists in determining the relation between air leakage flow rate in pipeline and the controlled air flow rate through control valve. Compressed air leakage flow rate in pipeline is calculated on the basis of pressure ratio measurements in two time periods - during leakage with or without the controlled flow in branch line. The proposed measurement methods of compressed air leakage in pipeline are independent of receiver and compressor parameters, which is not the case with traditional method measuring leaks by emptying the receiver. Branch connections of measurement device in compressed air pipeline do not require pipeline disassembly or modification. The methods can be used to measure compressed air leakage in any place of compressed air pipeline: in main line, distribution line and connection line.

A new measurement method of compressed air leakage in pipelines based on the controlled flow consists in determining the relation between air leakage flow rate q_L in leak point of pipeline and the controlled air flow rate q_{vc} directly measured by flow meter in branch line [3]. Practical formula for calculating the total volume leakage rate in the compressed air system using the indirect method with controlled flow in branch line then has the form [4]:

$$q_L = K_K\, K_T\, q_{vc} \frac{\ln\left(\dfrac{p_{Lu}}{p_{Ld}}\right) t_{Lc}}{\ln\left(\dfrac{p_{Lcu}}{p_{Lcd}}\right) t_L - \ln\left(\dfrac{p_{Lu}}{p_{Ld}}\right) t_{Lc}} \tag{1}$$

where:

q_L – total leakage volume flow rate by the leak points in pipeline,
p_{Lu}, p_{Ld} – upstream and downstream absolute pressure during leakage without the controlled flow,
t_L – measuring time during air leakage without the controlled flow,
p_{Lcu}, p_{Lcd} – upstream and downstream absolute pressure during leakage with the controlled flow,

t_{Lc} – measuring time during air leakage with the controlled flow,
q_{vc} – volumetric air flow rate measured by flow meter,
K_K – calibration factor dependent on the measurement conditions,
K_T – correction factor dependent on the temperature measurement,

$$K_T = \frac{T_i}{T_N} \tag{2}$$

where:
T_i – temperature recorded during the measurement,
T_N – temperature at conditions of reference ANR.

With a new indirect method compressed air leakage is estimated by measurement of pressure ratio in a two time intervals (Figure 1):

1. For the air leakage without the controlled flow the pressure drop ratio dp_{Lu}/dp_{Ld} in time intervals t_L is measured.
2. For the air leakage with the controlled flow the pressure drop ratio dp_{Lcu}/dp_{Lcd} in time intervals t_{Lc} is measured.

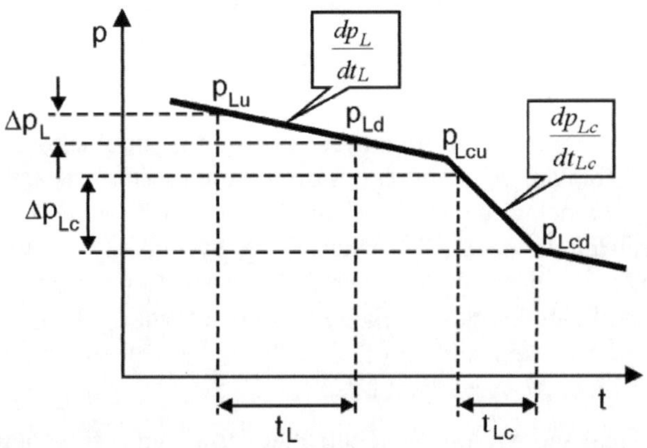

Figure 1. Pressure change in two time intervals in compressed air pipelines during the air leakage measurement based on the controlled flow in branch line.

The diagram of the air leakage measuring circuit in compressed air pipelines based on the controlled flow in branch line is shown in Figure 2. The measurement system MS is designed to work with a measuring device MD that reads the voltage input signals of the measuring transducer 1 (u_T, u_p) and the thermal flow meter 2 (u_q), and also generates a voltage output signal (u_k) to the proportional flow control 2/2 valve 3. The measuring system MS consists of a calculation block of leakage 4, a block forming the reference signal 5 a calculation block of flow rate 6, the feedback block 7, the controller block 8, and a block 9 that limits the set signal t_{Lc}. The way of measuring leakages that uses the measuring system MS includes: calibration, measurement without controlled flow and measurement of controlled flow. During the calibration of the measurement path of constant time of t_{Lc} of polytropic pressure drop through the control valve 3 is defined. On the basis of the measurement the pressure p_i in the pipeline, the critical pressure p_{cr} to the limit range of critical flow is determined. It is reasonable to agree the critical flow in the throttle valve 3, because the air flows into the atmosphere. By measuring the time constant τ, the pressure p_i in the pneumatic system and the critical pressure p_{cr} selected measurement time t_{Lc} controlled. Block 6 determines the flow rate q_{Lc} through the control valve 3 on the base of p_i measurement. It is possible to determine the flow rate q_{Lc} through the control valve 3 on the base of the measurement using the thermal flow meter 2.

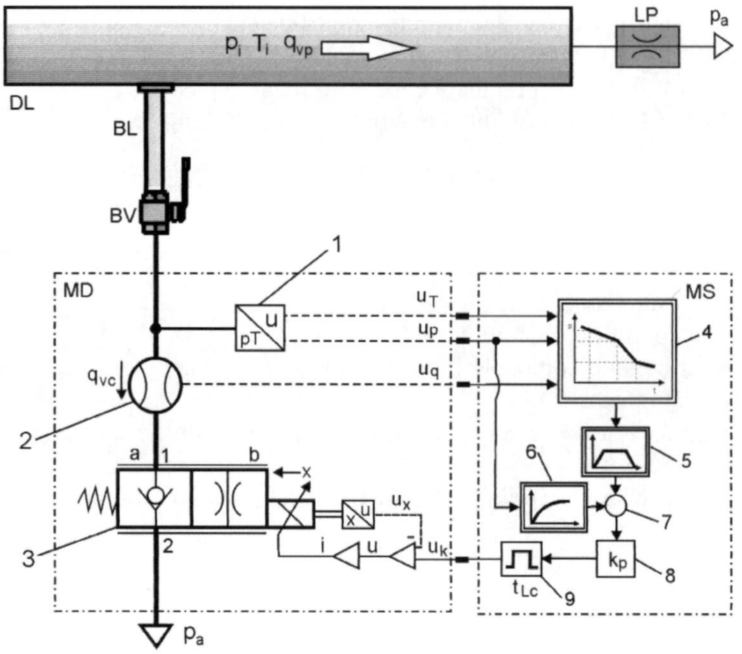

Figure 2. The schematic diagram of a measurement system connection to branch of compressed air pipeline: DL – distribution line, BL – branch line, BV – ball valve, LP – leak point, MD - measuring device, MS - measuring system, 1 - measuring transducer p/T, 2 - thermal flow meter, 3 - proportional flow control valve, 4 - calculation block of leakage, 5 - block forming the reference signal, 6 - calculation block of flow rate, 7 - feedback block, 8 - controller block, 9 - set signal block.

During measurements leakage without the controlled flow, when the control valve 3 is closed, the block 4 record the pressure p_{Lu} upper and lower pressure p_{Ld} during t_L, and on this basis the pressure drop ratio $\ln(p_{Lu}/p_{Ld})$ is determined caused by the leakage q_L of compressed air in the pipeline. Then during measurements leakage with the controlled flow, when the control valve 3 is opened, the block 4 record the pressure p_{Lcu} upper and lower pressure p_{Lcd} during t_{Lc}, and on this basis pressure drop ratio $\ln(p_{Lcu}/p_{Lcd})$ is determined due to leakage in the pipeline as well as the flow to the atmosphere. To carry out controlled flow, there was a block forming the reference signal 5, that generates a voltage of limited duration at the control valve input 3. To keep the measuring time a constant flow rate q_{lc} by the control valve 3, feedback block 7 determines the adjustment error, the controller 8 generates a voltage signal limited to the value of by the block 9. Time pulse voltage into the control valve 3 is determined in the range of t_{Lc} = 5 - 50 s.

3. Test of automatic measurement device

The view of portable compact measurement device type LT-I 200 for automatic measurement of leakage flow rate in compressed air pipeline is shown in Figure 3. This measuring procedure makes it possible to measure the determination of leakage flow rate in the test pipeline in a fully automatic way.
Standard features:

- automatic measurement of the leakage flow rate according to dedicated calculation procedure,
- measurement of the air pressure $p(t)$ to 0.8 MPa (8 bar),
- testing of pressure drops $\Delta p(t)$,
- measurement of volumetric flow rate of compressed air to 200 l/min.

IOP Conf. Series: Materials Science and Engineering **233** (2017) 012030 doi:10.1088/1757-899X/233/1/012030

Advanced features:

- compact device size 330x230x180 mm,
- integrated colour touch display WVGA (800x480), with a high-resolution resistive touch screen (16-bit colours),
- dual 10/100 Ethernet for Factory and World-wide Networking - supporting Modbus/TCP Master/Slave, FTP and HTTP,
- two RS232/RS-485 termination,
- USB 2.0 FLASH drive support - transfer files between the internal micro SD and the USB flash drive, USB Port for easy programming and application loading,
- 1MB memory, 32GB flash memory storage,
- power 230 V AC (24-30 VDC).

The device is configured with membrane keyboard and touch screen 1 on the front panel. The touch screen device has the basic functions for measuring the leakage in compressed air systems: F1 - measuring the flow rate, F2 - measuring the pressure, F3 - measuring the leakage in the pipeline. Pneumatic connection 2 allows direct measurement of leakage flow rate of compressed air in pipeline. Connection 3 is used for automatic measurement of the indirect method of leakage flow rate of compressed air in pipeline. Pneumatic control valve 5 is used to determine the controlled leakage flow rate q_{vc}. This measurement instrument after introducing the initial operating parameters such as: times t_L, t_{Lc}, the intensity of the controlled leakage flow rate q_{vc} performs automatic procedure for measuring and computing. As a result, we obtain automatic measurement the value of leakage flow rate q_L in compressed air pipeline.

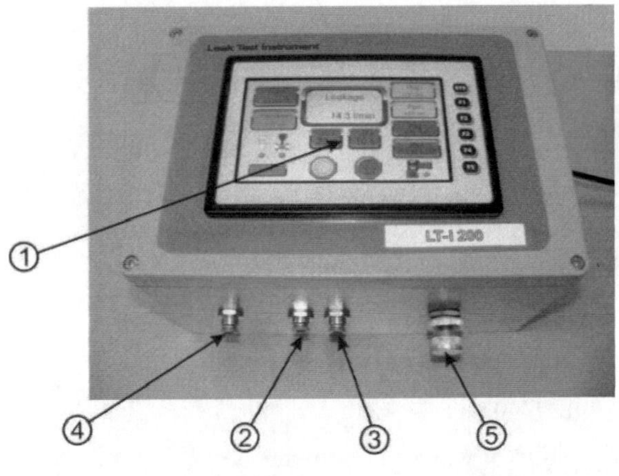

Figure 3. View of the device type LT-I 200 for automatic measurement of leakage flow rate in compressed air pipeline: 1 – touch panel, 2 - pneumatic connector for direct measurements of the air flow rate, 3 - pneumatic connector for indirect measurements of the leakage flow rate, 4 - output of pneumatic line, 5 - control valve.

After testing the portable measurement device type LT-I 200 two example of the measurement results of flow rate $q_{vc}(t)$ and pressure drop $p(t)$ on graph Figure 4 is show. During the measurements it was observed that after the operation of control valve, at the transition from the first to the second measuring interval, the pressure disturbances take place. Panel of the LT-I 200 device display measurement results of pressure p change in time intervals (according to Figure 1) and flow rate q_{vc}, and also calculation results of air leakage flow rate q_{vL} in compressed air pipeline. After completing the measurements parameters the leakage flow rate q_L was calculated automatically according to Equation (1).

IOP Conf. Series: Materials Science and Engineering **233** (2017) 012030 doi:10.1088/1757-899X/233/1/012030

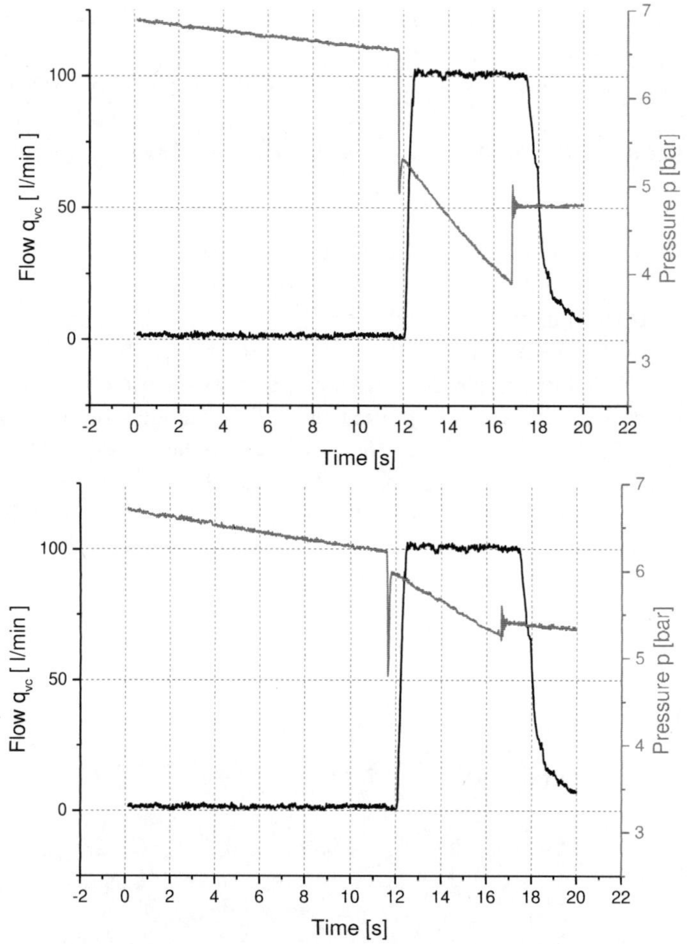

Figure 4. Graph of the measurement results of flow rate $q_{vc}(t)$ and pressure drop $p(t)$.

The results of the measurements of parameters $p(t)$, $q_{vc}(t)$ and calculations of leakage flow rate q_L for the analyzed example are summarized in Table 2.

Table 2. The results of the measurements and calculations.

Ex. 1.	Pressure p_{Lu}=6.8 [bar]	Time interval t_L=8[s]	Pressure p_{Lcu}=4.7 [bar]	Time interval t_{Lc}=2 [s]	Controlled flow rate q_{vc}= 100 [l/min]	Leakage flow rate q_L= 7.1 [l/min]
	Pressure p_{Ld}=6.6 [bar]		Pressure p_{Lcd}=4.2 [bar]			
Ex. 2.	Pressure p_{Lu}=6.7 [bar]	Time interval t_L=8[s]	Pressure p_{Lcu}=5.7 [bar]	Time interval t_{Lc}=2 [s]	Controlled flow rate q_{vc}=100 [l/min	Leakage flow rate q_L= 26 [l/min
	Pressure p_{Ld}=6.3 [bar]		Pressure p_{Lcd}=5.3 [bar]			

The demonstrative portable measurement device type LT-I 200 was honoured the medal for the innovative product on the 8th Exhibition of Pneumatics, Hydraulics, Drives and Controls PNEUMATICON'2015 in Kielce Trade Fairs (Figure 5).

Figure 5. The demonstrative portable measurement device type LT-I 200 in test pipeline.

4. Conclusions

The paper deals with the new automatic indirect measurement method of air leakage flow rate in compressed air system. Compressed air leakage flow rate in the pipeline is calculated automatically on the basis of pressure drop ratio measurements in two time periods - during leakage with and without the controlled flow. A portable measurement device for automatic measurement of leakage flow rate in compressed air pipeline is presented. With the use of this method, the leakages might be measured at any time and in any place of the pipeline: in the main pipeline, distribution line or connecting line the receivers (machines, equipment, tools).

References

[1] CS-Instruments News: *Energy analysis – flow measurement – leakage calculation.* 2016

[2] Dindorf R 2012 Estimating potential energy savings in compressed air systems *Procedia Engineering* **39** pp. 204-211

[3] Dindorf R, Wos P 2014 Indirect method of leakage flow rate measurement in compressed air pipelines *Applied Mechanics and Materials* **630** (57) pp. 57-62

[4] Dindorf R, Takosoglu J, Wos P 2017 *Development of pneumatic control systems* (Kielce University of Technology, Kielce)

[5] Dudić S P, Ignjatović I M, Šešlija D D, Blagojević V A, Stojiljković M M 2012 Leakage quantification of compressed air on pipes using thermovision *Thermal science* **16** pp. 621-631

[6] Liang H, Maolin C, Jiawei W 2010 Instantaneous leakage flow rate measurement of compressed air *Int. Conf. Mech. Autom. Control Eng. IEEE* pp. 2675–2679

[7] Marshall R 2011 Compressed Air System Leaks *Best Practices to Compressed Air. Compressed Air Challenge* (US DOE)

[8] Radgen P, Blaustein E 2001 Compressed Air Systems in the European Union LOG_X (Verlag GmbH, Stuttgart)

Acknowledgement

The demonstrative portable measurement device type LT-I 200 was realized within the project „The perspective Swietokrzyskie RSI - IV the stage", The task 5: „Creating the new forms of the cooperation of the science with the business", The Human Capital Operational Programme (HC OP) cofinanced from The European Social Fund (ESF).

IOP Conf. Series: Materials Science and Engineering **233** (2017) 012031 doi:10.1088/1757-899X/233/1/012031

Assessment of the state of the geometrical surface texture of seal rings by various measuring methods

S Adamczak[1], C Kundera[1], J Swiderski[1]

[1]Faculty of Mechatronics and Machinery Design, Kielce University of Technology, Aleja Tysiąclecia Państwa Polskiego 7, PL- 25314 Kielce, Poland

E-mail: kundera@tu.kielce.pl

Abstract. The present paper concerns the metrological measurements of the geometric structure of the faces of seal rings made of silicon carbide and carbon-graphite. Three different instruments, i.e. the stylus profilometer, the optical profilometer, and the atomic force microscope, were used to measure the geometric structure of surfaces. In the comparative analysis, an identical area of the ring surface which was mapped by three measuring instruments with different sampling densities resulting from their metrological characteristics was assumed. The measurements made show that for the silicon carbide ring, the surface texture measurements on an atomic force microscope and optical instruments more accurately represent the actual topography than the measurement determined by the stylus profilometer.

1. Introduction

The state of topography of surfaces many sliding connections of the mechanical parts (performing relative motion), determines their tribological and exploitation properties. In the case of mechanisms such as contact seals, sliding bearings or friction clutches, materials properties, technology and surface topography of the contacting surfaces are very important for their durability and reliability. The geometric texture of the surface determines the actual contact surface of co-operating elements and actual contact pressures [1,2]. In the case of face mechanical seals, the condition and microstructure of the surface significantly affect the tightness of the face contact of seal rings and the frictional resistance.

In the literature on face seals, for example [3-5], the theoretical models of the microroughness contact of the surface of seal rings and the estimates of the height of the gap and the flow rate (leakage) on their basis. Properties of materials and surface topography parameters of the co-operating seal rings are required as input parameters in these models. Hence, the reproduction accuracy of real conditions of the contact of seal rings depends on the physical assumptions adopted and the accuracy of input parameters, including surface roughness parameters.

In recent years, modern measuring techniques and specialized computer software have been developed to measure and analyze surface topography. Contact profilometry with a mapping blade, phase shifting interferometry, confocal microscopy, coherent scanning interferometry, focus variation microscopy, scanning tunneling microscopy, and atomic force microscopy, among others, belong to the methods of spatial topography [6]. The choice of the appropriate method for a given measurement task depends on the properties of a measured sample such as the hardness determining the ability to use contact methods or the degree of light absorption by the material of the measured element

IOP Conf. Series: Materials Science and Engineering **233** (2017) 012031 doi:10.1088/1757-899X/233/1/012031

determining the possibility of using optical methods. The state of surface topography characterized by the amplitude of roughness, the distance between the roughnesses and the inclination angle of roughness slopes are another important element that depends on the selection of the method [7].
The proper selection of the method of measuring very smooth surfaces that guarantees reliable results should consider an analysis of potential major sources of errors for each method affecting the measurement results [8].

This paper describes the results of roughness measurements of the faces of seal rings which were obtained from the measurements with three instruments: a stylus profilometer, an optical profilometer, and an atomic force microscope.

In the comparative analysis an identical surface area mapped by three used instruments was assumed, but with different densities of sampling resulting from their metrological characteristics.

In order to compare the measuring instruments used, five amplitude parameters and five surface parameters of the topographies of the faces of two seal rings were adopted.

2. Research object

The object of the study was the topography of faces of seal rings in the mechanical face seal. The figure shows a diagram of the mechanical face seal showing the gap formed by the actual (roughened) sealing surfaces of the two rings.

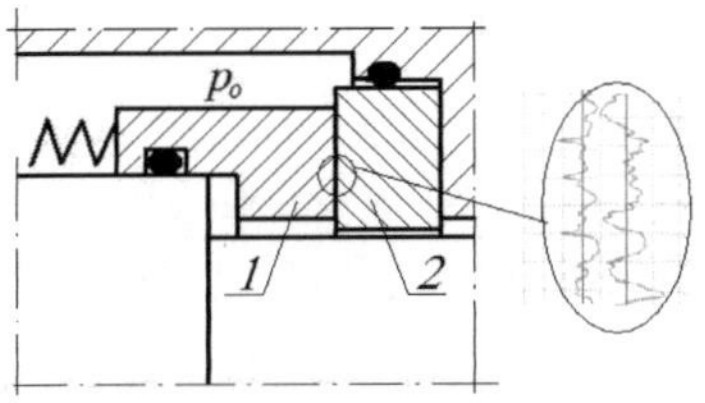

Figure 1. Mechanical face seal scheme: 1- rotating ring seal, 2- stationary ring seal

The stationary ring (2) (Figure 1) made of silicon carbide and the rotating ring (1) made of carbon-graphite and flexibly attached were subjected to metrological studies. The material kit of seal rings is a typical set used in modern face seal designs by many manufacturers. It is well known from the use of face seals that carbon-graphite is characterized by good sliding properties, whereas silicon carbide is a hard material resistant to wear and temperature. The hardness of the carbon-graphite ring was about 110 HR and that of the ring made of silicon carbide over 3000 HV. The surfaces of both rings (in the technological process) were grinded and then lapped to achieve values of surface roughness Ra between 0.03 and 0.06 μm. The finishing process (preparing faces) of the seal rings provides an isotropic roughness of the surface.

3. Methods of measurement

The topography of the faces of seal rings was measured by the contact method, the non-contact method, and on an atomic force microscope.

A Taylor Hobson Form Talysurf PGI 1230 profilometer was used to make measurements by the contact method. This instrument has an interferometer head with a measuring range of 12.5mm and a resolution of 0.8nm, which makes it possible to perform stereometric measurements of the surface while taking the contour, waveform and roughness of the surface into account. The instrument is equipped with a table to move the measured object in the Y axis with a maximum resolution of 1 μm. In the measurements presented in this paper, a conical measuring tip with the rounding radius of 2 μm and a vertex angle of 60° was used. Before the measurements 1mN pressure force, the traversing speed of the measuring tip v_{os}=0.5 mm/s was set on the device. Topographic measurements were made on surfaces of 1mm x 1mm, making 401 linear runs distant 1 μm apart.

Measurements by the non-contact method were performed on a Talysurf CCI instrument using coherent correlation interferometry which combines vertical scanning with optical interferometry. Thanks to this system, it is possible to make an analysis of the geometric surface with the vertical resolution up to 10pm and the constant in the measuring range of 2.2 mm. A Talysurf CCI has the ability to measure any surface of the reflectance between 0.3% and 100%. The measurement time is from tens of seconds to a few minutes and is dependent on the height of the roughness of the element being measured. The same measurement made on the contact profilometer can take from several dozen minutes to several hours. The x50 lens was used to scan the test surface of 0.33x0.33 mm. Measurements were recorded in the matrix of 1024 x 1024 points.

The third measuring instrument used to measure the surface topography of the examined rings was an AFM Dimension Icon atomic force microscope. The atomic force microscope illuminates the surface of the workpiece with a cantilever whose tip radius is a dozen or so nanometers. The force between the tip (needle) of the probe and the surface of the sample to be measured causes the lever to deflect. The detector measures the inclination of the lever as it moves over the surface of the element. The Dimension Icon atomic force microscope can work in three modes: contact, non-contact and intermittent contact. The measurements were made using an AFM instrument operating in the contact mode with a SCANASYST-AIR measuring probe with a 2nm tip radius. A maximum area of 90μm x 90 μm was used to measure the geometric structure of the surface. Measurement data were recorded in the matrix of 512 x 512 points.

4. Results of measurements and their analysis

Surface roughness measurements were carried out with the three above-described instruments on the faces of both seal rings in the examined face seal. The measuring areas for the measuring instruments used were different, i.e. for the Form Talysurf PGI stylus profilometer – 1mm x 1mm; for the Talysurf CCI optical profilometer – 0.33mm x 0.33mm, and for the AFM Dimension Icon atomic force microscope – 90 μm x 90 μm. Horizontal sampling densities (in x, y plane) used for each instrument are shown in the table below.

Table 1. Sampling densities in the x any y direction for the measuring instruments.

Instrument	Δx (μm)	Δy (μm)
Dimension Icon atomic force microscope	0.18	0.18
Talysurf CCI optical profilometer	0.33	0.33
Form Talysurf PGI 1230 stylus profilometer	0.125	1.00

Measurement data were recorded and then analyzed in a TalyMap Platinium software. An analysis of the measurement results was made for a uniform area of 90μm x 90μm. The surface area assumed for the analysis was determined by the technical feasibility of the measurement on the atomic force microscope. In order to compare and evaluate the values of the parameters obtained in the measurements on three of the above-mentioned instruments, the following geometrical surface texture (GST) parameters were chosen:

- amplitude parameters of the surface:
  – arithmetic mean height of the surface *Sa*,
  – root mean square height of the surface Sq,
  – maximum peak height of the surface Sp,
  –maximum pit height of the surface Sv,
  – maximum height of the surface Sz

and parameters of the areal material ratio curve - Abbott–Firestone curve:
  – core height Sk,
  – reduced peak height Spk,
  – reduced dale height Svk,
  – areal material ratio of the peak height Smr1,
  – areal material ratio of the dale height Smr2.

In addition, there are isometric images of the surface topography, material ratio curves and selected surface roughness profiles of the face of the stationary and rotary rings. As previously mentioned, an equal area of the surface was assumed to analyze the measurement results and the evaluation of the three measuring instruments. Figure 2 shows the isometric images of the face of the stationary ring (made of carbide) obtained from measurements made with the three instruments.

(a)

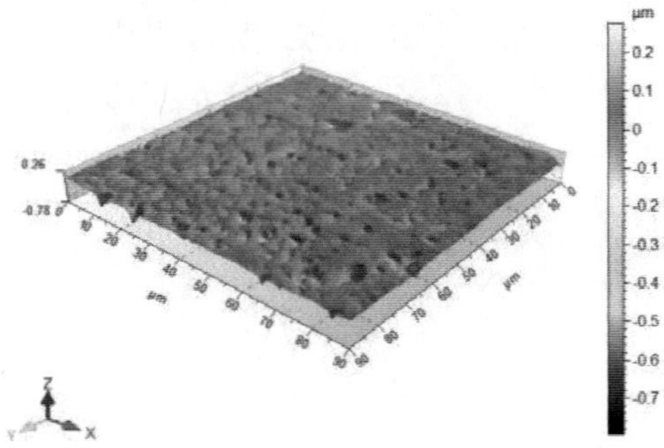

(b)

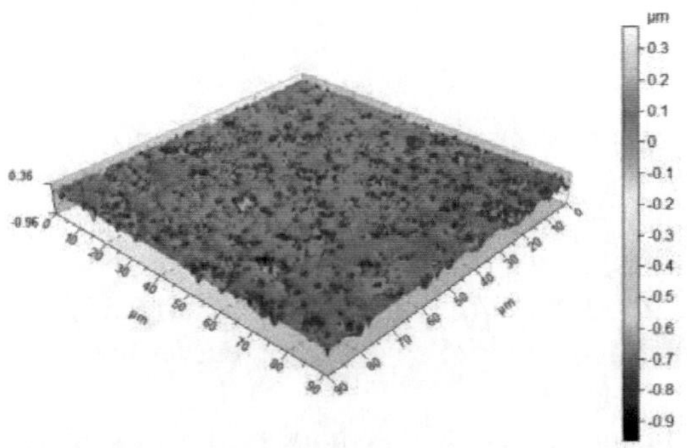

(c)

Figure 2. Isometric images of the topograhy face of the stationary ring obtained from measurements made with: (a) AFM; (b) PGI; (c) CCI

Exemplary roughness profiles isolated from the measuring surface of the stationary ring and material curves (Abbott-Firestone curve) are shown in Figure 3.

The isometric images of the surface topography (Figure 2), material contribution curves, and negative values of the asymmetry coefficient Ssk (skewness) indicate that the surface of the ring after the reach process has surface characteristics of the plateau type.

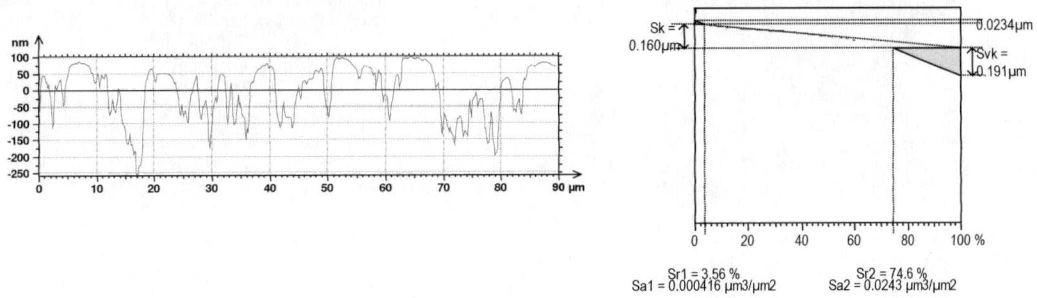

(a)

IOP Conf. Series: Materials Science and Engineering **233** (2017) 012031 doi:10.1088/1757-899X/233/1/012031

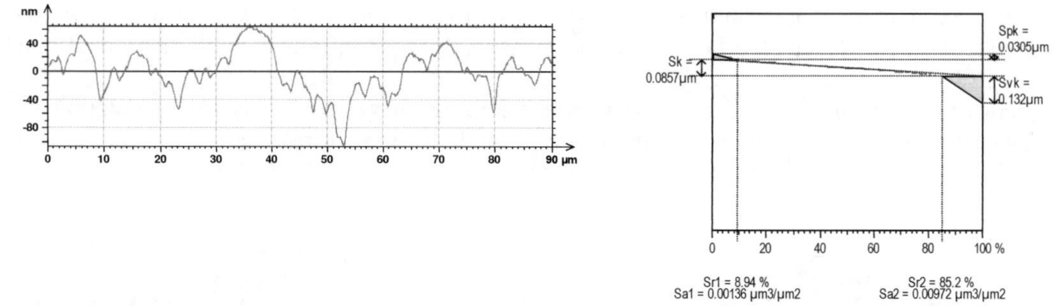

(b)

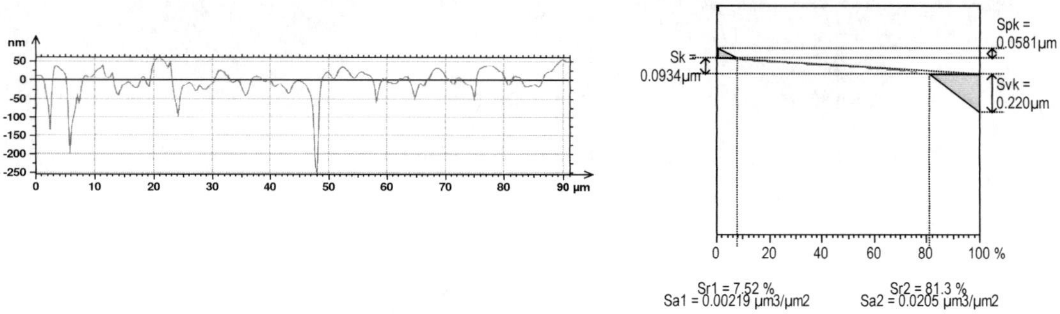

(c)

Figure 3. The extracted GST profile and parameters of material ratio curve determined: (a) AFM, (b) PGI, (c) CCI.

Similar measurements of GST were made for the carbon-graphite ring. Images of surface topography, profiles extracted from the analyzed surface and material contribution curves are presented in subsequent Figures 4,5.

A summary of the amplitude values of the GST parameters obtained from measurements made with three instruments: AFM, PGI, CCI, is presented in the diagrams: for the stationary ring (Figure 6) and for the rotary ring (Figure 7).

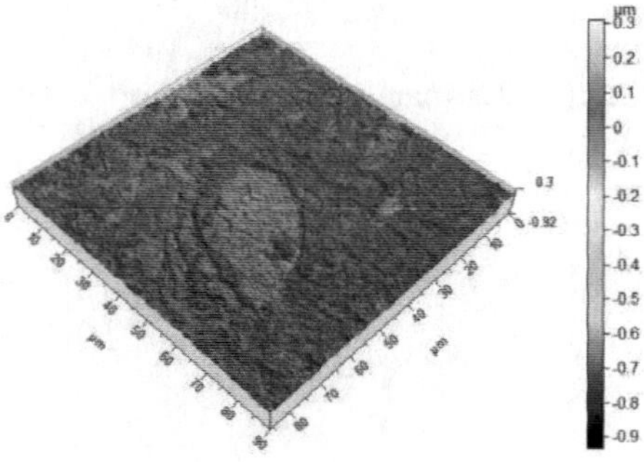

(a)

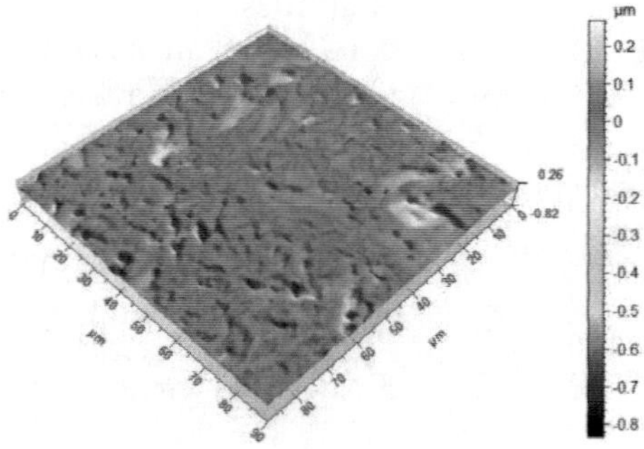

(b)

IOP Conf. Series: Materials Science and Engineering **233** (2017) 012031 doi:10.1088/1757-899X/233/1/012031

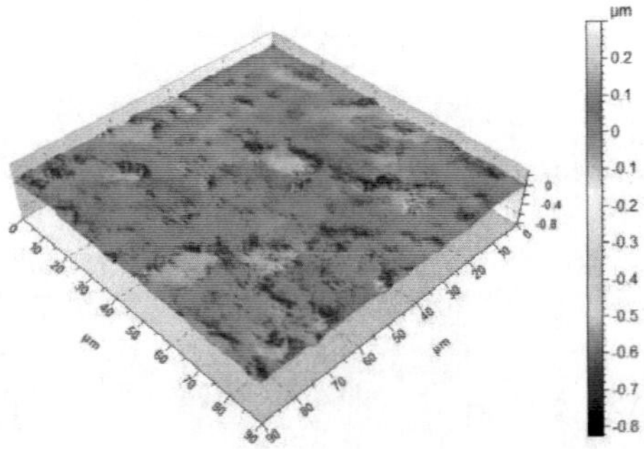

(c)

Figure 4. Isometric images topography of the face of the rotary ring obtained from the measurement made with the instruments: (a) AFM; (b) PGI; (c) CCI.

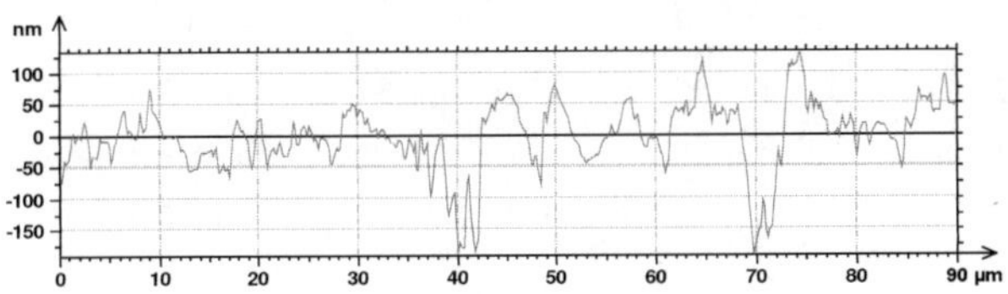

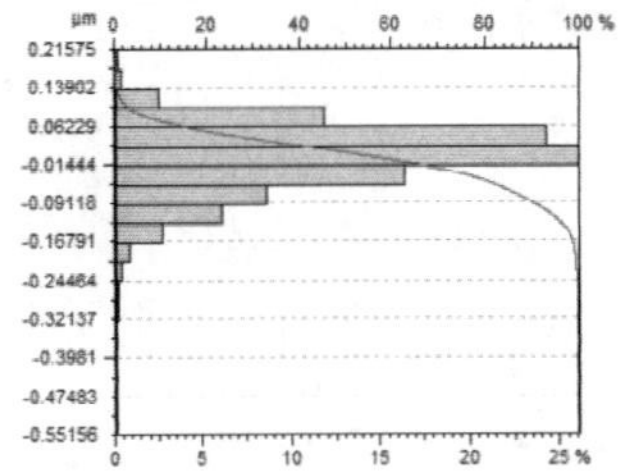

(a)

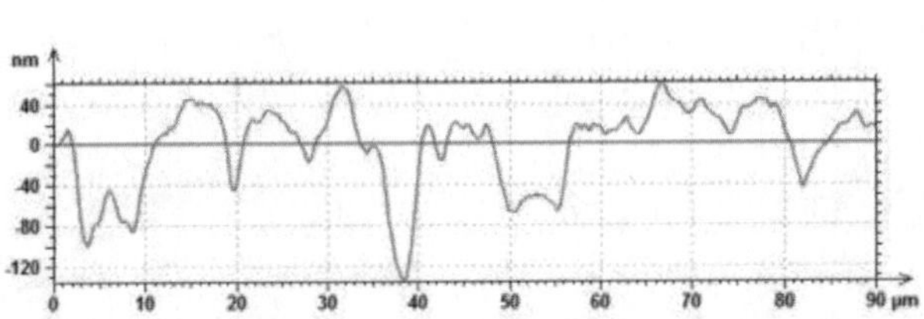

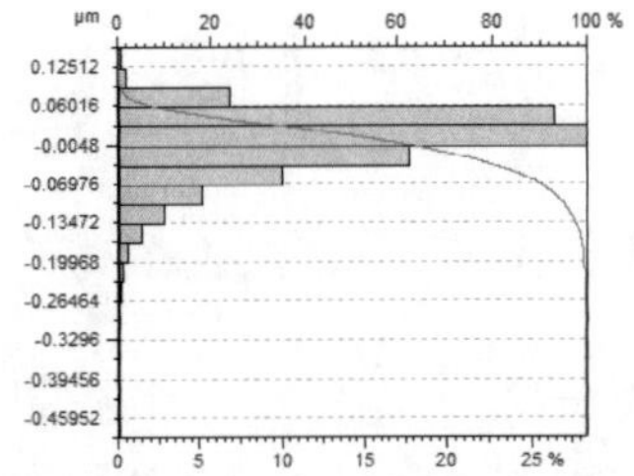

(b)

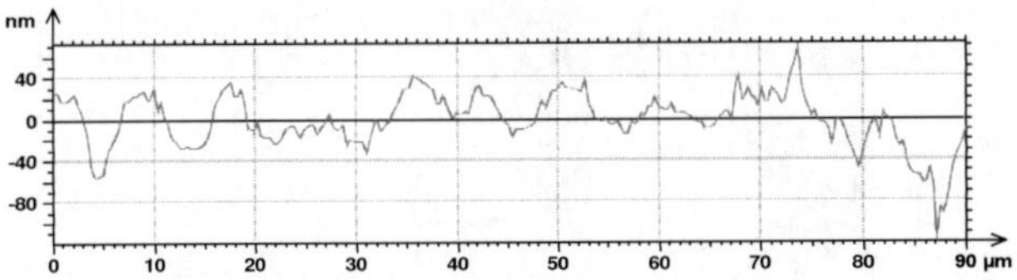

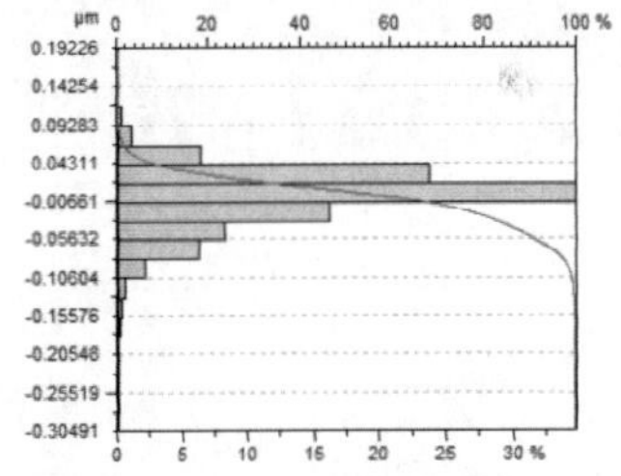

(c)

Figure 5. The extracted profile and the distribution of heights determined: (a) AFM, (b) PGI, (c) CCI.

IOP Conf. Series: Materials Science and Engineering **233** (2017) 012031 doi:10.1088/1757-899X/233/1/012031

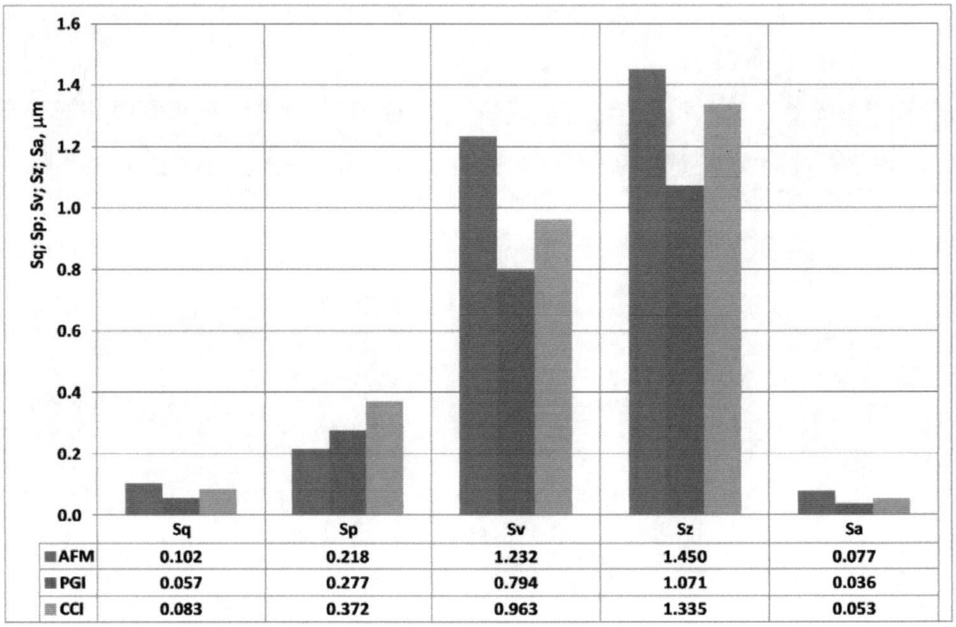

Figure 6. Summary of surface amplitude parameters GST for the stationary ring.

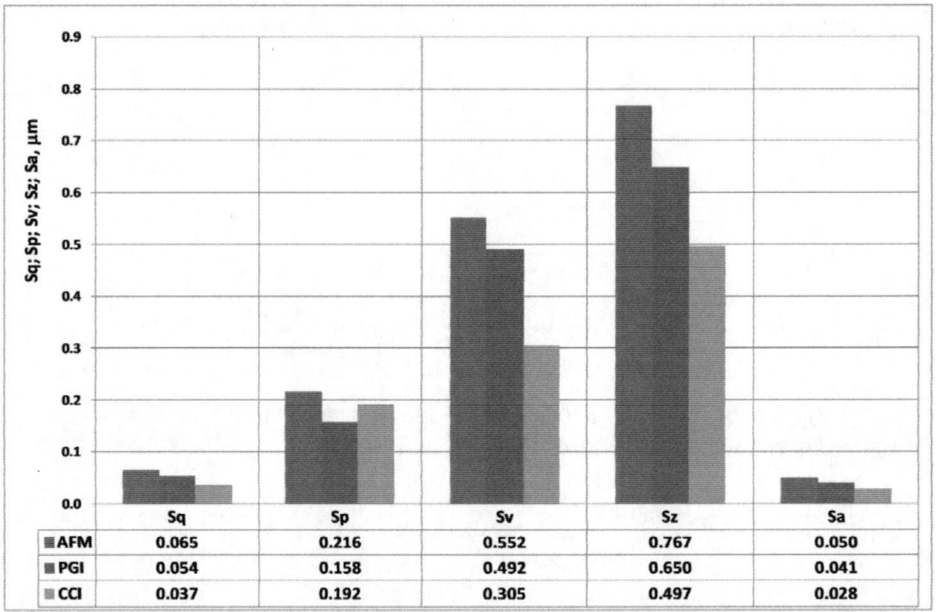

Figure 7. Summary of surface amplitude parameters SCP for the rotating ring.

In the presented charts (Figures 6, 7), it can be seen that for measurements performed on an AFM, the highest values of all surface amplitude parameters SGP were obtained. In the case of a rotary ring (Figure 7), values for amplitude parameters, i.e. Sa, Sq, Sz, Sv, increase for each instrument in the following order: AFM> PCI> CCI.

For the stationary ring (Figure 6), the values of the parameters Sa, Sq, Sz, Sv change for each instrument in the following order: AFM> CCI> PGI. The values of Sa, Sq, Sz parameters obtained for the atomic force microscope are nearly twice as high as those for the stylus profilometer.

The measurements made show that for the silicon carbide ring, the SGP measurements made with AFM and CCI instruments more accurately represent the actual topography than the measurement made with the PGI stylus profilometer, which is due to the filtering effect of the roughness by the mapping blade of the profilometer. The microstructure of the superfinished surface of this ring is more homogeneous and has less characteristic pores (Figure 2) than the superfinished surface of the carbon-graphite ring (Figure 4).

5. Conclusions

The following conclusions can be drawn from the analysis of the measurement results obtained from the three different measuring instruments:

1) the isometric images of the surface topography and the material contribution curves indicate that the surfaces of both rings after the reach process have surface characteristics of the plateau type;

2) on the basis of the graphs of structure directionality, it should be noted that the faces of the seal rings have an isotropic, random microroughness structure of 70 -80%;

3) the atomic force microscope, due to its metrological characteristics of horizontal sampling density, vertical resolution and the geometry of the mapping probe most accurately reproduces the actual stereometry of the very smooth elements, i.e. faces of the seal rings, which results in the highest values of amplitude parameters derived from this method;

4) differences in the values of geometric surface structure parameters obtained using the three different methods may also be due to the lack of homogeneity of surface topography in individual areas of the seal rings.

6. References

[1] Jackson R L and Green I 2011 On the modeling of elastic contact between rough surfaces *Tribology Transactions*, 54 pp 300–314

[2] Beheshti A and Khonsari M M 2012 Asperity micro-contact models as applied to the deformation of rough line contact *Tribology International* 52 pp 61–74

[3] Green I 2002 A transient dynamic analysis of mechanical seals including asperity contact and face deformation *Tribology Transactions* 45(3) pp 284–293

[4] Kundera C 2013 *Seals of rotating systems* Kielce University of Technology monograph M53, pp 286 (in Polish)

[5] Gacek J, Norymberczyk L and Płonka S 2015 Determination of a face seal's operational parameters on test bench *Measurement Automation Monitoring*, 61 12 pp 555–560

[6] Bhushan B 2001 *Modern Tribology Handbook*, chapter 2 pp 1–70

[7] Poon C Y and Bhushan B 1995 Comparison of surface roughness measurements by stylus profiler, AFM and non-contact optical profiler *Wear* 190 pp 76–88

[7] Miller T, Adamczak S, Świderski J, Wieczorowski M, Łętocha A and Gapiński B. 2017 Influence of temperature gradient on surface texture measurements with the use of profilometry *Bulletin of the Polish Academy of Sciences Technical Sciences* 65 (1)

[8] ISO 25178-2:2012 (E) Geometrical product specifications (GPS) – Surface texture: Areal – Part 2: terms, definitions and surface texture parameters.

IOP Conf. Series: Materials Science and Engineering **233** (2017) 012032 doi:10.1088/1757-899X/233/1/012032

Numerical analysis of the non-contacting gas face seals

S Blasiak

Faculty of Mechatronics and Mechanical Engineering, Kielce University of Technology, Aleja Tysiąclecia Panstwa Polskiego 7, PL- 25-314 Kielce, Poland

E-mail: sblasiak@tu.kielce.pl

Abstract. The non-contacting gas face seals are used in high-performance devices where the main requirements are safety and reliability. Compliance with these requirements is made possible by careful research and analysis of physical processes related to, inter alia, fluid flow through the radial gap and ring oscillations susceptible to being housed in the enclosure under the influence of rotor kinematic forces. Elaborating and developing mathematical models describing these phenomena allows for more and more accurate analysis results. The paper presents results of studies on stationary ring oscillations made of different types of materials. The presented results of the research allow to determine which of the materials used causes the greatest amplitude of the vibration of the system fluid film-working rings.

1. Introduction

The stricter standards for emissions to the environment impose on the constructors of sealing nodes extremely strict requirements for the design of non-contacting face seals. It should be emphasized that these seals are used practically in all branches of industry and work in diametrically different operating and environmental conditions. The non-contacting face seals have been widely used in rotary machines for many years, including compressors and pumps, mixers, etc., where they play an extremely important role in the separation of sealed media from the environment. Main task for the seals is to maintain maximum tightness, but maintaining this task proves extremely difficult, due to a number of external and internal factors that affect the behavior of the adopted in construction height of the radial gap.

In case of the non-contacting face seals lubricated with gas, the methods enabling to maintain a stable layer of medium separating working rings is the use of various types of geometric modifications (microstructures) on the race of the working rings. Introduction of modifications in the form of conical, radial or helical channels and texturing of the face or impulse seals results in an increase in the force generated in the fluid film. The increase in ring separating force allows to maintain a stable the fluid film and eliminates the contact of the sealing rings during operation. Another important aspect of the research on non-contact gas face seals is the study of vibration analysis of the sealing rings. Depending on the construction, the con-contacting face seals can be divided into two groups: FMS - *Flexibly-Mounted Stator* and FMR - *Flexibly-Mounted Rotor*. In both of these design variants, kinematic constraints are derived from the anti-ring rigidly attached to the shaft of the rotor machine.

The analyses literature presents the results of numerical calculations of a complex mathematical model including: radial gap height function, nonlinear Reynolds equation for compressible medium, and vibration dynamic equation for ring vibration in the housing. To correct the system of equations describing stationary ring oscillations, it is necessary to know the pressure distribution in fluid film. Practically in all scientific publications, the two-dimensional Reynolds equation is solved using numerical methods, usually Finite Volume Method (FVM) or Finite Element Method (FEM) [1], rarely, authors use a finite difference method [2] or meshfree methods [3].

While reviewing the literature of recent years, it can be concluded that practically all types of applied race modifications have a positive influence on the non-contacting gas face seals [4–9]. This has influence on the reduced wear on the work surface (sliding) of the rings while minimizing leakage, friction temperature and thermal distortions [10–15]. Achieving these results is possible in the case of a proper design of a sealing node containing non-contacting gas face seals. However, this requires considerable financial resources to carry out a large number of experiments. An alternative way to achieve the desired effect is to develop a computational aided design apparatus for this type of non-contacting seals with respect to operating requirements.

This paper presents the results of ring vibration analyzes, being housed in a housing for a non-contacting face seal based on the author's computer program. Simulation studies were performed for a ring with conicity for four types of materials, i.e. carbon-graphite, carborundum, stellite, tungsten carbide. The subject matter was aimed at determining the influence of the physico-chemical properties of the material used on the ring [16–18], which was flexibly attached to the housing for correct operation of gas non-contacting face seal.

2. Gas face seal physical model

The design assumption concerning the non-contacting gas seals permits the operation of such seals with minimal leakage.

This means that between the front surfaces of the cooperating rings, there is a small radial piston of with height of up to several micrometers, choking leakage. Based on the work [4] and [19, 20], Figure 1 presented a scheme of the non-contacting FMS gas face seal.

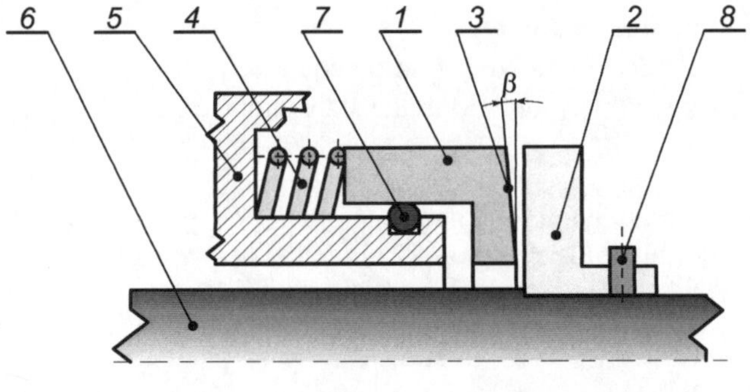

Figure 1. Schematic cross section drawing of a non-contacting gas face seal; 1 – stator, 2 – rotor, 3 – taper angle, 4 – spring, 5 – casing, 6 – shaft, 7 – O-ring, 8 – steady pin.

The seal consists of two rings, of which one stator (1) is mounted flexibly in the housing, and the other rotor (2) rotates with the shaft (6) of the rotor machine. The radial gap, which occurs between the cooperating faces, is filled with a sealed medium and forms a gas cushion. Fluid film has certain elastic-damping properties and induces the transfer of vibrations from the rotor to the stator.

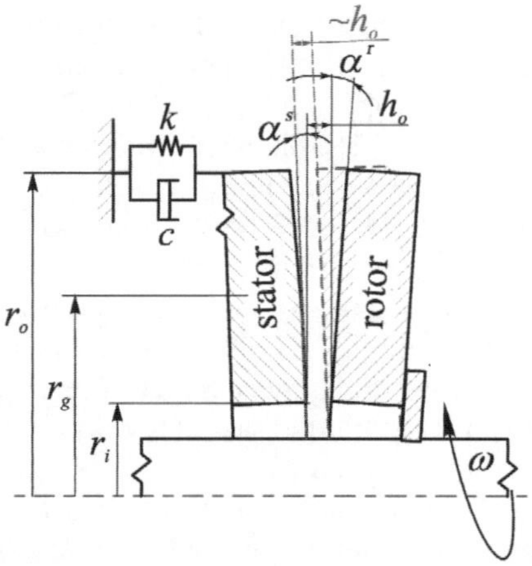

Figure 2. Schematic cross section of a non-contacting mechanical face seal.

Typically, execution errors, mounting errors or shaft defects during operation cause an unbalanced alignment of the sealing rings (Figure 2) and generation of additional axial and angular inducing oscillations from the rotor. With high rigidity of the layer separating the rings or under stable operating conditions, the sealing faces may be parallel to each other, meaning that they move synchronously and the gap is practically constant near the nominal value h_o.

3. Gas face seal mathematical model

The mathematical model of a non-contacting gas face seal is a discrete-continuous system of differential equations, i.e.:

- equations describing fluid motion in a film,
- equation of stream continuity (mass conservation equation)
- ring vibration equations flexibly mounted in the stator housing.

3.1 Lubrication analysis

In the works concerning the non-contacting gas face seals, the pressure distribution in the gap is determined on the basis of the Reynolds equation. [21, 22, 19] described by the dependency:

$$\vec{\nabla}\left[ph^3\vec{\nabla}p - 6\mu\omega r\, ph\vec{e}_\theta\right] = 12\mu\frac{\partial(ph)}{\partial t} \tag{1}$$

where $p = p(r,\theta)$ – pressure, $h = h(r,\theta)$ – film thickness function, μ – gas viscosity, ω – shaft angular velocity.

Along with the boundary conditions and periodicity condition:

$$p(r,\theta)\big|_{r=r_i} = p_i;\; p(r,\theta)\big|_{r=r_o} = p_o \tag{2}$$

$$p(r,\theta)\big|_{\theta=0} = p(r,\theta)\big|_{\theta=2\pi} \tag{3}$$

As mentioned earlier in the operation of the device, both the stator and the rotor perform axial and angular vibrations, which has influence on the function describing the height of radial gap:

$$\begin{aligned} h(r,\theta) &= h^r - h^s = h_o + \left(z^r - z^s\right) + \\ &+ r\left(\left(\alpha_x^{\,r} - \alpha_x^{\,s}\right)\sin(\theta) - \left(\alpha_y^{\,r} - \alpha_y^{\,s}\right)\cos(\theta)\right) + \left(r - r_i\right)\beta \end{aligned} \tag{4}$$

where z^s α_x^s, α_y^s – stator axial and angular degrees of freedom, z^r α_x^r, α_y^r rotor axial and angular degrees of freedom.

In the model in question, it was assumed that the rotating ring moving along the machine shaft performs harmonic angular vibrations α_x^r, α_y^r – in the form of:

$$\alpha_x^r = \alpha^r \cos(\omega t),\ \alpha_y^r = \alpha^r \sin(\omega t) \tag{5}$$

and their derivatives over time:

$$\dot{\alpha}_x^r = -\omega \alpha^r \sin(\omega t),\ \dot{\alpha}_y^r = \omega \alpha^r \cos(\omega t) \tag{6}$$

On the other hand, the parameters α_x^s, α_y^s are the answer in the form of angular vibrations of the stator. The mutual misalignment of a pair of rings can be represented graphically as on Figure 2.

$$\alpha_x^r(0) = \alpha^r,\ \alpha_y^r(0) = 0,\ \dot{\alpha}_x^r(0) = 0,\ \dot{\alpha}_y^r(0) = \omega \alpha^r \tag{7}$$

From dependencies (5) and (6) result also the initial conditions for angular and axial displacements at $t = 0$ (eq. (7).

3.2 Kinematic of face seal

The ring, which is flexibly attached to the housing, performs a complex movement that can be described by the system of differential equations of ordinary form:

$$\begin{aligned} m\ddot{z}^s + c\dot{z}^s + k z^s + F_{zeq} &= F_z \\ J\ddot{\alpha}_x^s + c_k \dot{\alpha}_x^s + k_k \alpha_x^s - M_{xi} - M_{xeq} &= M_x \\ J\ddot{\alpha}_y^s + c_k \dot{\alpha}_y^s + k_k \alpha_y^s - M_{yeq} &= M_y \end{aligned} \tag{8}$$

where c – axial damping coefficient, c_k – angular damping coefficient, F – force, J – transverse moment of inertia, k – axial stiffness coefficient, k_k – angular stiffness coefficient, m – stator mass, M – moment, M_{xi} – moment due to stator initial misalignment, M_{xeq} – equilibrium moment.

The equations of the movement flexibly attached in the stator housing are stored in an inertial system *OXYZ* (Figure 3) similarly as in the paper [1].

The constants c, k mean the damping factor and the axial elasticity, respectively c_k, k_k – coefficient of damping and angle elasticity. The coefficients depend on the properties of the elastic-damping elements of the stator mounting (Figure 4) and are linked together according to the following relations [19]:

$$c_k = 0.5 c r_{s1}^2,\ k_k = 0.5 k r_{s2}^2 \tag{9}$$

r_{s1} and r_{s2} respectively the radius on which the spring is mounted and the secondary seal – o-ring. $r_{s1} = (r_o + r_b)/2$, $r_{s2} = r_b$. Where r_o - stator outer radius, r_b - stator balance radius.

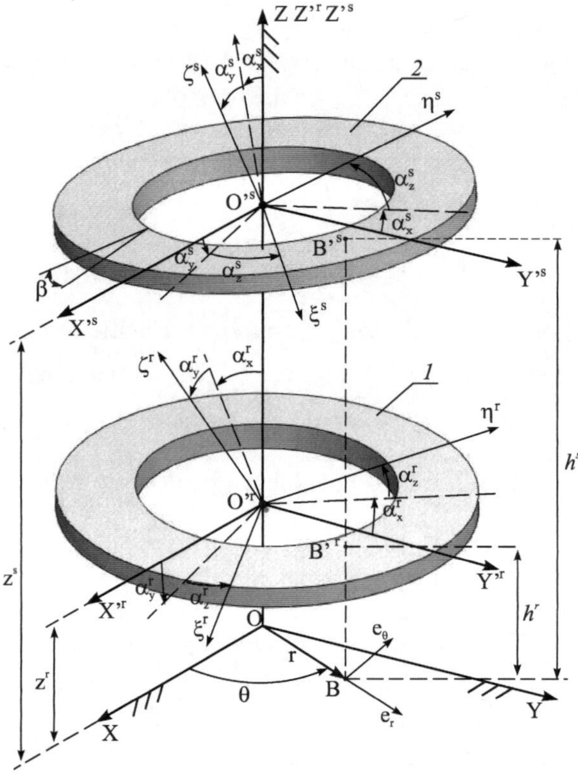

Figure 3. Kinematic model of the non-contacting face seal: 1 – rotor, 2 – stator.

In the system of equations (8), it was assumed that: m – ring weight, was determined according to dependency [20]:

$$m = \left(1/3 + 2B/3\right)\pi\rho l\left(r_o^2 - r_i^2\right) \tag{10}$$

The transverse moment of inertia is described by the equation: $J_x = J_y = J = \int r^2 dm$.

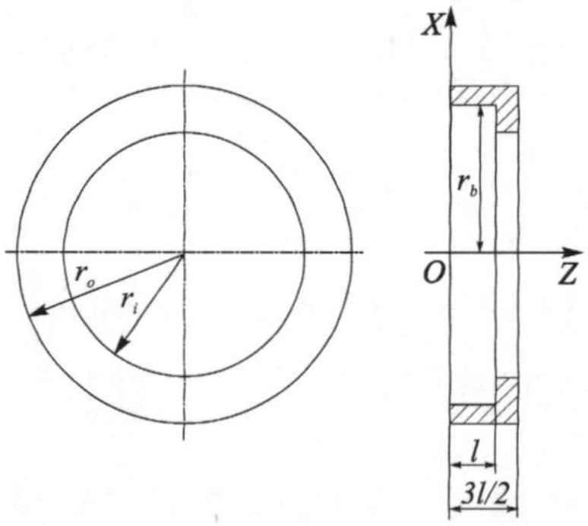

Figure 4. Geometry of the stator.

The values of force $-F_z$ and moments – M_x and M_y, occurring on the right side of the system of equations (8) depend directly on the pressure distribution in the gap. The pressure, however, depends on the height of the gap, the assumed flow pattern, and the operating conditions of the seal.

$$\begin{aligned} F_z &= -\iint_{\Omega} p(r,\theta)\ d\Omega, & F_{zeq} &= -\iint_{\Omega} p_{eq}(r,\theta)\ d\Omega \\ M_x &= -\iint_{\Omega} p(r,\theta) r\sin(\theta)\ d\Omega, & M_{xeq} &= -\iint_{\Omega} p_{eq}(r,\theta) r\sin(\theta)\ d\Omega \\ M_y &= \iint_{\Omega} p(r,\theta) r\cos(\theta)\ d\Omega, & M_{yeq} &= \iint_{\Omega} p_{eq}(r,\theta) r\cos(\theta)\ d\Omega \end{aligned} \quad (11)$$

Static and dynamic forces and moments are calculated by numerical integration of the gas film pressure p and p_{eq} (in a state of static equilibrium) in the area Ω of the ring. Based on such formulated mathematical model, a computational algorithm and a C ++ computer application has been developed allowing for simulation studies.

4. Results and discussion

Parameters used to perform numerical calculations for non-contacting gas face seal, were compared in Table 1. One adopted, as a working medium, the air with dynamic viscosity $\mu = 1.8 \cdot 10^{-5} (Pa \cdot s)$.

One considered the sealing structure which working rings have an outer radius of 60 mm and an internal radius of 48 mm. It has also been assumed that the rotor performs rotational movement at an angular speed of 2094 rad/s, just as in operation . One assumed the stator material density respectively for Carbon-graphite – $\rho = 1.8 \cdot 10^3 (kg \cdot m^{-3})$, Carborundum – $\rho = 3.1 \cdot 10^3 (kg \cdot m^{-3})$, Stellite – $\rho = 8.4 \cdot 10^3 (kg \cdot m^{-3})$, Tungsten carbide – $\rho = 1.5 \cdot 10^4 (kg \cdot m^{-3})$. The calculations were made similarly as in the paper [5] the height of radial gap at level h_o=6 μm.

Table 1. Characteristic operation parameters of the gas face seal.

Parameter	Value	Parameter	Value
Stator inner radius $r_i (mm)$	60	Stator density $\rho(kg \cdot m^{-3})$	var
Stator outer radius $r_o (mm)$	48	Angular excitation frequency $\omega(rad/s)$	2094
Pressure at inner radius $p_i (MPa)$	0.101	Support axial stiffness $k(N \cdot m^{-1})$	5 10^5
Pressure at outer radius $p_o (MPa)$	0.2	Support axial damping $c(N \cdot s \cdot m^{-1})$	300
Dynamic viscosity $\mu(Pa \cdot s)$	1.8 10^{-5}	Nominal clearance height $h_o (m)$	6 10^{-6}
Balance ratio $B = (r_o^2 - r_b^2)/(r_o^2 - r_i^2)$	0.6	Angular excitation amplitude $\alpha^r (rad)$	150 10^{-6}
Stator length $l(mm)$	50	Taper angle $\beta(rad)$	$\frac{0.5 \cdot h_o}{(r_o - r_i)}$

Figure 5 (a) shows the changes in the minimum radial gap height depending on the number of revolutions for four different stator densities ρ. By analyzing graphically presented results, it can be seen that for the carbon graphite and carborundum, the period of transient vibration is elongated and the stabilization of the vibration amplitude of the ring flexibly attached to the housing occurs only after rotation of the shaft of the rotor machine.

The increase in inertia force occurs together with increase of the density ρ, and greater inertia forces the stator to respond more rapidly to the sudden motion caused by the rotor displacement. This is perfectly visible to the ring made of stellite. The transition period is significantly shorter (up to 3 revolutions), but the amplitude of the angular vibration increases, which has influence on a significant reduction in the minimum radial gap height.

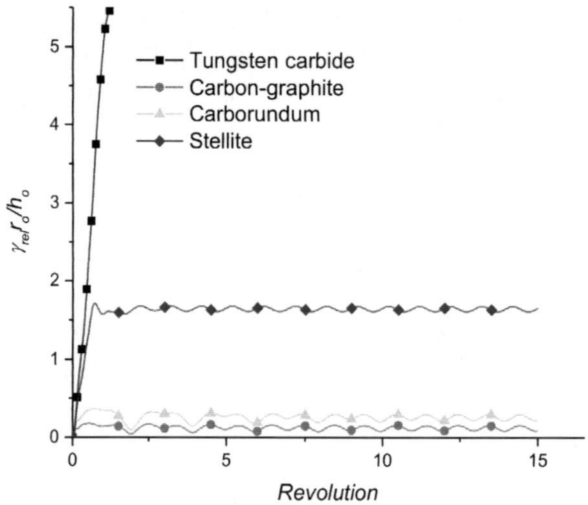

(a)

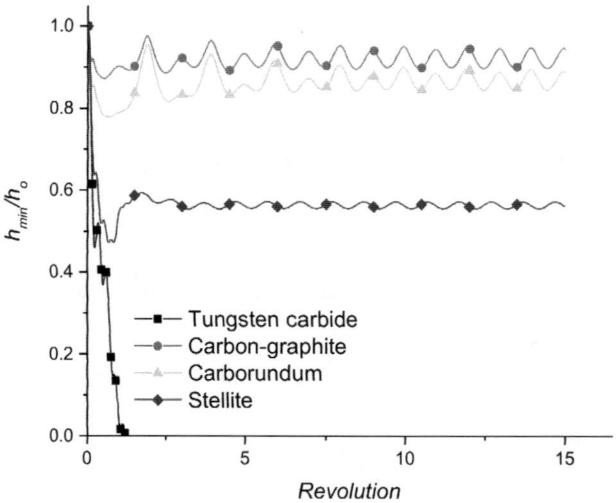

(b)

Figure 5. (a) change of minimum radial gap height, (b) transient dynamic response.

The use of stator material with even higher density ρ, destabilizes the operation of the working rings system. For the parameters collected in Table 1 for the non-contacting gas seal, the amplitude of the angular vibration increases drastically, leading to the discontinuity of the gaseous film and the contact of the working ring races.

5. Conclusions

The presented paper presents simulations of the non-contacting face seal, lubricated with gas, with flexibly mounted stator. Numerical calculations consisted in solving a mathematical model involving the nonlinear Reynolds equation and the equation of motion for the stationary ring. The demonstrated example of numerical calculations proves that the type of material used on working rings in non-contacting face seal can influence its proper non-contact operation. Sealing operation destabilization can be caused by too large mass of the ring flexibly attached to the housing, and may also lead to excessive wear of the cooperating working rings. Another undesirable occurrence is the occurrence of dry friction and excessive temperature increase of the sealing ring leading to the occurrence of thermoelastic deformation. Increasing the amplitude of the angular vibration also causes an increase in leakage. In order to avoid costly repairs and increase the life of the non-contacting seals used, computer programs are developed that can be useful tools to assist in selecting seals for specified operating conditions.

References

[1] Ruan B 2000 Finite Element Analysis of the Spiral Groove Gas Face seal at the slow speed and the low pressure conditions — slip flow consideration *Tribology Transactions* **43** (3) pp 411–418
[2] Blasiak S, Zahorulko A V 2016 A parametric and dynamic analysis of non-contacting gas face seals with modified surfaces *Tribology International* **94** pp 126–137
[3] Zhang H, Huang W, Wang Y, Lu L, Lyu S 2017 Research on meshfree method for analyzing seal behavior of a T-DGS *International Journal of Precision Engineering and Manufacturing* **18** (4) pp 529–536
[4] Miller B A, Green I 2001 Numerical formulation for the dynamic analysis of spiral-grooved gas face seals *Journal of Tribology* **123** (2) pp 395–403
[5] Miller B A, Green I 2003 Semi-analytical dynamic analysis of spiral-grooved mechanical gas face seals *Journal of Tribology* **125** (2) pp 403–413
[6] Ma C, Bai S, Peng X 2016 Thermoelastohydrodynamic characteristics of T-grooves gas face seals International *Journal of Heat and Mass Transfer* **102** pp 277–286
[7] Zahorulko A 2015 Theoretical and experimental investigations of face buffer impulse seals with discrete supplying *Eastern-European Journal of Enterprise Technologies* **4** (7(76)) pp 45-52
[8] Martsynkovskyy V, Zahorulko A, Gudkov S, Mischenko S 2012 Analysis of buffer impulse seal *Procedia Engineering* **39** pp 43-50
[9] Gaft J, Martsynkovskyy V, Gromyko B, Zahorulko A 2003 Design and calculation of mechanical seals with self-adjusting clearance *Proc. XVII Int. Conf. on Fluid sealing* (BHR Group, York, England) pp 505-520
[10] Blasiak S, Kundera C 2012 A numerical analysis of the grooved surface effects on the thermal behavior of a non-contacting face seal *Procedia Engineering* **39** pp 315–326
[11] Blasiak S, Kundera C, Bochnia J 2012 A numerical analysis of the temperature distributions in face sealing rings *Procedia Engineering* **39** pp 366–378
[12] Blasiak S, Takosoglu J E, Laski P A 2014 Heat transfer and thermal deformations in non-contacting face seals *Journal of Thermal Science and Technology* **9** (2)
[13] Ding S, Bai S 2017 Thermoelastohydrodynamic behaviour of inclined-ellipse dimpled gas face seals *Science China Technological Sciences* **60** (4) pp 529–537
[14] Ding X, Lu J 2016 Theoretical analysis and experiment on gas film temperature in a spiral groove dry gas seal under high speed and pressure *International Journal of Heat and Mass Transfer* **96** pp 438–450

[15] Thomas S, Brunetière N, Tournerie B 2007 Thermoelastohydrodynamic behavior of mechanical gas face seals operating at high pressure *Journal of Tribology* **129** (4) pp 841–850
[16] Bochnia J, Blasiak S 2016 Anisotrophy of mechanical properties of a material which is shaped incrementally using polyjet technology *Engineering Mechanics 2016* ed I Zolotarev and V Radolf pp 74–77
[17] Bochnia J 2012 Ideal material models for engineering calculations *Procedia Engineering* **39** pp 98–110
[18] Kundera C, Bochnia J 2014 Investigating the stress relaxation of photopolymer O-ring seal models *Rapid Prototyping Journal* **20** (6) pp 533–540
[19] Green I, Barnsby R M 2001 A simultaneous numerical solution for the lubrication and dynamic stability of noncontacting gas face seals *Journal of Tribology* **123** (2) pp 388–394
[20] Chen Y, Jiang J, Peng X 2016 Gas film disturbance characteristics analysis of high-speed and high-pressure dry gas seal *Chinese Journal of Mechanical Engineering* **29** (6) pp 1226–1233
[21] Wang B 2011 Numerical analysis of a spiral-groove dry-gas seal considering micro-scale effects *Chinese Journal of Mechanical Engineering* **24** (01) pp 146–153
[22] Kundera C 2003 Determination and analysis of cross-couplings of axial and angular vibrations of a flexibly mounted ring in a non-contacting face seal *Journal of Tribology* **125** (4) pp 797–803

IOP Conf. Series: Materials Science and Engineering **233** (2017) 012033 doi:10.1088/1757-899X/233/1/012033

Influence of feeders on operating characteristics of the impulse seals

V A Martsynkovskyy[1], S Hudkov[2], C Kundera[3]

[1] Department of General Mechanics and Dynamics of Machines, Sumy State University, 2, Rimsky-Korsakov Street, Sumy 40007, Ukraine
[2] Department of General Mechanics and Dynamics of Machines, Sumy State University, 2, Rimsky-Korsakov Street, Sumy 40007, Ukraine
[2] Faculty of Mechatronics and Machine Building, Kielce University of Technology, Aleja Tysiąclecia Państwa Polskiego 7, PL- 25314 Kielce, Poland

E-mail: s.gudkov@omdm.sumdu.edu.ua

Abstract. The paper presents an refined calculation of impulse face seals with a self-regulated gap. Unlike the existing methods of calculation, wherein, for the sake of simplicity, influence of the feeder conductance is neglected, this article describes conductance effect on the operating characteristics of the seal. Static calculation method of a face impulse seal of the high-speed pump is presented.

1. Design and working principle of the seal

The simplest design of a single-stage impulse seal (Figure 1) differs from the mechanical face seal only by the fact that closed chambers 2 are located on the end face of the axially movable ring 1, and several radial feeding channels 5 are made on the rotating support ring 6, and these channels are opened towards the sealing hole [1]. Sealing medium under sealing pressure p_1 is injected through these channels into the chambers during the short time intervals $t_c = \beta_c / \omega$ while the rotating channels 5 pass by the chambers 2. At these moments, pressure p_2 in the chambers increases in an abrupt manner to a value $p_{2\max} = p_1$.

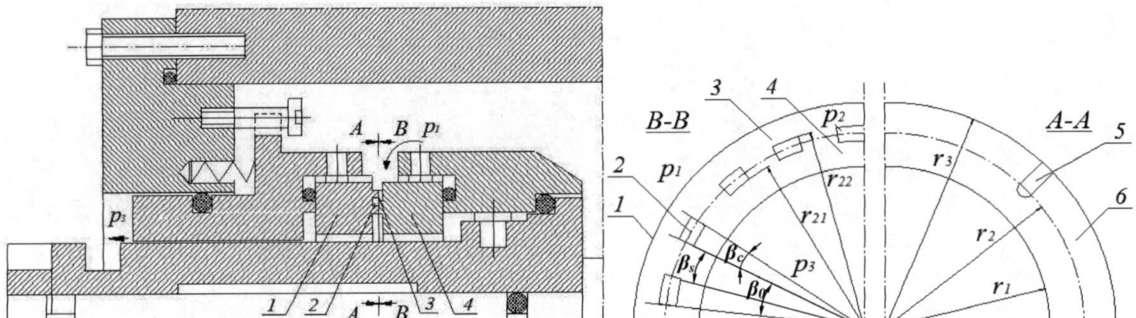

Figure 1. Scheme of the impulse face seal.

Nature of the pressure change in the chamber over a period of $T = 2\pi/\omega n_i$ (n_i -number of feeders) between two next injections depends on the hydraulic resistance or conductance g_i of the feeders and conductance of the internal (on the side of the condensed pressure) 3 and outer 4 face gap throttles $g_1(z)$ and $g_3(z)$. Figure 2 (a) [2] shows approximate graphs of pressure variation in a separate chamber: the larger is the gap, the less is $p_{2\min}$ and the averaged pressure $\overline{p}_2$ over a period $T = 2\pi/\omega n_i$.

A similar picture takes place during the expansion process, when a feeder is outside the sector β_c on a time interval $T - t_c$. After a feeder leaves the sector β_c occupied by the chamber, pressure p_2 in the chamber begins to decrease due to the outflow of the compressed medium through the outer flat gap 4; A_1, A_3 - areas of flat annular gaps walls (Figure 2, (b)). Pressure drop continues until the next injection (Figure 2 (a)), and depth of reduction, i.e. the value $p_{2\min}$, depends on the face gap: the larger is the gap, the less is $p_{2\min}$. A combination of parameters wherein pressure reaches the minimum value p_3 within the time $t < T = 2\pi/\omega n_i$ is possible. If a gap value decreases, amplitude of pressure changes in the chamber decreases, and the average pressure increases.

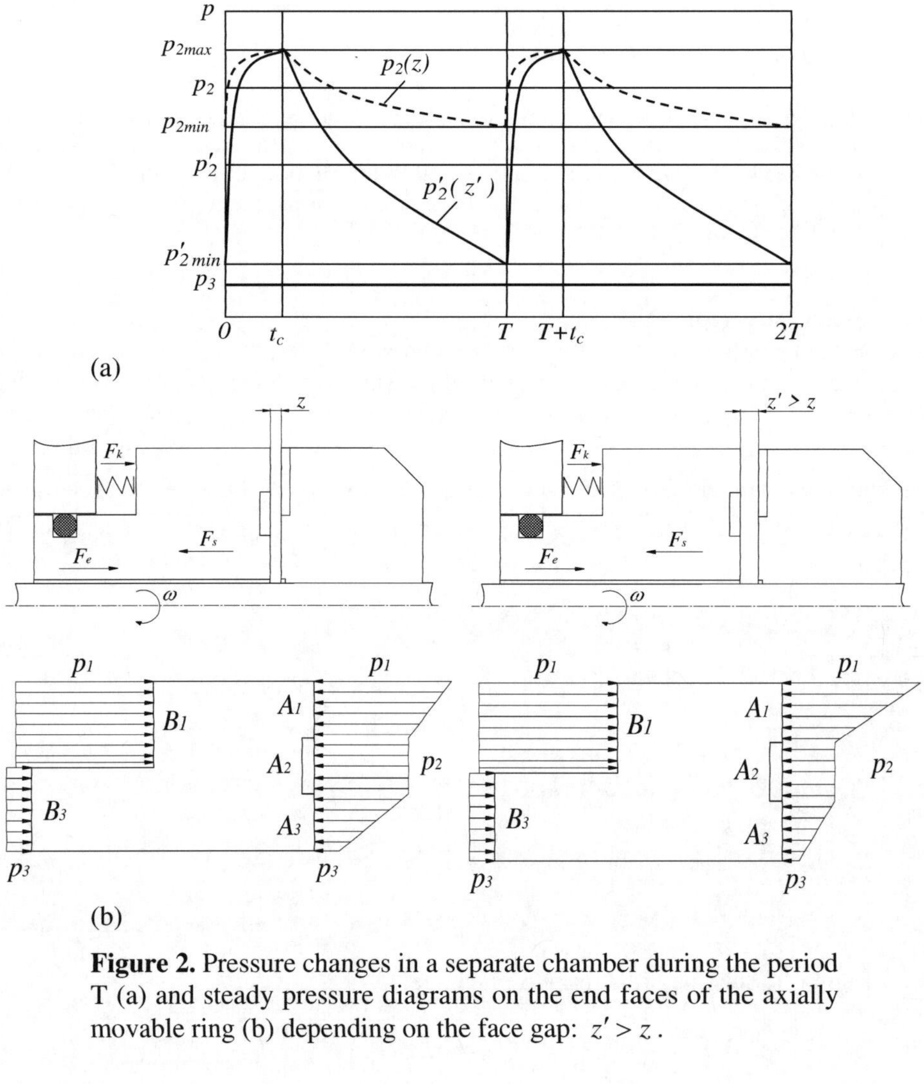

Figure 2. Pressure changes in a separate chamber during the period T (a) and steady pressure diagrams on the end faces of the axially movable ring (b) depending on the face gap: $z' > z$.

The shorter is the time between injections, the lower is the depth of pressure drop $p_{2\min}$ in the chambers, the higher is the averaged pressure p_2 in the chambers the greater is the force F_s opening the end joint.

Pressure force $F_s(z)$ opening the face joint depends on the pressure $\bar{p}_2(z)$ and, consequently, on the gap size. As the gap size is reduced, pressure increases, and its balance with the external force F_e (Figure 3), not depending on the gap size, is upset. Under the influence of positive difference $F_s - F_e > 0$ on the axially movable ring, gap size decreases (Δz) so that the balance $F_s = F_e$ is recovered. The sharper the dependence $F_s(z)$ drops, the less the new equilibrium position M' deviates from the original position of M.

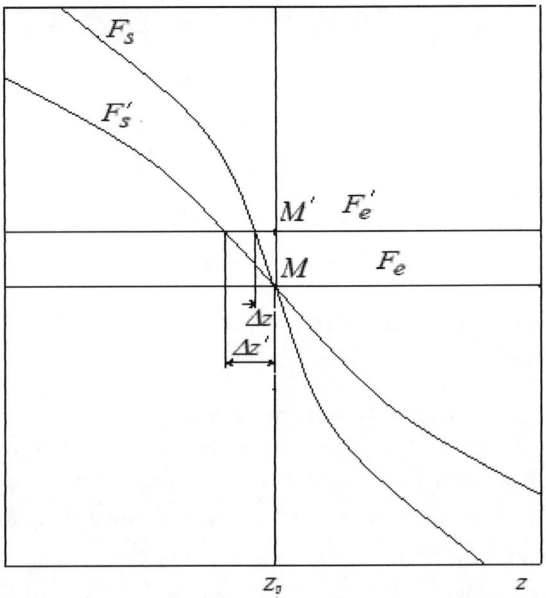

Figure 3. Effect of the slope of a curve $F_s(z)$ on static deflection under the influence of an external load F_e.

Thus, there is a negative feedback between the face gap z (adjustable value) and the force F_s (regulating effect), which provides self-regulation of the face gap (Figure 4).

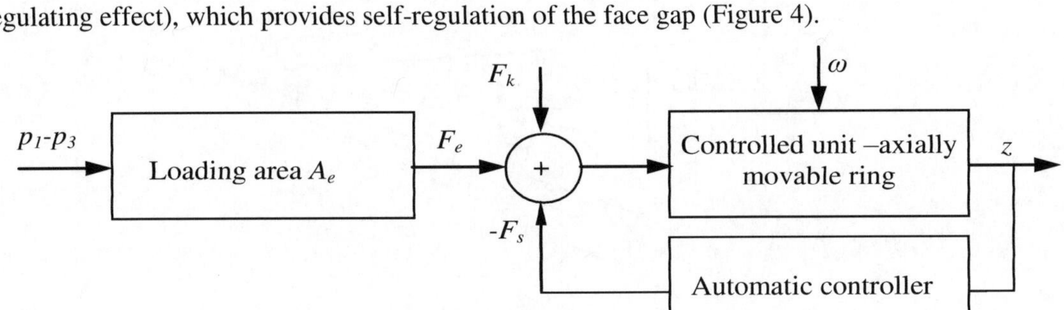

Figure 4. Structural scheme of the impulse seal as an automatic control system.

Seal performance is based on the developing of high-frequency pressure impulses in the discharge chambers, so it is called the impulse seal. Impulse frequency of the total axial force of pressure in the gap, acting on the axially movable ring – $\omega n_i n_c$ ($-n_c = 2\pi/\beta_0$ number of chambers) is large in comparison with rotation speed. Because of large damping in the micron face gap and inertia of the axially movable ring, the ring hardly reacts to such high-frequency impulses. For these reasons, high-frequency forced oscillations of the ring in the static calculation are not taken into account.

Static characteristic is found from the equation of axial equilibrium of an axially movable ring. In the equilibrium position, pressure force F_s, opening the face gap, balances external pressure force F_e and force of elastic elements F_k : $F_s = F_e + F_k$. Averaged pressure p_2 in the chambers over a period between successive injections determines the force F_s. For simplicity, a linear pressure change along the face gap radius is assumed (Figure 5).

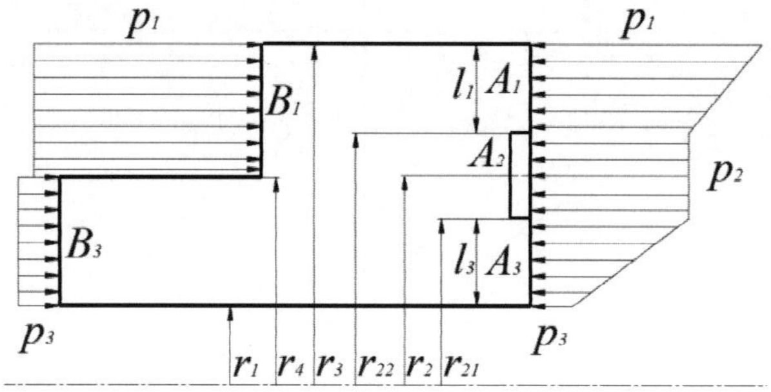

Figure 5. Pressure distribution on the end faces of the axially movable ring.

2. Calculation of the averaged pressure in the chambers

To estimate pressure in the chambers, it is necessary to consider a radial flow of viscous compressible fluid in a flat channel, which looks like a sector with a central angle β_c and a radial size $r_3 - r_1$, formed by sealing surface elements and separated by a flow chamber (Figure 1, (b)).

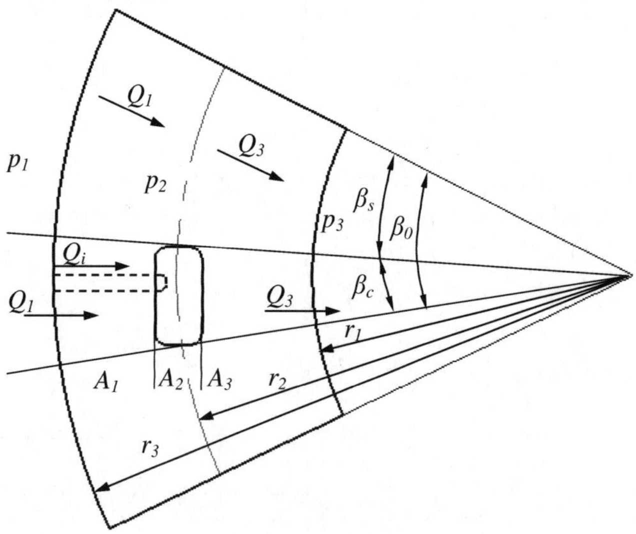

Figure 6. Segment of a flat channel with a flow chamber.

Right wall of the gap wherein feeders are located, rotates; the left one has a freedom of axial displacement and moves inside the micron-sized face gap. Circumferential component of the flow is not taken into consideration. The flow in the channels is nonstationary. A feeder (shown in dashed

IOP Conf. Series: Materials Science and Engineering **233** (2017) 012033 doi:10.1088/1757-899X/233/1/012033

lines), passing over time $t_c = \beta_c/\omega$ through a liquid-filled chamber, steplike supplies pressure p_1 to it. As a result, pressure rises to the maximum value p_1, compressing liquid in the chamber. After a feeder leaves the sector β_c, liquid volume compressed in the chamber flows out and pressure decreases to the initial minimum value. The expansion process takes place over a period of time $T - t_c$. After that, compression takes place again, and the process repeats (Figure 2, (a)).

During compression, difference in the liquid volume inflowing into the feeder and inner gap throttle $(Q_i + Q_1)(dt)_c$ and outflowing $Q_3(dt)_c$ from the outer one, is compensated by the liquid, which fills the volume $-dV$ released in the chamber because of previous liquid compression:

$$(Q_i + Q_1 - Q_3)(dt)_c = -dV \tag{1}$$

In the process of expansion, volume of the outflowing liquid $Q_3(dt)_p$ is greater than the volume of the inflowing one- $Q_1(dt)_p$ by the amount dV with the opposite sign:

$$(Q_3 - Q_1)(dt)_p = dV \tag{2}$$

Equations (1) and (2) of the volume balance differ only by flow rate Q_i through the feeder and the initial condition: compression begins with the minimum pressure, and expansion with the maximum one.

As defined by [5], the modulus of volume elasticity of a liquid is defined $-E = -V_0 \dfrac{dp}{dV}$, from which

$$dV = -\frac{V_0}{E} dp \tag{3}$$

Here $dV = V - V_0$- difference between the final V and initial V_0 volumes of the liquid, V_0 - unchanged volume of the chamber, E - isothermal modulus of liquid volume elasticity; Q_i- flow rate through the feeder, Q_1, Q_3- flow rates outflowing from the internal and external face throttles of the sector β_c.

Modulus of water elasticity is $E \approx 2200\ MPa$. To seal the pump with discharge pressure $p_1 = 18 МПа$, $p_2 \approx 0{,}5 p_1$ the relative volume of compression is $\Delta V/V_0 \sim 10^{-2}$, i.e. it is approximately equal to 1% of the chamber volume. Therefore, there is no need to expect a large effect from compressibility on the seal performance. The change of volume ΔV during compression and expansion differs only in sign.

Adding to term the equations (1) and (2), we obtain an equation for the entire period T

$$Q_i(dt)_c + (Q_1 - Q_3)\left[(dt)_c + (dt)_p\right] = \frac{V_0}{E}\left[(dp)_c + (dp)_p\right]$$

The sums of incremental components of time and pressure cover the entire period T between the next injections, therefore, one can indicate: $(dt)_c + (dt)_p = dt,\ (dp)_c + (dp)_p = dp$:

$$Q_i(dt)_c + (Q_1 - Q_3)dt = \frac{V_0}{E} dp \tag{4}$$

Differentials of independent variables are constant quantities:

$$(dt)_c \approx \Delta t_c = t_c - 0 = \beta_c/\omega,\ dt = \Delta T = T - 0 = 2\pi/n_i\omega \tag{5}$$

A pressure differential, accurate to a second-order quantity, can be presented as following:

$$dp = (dp)_c + (dp)_p \approx \Delta p = \Delta p_c + \Delta p_p = p_{2max} - p_{2min} + p_{2min} - p_{2max} = 0$$

Taking into account the obtained relations, an equation (4) after dividing by dt represents an approximate, time-independent equation of flow rate balance:

$$Q_i \frac{t_c}{T} + Q_1 - Q_3 = 0 \quad (6)$$

For a laminar flow, the flow rate is linearly dependent on pressure drops:

$$Q_i = g_i(p_1 - p_2),\ Q_1 = g_1(p_1 - p_2),\ Q_3 = g_3(p_2 - p_3) \quad (7)$$

where

p_2 – averaged value of pressure in the chambers.

Conductance values of the face throttles for laminar flows are proportional to the cube of the gap z [3,7] and are expressed by the equations:

$$g_1 = g_{1n}u^3,\ g_3 = g_{3n}u^3,\ u = z/z_n,$$

$$g_{1n} = \frac{\beta_c z_n^3}{12\mu \ln(r_3/r_{22})} \approx \frac{\beta_c z_n^3 r_3}{12\mu l_1},\ l_1 = r_3 - r_{22}, \quad (8)$$

$$g_{3n} = \frac{\beta_c z_n^3}{12\mu \ln(r_{21}/r_1)} \approx \frac{\beta_c z_n^3 r_{21}}{12\mu l_3},\ l_3 = r_{21} - r_1$$

z_n = (10...20) μm is a nominal face gap accepted for this device design, $u = z/z_n$ - dimensionless current gap, μ - dynamic viscosity of the fluid to be sealed.

Feeder conductance g_i depends on the shape and dimensions of the cross-section and it is determined experimentally, as a rule. Conductance estimate of an open radial channel can be obtained using the Hagen-Poiseuille equation for tubes of the circular cross section [6]:

$$g_i = \pi d_i^4 / 128\mu l_i \quad (9)$$

where

d_i, l_i – diameter and length of an equivalent tubular feeder.

Having substituted a equation for flow rate in (6) and using (5), it is possible to obtain the general equation of balance for volumetric flow

$$\left(g_i \frac{t_c}{T} + g_1\right)(p_1 - p) - g_3(p - p_3) = 0$$

from which the averaged pressure in the chambers can be found, depending on the dimensionless face gap:

$$p = \frac{\left(g_i \frac{t_c}{T} + g_1\right)p_1 + g_3 p_3}{g_i \frac{t_c}{T} + g_1 + g_3}$$

Presents the resulting expression to the dimensionless form by introducing the following description:

$$\psi_2 = \frac{p_2}{p_n}, \psi_1 = \frac{p_1}{p_n}, \psi_3 = \frac{p_3}{p_n}; \alpha_{1i} = \frac{g_{1n}}{G_i}, \alpha_{3i} = \frac{g_{3n}}{G_i}, G_i = g_i \frac{t_c}{T} = g_i n_i \frac{\beta_c}{2\pi} \tag{10}$$

p_n - nominal value of sealing pressure. As a result, one gets

$$\psi_2 = \frac{\psi_1 + (\alpha_{1i}\psi_1 + \alpha_{3i}\psi_3)u^3}{1 + (\alpha_{1i} + \alpha_{3i})u^3} \tag{11}$$

3. Regulating effect and hydrostatic stiffness

Regulating effect is a total axial force of pressure influencing on the contact surface of the axially movable ring, which depends on the face gap. Using a linear pressure diagrams shown in Figure 5, we calculate the pressure force F_s on the contact surface, opening the face gap (regulatory impact), pressure force F_e which presses a ring to the support disk (external load) and the force F_k of the elastic elements (control input):

$$F_s = 0{,}5(p_1 + p_2)A_1 + p_2 A_2 + 0{,}5(p_2 + p_3)A_3 = F_{s0} + Ap_2(u)$$

$$F_{s0} = 0{,}5(A_1 p_1 + A_3 p_3) \tag{12}$$

$$A_1 = \pi(r_3^2 - r_{22}^2) \quad A_2 = \pi(r_{22}^2 - r_{21}^2) \quad A_3 = \pi(r_{21}^2 - r_1^2) \tag{13}$$

$$A = 0{,}5(A_1 + 2A_2 + A_3) -$$

- effective area of contact relative to pressure p_2;

$$F_e = B_1 p_1 + B_3 p_3, F_k = k(\Delta + z) \quad B_1 = \pi(r_3^2 - r_4^2), B_3 = \pi(r_4^2 - r_1^2) \tag{14}$$

As it can be seen from expressions (12) and (14), only the component Ap_2 of the force F_s depends on the face gap value, therefore, this component causes regulatory impact.

The equation of axial equilibrium $F_s = F_e + F_k$, taking into account the fact that the face gap value is negligibly small in comparison with the preliminary deformation of the elastic elements $z << \Delta$, after substitution of the forces (12) and (14) takes the form:

$$Ap_2 = -0{,}5(A_1 p_1 + A_3 p_3) + B_1 p_1 + B_3 p_3 + k\Delta$$

where k is a reduced coefficient of axial stiffness of the elastic elements. It is necessary to divide this equation term-by-term by Ap_n and pass on to dimensionless forces, introducing the following notations:

$$\chi = F_k / Ap_n = k(\Delta + z)/Ap_n \approx k\Delta/Ap_n \quad \varphi_e = F_e / Ap_n = \frac{1}{A}(B_1\psi_1 + B_3\psi_3)$$

$$\varphi_s = \frac{F_s}{Ap_n} = \varphi_{s0} + \psi_2(u) \qquad \varphi_{s0} = \frac{F_{s0}}{Ap_n} = \frac{1}{2A}(A_1\psi_1 + A_3\psi_3) \tag{15}$$

As a result, we obtain a dimensionless regulatory impact effect:

$$\psi_2(u) = \frac{B_1 - 0{,}5A_1}{A}\psi_1 + \frac{B_3 - 0{,}5A_3}{A}\psi_3 + \chi \tag{16}$$

For further transformations, one can use the obvious equality $B_1 + B_3 = A_1 + A_2 + A_3$ (Figure 5), introduce the notation of dimensionless areas K, σ, and open the difference $\varphi_e - \varphi_{s0}$:

$$K = (B_1 - 0{,}5A_1)/A,\ \sigma = (B_3 - 0{,}5A_3)/A = 1 - K;\ \varphi_e - \varphi_{s0} = K\psi_1 + (1-K)\psi_3 \quad (17)$$

In this case, axial equilibrium equation (16) of an axially movable ring is:

$$\psi_2(u) = K\psi_1 + (1-K)\psi_3 + \chi \quad (18)$$

where $\psi_2(u)$ is also determined by the equation (11), which is the consequence of a balance equation of the flow rate. Joint solution of a balance equation of the flow rate and a equation of axial equilibrium of an axially movable ring, makes it possible to express a dependence of the gap value on external disturbances ψ_1, ψ_3 and the driving force χ, i.e. determine the static characteristics.

The steepness of curve $F_s(z)$ (Figure 3) is determined by the slope of the tangent, i.e. by means of a derivative $k_s = \partial F_s/\partial z$, which is called a coefficient of hydrostatic stiffness. In the dimensionless form, stiffness coefficient is expressed by the equation $\kappa_s = \partial\varphi_s/\partial u$. Since in the expression φ_s (15) only the second additive $\psi_2(u)$ depends on the gap value, using the equation (11), one obtains

$$\kappa_s = \frac{\partial\varphi_s}{\partial u} = \frac{\partial\psi_2}{\partial u} = -\frac{\alpha_{3i}}{\left[1 + (\alpha_{1i} + \alpha_{3i})u^3\right]^2}\cdot 3u^2\Delta\psi \quad (19)$$

The greater is the steepness $F_s(z)$, i.e module κ_s, the smaller is the gap position deviation from the initial equilibrium state when the external load F_e changes (Figure 3), the minor is the static error. Thus, in order to improve static characteristics, it is necessary to increase modulus of the hydrostatic stiffness coefficient, i. e. increase conductance g_{3n} and reduce G_i. Attempts to reduce α_{3i} are rarely successful, since conductance of the face gap is determined by the micron-sized gap, and conductance of feeders - by the diameter of a channel - 0.4 ... 1.0 *mm*. Many different designs of throttles have been developed for bearing supports with gas lubrication [8,9]. Such throttles can be also used for impulse seals. Coefficient of hydrostatic stiffness (19) is less than zero, and it provides stability of the equilibrium position. Hydrostatic stiffness makes it possible to estimate minimum value ω_0 of the own frequency of axially movable ring vibrations and, if necessary, to escape possible resonance. In the equilibrium position:

$$\omega_0 = \left(\frac{k + k_s}{m}\right)^{1/2} = \left[\frac{k}{m}\left(1 + \frac{Ap_n}{kz_n}\kappa_s\right)\right]^{1/2} \quad (20)$$

4. Static and discharge characteristics

Static characteristic is a dependence of the steady-state gap value on external disturbances level. It is determined from the joint solution of equations (18) and (11):

$$K\Delta\psi + \psi_3 + \chi = \frac{\psi_1 + (\alpha_{1i}\psi_1 + \alpha_{3i}\psi_3)u^3}{1 + (\alpha_{1i} + \alpha_{3i})u^3}$$

From this equality one can find:

$$u = \left\{\frac{\alpha_{i1}[(1-K)\Delta\psi - \chi]}{[K(1+\alpha_{31}) - 1]\Delta\psi + (1+\alpha_{31})\chi}\right\}^{1/3},\quad \alpha_{31} = g_{3n}/g_{1n} \quad (21)$$

Equation (21) enables to make some important practical conclusions and recommendations.

1. Working range of the seal is limited by positive values of the face gap $u > 0$ in the entire variation range of the external influences $\Delta\psi$ and χ. In addition, these equations are reduced to inequations:

$$\frac{1}{1+\alpha_{31}} < K < 1, \ \Delta\psi > \frac{\chi}{1-K} \tag{22}$$

$\Delta\psi_{\min} = \chi/(1-K)$- minimum value of the permissible sealed differential pressure, upper value in the indicated range K (22) is not limited. Operating range does not depend on the conductance of the feeders g_i.

2. In the nominal mode $\Delta\psi \approx 1$, parameters included in (21) must provide face gap value close to the optimal one: $z \approx z_n$, $u \approx 1$ It is possible if numerator and denominator of the equation (21) are equal. From this equality, we find a load factor providing nominal gap parameters at $\Delta\psi \approx 1$:

$$K_n = \frac{1+\alpha_{i1}}{1+\alpha_{i1}+\alpha_{31}} - \chi \tag{23}$$

The load factor determined by this equation complies with the condition (22) $K_n < 1$.

Geometric parameters and preliminary compression force of elastic elements included in (23) are selected from constructive and technological considerations. Required value of the load factor K_n is provided by the appropriate choice of the internal radius r_4 (Figure 5) of the loading area B_1 (14). The last one is related to the load factor by the equation (18), whereof it follows:

$$B_1 = K_n A + 0{,}5A_1 = \pi\left(r_3^2 - r_4^2\right), \quad r_4 = \left\{r_3^2 - \frac{1}{\pi}\left(K_n A + 0{,}5A_1\right)\right\}^{0,5} \tag{24}$$

Equation (21) shows that the load factor K_n and the force of elastic elements χ reduce the gap value, and pressure drop increases it. If $\chi = 0$, a gap (21) does not depend on the condensed pressure difference and maintains a constant value. Limiting value of the gap when the condensed differential pressure increases is expressed by the quation:

$$\lim u_{\Delta\psi\to\infty} = \left\{\frac{\alpha_{i3}(1-K)}{K-(1-K)\alpha_{13}}\right\}^{1/3} \tag{25}$$

To calculate the flow rate, an expression for the total flow rate through the outer (output) flat annular throttle can be used. The flow rate $Q_3 = g_3(p_2 - p_3)$ and conductance (4) were determined for a sector with a central angle β_c (Figure 6). For the entire gap, the angle β_c in equations (4) is to be replaced by 2π:

$$Q = \frac{2\pi}{\beta_c} Q_3 = \frac{2\pi}{\beta_c} g_{3n} p_n u^3 (\psi_2 - \psi_3)$$

It is necessary to assign it to the basic flow rate equation: $Q_n = \frac{2\pi}{\beta_c} g_{3n} p_n$ and the required flow rate in a dimensionless form is obtained as following:

$$\overline{Q} = \frac{Q}{Q_n} = u^3(\psi_2 - \psi_3) \tag{26}$$

We substitute the value- $\psi_2(u)$ (11) in the last equality:

$$\overline{Q} = u^3 \Delta\psi \left[\frac{1+\alpha_{1i}u^3}{1+(1+\alpha_{31})u^3}\right] \tag{27}$$

If the expression for the gap (21) is used, then from equation (27) it is possible to obtain an analytical dependence of the flow rate on the compressed pressure and on the force of the spring preliminary compression. However, this dependence turns out to be too lengthy, so to calculate the flow rate it is more convenient to use a numerical calculation of the face gap by equation (21). A simple estimate is presented by the equation (27) for the nominal gap $u = 1$:

$$\overline{Q} = \Delta\psi\left[\frac{1+\alpha_{i1}}{1+\alpha_{i1}+\alpha_{31}}\right] \quad (28)$$

It is necessary to consider an important case of the compressible pressure dependence on the rotor speed. In centrifugal machines with an adjustable drive, pressure to be compressed is proportional to the square of the rotor speed $p_1 = C\omega^2$ [3]. Here C is a generalized parameter determined by geometry of the machine flow section and preserving an approximately constant value at all rotational frequencies. In the nominal mode $p_{1n} = C\omega_n^2$, therefore

$$C = \frac{p_{1n}}{\omega_n^2},\ p_1 = C\omega^2 = p_{1n}\frac{\omega^2}{\omega_n^2} = p_{1n}\Omega^2,\ \psi_1 = \frac{p_1}{p_{1n}} = \Omega^2;$$

$$\Delta\psi = \frac{p_1 - p_3}{p_n} = \left(\Omega^2 - \psi_3\right),\ p_n = p_{1n} \quad (29)$$

Thus, in static (21) and discharge (27) characteristics, a replacement $\Delta\psi = \Omega^2 - \psi_3$ is to be done:

$$u = \left\{\frac{\alpha_{i1}\left[(1-K)\left(\Omega^2-\psi_3\right)-\chi\right]}{\left[K(1+\alpha_{31})-1\right]\left(\Omega^2-\psi_3\right)+(1+\alpha_{31})\chi}\right\}^{1/3} \quad (30)$$

$$\overline{Q} = u^3\left(\Omega^2-\psi_3\right)\left[\frac{1+\alpha_{1i}u^3}{1+(\alpha_{1i}+\alpha_{3i})u^3}\right] \quad (31)$$

In this case, independent external influence is not the condensed pressure drop, but the rotor speed.

5. Calculation procedure and numerical example

Static calculation procedure will be considered using engineering calculation example of an impulse mechanical seal for a high-pressure feed pump of a smaller size and higher rotational speed.

1. Initial data: $p_3 = 0{,}9 MPa$, $p_n = p_{1n} = 18\ MPa$, $\omega = \omega_n = 600 1/c$

Sealing medium – water, $\mu = 0{,}653\cdot 10^{-3}\ Pa\cdot s$, $E = 2{,}2\cdot 10^3 MPa$.

2. For design reasons, we choose dimensions of the contact surface: $r_1 = 0{,}05 m$, $r_3 = 0{,}07 m$, $r_2 = 0{,}5(r_1 + r_3) = 0{,}06\ m$, $r_{21} = 0{,}057 m$, $r_{22} = 0{,}063 m$, $l_1 = l_3 = 0{,}007 m$, and also sizes and number of chambers and feeders: $n_i = 2$, $n_c = 18$, $V_0 \approx 3\cdot 10^{-7} m^3$, $d_i = 0{,}4\cdot 10^{-3}$ m, $l_i = 8\cdot 10^{-3}\ m$, $\beta_s = 2\pi/n_c = 0{,}349$, $\beta_c = b_c/r_2 = 0{,}25\ rad$.

3. To calculate areas of face bands we use the following equations:

$A_1 = \pi(r_3^2 - r_{22}^2) = 2{,}92\cdot 10^{-3} m^2$, $A_2 = \pi(r_{22}^2 - r_{21}^2) = 2{,}62\cdot 10^{-3} m^2$, $A_c = 64\cdot 10^{-6} m^2$.

$A_3 = \pi(r_{21}^2 - r_1^2) = 2{,}35\cdot 10^{-3} m^2$, $A = 0{,}5(A_1 + 2A_2 + A_3) = 4{,}9\cdot 10^{-3} m^2$,

4. We determine conductance of gap throttles (8) and their dimensionless values (10).

Calculation results are given in Tables 1-3 for 3 nominal gap values. It makes possible to evaluate the effect of z_n on static characteristics.

Table 1. Conductance values of gap throttles

z_n, μm	G_i	g_{1n}	g_{3n}	α_{i1}	α_{i3}
10		$3,19\cdot10^{-13}$	$2,6\cdot10^{-13}$	30	36,8
15	$9,57\cdot10^{-12}$	$1,077\cdot10^{-12}$	$8,768\cdot10^{-13}$	8,89	10,91
20		$2,552\cdot10^{-12}$	$2,078\cdot10^{-12}$	3,75	4,6

5. We calculate coefficients of hydrostatic stiffness (20).

Table 2. Coefficients of hydrostatic stiffness

z_n, μm	10	15	20
κ_s	-0,069	-0,18	-0,281

$\kappa_s < 0$, i.e. the stability condition holds for all 3 values of z_n.

6. We calculate load factor K_n (23), which provides nominal gap value at the nominal mode, loading area B1 and inner radius r4 (24). To estimate effect of the pre-compression force of elastic elements, we calculate three of its values: x = 0.004, 0.01, 0.03.

Table 3.

z_n, μm	K_n			B_1, m^2			r_4, m		
	χ=0,004	χ=0,01	χ=0,03	χ=0,004	χ=0,01	χ=0,03	χ=0,004	χ=0,01	χ=0,03
10	0,97	0,964	0,944	$6,21\cdot10^{-3}$	$6,19\cdot10^{-3}$	$6,09\cdot10^{-3}$	0,054	0,054	0,054
15	0,92	0,914	0,894	$5,97\cdot10^{-3}$	$5,94\cdot10^{-3}$	$5,84\cdot10^{-3}$	0,055	0,055	0,055
20	0,85	0,844	0,824	$5,62\cdot10^{-3}$	$5,6\cdot10^{-3}$	$5,5\cdot10^{-3}$	0,056	0,056	0,056

7. According to the equations (21) and (27) static and consumption characteristics are constructed (Figures 7, 8).

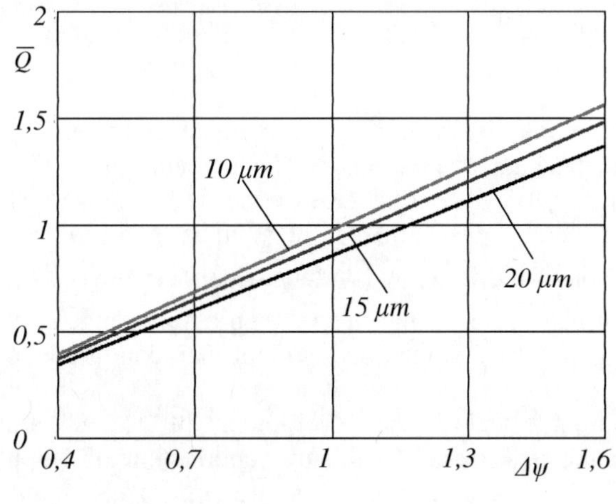

Figure 7. Discharge characteristics for various z_n.

IOP Conf. Series: Materials Science and Engineering **233** (2017) 012033 doi:10.1088/1757-899X/233/1/012033

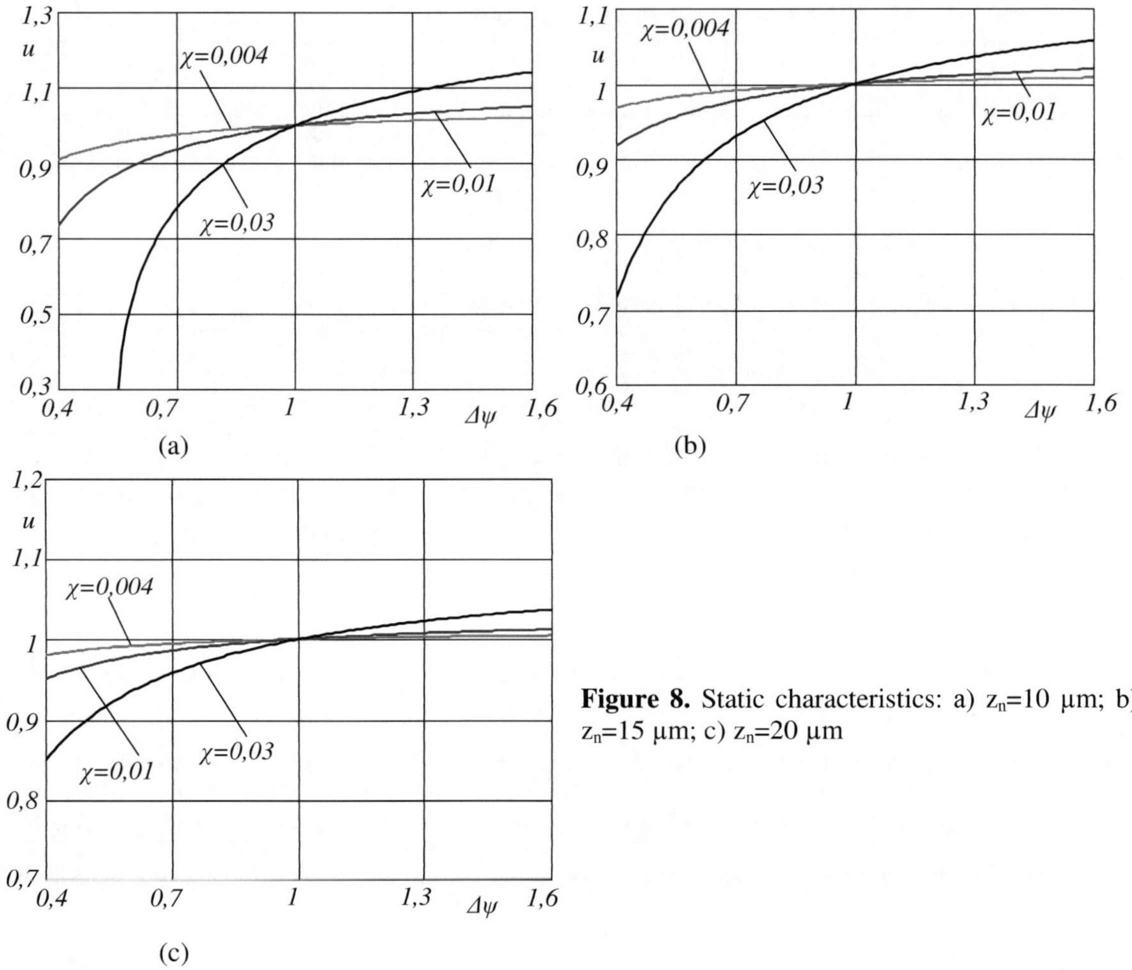

Figure 8. Static characteristics: a) z_n=10 μm; b) z_n=15 μm; c) z_n=20 μm

6. Conclusion

The paper deals with a face impulse seal of a high-speed centrifugal pump, which has several advantages over traditional mechanical face seals. The extended analysis described above made it possible to estimate influence of feeder conductance on the static characteristics of the impulse seal. In particular, the steady-state value of a dimensionless face gap under equal external forces is proportional to the cubic root of the relative conductance α_{i1}: $u \sim \sqrt[3]{G_i/g_{1n}} \sim \sqrt[3]{g_i n_i \beta_c / g_{1n} 2\pi}$. From the derived characteristics it can be seen, that in the set pressure range of the sealed liquid, face gap value slightly differs from the base value, and so optimal conditions of the seal operation are guaranteed.

The main problem of impulse seals is the development of structures and calculation of feeding channels with high hydraulic resistance (with low conductance). Impulse seal performance is significantly affected by the force of preliminary compression of elastic elements. Increasing of this force limits working area and increases static error, so it is necessary to monitor the load value and uniformity of its distribution along the circumference.

References

[1] Martsynkovskyy V, Zahorulko A, Gudkov S, Mischenko S 2012 Analysis of buffer impulse seal *Procedia Engineering* **39** pp 43–50

[2] Zahorulko A, Lisovenko D, Martsynkovskyy V, 2016 Development and investigation of face buffer impulse seal of centrifugal compressor *Eastern European Journal of Enterprise Technologies* **1** 7(79) pp 30–39
[3] Marcinkowski W A, Kundera C, Gaft J Z 2005 Podwojne uszczelnienia impulsowe *Hydraulika i Pneumatyka* **5** pp 8–11
[4] Martsynkovskyy V A 1980 *Contact free seals of the rotor machines* (in Russian) (Moscow: Machinebuilding) p 200
[5] Emtsev B T 1978 *Technical hydraulic mechanics: a textbook* (in Russian) (Moscow: Machinebuilding) p 463
[6] Martsynkovskyy V A 2005 *Annular seals: theory and practice* (in Russian) (Sumy: SSU) p 416
[7] Martsynkovskyy V A 2005 *Hydrodynamics of the throttling channels* (in Russian) (Sumy: SSU) p 337
[8] Luchin G A 1989 *Gas supporting bearings of the turbo machines* (in Russian) (Moscow: Machinebuilding) p 240
[9] Maksimov V A 2002 *Gas lubrication: prospects of application in turbomachinery* (in Russian) (Kazan: ZAO NIIturbokompressor named after VB Shnepp) p 154

IOP Conf. Series: Materials Science and Engineering **233** (2017) 012034 doi:10.1088/1757-899X/233/1/012034

Remarks on modeling the flexible seal ring housing

C Kundera[1], V Martsynkovskyy[2], A Zahorulko[2]

[1] Kielce University of Technology, Faculty of Mechatronics and Machinery Design, Al. 1000-lecia PP 7, 25-314 Kielce, Poland
[2] Sumy State University, Department of General Mechanics and Machine Dynamics, R-Korsakova 2, 40007 Sumy, Ukraine

E-mail: kundera@tu.kielce.pl

Abstract. When formulating models of dynamic face seals, two issues are of essential importance. The first concerns the determination of the forces formed in the medium film in the gap formed by the faces of two rings, while the second refers to the model of the flexible housing of one of the seal rings. The first issue includes the analysis of the medium flow through a narrow slit, taking thermal phenomena and deformations of slit-forming surfaces into account. The second issue concerns the modeling of properties of seal structural components, especially the modeling of elastic-damping properties of the flexible seal ring housing. This paper presents the results of simple relaxation tests of elastomeric rings and the procedure for analyzing these results and evaluating the elastic-damping properties of the rings tested. Finally, experimental results will be compared with theory.

1. Introduction

A number of theoretical and experimental works on the dynamics of seals, especially face non-contact seals, e.g. [1-6] have been published in recent years. Their review shows that dynamic properties of seals can have a significant impact on the dynamics of the whole rotor machine of a pump type, compressor. When formulating models of dynamic face seals, two issues are of essential importance. The first concerns the determination of the forces formed in the medium film in the gap formed by the faces of two rings, while the second refers to the model of the flexible housing of one of the seal rings. The first issue includes the analysis of the medium flow through a narrow slit, taking thermal phenomena and deformations of slit-forming surfaces into account. The problem is complex and still relevant, both in analytical solutions and in experimental research. The second issue concerns the modeling of properties of seal structural components, especially the modeling of elastic-damping properties of the flexible seal ring housing.

In many works with the analysis of the dynamics of face seals, the Kelvin-Voita (K-V) model [1-3] is most often used to describe elastic-damping properties of a flexible housing. Adopting the K-V model greatly simplifies the formulation of equations on the motion of a flexibly attached ring, and, then, the performance of the dynamic analysis of the system composed of a seal and a machine shaft.

Mechanical properties of elastomeric rings of the circular cross-section, the so called "O-rings", have been the subject of works not only on face seals, but also on plain bearings, pneumatic and hydraulic cylinders. The characteristics determining these properties are generally nonlinear and depend on deformation and deformation rates. They are made of polymer, gum-like and other materials.

This paper was motivated by works containing more complex rheological models used to describe properties of viscoelastic dampers, such as [7] and the properties of a gas film, for example, in a gas thrust bearing [8], using a model in the dynamic analysis [9]. Considering the theoretical models described in the above-mentioned papers, this paper presents the results of simple relaxation tests of elastomeric rings and the procedure for analyzing these results and evaluating the elastic-damping properties of the rings tested.

2. Presentation of the problem

The present analysis focuses on the flexible housing of a seal ring in the mechanical face seal. Figure 1 shows an example of the classic design of a face seal with a flexibly attached ring and, additionally, rotating together with the shaft of the machine i.e. the pump. The seal ring flexibly attached can be axially and angularly displaced relative to its transverse inertia axes.

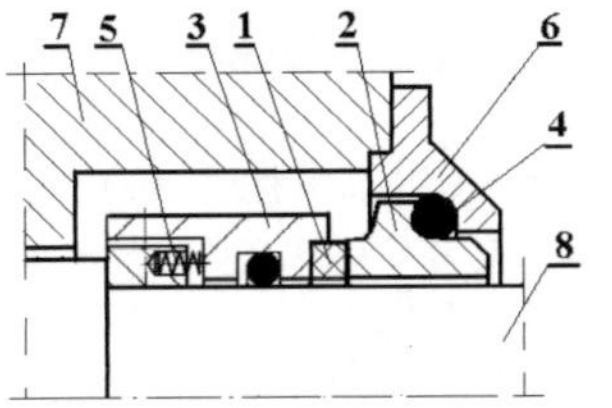

Figure 1. Diagram of the face seal design: 1- seal ring flexibly attached; 2- seal ring, stationary; 3,4- elastomeric O-rings; 5- spring, 6-cover; 7- pump packing; 8- shaft.

The elastomeric ring and different types of bellows (Figure 2) fulfill the function of a secondary seal in the flexible seal housing. Elastic bellows also act as a compression spring, and can be made of elastomeric material, Teflon, and thin stainless steel with good elastic properties and resistance to corrosion. All connections in bellow secondary seals are inseparable.

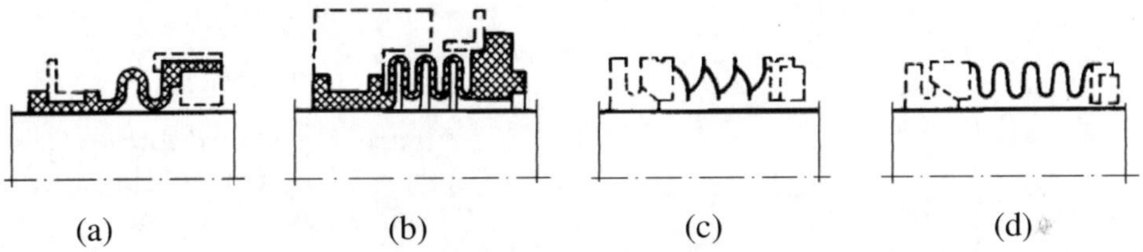

Figure 2. Diagram of secondary seals with elastic bellows: (a) elastomeric bellow; (b) PTFE bellow; (c), (d) metal bellows.

Secondary seals play an important role in the reliable operation of the entire face seal. Therefore, their elastic-damping properties and friction force at the contact surface, e,g. with the shaft of the machine, should be considered in design calculations and, in particular, in dynamic analyses.

The frictional force on the secondary seal depends on its design (shape, mounting), load system (pre-clamp), lubrication conditions of contact with the seal surface. At the contact of the ring with the shaft, a friction force which, depending on the direction of the displacement, increases or decreases the pressing force of the elastic element, appears when axial displacements of the seal ring are involved. Thus, the friction force in the secondary seal can have a significant effect on the balance of forces acting on the main seal ring and determine the tightness. Under dynamic conditions, especially at the low amplitude of vibrations of the ring flexibly attached, the elastomeric ring will only be subject to internal deformations that depend on its material properties.

3. Identification of the properties of elastomeric rings

Different rheological models are used to describe elastic-damping properties of elastomeric rings, and their parameters must be experimentally determined. Such experiments are generally very complex, because they must take not only deformations, but also deformation rates at different times into account. These determinants (circumstances) make us try to simplify both the deformation model itself and the experiment leading to the determination of its parameters.

IOP Conf. Series: Materials Science and Engineering **233** (2017) 012034 doi:10.1088/1757-899X/233/1/012034

In the earlier work [10] an experiment was carried out on the "Ispect mini" strength machine using the appropriately designed construction of the elastomeric ring (Figure 3). The construction used is characteristic for the O-ring working in the conditions of the dynamic reciprocating motion load, e.g. in the face seal or servomotor.

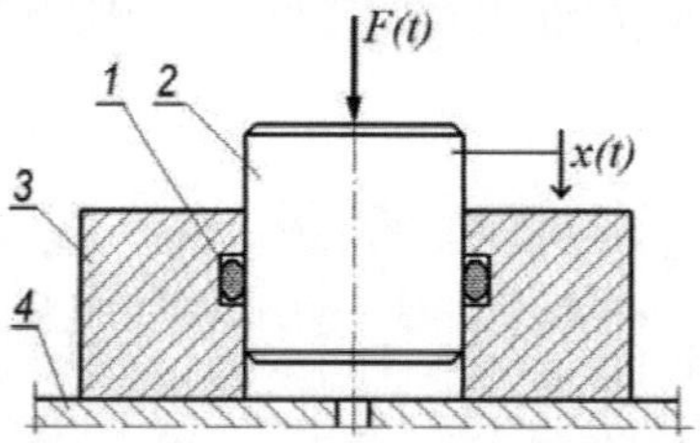

Figure 3. Diagram of the construction at the post: 1 – O-ring, 2 – piston, 3 – housing, 4 – base.

During the experiments, the following actions of the load force were used to determine the relaxation characteristics:

- force increment at 1 mm/s until the desired displacement value is reached, eg 0.5 mm;
- maintenance of the set displacement value for a specified period of time during the simultaneous recording of the decrease in the load value due to the relaxation of the O-ring material. The maintenance time of the set displacement value was between 90 and 300 s.

The above-mentioned test procedure was carried out for elastomeric rings of dimensions: d = 3.5 mm, D = 54.5 mm, made of nitrile mixture (NBR). An example of the ring load change graph obtained directly from the computer controlling the test machine is shown in Figure 4.

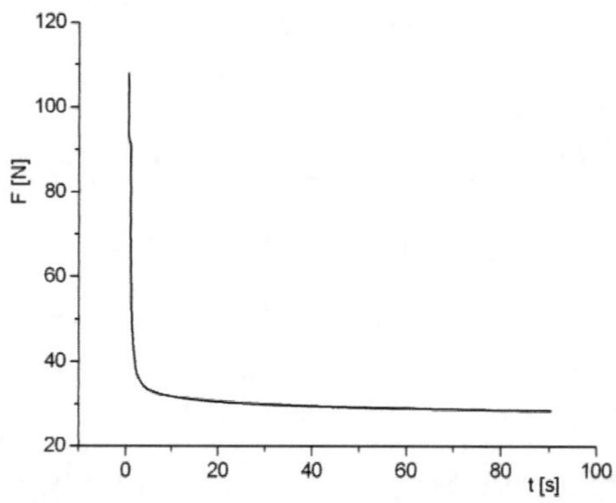

Figure 4. The load force relaxation graph for the O-ring - step displacement s=0.5 mm.

The course of the load force change (relaxation force) describes the properties of the material the test ring was made of. By dividing the load force $F(t)$ by the set value of displacement x_o we obtain the characteristics of the replacement coefficient of elasticity K_{rel}, often referred to as the coefficient of relaxation stiffness or the relaxation function. Figure 5 shows the characteristics of the replacement coefficient of elasticity K_{rel} corresponding to the course of the relaxation force shown in Figure 4.

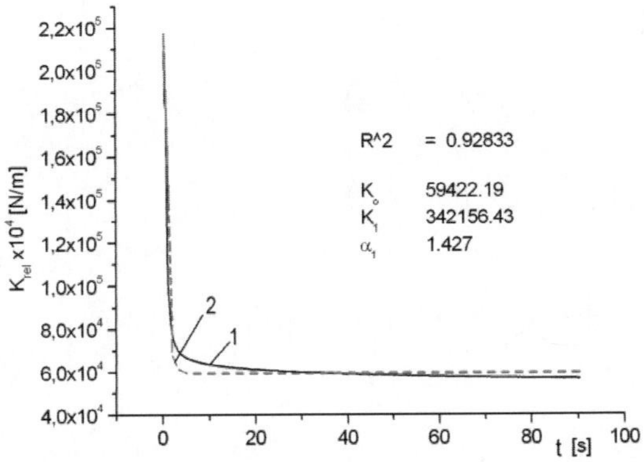

Figure 5. The relaxation stiffness: 1- from the experiment; 2- approximation.

Considering the results obtained from the experiment (Figure 5), the replacement coefficient of elasticity (the relaxation stiffness) can be approximated by the three-parameter function:

$$K_{rel}(t) = K_o + K_1 e^{-\alpha t} \tag{1}$$

where:

K_o, K_1, α – parameters obtained by approximation of the experimental characteristics.

Considering Figure 5, it can be seen that the approximation of experimental data with a three-parameter function is not very accurate. It can be improved by using the relaxation function with more parameters.

In the studies of photopolymers described in the above-mentioned paper [10], to approximate the results of the experimental tests, a five-parameter relaxation function was adopted in the form of:

$$K_{rel}(t) = K_o + K_1 e^{-\alpha_1} + K_2 e^{-\alpha_2} \tag{2}$$

The relaxation function assumed corresponds to the rheological model (Figure 6) consisting of two Maxwell models connected in parallel with the Hooke model.

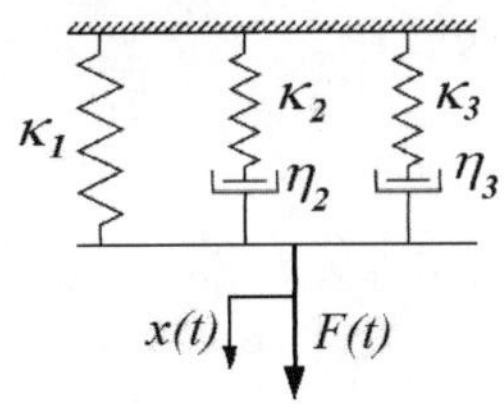

Figure 6. Five-parameter rheological model.

If we adopt the five-parameter relaxation function to fit the above-presented experimental results (Figure 5), then we obtain very good accuracy, which shows the value of the R^2 test (Figure 7).

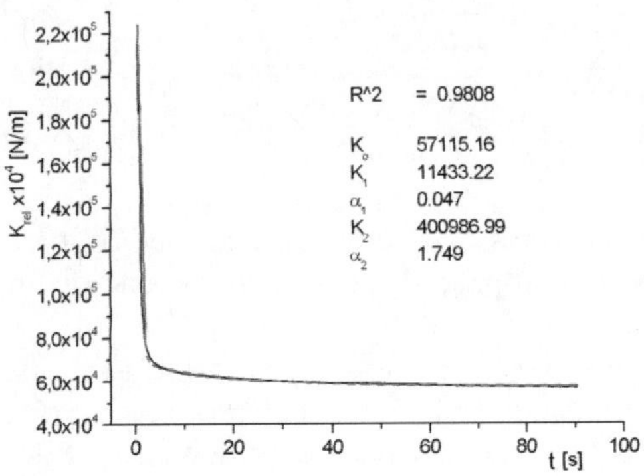

Figure 7. Approximation of the five-parameter relaxation function.

IOP Conf. Series: Materials Science and Engineering **233** (2017) 012034 doi:10.1088/1757-899X/233/1/012034

The evaluation of the elastic-damping properties of the system, in our case of a flexible ring housing, can also be carried out in the frequency domain (or the Laplace domain).

The force-displacement relationship can be expressed in analogy to the stress-strain constitutive law for a viscoelastic material [11]:

$$P(t) = \int_0^t K_{red}(\vartheta)\dot{x}(t-\vartheta)d\vartheta \tag{3}$$

where: $K_{rel}(t)$ – function of relaxation (relaxation stiffness).

After the Fourier transformation of the constitutive equation (3), we obtain:

$$F_p(j\omega) = j\omega F_{K_{red}}(j\omega)F_x(j\omega) \tag{4}$$

The frequency dependent complex impedance is:

$$H(j\omega) = j\omega F_{K_{red}}(j\omega) \tag{5}$$

where: $F_{Krel}(t)$ – transform of the relaxation function.

For the relaxation function in the form (1), its transform is respectively:

$$F_{K_{red}}(j\omega) = \frac{K_o}{j\omega} + \frac{K_1}{\alpha_1 + j\omega} \tag{6}$$

and the frequency dependent complex impedance (5) takes the following form:

$$H(j\omega) = K_o + K_1\frac{j\omega}{\alpha_1 + j\omega} \tag{7}$$

The real part of the complex impedance (7) determines the elastic properties of the system, while the imaginary part divided by ω - damping properties of the system.

$$K(\omega) = H^{re}(\omega) = K_o + K_1\frac{\omega^2}{\alpha_1^2 + \omega^2} \tag{8}$$

$$D(\omega) = \frac{H^{im}(\omega)}{\omega} = K_1\frac{\alpha_1\omega}{\alpha_1^2 + \omega^2} \tag{9}$$

If we take the five-parameter relaxation function (2) then the complex impedance and its real and imaginary parts take the following form:

$$H(j\omega) = K_o + K_1\frac{j\omega}{\alpha_1 + j\omega} + K_2\frac{j\omega}{\alpha_2 + j\omega} \tag{10}$$

$$K(\omega) = H^{re}(\omega) = K_o + K_1\frac{\omega^2}{\alpha_1^2 + \omega^2} + K_2\frac{\omega^2}{\alpha_2^2 + \omega^2} \tag{11}$$

$$D(\omega) = \frac{H^{im}(\omega)}{\omega} = K_1\frac{\alpha_1\omega}{\alpha_1^2 + \omega^2} + K_2\frac{\alpha_2\omega}{\alpha_2^2 + \omega^2} \tag{12}$$

The experimental test of determining the relaxation functions (1), (3) was performed at the axial load (Figure 3), hence we obtain the elasticity and the axial damping of the system, depending on the frequency.

The stiffness $K(\omega)$ and damping $D(\omega)$ characterize the linear reaction of the–force to the small sinusoidal seal ring (or shaft) displacement about equilibrium and to the corresponding velocity, respectively. The stiffness and damping together make up the complex frequency response (stiffness).

$$G(j\omega) = K(\omega) + j\omega D(\omega) \tag{13}$$

The real part of the frequency response is also called the storage modulus, and the imaginary part is called the loss modulus. After inserting values of the relaxation function parameters, i.e. K_o, K_1, K_2, α_1, α_2, experimentally determined (in the above-described experimental or relaxation tests) into the relationship (7÷12), we obtain fully defined characteristics of the elasticity and damping of the flexible housing shown in Figure 3. Figs 8 and 9 show the characteristics of elasticity (8, 11) and damping (9, 12) of the flexible ring housing as a function of the frequency of induction.

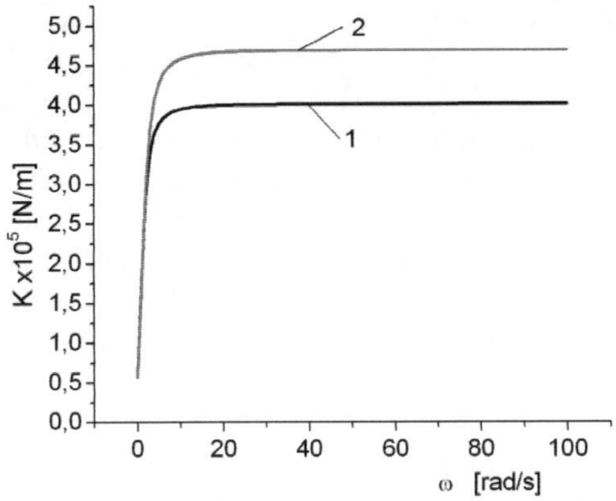

Figure 8. Elasticity characteristics of the flexible housing for two rheological models: 1-three-parameter and 2-five-parameter.

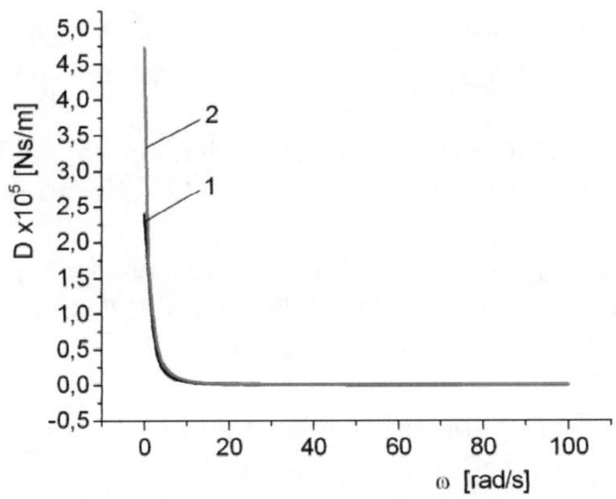

Figure 9. Damping characteristics of the flexible housing for rheological models: 1-three-parameter and 2-five-parameter.

Considering the characteristics obtained, it can be seen that with the increase of the induction frequency, the elasticity of the flexible housing increases and the damping properties decrease to the fixed (stabilized) values.

Comparing the two relaxation approximation models assumed shows that for a five-parameter model, higher values of elasticity and damping coefficients are obtained for a fixed induction frequency.

In the range of low induction frequencies, the flexible housing is characterized by low elasticity but high damping properties. Such characteristics of the flexible housing can be important in the analysis of the dynamics of mechanical face seals, the so-called "low-speed".

The above-mentioned description of the elastic-damping properties of the flexible housing has been carried out in the frequency domain. The dependencies determined can be directly used in a further analysis of the dynamics of a mechanical face seal, but also performed in the frequency domain. In addition, it should be borne in mind that the elastic-damping properties of a flexible housing depend on its design and the material of which the elastomeric ring was made. A relaxation test should be performed and the appropriate parameters of the relaxation function should be determined for another design of the housing and another elastomeric ring material.

4. Conclusions

The procedure for identifying elastic-damping properties of a flexible housing of the seal ring in a mechanical face seal described in this article includes simple experimental relaxation tests and an analysis of their results in the frequency domain. The five-parameter relaxation model of the relaxation function adopted for the analysis showed to be very well adjusted to the experimental results. The determined characteristics of elasticity and damping can be directly used in the analysis of dynamics and the design of face seals.

References

[1] Wileman J and Green I 1999 Parametric investigation of the steady-state response of mechanical seal with two flexibly mounted rotors *ASME Journal of Tribology* **121** (1) pp 69-76
[2] Brunetiére N, Tournerie B and Frêne J 2003 TEHD lubrication of mechanical face seals in stable tracking mode: Part 2 Parametric study *ASME Journal of Tribology* **125** (3) pp 617–627
[3] Kundera C 2003 Determination and analysis of cross-couplings of axial and angular vibrations of a flexibly mounted ring in a non-contacting face seal *ASME Journal of Tribology* **125** (4) pp 797–803
[4] Zahorulko A 2015 Theoretical and experimental investigations of face buffer impulse seals with discrete supplying *Eastern-European Journal of Enterprise Technologies* **4** (7(76)) pp 45-52
[5] Martsynkovskyy V, Zahorulko A, Gudkov S, Mischenko S 2012 Analysis of buffer impulse seal *Procedia Engineering* **39** pp 43-50
[6] Gaft J, Martsynkovskyy V, Gromyko B, Zahorulko A 2003 Design and calculation of mechanical seals with self-adjusting clearance *Proc. XVII Int. Conf. on Fluid sealing* (BHR Group, York, England) pp 505-520
[7] Lewandowski R and Chorążyczewski B 2007 Remarks on modelling of passive viscoelastic dampers *Proceedings of the 9th International Conference Modern Building Materials Structures and Technique 16-18 May 2007 Vilnius Lithuania*
[8] Miller B and Green I 1998 Constitutive Equations and the Correspondence Principle for the Dynamics of Gas Lubricated Triboelements *ASME Journal of Tribology* **120** (2) pp 345–352
[9] Gudkov S, Martsynkovskyy V and Kundera C 2016 An analysis of axial vibration in face packing seals *Solid State Phenomena* **248** pp 142-151
[10] Kundera C and Bochnia J 2014 Investigating the stress relaxation of photopolymer O-ring seal models *Rapid Prototyping Journal* **21** (6) pp 533-540
[11] Kundera C 2013 Seals of rotating systems *Kielce University of Technology monograph M53*, pp 286

IOP Conf. Series: Materials Science and Engineering **233** (2017) 012035 doi:10.1088/1757-899X/233/1/012035

Dynamic analysis of the mechanical seals of the rotor of the labyrinth screw pump

A Y Lebedev[1], P M Andrenko[1] and A L Grigoriev[1]

[1]National Technical University "Kharkiv Polytechnic Institute", Ukraine, Kharkiv

E-mail: andrenko1947@gmail.com

Abstract. A mathematical model of the work of the mechanical seal with smooth rings made from cast tungsten carbide in the condition of liquid friction is drawn up. A special feature of this model is the allowance for the thermal expansion of a liquid in the gap between the rings; this effect acting in the conjunction with the frictional forces creates additional pressure and lift which in its turn depends on the width of the gap and the speed of sliding. The developed model displays the processes of separation, transportation and heat removal in the compaction elements and also the resistance to axial movement of the ring arising in the gap caused by the pumping effect and the friction in the flowing liquid; the inertia of this fluid is taken into account by the mass reduction method. The linearization of the model is performed and the dynamic characteristics of the transient processes and the forced oscillations of the device are obtained. The conditions imposed on the parameters of the mechanical seal are formulated to provide a regime of the liquid friction, which minimizes the wear.

1. The introduction and the statement of task

The labyrinth screw pump (LSP, sealing assembly is shown in Figure 1 [1]) is installed in an oil well at a depth of about 1 km at the lower end of the pipeline, through which the vane pumps lift the oil to the surface of the earth; up to the specified mark, it flows under the influence of layer's pressure.

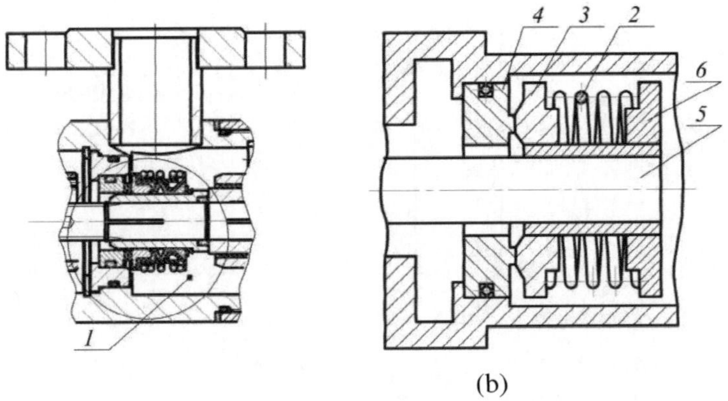

(a) (b)

Figure 1. Mechanical seal of the rotor of the LSP: (a) – design; (b) – scheme. 1– outlet chamber; 2 – spring; 3,4 – clips of rings; 5 – shaft; 6 – spring support.

The oil in the cavity of the LSP has the temperature of 120^0C and the pressure of about 10 MPa. It is known [2] that there is a lot of gas which is dissolved in the layer's oil and during the lifting along the well due to the pressure drop, about 2/3 of its mass is released in the form of bubbles and thus the oil becomes a two-phase medium. The task of LSP is to crush the bubbles and mix them with the liquid, which is necessary for a reliable operation of the vane pumps.

During the operation the LSP raises the oil pressure by 0.2 MPa, and this difference acts on the mechanical seal of the output chamber, shown in Figure 1 The presence of the pressure drop, as well as the specific physical properties of the working body and a very high purity of the sealing surfaces (roughness of the rings made of the wear-resistant alloy of two tungsten carbides – is only 0.03 micro mill.), create the prerequisites for the maintaining the liquid friction regime in the gap what minimizes the wear. The task of the research was to determine the compaction parameters which ensure this mode.

2. The features of the sealed fluid

For every 100 m increase in depth, the temperature of the earth's rock rises by 3 degrees. The depth of the layer from which oil is extracted reaches about 3 km and the pressure in the layer is close to 30 MPa. When the oil is being lifted along the well its temperature changes slightly and the pressure decreases threefold which is a result of the release of a large part of the dissolved hydrocarbon gas and nitrogen into the gas phase. In this case the initial viscosity of the liquid phase increases by 1.5 ... 2 times.

The rotation of the rotor leads to a separation of the oil in the outlet chamber in terms of density: the water which is present in the layer's oil is pressed to the walls of the chamber and the gas phase remains in the center. As a result the liquid hydrocarbon fractions are fed to the compact but their viscosity does not remain constant and varies within rather wide limits.

It is generally accepted that the average kinematic viscosity of the degassed and separated oil at a temperature of 120^0 C is about 4 cSt, but in the layer's oil where there is a lot of gas is dissolved the viscosity is much lower; besides, some types of the oil have a higher viscosity [2].

In the layer, the viscosity of oil ν (in cSt) obeys the Mayer equation

$$\ln(\nu+0.8)/\ln(\nu_{20}+0.8)=\left(293/(273+t_{lay})\right)^{3.5} \tag{1}$$

ν_{20} – the viscosity of the same oil at a temperature of 20^0C; $t_{lay}=120^0$C

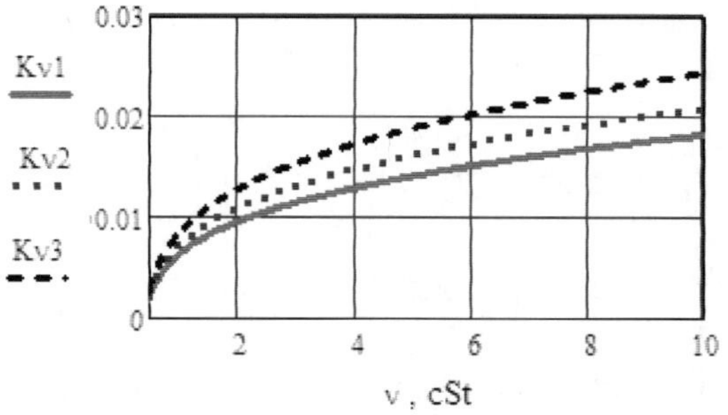

Figure 2. Changing the index K_ν depending on the viscosity ν of the layer's oil and temperature: $K\nu1 - t_{lay}=120^0$C, $K\nu2 - t_{lay}=70^0$C, $K\nu1 - t_{lay}=20^0$C.

Differentiating the both sides of equality (1) with respect to the temperature the coefficient of change (index) of the viscosity is obtained:

$$K_\nu=\nu^{-1}\,d\nu/dt_{lay}=-3.5\,(1+0.8/\nu)\,\ln(\nu+0.8)/(273+t_{lay})$$

In order to take into account the partial degassing during the lifting of the oil through the well and its separation in the pump, the viscosity in the right side of this formula should be reduced by 2 times. As a result, the dependencies presented in Figure 2 are obtained. The kinematic viscosity ν of oil in the pump varies within the limits of 1 ... 4 cSt; for further analysis an average value $\nu_{med} = 2$ cSt will be taken.

3. The changes in the temperature of the liquid and the material of the ring along the length of the gap

It was noted in the work [3] that a radial temperature drop leads to a conicity of the gap, which in its turn entails a change in the lift force. To calculate the conicity, the heat conduction problem for the radial section of the ring is solved (Figure 3). The release of the heat is caused by the operation of the fluid friction force and the heat flux density q is described by the formula

$$q(x) = \rho\nu((R-x)\omega)^2 / (h_0 - \Delta h(x)) \quad (2)$$

where

ρ, ν – is the density and kinematic viscosity of the oil;

R, ω – is the mean radius and the angular velocity of the rotation of the ring;

h_0 – is the gap's height on the outer radius of the ring;

Δh – is the change of the gap;

x – is the radial coordinate, directed along the flow of the fluid;

$x \in [0, H]$;

$\Delta h(0) = 0$;

H – is the width of the ring.

The task, after neglecting the radial contraction of the ring and the flow of heat from the pump shaft, takes the form of:

$$\nabla^2 T = 0;\ \partial T / \partial x\ \big|_{x=0} = 0;\ \lambda_c \partial T / \partial x + \alpha T\big|_{x=H} = 0;\ \partial T / \partial y\ \big|_{y=0} = 0;\ \lambda_c \partial T / \partial y\ \big|_{y=H_Y} = 0.5q \quad (3)$$

and its solution was obtained by Fourier's method of separation of variables [4] in the form of a series

$$T(x, y) = 0.5H\lambda_c^{-1} \sum\nolimits_{j=0,1,...} A_j [\gamma_j \operatorname{sh}(\gamma_j H_Y / H)]^{-1} \operatorname{ch}(\gamma_j y / H) \cos[\gamma_j (1 - x / H)] \quad (4)$$

where the separation constants γ_j and the Fourier coefficients A_j are defined by the equalities:

$$\gamma_j tg\, \gamma_j = \alpha H \lambda_c^{-1}, \quad A_j = 2H^{-1} \int_0^H q(x) \cos(\gamma_j (1 - x / H)) dx;\ \Delta\alpha = \alpha H \lambda_c^{-1}$$

$\nabla^2[.]$ – is Laplace operator;

$\nabla^2 T = \partial^2 T / \partial x^2 + \partial^2 T / \partial y^2$;

α – is the coefficient of heat transfer to the liquid in the output chamber of the LSP (its temperature in the model (3) is considered zero);

λ_c – is the average thermal conductivity of the material of the ring (in the calculation the thermal conductivity of the cast tungsten carbide $\lambda_r = 29.5$ W m^{-1} K^{-1} is used);

H_Y – is the height of the ring.

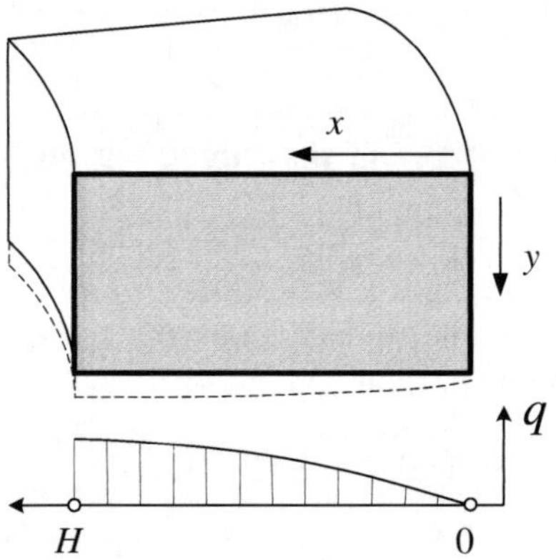

Figure 3. The section and the deformation of the ring and the heat flow in the gap.

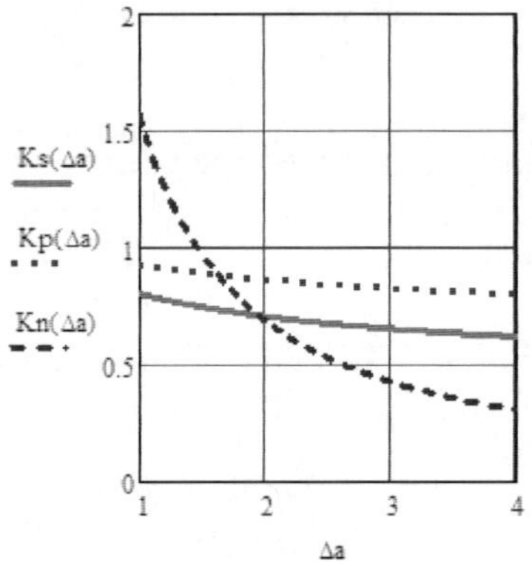

Figure 4. The graphs of the coefficients of the proportionality for the temperatures.

Integrating the series (4) over the height of the ring the following data is obtained:

$$\int_0^{H_Y} T(x,y)\,dy = 0.5H^2\lambda_c^{-1}\sum_{j=0,1,...} A_j\gamma_j^{-2}\cos(\gamma_j x/H) \tag{5}$$

and, it is typical, that this result does not depend on the height of the ring H_Y. The change in the temperature of the liquid in the gap (i.e., the temperature of the channel wall) is determined by the equality

$$T_l(x) = 0.5H\lambda_c^{-1}\sum_{j=0,1,...} A_j[\gamma_j\,\mathrm{th}(\gamma_j H_Y/H)]^{-1}(1-\cos[\gamma_j(1-x/H)]) \tag{6}$$

and when the ratio $H_Y > 0.5H$ is weakly dependent on the height H_Y.

The analysis showed that in the sums (4) - (6) one can confine the first term; then all the thermal characteristics of the ring are proportional to one temperature T_0:

$$\Delta T_r = Kp(\Delta a)\,T_0;\ \Delta T_l = Ks(\Delta a)\,T_0;\ \Delta T_{in} = Kn(\Delta a)\,T_0;\ \mathrm{grad}_x T = 2Kp(\Delta a)\,T_0/H$$

where

$T_0 = 0.5q_{med}H/\lambda_r$ – is the calculated value of the temperature obtained under the condition of the transfer of the heat with density $q_{med} = (\rho\nu(R_{med}\,\omega)^2/h_0)$ from the inner radius of the ring to the outer radius;

$R_{med} = R - 0.5H$;

ΔT_r , ΔT_l – are the radial temperature differences of the ring and the liquid;

T_{in} – is the temperature of the material in the initial section of the ring; the corresponding thermal elongations in the inlet ΔH_{in} and the outlet sections ΔH_{out} are determined by the formulas

$$\Delta H_{in} = \alpha_c T_{in} H;\quad \Delta H_{out} = \alpha_c\,\Delta T_{out}\,H, \tag{7}$$

α_c – is the coefficient of thermal linear expansion of the cast tungsten carbide, $\alpha_c = 5.8\cdot10^{-6}\ \mathrm{K}^{-1}$; the dependence of the coefficients Ks, Kp, Kn on the parameter $\Delta\alpha$ is given in the graphs of Figure 4.
With a good organization of the heat removal from the rings, the values $\Delta\alpha$ are located in the interval 2 ... 3.

In [3], the restriction on the gradient of the expansion of the material of the ring is indicated; it is $6\cdot10^{-5}$, that is why, the condition $Kp(\Delta a)\alpha_c T_0 < 6\cdot10^{-5}$ must be satisfied. From this point, as well as from the graphs in Figure 4, it follows that for a ring from the cast tungsten carbide, the value ΔT_l of the rise in the temperature of the liquid in the channel should not exceed $6.5\,\mathrm{K}$.

4. The change in ring's temperature in the transient process

If the gap or other parameters used in the right side of the equation (2) are being changed during the time t, then a stationary task (3) is transformed into a non-stationary task:

$$\lambda_c \nabla^2 T = c_c \rho_c \partial T/\partial t;\ \partial T/\partial x\ \Big|_{x=0} = 0;\ \ ....\ ;\ \lambda_c \partial T/\partial y\ \Big|_{y=H_Y} = 0.5 q(x,t)$$

where

c_c, ρ_c – is the specific heat and the density of the material of the rings.

Applying the Laplace transformation and the Fourier method, the following equalities are obtained:

$$\int_0^{H_Y} T(x,y,s)dy = 0.5H^2\lambda_c^{-1}\sum\nolimits_{j=0,1,...} A_j(s)(\gamma_j^{\,2} + \alpha_p H^2 s)^{-1}\cos[\gamma_j(1-x/H)]$$

$$A_j(s) = 2H^{-1}\int_0^H q(x,s)\cos[\gamma_j(1-x/H)]dx$$

where

s – is the Laplace variable; as a result for thermal deformations $\Delta H_{in}, \Delta H_{out}$ the solutions for images in the form of generalized Fourier series are obtained:

$$\Delta H_{in} = 0.5H^2\lambda_c^{-1}\sum\nolimits_j A_j(s)(\gamma_j^{\,2} + \alpha_c H^2 s)^{-1}\cos\gamma_j,\ \Delta H_{out} = 0.5H^2\lambda_c^{-1}\sum\nolimits_j A_j(s)(\gamma_j^{\,2} + \alpha_c H^2 s)^{-1}$$

$$A_j(s) = 2H^{-1}\int_0^H q(x,s)\cos(\gamma_j x/H)dx,\quad \alpha_c = c_c\,\rho_c\lambda_c^{-1}$$

Returning to the originals and using the first terms of the series (the others are small), the previous proportionalities are obtained

$$\Delta H_{in}(t) = K_{in}HT_0(t);\quad \Delta H_{out}(t) = K_{out}HT_0(t) \tag{8}$$

and the differential equation for the design temperature $T_0(t)$:

$$\alpha_c H^2\gamma_0^{-2}\ dT_0(t)/dt + T_0(t) = 0.5q_0(t)H\lambda_c^{-1};\quad q_0(t) = \rho\nu(R_{med}\omega(t))^2/h_0(t) \tag{9}$$

5. The condition of the static equilibrium of the sealing ring

If the gap exceeds the size of the roughness by 3 times or more then the liquid friction regime is observed. Here, the rotating ring does not have a contact with the stationary ring and its equilibrium is achieved when the oppositely directed forces are equal $F_{spr}(h_0) = F_p(h_0) + F_{gr}$ – the force of the spring $F_{spr}(h) = F_{spr.0} + z_{spr}\,h$, the force of gravity F_{gr} (in which Archimedes' law is taken into account, as well as the third part of the mass of the moving coils of the spring) and the fluid pressure force in the gap

$$F_p(h_0) = \pi\int_0^H (R-x)\,p(x)dx \tag{10}$$

where

h_0 – is the initial clearance on the outer radius R of the ring;

z_{spr} – the coefficient of the spring's stiffness;

x – the radial coordinate of the section;

$p(x)$ – the diagram of the oil overpressure in the gap, which satisfies the known differential equation [5] $dp/dx = -12\rho_{oil}\,\nu u/h^2$ or if the flow velocity u passes to the mass flow $\dot{m} = 2\pi(R-x)hu$,

$$dp/dx = -[6\dot{m}\nu_0/(\pi R h_0^3)]\,[\delta_\nu(x)/(\delta_r(x)\delta_h{}^3(x))] \tag{11}$$

$\delta_r = (R-x)/R$, $\delta_h = (h-h_0)/h_0$, $\delta_\nu = \nu/\nu_0$ – the relative values of the radius, the gap and the kinematic viscosity in the section x; ν_0 – is the value of the viscosity at the inlet to the seal.

For the differential equation (11), the boundary value task with boundary conditions is solved $p(0)=\Delta p$, $p(H)=0$, where Δp – is the preset pressure drop on the slits, from which the correction for the centrifugal force of Coriolis is subtracted $\Delta p_C = 0.5\rho_{med}V_{t.med}^2$; $V_{t.med}$ – is an average tangential velocity of fluid in the gap (in the investigated LSP at the operating mode $V_{\grave{o}.\tilde{n}\eth} \approx 5.5\text{mc}^{-1}$); ρ_{med} – is its average density. Then the result of the solution is substituted into the integral (10).

If it is assumed that the gap and the viscosity along the length of the gap do not change, then integration gives the value of the force

$$F_p(h_0) = f_R\,k_R\,\Delta p/2$$

where

f_R – is the area of the ring; and the correction $k_R \approx 1 + 0.5H/R_{med}$ takes into account the taper, caused by the annular form of the gap; it is present in all further results.

It is taken into account that the values $\delta\nu(x)$ and $\delta h(x)$ vary with the change in the temperature:

$$\delta\nu(x) = 1 + K_\nu(T_l(x)/\Delta T_l)\,;\ \delta h(x) = 1 + \alpha_p(2H/h_0)k_H(T_l(x)/\Delta T_l))$$

where

$$k_H = Kp(\Delta a)/Ks(\Delta a) \approx 0.8\,; \quad \Delta T_l = T_l(H)$$

K_ν – is the viscosity index at $t_{lay} = 120^0\text{C}$ (Figure 2).

And the amendment $\delta\nu(x)$ differs slightly from 1, therefore $\delta_\nu(x)/(\delta_h(x))^3 \approx \Psi(x)$, where

$$\Psi(x) = (1+\delta T_l(x)/\Delta T_l)^{-3}; \qquad \delta = \Delta T_l(\alpha_c(2H/h_0)k_H - K_\nu/3)$$

The use of the function $\Psi(x)$ in equation (11), and then in the integral (10), leads to the following result:

$$F_p(h_0) = f_R k_R(\Delta p/2)\,[1 + 0.45\delta/(1-0.4\delta)] \tag{12}$$

In particular, if $\delta \approx 1$, i.e. the taper of the gap leads to an almost complete overlap of the flow near the inner radius of the ring, then $F_p(h_0) \approx f_R\Delta p$ and this is the maximum tearing force which can be obtained during the liquid friction regime. However, if in the operating mode of the LSP this result is obtained (due to the corresponding tightening of the spring), this can lead to overheating of the sealing ring.

For the test compaction, the connection between the viscosity ν, the outer gap h_0 and the maximum oil heating temperature ΔT_l in the gap is determined by the equation:

$$\Delta T_l = Ks(\Delta a)[\rho\nu(R_{med}\omega)^2/(\lambda_c/H)]/(h_0 - 2k_H\alpha_c H\,\Delta T_l) \tag{13}$$

hereof it is easy to express the height of the gap through the values of the viscosity and the temperature:

$$h_0 = Ks(\Delta a)[\rho\nu(R_{med}\omega)^2 H/\lambda_c]/\Delta T_l + 2k_H\alpha_c H\,\Delta T_l$$

Then the formula (12) was used, which was rewritten in the following form:

$$F_p(h_0) = f_R k_R(\Delta p/2)[1+\delta F]$$

where

$$\delta F = \Delta T_l A_T; \quad A_T = [0.9(\alpha_c H k_H / h_0 - K_\nu / 6)]/(1 - 0.8\alpha_c H k_H \Delta T_l) \tag{14}$$

The coefficient δF indicates the fraction of the repulsive force which was acquired after the ring was heated. In Figure 5 (a), the graphs corresponding to the dependences (13), (14) for the parameters of the analyzed compaction are plotted at the maximum permissible value $\Delta T_l = 6.5^0$Ñ, and also at 5^0C and 8^0C. With a viscosity $\nu = \nu_{med} = 2$ cSt, the gap h_0 is 0.65 μm, and at the inner radius it decreases to 0.45 μm, 0.5 μm and 0.4 μm accordingly.

If the tightening of the spring is selected and remains unchanged then the coefficient δF is not changed. As we can see, for equal values δF an increase in the viscosity leads to an increase in the steady temperature drop and following this rule the tightening must be made at the highest viscosity, where $\delta F \approx 1$ and it is difficult to make such a fine adjustment. But if the viscosity fluctuation takes a short time, then, as we shall show later the temperature difference will also vary little. Therefore the tightening of the spring should be calculated based on the average viscosity of the oil.

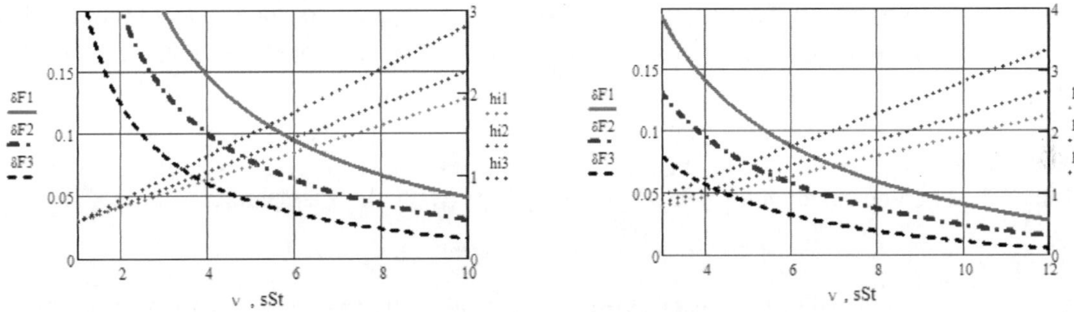

(a) (b)

Figure 5. The coefficients δF of increase in the lift force and the initial gaps hi, μm with temperature difference 8^0C (curves 1), 6.5^0C (curves 2) and 5^0C (curves 3) and the oil temperature: (a) – 120^0C, (b) – 20^0C.

In the formula (14), two factors affect the change in the lift force: the thermal expansion of the rings increases this force and the decrease in viscosity with the increasing of the temperature reduces on the contrary. With the viscosity $\nu_{med} = 2$ cSt, the first factor is stronger than the second one by an order of magnitude, but at high viscosities (due to the increase in the gap) this gap is reduced. If the temperature is significantly reduced (Figure 5 (b)), then with increased viscosities $\nu > 6$ cSt, it is going to be difficult to adjust the seal to work in liquid friction mode.

6. The modeling the movement of the ring in the transient modes

As it follows from (9) the time of the thermal inertia of the rings is $\tau_T = \alpha_c H^2 \gamma_0^{-2} \approx 1.6$ s, but when the ring moves the stronger factors which make slower the transient processes are acting. It is considered to observe the one dimensional non stationary motion of a the viscous incompressible fluid along the gap of the variable and the varying height $h(x,t)$, described by the well-known equations (here $S = 2\pi(R-x)h(x)$ – this is the area of the annular section of the gap):

$$\partial \dot m / \partial t + u \partial \dot m / \partial x + S \partial p / \partial x + 12 \nu \dot m / h^2 = 0; \; \partial(\rho S)/\partial t + u \partial(\rho S)/\partial x + \partial \dot m / \partial x = 0$$

The velocity of the flow is small, so the convective terms are not considered:

$$\partial p/\partial x = -12\nu \dot{m}/(S h^2) - \rho \dot{u}; \quad \partial \dot{m}/\partial x = -(\dot{\rho} + \dot{h}/h) S$$

And the acceleration $\dot{u}$, by virtue of the non-compressibility of the liquid is considered to be proportional to the axial acceleration of the ring:

$$\dot{u} = [(x - 0.5H)/h] d^2 h/dt^2$$

The integration of the equation

$$dp/dx = -\rho[(x - 0.5H)/h]\, d^2 h/dt^2$$

under the boundary conditions $p(0) = 0$, $p(H) = 0$ and substitution of the result into the integral (10) leads to the appearance of inertia force $F_{in} = m_{red}\, d^2 h/dt^2$, where m_{red} – is the mass of fluid in the gap, reduced to the ring,

$$m_{red} = (1/3)\, \rho_{oil} f_R\, k_R\, h\, (0.5H/h)^2$$

Note that the mass m_{red} increases in the proportion to the reduction of the gap and in the working modes of the compaction exceeds the mass m_R of the moving ring (to which, according to the known rule, the inertial mass of the spring's turns is shown) in several times.
The equation $\partial p/\partial x = -12\nu \dot{m}/(S h^2)$ coincides with (11), but the magnitude of the mass $\dot{m}$ flow due to non-stationarity of the process for different cross sections can be different. Note that on the right-hand side of the second equation the density of the liquid varies with its temperature, and

$$\dot{\rho} = -[\alpha_{oil} + \alpha_c (2H/h)(k_H\, T_l(x)/\Delta T_l + K_{in})]\rho \Delta \dot{T}_l$$

where α_{oil} – is the coefficient of thermal expansion of oil, $\alpha_{oil} \approx 0.012\text{K}^{-1}$, and $\dot{h}(0,t) = dh/dt$ – is the speed of the axial movement of the ring; $K_{in} = Kn(\Delta a)/Kp(\Delta a) \approx 0.6...0.8$ (Figure 4).

In order to determine the value $\dot{m}$ the second equation was integrated:

$$\dot{m}(x) = \dot{m}(0) - \int_0^x [\dot{\rho}(z) + \dot{h}(z)/h(z)] S(z)\, dz$$

then this function was inserted into the equation (11) and the integral (10) using the established procedure. As a result the following formula for detachment strength was obtained:

$$F_p = f_R k_R (\Delta P/2)[1 + A_T \Delta T_l] - m_{red} d^2 h/dt^2 - 12 m_{red} (\nu/h^2) dh/dt + 12 m_{red} (\nu/h^2) B_{\dot{T}} \Delta \dot{T}_l$$

where

$B_{\dot{T}} = K_{\dot{T}} (\alpha_{oil} + \alpha_c (2H/h) k_H)$;

$K_{\dot{T}} = 1/15 + K_{in}/12$.

Taking into account the forces of the spring and the inertia of the ring, the equation of axial oscillations is obtained:

$$(m_R + m_{red}) d^2 h/dt^2 + (12\nu/h^2) m_{red}\, dh/dt + z_{spr} h = f_R k_R (\Delta p/2)[1 + A_T \Delta T_l] + \\ + (12\nu/h^2) m_{red}\, B_{\dot{T}} \Delta \dot{T}_l + F_{gr} - F_{spr.0} - 2 z_{spr}\, K_{in}\, \alpha_c H\, \Delta T_l. \quad (15)$$

The last term on the right-hand side of (15), which takes into account the deformation of the spring caused by the thermal expansion of the sealing rings can be neglected.

Note that the mass m_{red} increases in the proportion to the change in the gap and on the workers modes of compaction exceeds the mass m_R in several times.

The varying part of the force depends on the temperature and the rate of the change of the temperature, which is determined from the equation (9). Equations (9), (16) are supplemented by an equation describing the angular oscillations $\psi(t)$ of the sealing ring relatively to the shaft:

$$j_{m.\,tors}\, d^2\psi/dt^2 + f_R(\rho\nu/h)R_{med}^2\, d\psi/dt + z_{spr.\,tors}\,\psi = f_R(\rho\nu/h)R_{med}^2\,\omega \tag{16}$$

where $j_{m.\,tors}, z_{spr.\,tors}$ – is the moment of the inertia of the ring and the torsional rigidity of the spring and the formula (9) for the heat flux is refined:

$$q_0(t) = \rho\nu(R_{med}(\omega(t) + d\psi/dt))^2 / h_0(t)$$

Thus, the dynamics of the compaction is described by a system of three nonlinear differential equations and is of the 5th order.

7. The frequency analysis and the model simplification

The linearization of the system of equations (9), (15), (16) and the analysis of the frequencies of its free oscillations showed that in the transient condensation process there are damped periodic and aperiodic oscillations. The periodic angular vibrations of the rotating ring have a frequency of about 12 Hz and the attenuation decrement that increases as the gap decreases; these quantities are mainly determined by the coefficients of equation (16). The aperiodic process of changing the radial temperature difference in the gap has a time constant $\tau_T \approx 1.6$ s which is mainly determined by the coefficients of equation (9).

The axial movements of the ring have two time constants:

$$\tau_{h.1} = (1 + m_R/m_{red})^{-1} h^2/(12\nu) \approx h^2/(12\nu) < 10^{-6}\text{s}, \quad \tau_{h.2} = 12(m_{red}/z_{spr})\,\nu/h^2 \approx 10^2...10^3\text{s}$$

As one can see, the first process is too short-lived and it is not necessary to take it into account when modeling it. To eliminate this process it is sufficient to exclude the force of inertia from equation (15). As a result, the equation of the motion has the first order:

$$dh/dt = [h^2/(12\nu\, m_{red})]\left[f_R k_R(\Delta p/2)[1 + A_T\Delta T_l] + F_{gr} - F_{spr.0} - z_{spr}h\right] + B_T\Delta\dot{T}_l \tag{17}$$

8. The results of solving the test problems

Figures 6-9 show the results of the calculation of transient processes caused by the various factors. The change in values is shown in a relative form:

$$\delta h = h(t)/h(0),\ \delta\psi = \psi(t)/\psi(0),\ \delta T = \Delta T_l(t)/\Delta T_l(0),\ \delta Fd = Fd(t)/F_{spr.0}$$

where Fd – is the sum of the active forces applied to the ring (without accounting the frictional force); the argument t as it was before designates the time and is measured in seconds.

In Figure 8, the values δh and δT are shown in the percentages and the fractions of a percent. Analyzing the graphs, one can come to the conclusion that the frictional forces effectively dampen the action of the perturbations so that the integrity remains. The leakage of oil through the end gap does not exceed $0.5\ \text{G h}^{-1}$ what is considered permissible for this device.

IOP Conf. Series: Materials Science and Engineering **233** (2017) 012035 doi:10.1088/1757-899X/233/1/012035

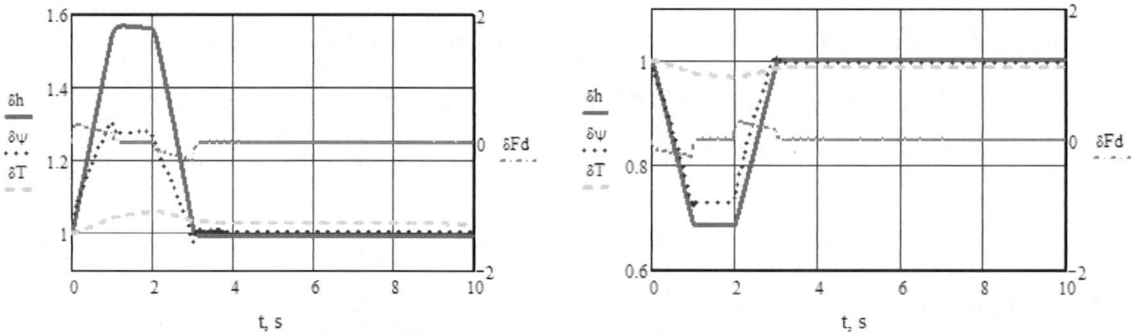

(a) (b)

Figure 6. Fluctuation of viscosity (in 2 times): (a) – into the large, (b) – into the smaller side.

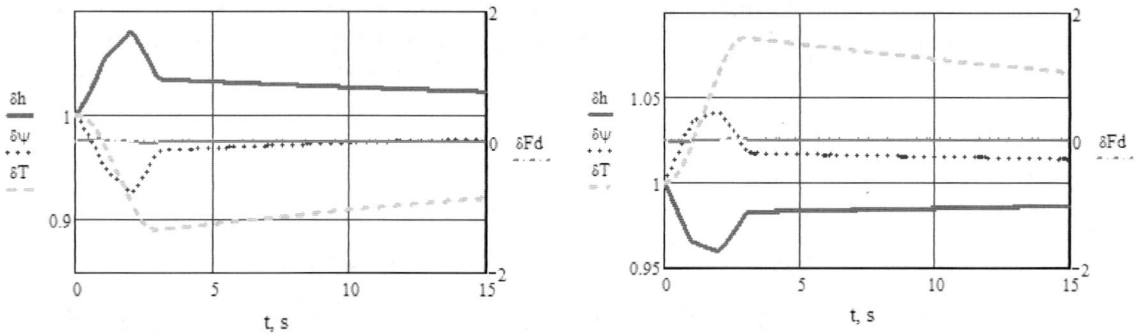

(a) (b)

Figure 7. Fluctuation of the pressure drop (in 2 times): (a) – into the large, (b) – into the smaller side.

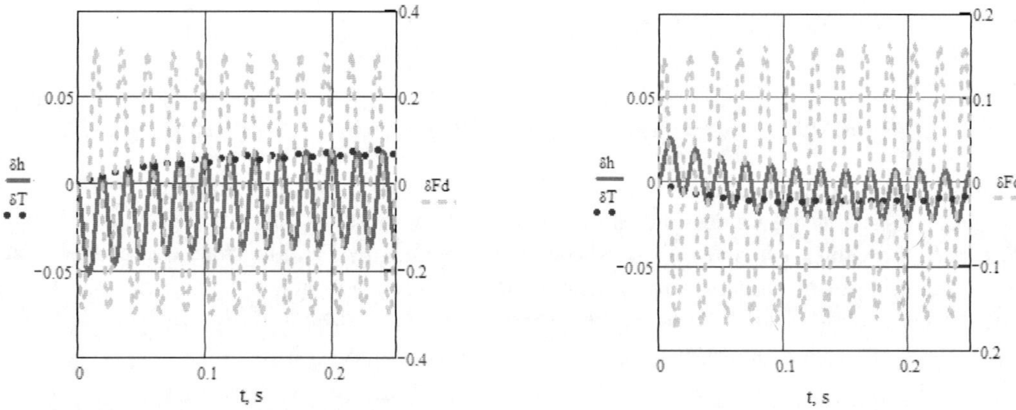

(a) (b)

Figure 8. Forced oscillations with a frequency of 50 Hz: (a) – axial vibration of the housing with an amplitude of 0.2 mm, (b) – unevenness of the pump feed 2.5%. The value δh is shown as a percentage, and δT – in tenths of a percent (Ppm). Time t is counted from the beginning of the cause of the oscillations.

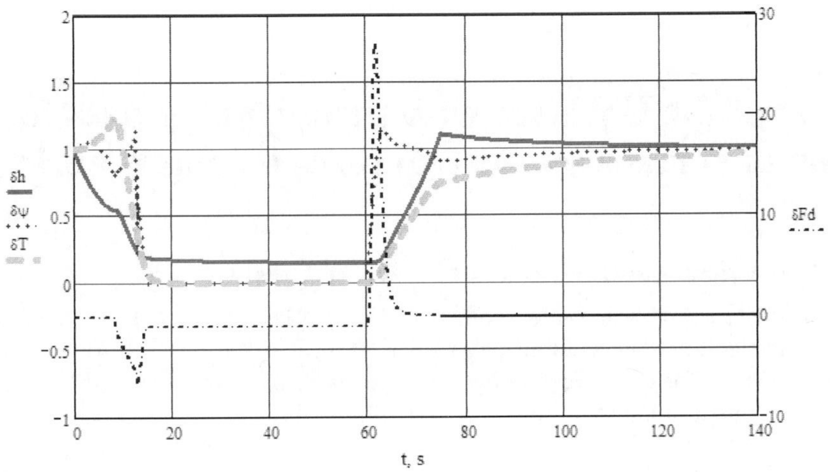

Figure 9. Simulation of power failure with a duration of 60 s.

Moreover, there is no significant overheating of the rings in these processes.

9. Conclusion

In order to ensure a stable operation of the mechanical seal in the liquid friction mode the designer needs to create an additional hydrodynamic force which tends to increase the gap between the rings by the increasing of the speed of the shaft or the pressing force. The authors have left the traditional view which says that the cause of this force is a roughness or and undulation of the surfaces, and there are explanations about this case.

The maximum size of irregularities on the ground surface of a ring made from cast tungsten carbide does not exceed 0.03 μm, which is an order of magnitude smaller than the working gap. In addition, cast tungsten carbide – this is 96% (by weight) wolfram and its high firmness prevents an increase in roughness during the usage what occurs on the contrary with the seals which made from soft composite materials.

It turned out that the appearance and the change of the additional lifting force can be explained by the uneven temperature deformation of the rings which lead to the appearance of the taper of the gap. The lack of empiricism in the formula of the lifting force made it possible to explain the static and the dynamics of the compaction from a unified position. The calculated equations reflect the factors known from the literature but in a closed form they are considered and used for the first time.

Further on, the authors plan to specify the model: take into account the heat flowing from the shaft as well as the regular plastic deformations of the sealing surfaces arising during operation. Then they will look for the ways to identify the model but if to take into account the extreme values of oil pressure and temperature it will be difficult to do this.

References

[1] Andrenko P M and Lebedev A Y 2017. *Labyrinth screw pumps* (Kharkiv: Panov) p 156
[2] Eygelson A S and Shaykhali D M 1989. *Calculation of density and viscosity of layer oil by data of surface degassing.* (Moscow: *Geology of oil and gas*) **11**
[3] Mayer E 1978 *Mechanical seals* (Moscow: Mechanical Engineering) p 288
[4] Farlow S 1983 *Partial differential equations for researchers and engineers* (Moscow: Mir) p 381
[5] Loitsyansky L G 1978 *Mechanics of liquids and gases* (Moscow: Nauka) p 736

IOP Conf. Series: Materials Science and Engineering **233** (2017) 012036 doi:10.1088/1757-899X/233/1/012036

Utilization of the UV laser with picosecond pulses for the formation of surface microstructures on elastomeric plastics

B Antoszewski[1], S Tofil[2], M Scendo[3], W Tarelnik[4]

[1] Centre for Laser Technologies of Metals, Kielce University of Technology, Aleja Tysiąclecia Państwa Polskiego 7, PL- 25314 Kielce
[2] Centre for Laser Technologies of Metals, Kielce University of Technology, Aleja Tysiąclecia Państwa Polskiego 7, PL- 25314 Kielce
[3] Institute of Chemistry, Jan Kochanowski University in Kielce, ul. Żeromskiego 5, PL- 25369 Kielce
[4] Sumy National Agrarian University, 160 G. Kondrateva Str., 40021, Sumy, Ukraine

E-mail: ktrba@tu.kielce.pl

Abstract. Elastomeric plastics belong to a wide range of polymeric materials with special properties. They are used as construction material for seals and other components in many branches of industry and, in particular, in the biomedical industry, mechatronics, electronics and chemical equipment. The micromachining of surfaces of these materials can be used to build micro-flow, insulating, dispensing systems and chemical and biological reactors. The paper presents results of research on the effects of micro-machining of selected elastomeric plastics using a UV laser emitting picosecond pulses. The authors see the prospective application of the developed technology in the sealing technique in particular to shaping the sealing pieces co-operating with the surface of the element. The result of the study is meant to show parameters of the UV laser's performance when producing typical components such as grooves, recesses for optimum ablation in terms of quality and productivity.

1. Introduction

The widespread trend toward miniaturization also applies to elements made of polymers. The wide use of elastomeric plastics results from the special properties of these materials such as deformability, elasticity, chemical stability, weather resistance, good electrical insulation and sometimes biocompatibility and heat resistance. They are used in biomedical, mechatronic, chemical, and electronic industries where, among others, such elements are made as micro-flow systems, surfaces with special adhesive properties, chemical and biological microreactors. Practically, mechanical methods of shaping surface structures of elastomers are ineffective in this case. To meet these needs, laser micromachining having such features as high precision, controlled heat-affected zone (HAZ), automation of the process ensuring treatment repeatability, non-contact machining, minimal range of finishing work are used. Lasers emitting radiation with ultra-short pulses contributing to removing the material without heat effects due to cold ablation are particularly useful here. In the scientific literature more and more articles on laser treatment of elastomeric and generally polymeric plastics appear [1-4]. The laser micromachining of elastomeric plastics is an alternative to lithographic methods. It is one of the direct methods utilizing the most common sources of laser radiation: CO_2 (wavelength 10.6 μm - close to infrared) and Nd-YAG (successive wave harmonics: 1064, 532, 326 and 266 Nm - from near infrared to ultraviolet) including ultrashort pulse lasers.

The constant optimization of production processes results in a continuous increase in requirements for elastomer seals. These requirements can be very different and depend not only on application but also on industry. In general, the suitability of sealing material for a given application is determined by three groups of factors: working temperature, chemical resistance and mechanical properties. Interactions such as working temperature and mechanical properties should also be considered. In addition to the correct material selection, sealing performance can also be determined by issues such as design, construction geometry and physical, chemical and stereoscopic surface properties. In particular, this applies to those parts of the sealing surface on which the sealing at the point of contact of the seal and the surface of the machine element or the housing depend. These are mainly small sealing parts such as sealing lips of different shapes, and cylindrical and flat surfaces. The shaping of surface properties, including adhesive ones, of these parts is essential for processes occurring in the sealing gap. Laser micromachining makes it possible. In many cases it is necessary for PUWO seals of harder materials such as PTFE to use hardened rollers up to 58 - 62 HRC to prevent the fast cutting of a groove under the lip. For dry seals, it is recommended to lubricate the seals with suitable grease or use air lubricants when installing. Due to lack of lubrication the seal may be jammed in the housing. These problems can be solved to a large extent by the use of micro-machining of the elastomeric material at the point of contact of the sealing portion with the surface of the machine element or housing. The shaping of micromachining may be of a different nature. They may be casket-shaped grease reservoirs, groove profiles that form microscopic blades, and assemblies of channels dosing the flow. Such solutions may be used, for example, in the elimination of the *stic slip* phenomenon and the prospect of using micromachining in combination with the elasticity and dynamics of the seal seems very promising.

With this in mind, the paper explores the potential of micromachining typical materials used for seals.

2. Theoretical basics

As a direct effect of the laser spot on the surface of the material, combustion, melting, sublimation, depolymerization (PMMA), ablation (in particular, desorption),foaming, color change, annealing (metals) occur depending on the type of material and laser beam parameters (wavelength, pulse duration, repetition frequencies, etc.). The classic method of machining directly with the CO_2 laser beam has been used for several years now. Unfortunately, it has serious disadvantages: low quality of the obtained structure - high surface roughness strongly dependent on the material used and the minimal spot diameter of 100 μm, which significantly limits the application area. The improved TC & T (*Laser Through-Cutting and Pattern Transfer*) method of direct CO_2 laser machining was developed by Liu and Gong [5]. Thanks to its application, it was possible to obtain micro-channels with a minimum diameter of 30 μm, with smooth surfaces of the bottom and side walls.

Laser methods of surface treatment of polymeric materials [1,2] using radiation to generate thermal effects consist in melting micro areas of the surface with a thickness of several micrometers, in increasing the surface roughness and removing its fragments, and if the process is carried out in the presence of oxygen, oxidation on the surface also occurs. This method is suitable for polyamides, polyesters and aramid materials, or for polyolefines and PTFE if ultraviolet absorbers are introduced to them because the absorption coefficient of laser radiation by the surface of the polymeric material determines the effectiveness of this technology.

Laser radiation is used for very fine processing of small surfaces of complex geometrical form, e.g. in electronics, or in the manufacture of medical devices where they are additionally provided with sterilization. Laser radiation affects the improvement of wetting and free surface energy values by the implementation of polar functional groups, mainly in the oxidation process, the degree of the polymer crosslinking, and the type of the geometrical structure of the surface without changing the core properties.

When polycarbonate, polytetrafluorate or polystyrene are processed, the use of the laser radiation wavelength less than 308 nm is required, which is provided by an ArF excimer laser. If the radiation energy lower or higher, respectively, than the ablation threshold of the mentioned polymeric materials

IOP Conf. Series: Materials Science and Engineering **233** (2017) 012036 doi:10.1088/1757-899X/233/1/012036

is used, an improvement in hydrophilic or hydrophobic properties is provided, which is accompanied by an increase in roughness [2, 6].

3.Theoretical basics

Laser ablation is the process by which the chemical bonds of macromolecules of polymeric material are destroyed by laser light. There are two mechanisms of ablation which can occur separately or simultaneously: photochemical and photothermal. The thermal ablation of a polymer occurs as a result of the excitation of molecules into the high energy state by laser radiation. This results in generation and, at a later stage, in heat accumulation and an increase in material temperature. The thermal ablation occurs after exceeding the temperature value of the material called the threshold ablation temperature (T_D). The energy initiating the thermal ablation is expressed by the following relation:

$$E_j^{th} = c_v \frac{(T_D - T_R)}{\alpha(1-R)}$$

where:

T_R - initial temperature of polymer material, c_v - specific heat of polymeric material, α - radiation absorption coefficient, R - reflectivity factor of laser radiation.

The mechanism of photochemical ablation consists in the photolytic breaking of chemical bonds, especially C-H bonds. Ablation achieves its full degree of development when a large number of (n) chemical bonds simultaneously break as a result of a laser pulse. The value of the energy that initiates ablation (the threshold value of ablation energy) describes the dependence:

$$E_j^{th} = n \frac{h\nu}{\phi\alpha(1-R)}$$

where;

ϕ - quantum yield of breaking bonds (value from 0-1 range), $h\nu$ - photon energy.

4. Experimental

Experimental research was carried out on a test bench existing at the Center for Laser Technology of Metals of the Świętokrzyski University of Technology.

A test bench is a laser machine for micro-machining. The test bench layout is shown in Figure 1.

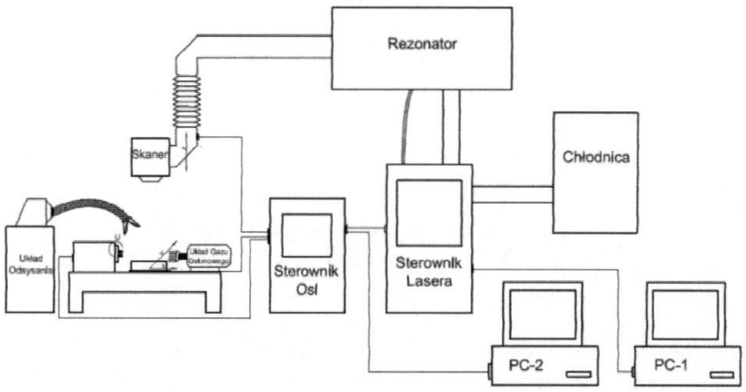

Figure 1. TruMicro5325c laser test bench layout.

The characteristics of the basic units of the laser machine for micromachining are as follows:

Laser TruMicro 5235c - type of laser: pulsed diode impulse laser disk with 3 harmonic generation, - wavelength: 343 nm, - average power: 5W, - minimum pulse duration: 6,2ps, - 400kHz pulse frequency with the possibility of dividing by natural numbers from 1 to 10000, - maximum pulse energy: 12,6μJ, - mod: TM_{00}, - M^2=1,3 - maximum fluency: 4,8 J/cm^2

Scanner IntelliSCAN 14 - biaxial scanner (XY) for ultraviolet light 343 nm, - aperture: 14 mm, - standard F-Theta lens with focal length: f = 160 mm, - Rayleigh length - 0,76 mm, - spot diameter - 18,2 µm.

XYZ coordinate table - XYZ movement range: 200 mm, 200mm, 275 mm, - bearing capacity: 70kg spread evenly distributed load, 10kg for the Z axis, unevenly distributed load, - repeatability of positioning: +/- 0.0001 mm, - resolution of measurement of displacement: 0.5 micrometer, - max speed: 500mm/s.

Experimental

Three materials used in the construction of seals with distinctly different properties were selected for the study. NBR nitrile rubber is the most popular material for general purpose seals. PEBAX is used for precision seals in laboratory and medical equipment. Polamid PA6 is characterized by its high stiffness, high temperature resistance and durability. It is used, among other things, for flat seals and support rings.

Table 1. Materials tested - selected properties.

Name of the material	Density of the material kg/cm^3	°Shor hardness
NBR nitrile rubber	1.1	50
Poliamid PA 6	1.13	74 (skala D)
PEBAX	0.59	54

The purpose of the conducted research was to determine the possibility of micro-surface machining of selected elastomer plastics for shaping structures useful in sealing technology.

The attempt consisted in making longitudinal micro recesses on the materials tested with a maximum laser power of 400 kHz and variable scan speeds of 1000, 2000, 3000, 4000 and 5000 mm / s. The treatment marks were identified by analyses done using HIROX microscopy.

An exemplary 3D view of the cut and section profile is shown in Figures 2 and 3 HIROX microscope software allows you to determine the depth and width of the groove as well as its cross-sectional area. Knowing the scan speed and material density, the machining efficiency was determined. A summary of the measurement results is shown in Table 2

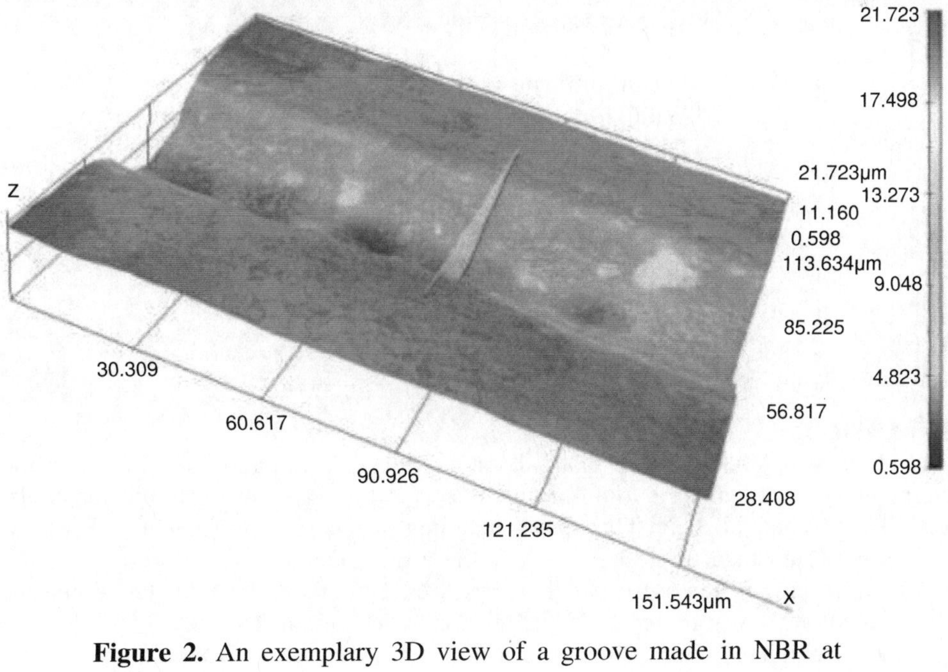

Figure 2. An exemplary 3D view of a groove made in NBR at 1000 mm / s.

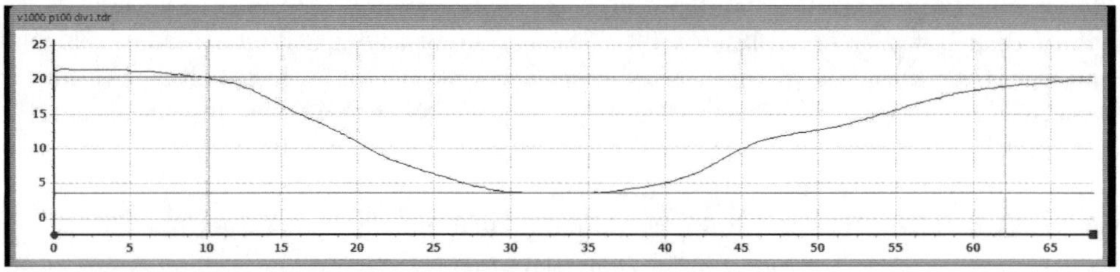

Figure 3. Profile of groove section in NBR at 1000 mm / s. mm/s.

Table 2. Summary of research results.

Scan speed [mm/s]		NBR	PA 6	PEBAX
1000	Depth [μm]	16.861	11.545	24.004
	Width [μm]	51.936	39.906	58.956
	Machining efficiency [g/s] x 10^{-6}	349.277	224.353	369.026
2000	Depth [μm]	7.883	5.441	11.495
	Width [μm]	53.551	25.494	37.438
	Machining efficiency [g/s] x 10^{-6}	294.382	126.985	163.947
3000	Depth [μm]	4.617	2.970	8.446
	Width[μm]	40.756	15.926	36.593
	Machining efficiency [g/s] x 10^{-6}	219.453	113.943	208.920
4000	Depth [μm]	4.503		6.823
	Width [μm]	41.897		38.194
	Machining efficiency [g/s] x 10^{-6}	291.540		240.625
5000	Depth [μm]	3.955		2.803
	Width [μm]	37.555		24.074
	Machining efficiency [g/s] x 10^{-6}	271.277		107.391

5. Analysis of results

Surveys and observations have shown that a laser emitting UV radiation at 343 nm in picosecond pulses can be recommended as a tool for the micromachining of elastomeric materials and, in particular. NBR, PA6 and PEBAX. The micromachining of recesses is efficient and precise and the workpiece does not lose its elasticity. At the site of the treatment and its surroundings, no charring and other signs of overheating were observed. These recesses retain their characteristic dimensions with a +/- 15% deviation along the entire length of the groove, which demonstrates the high repeatability and accuracy of the machining. The estimated removal efficiency of the material is in the range of 369.026 10-6 g / s to 107.391 10-6 g / s and was the highest for PEBAX. The hardest material to work was PA6 which is the material of the highest density and hardness of the tested materials. As the scanning speed increases when the recesses are made, the process yield and depth of grooves are reduced. The greatest changes in depth occur when passing from 1000 mm / s to 2000 mm / s. Detailed microscopic examinations are required to evaluate structural changes at the work site. The conducted research confirms the possibility of applying the developed technology to shaping the surface parts of the seals cooperating with the surface of the machine element or the sealing housing. This opens new prospects for improving the performance of seals such as reducing friction and damage probability, improving lubrication conditions, and increasing durability.

6. References

[1] Masayuki Okoshi, Jianzhao Li, Peter R Herman 2005 F-laser writing of silica optica waveguides in silicone rubber *Photon Processing in Microelectronics and Photonics IV*, ed J Fieret, P R Herman, T Okada, C B Arnold, F G Bachmann, W Hoving, K Washio, Y Lu, D B Geohegan, F Träger, J J Dubowski (SPIE, Bellingham, WA) **5713**

[2] Anil Kurella, Narendra B Dahotre 2005 Review paper: Surface Modification for Bioimplants: The Role of Laser Surface Engineering *Journal of Biomaterials Applications* **20**

[3] N E Stankova et al. 2015 f- and ns-laser processing of polydimethylsiloxane (PDMS) elastomer: Comparative study *Applied Surface Science* **336** pp 321-328

[4] C Dupas-Bruzek et al. 2009 Transformation of medical grade silicone rubber under Nd: YAG and excimer laser irradiation: First step towards a new miniaturized nerve electrode fabrication process *Applied Surface Science* **255** pp 8715-8725

[5] Liu H.B., Gong H.Q. 2009 Templateless prototyping of polydimethylsiloxane microfluidic structures using a pulsed CO2 laser *J. Micromech. Microeng.* **19** (3) pp 1-8

[6] Radek N, Pietraszek J, Antoszewski B 2017 The Average Friction Coefficient of Laser Textured Surfaces of Silicon Carbide Identified by RSM Methodology *Advanced Materials Research* **874** pp 29-34

IOP Conf. Series: Materials Science and Engineering **233** (2017) 012037 doi:10.1088/1757-899X/233/1/012037

Problems and criteria of quality improvement in end face mechanical seal rings through technological methods

V Tarelnik[1], A Belous[1], B Antoszewski[2], A Zukov[3]

[1] Sumy National Agrarian University, 160 G. Kondrateva Str., 40021, Sumy, Ukraine
[2] Centre for Laser Technologies of Metals, Kielce University of Technology, Aleja Tysiąclecia Państwa Polskiego 7, PL- 25314 Kielce
[3] PJSC "Sumy NPO", 58 Gorkogo Str., Sumy, 40004, Ukraine

E-mail: tarelnik@i.ua

Abstract. In this paper are presented the recommendations for material's selections of the mechanical seals rings and basic productive and operating requirements. The system of a directional selection of technology that ensures the required quality of working surfaces of the mechanical seals rings covers their entire life cycle. The mathematical frictional model is proposed as an instrument for calculating a linear and weighing abrasion of the mechanical seals rings and helps to improve selection's criteria and the most rational method of strengthening.

1. Introduction

The creation of reliable seal assemblies, that ensure tightness for a long time in a wide range of temperatures and pressures, is one of the major problems in the design of pump and compressor machines and units.

One of the most common sealing elements used in pumps, compressors and various chemical apparatus (reactors, mixers, etc.) is a mechanical seal (MS). It is a sealing device consisting of two parts in the form of rings: one of which rotates with the shaft and the other is fixedly connected to the body (Figure 1).

Figure 1. Rings of friction pairs of mechanical seals.

Seal performance is influenced by design, technological and operational factors. The most important ones are the properties of working environment, modes of operation, material properties of a sealed connection, limits on the leakage, the resource, the total period of use, toxicity and chemical environments.

Seal rings are a friction pair, which serves as the main sealing element. Hence they are made of special materials, whose selection depends on the operating conditions. The properly selected material of the rings provides reliable, safe and trouble-free operation of the seal assembly and, therefore, the entire unit.

2. Analysis of the main achievements and publications

According to [1], the design choice is largely determined by the physical and chemical properties of the medium, its state of aggregation (gas, liquid), pressure, temperature, viscosity, content of suspended solid particles and salts, chemical aggressiveness, flammability (being heated in contact with the atmosphere), the degree of hazard to humans and the environment.

It should be noted that the correct choice of material and its mechanical and physical characteristics plays the significant role in selecting design of the sealing rings. It is the combination of the materials for friction pairs what is more crucial rather than materials per se.

In the selection of the optimal sliding couples it has to be considered the corrosion resistance and durability of the materials, the possibility of heat removal from the zone of friction and materials compatibility i.e. the ability to work without setting and jams [2].

For each case the choice of the most suitable materials of rings can be made only by a careful comparison of their working conditions, raw material properties, the changes, which occur on the friction surfaces, etc.

Some rules are proposed in [3] for the combination of materials. For example, it is recommended to combine a solid material with a solid material (a combination of the nitrided, chrome and hardened steels). Such friction pairs have a high wear resistance due to the small mutual penetration of their surfaces. The application of the running-in coating improves the reliability of pairs in the most dangerous period of work - during running-in. The application of these pairs is limited by sliding velocities. High precision manufacturing and assembly, considerable structural rigidity, rigorous running-in, improved lubrication conditions greatly expand the scope of the friction pairs made of hard materials.

In [4] a parameter is used to describe the intensity of working conditions of mechanical seals, which is the product of the pressure p in the fluid average circumferential velocity slip ν in a friction pair. Depending on the parameters p and ν seals are subdivided into four categories (Table. 1).

Table 1. The classification of seals by the parameter $p\nu$

Seal category	P, MPa	ν, m/sec	$p\nu$, MPa· m/sec	Parameter definition
I	≤ 0,1	≤ 10	≤1	Low
II	≤ 1,0	≤ 10	≤ 5	Average
III	≤ 5,0	≤ 20	≤ 50	High
IV	>5,0	>20	>50	Highest

In [5] it is summarizes the results of studies [6-9] and presented requirements on the materials of friction pairs, which are designed to work as the mechanical seals in contact with aggressive media:

- Resistance to aggressive environments;
- Impermeability to fluids in contact with the rings;
- Resistance to temperature gradient;
- Absence of setting and seizure;
- Sufficient frictional heat resistance;
- The maximum possible service life;
- Time stability of the coefficient of friction and its minimum value.

Depending on the operating conditions, the following materials are proposed: plastic antifriction carbon-based materials, metals (iron, steel alloys, non-ferrous metals), and other hard alloys. [1-9].

According to [10], the seals for rotors have to meet two main requirements: have the desired tightness and high reliability at a given pressure drop, rotating speeds, temperatures and physical properties of the sealed medium.

It should be noted that the mechanical seal shaft requires a minimum leakage for the operation. As a result, the gap between the rings must be very small, lubricating film has to be very thin and the surface of sealing materials must withstand the friction against each other, often at high speeds and loads. Therefore, the materials of mechanical seal surfaces must have a low coefficient of friction, sufficient hardness, good corrosion resistance and high thermal conductivity.

Mechanical seals of non-contact type (Figure 2) are widely applied in the high-speed pumps and compressors for high pressure, where the end faces of the rings are in contact for a very short time in moments of starting and stopping the machine. Guaranteed clearance size 0.003-0.004 mm between the ends of a pair leads to the fact that the sealing surfaces of the rings almost do not wear out. Despite that, their sealing rings are made entirely of wear-resistant materials such as tungsten carbide, silicon carbide, various kinds of graphites. The cost of such rings is hundreds and thousands US dollars, resulting in the high cost of seal assemblies in general.

The seal (Figure 2) comprises an axially movable ring 1, set in the housing 2, which is preloaded with spring 3 and the metal supporting ring 4, fixed to shaft 5. At the end of the working surface 6 of the ring 1 there is a set of chambers 7 intermittently arranged along the circumference of the ring . On the ring 4 there are several tangential channels 8, with middle portions of which located on the axis at the same distance as the cameras 7. The number of channels 8 is less than the number of cameras 7, so that they can communicate with the cavity of hardening medium only a part of cameras. Secondary seal 9 of the axially movable ring 1, which is shaped in the form of a thin metal shell, is made as one piece with the ring 1. This shell is shaped in a conical whisker, thickening towards the connection to the axially movable ring 1. This conical whisker has a sealing surface 10 in place of the contact with the sealing surface of the bushing 11.

Note that when the whisker (the secondary seal) contacts with the sleeve, fretting process occurs, i.e. whisker surface 10 reciprocating (with an amplitude of 3 - 4 mm) on the surface of the sleeve 11. As a result of the fretting corrosion the destruction of the contact surfaces happens. Thus, the contact surface whiskers except watertight connections should have high wear resistance.

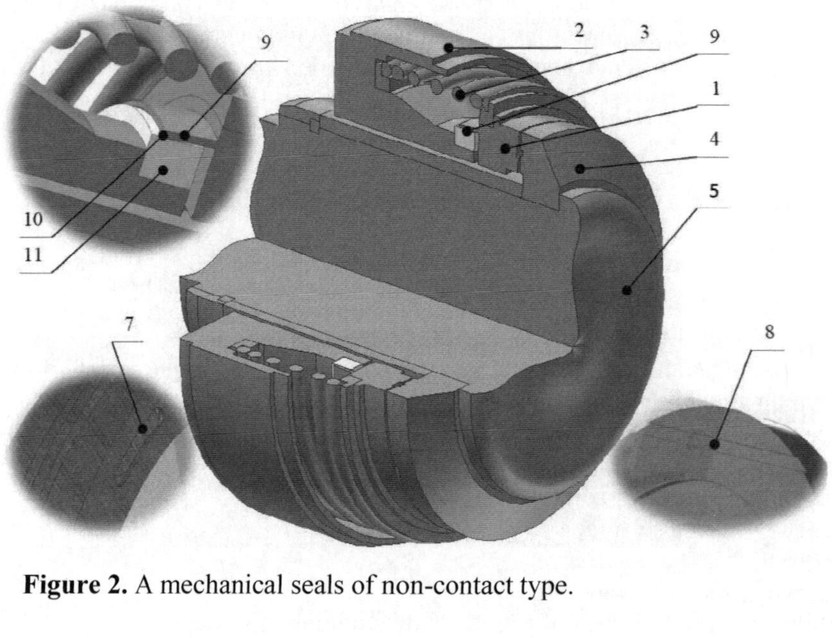

Figure 2. A mechanical seals of non-contact type.

At the same time, the improvement of the technique demands a high and ultrahigh pressures, temperature extremes (from highest to cryogenic), corrosive environments and the like. The impact of such conditions on the tightness of detachable joints is very huge. Therefore, the use of metal composite materials, such as "basis-coating" that combine the protective properties of the coating with the mechanical strength of its foundations [11], becomes advantageous to the manufacture of mechanical seas rings.

In [12] it is analyzed the work of the gate pulse compression. It is noted that the use of such seals can save energy and resources and increase the environmental safety of the pumping and compression equipment.

The change in protective and tribological properties of the surfaces of the parts can be achieved through the formation of a special relief surfaces of friction pairs [13,14].

As a reserve of the improvement of the quality specifications can be the use of composite materials in the plated steel parts. The usage of the cladding parts with special material properties allows to create units with the necessary strength, reliability and durability [15].

Given that not all materials can be used for surfaces of mechanical seals, their choice is, therefore, crucial.

In the literature, there are no comprehensive studies aimed at developing the technology of the selection of the material for rings and their elements, the use of which will provide the maximum life of the specification. None of the available literature recommendations on the choice of materials for mechanical seals is universal.

The aim of this work is to develop the system and criterions of a directional selection of the technology that ensures the required quality of working surfaces of the mechanical seal rings by the analysis and synthesis of existing counterparts, industry experience and suggestions in domestic and foreign literature.

3. Statement of the basic material research

The system of a directional selection of technology that ensures the required quality of working surfaces of the mechanical seals rings covers their entire life cycle (Figure 3), which includes the material of rings and their elements, the technology of manufacture, maintenance etc. They are viewed through special directed choice techniques. It is necessary to consider the mutual impact of selected methods, which will ultimately affects the quality of the whole product.

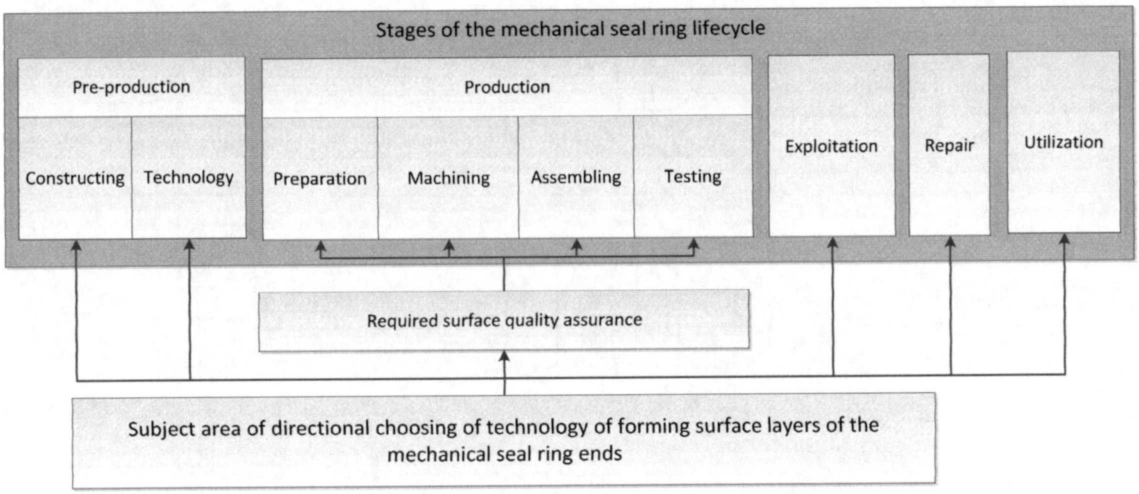

Figure 3. Using of directional choosing of technology of forming surface layers of the mechanical seal ring ends at different lifecycle stages.

The need for a systemic approach to research requires analysis of the expedient use of directional selection technology to ensure the required quality of the surface layers of the rings at all stages of their life cycle. It is now obvious that the problems of improving the wear resistance of details of friction units should be carried out in close co-operation design, technological and tribological solutions. Proper selection of materials is only possible if the analysis of structural and tribological characteristics of friction unit and the conditions of its operation is carried out.

At the stage of pre-production design the design of the mechanical seal rings, which perform certain functions, it is important to know the techniques, the use of which may provide the required characteristics of the surface and in accordance with the quality parameters to assign it (technological rationality of design). Production experience shows that the range of such methods can be wide.

At the pre-production technological stages the knowledge of the methods to improve the quality of the surface layers of machine parts allows to plan a rational technology of obtaining the desired properties.

As a result of research, it is possible to choose the most efficient method of producing ring billets of the required quality. Perhaps they will be made of cheaper materials, with less allowance for processing etc. It is possible a more rational use of thermal machining, reducing the number and duration of the individual stages.

Knowing the requirements that apply to the surfaces of the rings, it is possible to select those methods of machining that will be most suitable and economically feasible.

It is also necessary to know the obtained results of studies when planning and implementing the assembly process. The choice of assembly operations: welding, assembly, thermo-assembly etc. - is dependent on the quality of the surface of the previous layer. This leads to a deeper analysis of the assembly process, as in the final stages of the production process the necessary characteristics of the product are finally formed.

During the formation of the surface layer with the desired characteristics control methods and testing of the seal are applied. Knowing the material and surface quality rings, it can be predicted the conditions under which mechanical seals will work better or worse and in this regard, using the obtained results, it is possible to manage the process of rational exploitation.

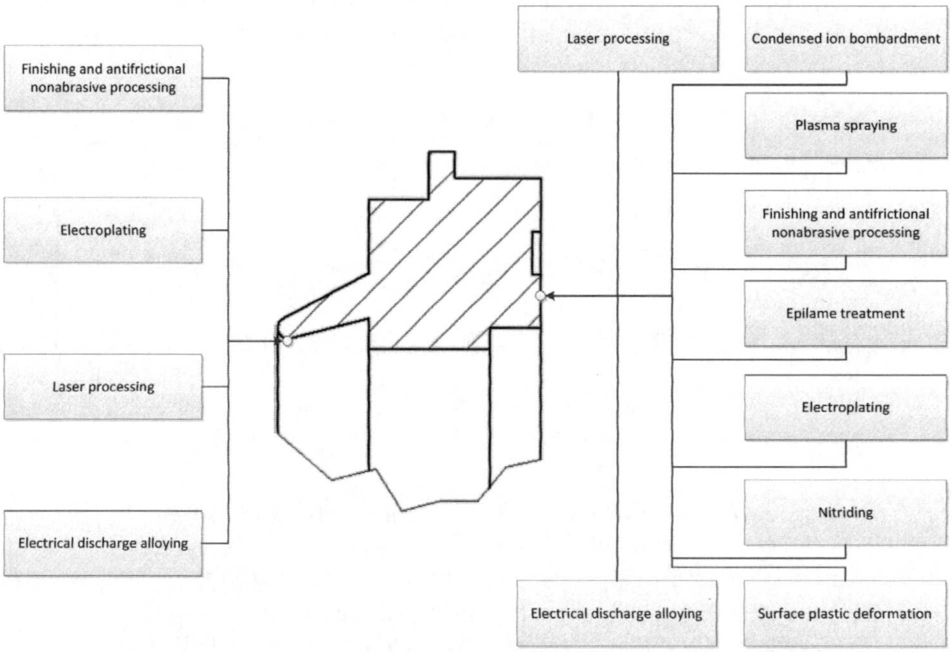

Figure 4. Methods to achieve the required quality of mechanical seas rings surfaces.

Using the directional choice of technology to ensure the required quality specifications for rings at the stage of repair allows to solve the problem to restore their state more efficiently. Practice shows that at this stage the use of research results has a significant economic impact.

The results also can be used for the rational utilization of the rings, since processing thereof largely depends not only on the material composition and structure, but also on their sealed environment (acid, venoms, foods, etc.).

Figure 4 shows a set of methods to achieve the desired quality of the contacting surfaces of the rings 1 and 4 specifications and the sealing surface of the whisker (Figure 2). This set is the practical field application of the results of academic research undertaken.

While solving the problem of improving the quality of ring surface layer, it is important to consider not only the cost but also the environmental characteristics of the process (Figure 5).

The costs of maintaining the environmental characteristics of rings are added to the overall costs. It is worth to be noted that environmental characteristics may be used as an independent optimization criteria in choosing pre-selected cost-effective options. A set of methods of forming the surface of a required quality, a number of layers in coating and equipment modes of operation can vary during optimization.

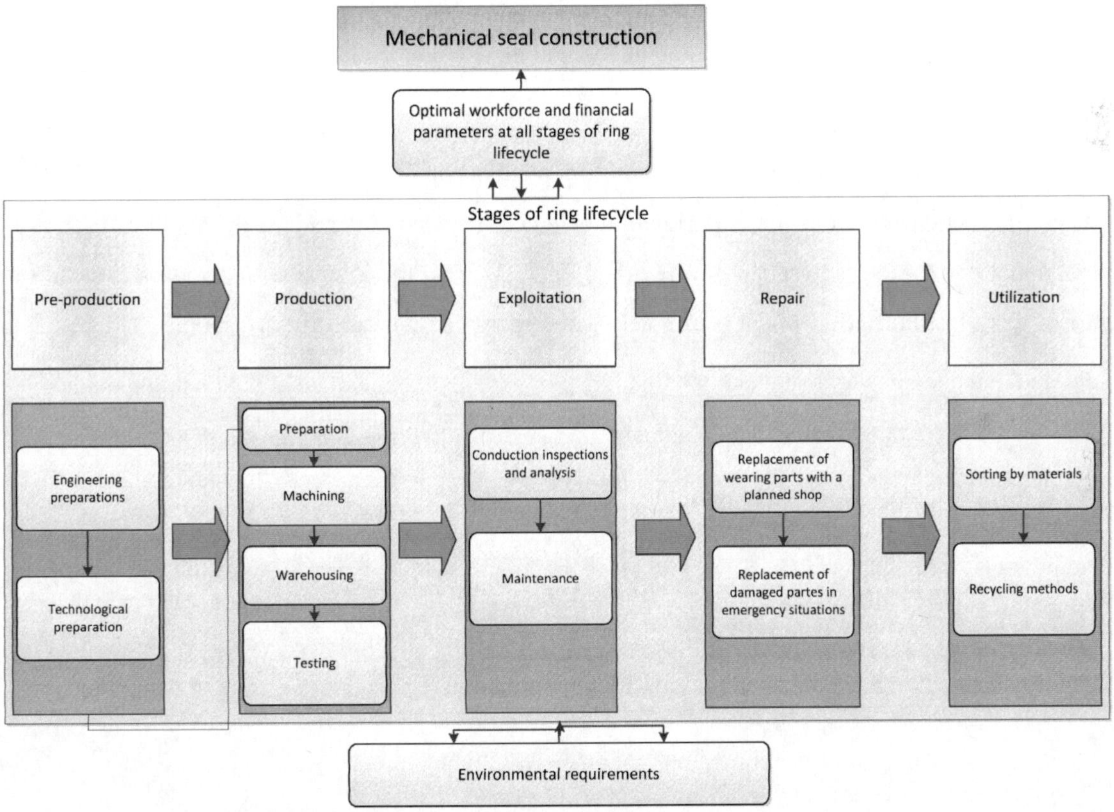

Figure 5. Relationship between ring technological parameters at different stages.

There are a lot of criterions for measuring parameters of quality of surface layers such as: full workload, the fatigue cycles limit, surface hardness, creep limit, constructional parameters of details etc. Different varieties of them can be used depend on the operating condition.

During the exploitation period, energy transformation from outside energy to energy of inside processes is occurred by temperature and diffusion mechanism. As a result, structural properties of materials could be reversed and lifecycle of whole machine will be reduced.

IOP Conf. Series: Materials Science and Engineering **233** (2017) 012037 doi:10.1088/1757-899X/233/1/012037

The main benefit of using the energy as the base criteria in technological process is the creature of mathematical model which includes all operating, technological and constructional parameters and helps to provide required characteristics of details.

Intensive mechanical loads, which are the result of outside friction processes, lead up to sharp change in mechanical and physical characteristics of detail's working surfaces therefor the main criteria of abrasion is the energy of activation [16].

The disconnection of links is a discrete stage of activation whereby the energy of activation E_A is a difference between the link's energy U_0 and the energy of deformation $\gamma \cdot \sigma$.

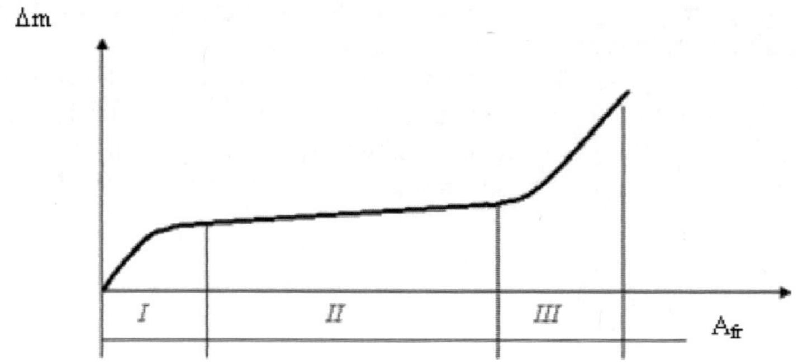

Figure 6. The stages of abrasion:
1 – cold burnishing;
2 – stable abrasion;
3 – destructive abrasion.

Figure 6 shows that in the first and second stages there is the exponentially increasing reliance between work lost in friction (A_{fr}) and weight loss by abrasion (Δm).

It can be concluded from the experimental reliance between Δm and $(-A_{fr})^{-1}$ that there is the proportional correlation between $\ln \Delta m$, $(-A_{fr})^{-1}$ and E_A. So the intensive of abrasion depend on a value of work lost in friction which is as much as the energy of activation $\ln \Delta m \approx (-A_{fr})^{-1}, E_A$.

$$\Delta m = C \cdot \exp\left(\frac{-E_A}{A_{fr}}\right) \tag{1}$$

$C = \Delta m_{max}$ – value of maximal tolerable abrasive. Than,

$$\Delta m = \Delta m_{н} \cdot e^{\frac{-E_A}{A_{mp.}}} \tag{2}$$

The correlation (2) is the equation of abrasion.

If we assume that $E_A = A_{fr}$ than (2) can be transformed in (3).

$$\frac{\Delta m}{\Delta m_{max}} = e^{-1} \tag{3}$$

Consequently, the energy of activation E_A is the physical quantity which shows what work lost in friction is needed to serve the condition (3). E_A [J] is a constant of abrasion.

In order to measure linear abrasion (the alteration of a detail's dimensions) the equation (2) has to be transformed in (4)

$$\Delta h \cdot S \cdot r = \Delta h_{н} \cdot S \cdot r_{max} \cdot \exp(\frac{-E_A}{A_{fr}}) \tag{4}$$

S – the area of wearing surface;
r – density of surface layer (in the first and second stages Figure 6);
$r_{\max}$ – density of surface layer (in the beginning of the third stage Figure 6).
$\Delta h_{\max}$ - value of maximal depth of abrasive.

After transformation:

$$\Delta h = \Delta h_{\max} \cdot \frac{r_{\max}}{r} \cdot \exp\left(\frac{-E_A}{A_{fr}}\right) \tag{5}$$

If we assume that $E_A = A_{fr}$ than (5) can be transformed in (6) and (7).

$$\Delta h = \Delta h_{max} \frac{r_{max}}{r} \cdot e^{-1} \tag{6}$$

$$\frac{\Delta h \cdot r}{\Delta h_{\max} \cdot r_{\max}} = e^{-1} \tag{7}$$

Consequently, the energy of activation E_A is the physical quantity which shows what work lost in friction is needed to serve the condition (7) if densities r and $r_{\max}$ are equal.

In order to measure a value of work lost in friction for weight loss by abrasion ore for linear abrasion the equation (2) and (5) have to be transformed in (8) and (9) accordingly.

$$A_{fr.} = \frac{E_A}{\ln \frac{\Delta m_{\max}}{\Delta m}} \tag{8}$$

$$A'_{fr} = \frac{E_A}{\ln \frac{\Delta h_{\max} \cdot r_{\max}}{\Delta h \cdot r}} \tag{9}$$

Using this mathematical system to ensure the required quality specifications for rings at the operation stage allows to solve the direct task (quantification of weight and linear abrasion when the work lost in friction is known) and the indirect task (quantification of work lost in friction when weight ore linear abrasion is known). In addition, if duration of abrasion is known the rational exploitation of machine will be possible.

The results also can be used for reliable prediction of surface abrasion for wide variety of technological methods. The main criterions for system of directional choosing of technology of forming surface layers are the constants of the equation of abrasion ($E_A, \Delta h_{\max}, \Delta m$). If the basic input dates (value of abrasion, time and operating costs) are known it will be easy to choose the optimal technology which allows to achieve pre-set surface properties (Figure 7).

IOP Conf. Series: Materials Science and Engineering **233** (2017) 012037 doi:10.1088/1757-899X/233/1/012037

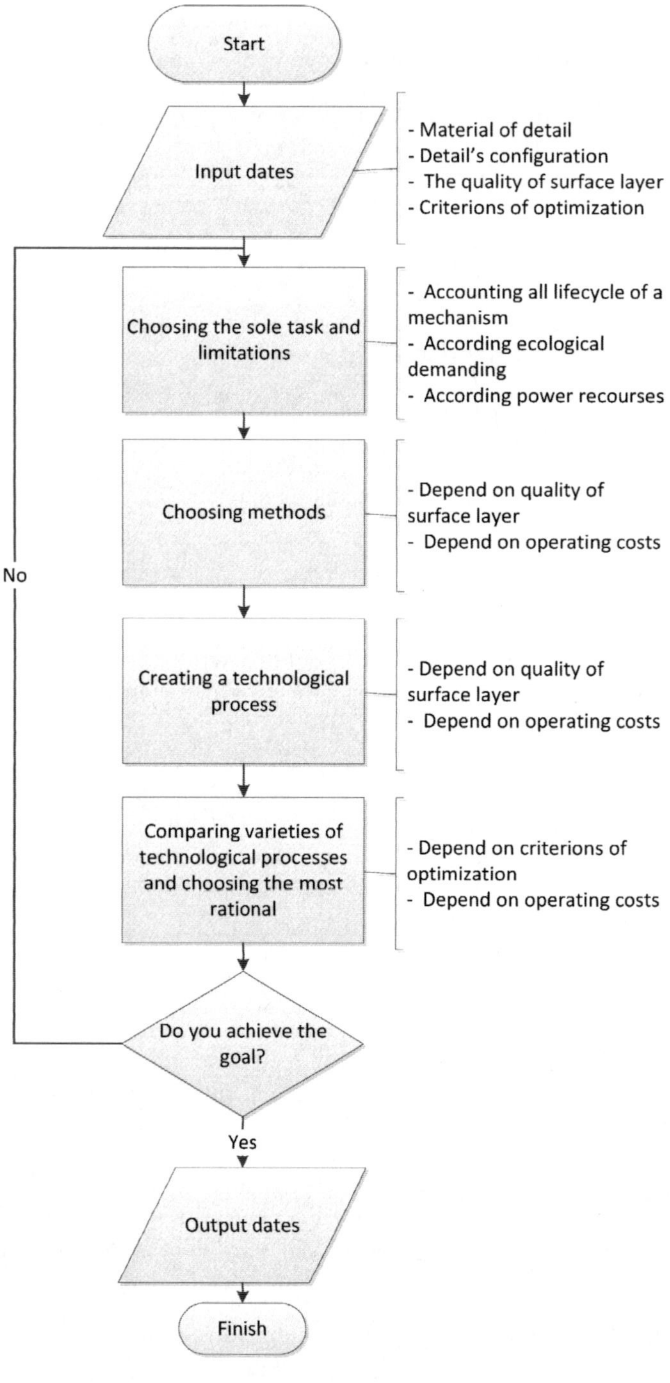

Figure 7. Algorithm of the directional choice of technology to ensure the required quality specifications for details.

4. Conclusions

This research allows to develop a common point of improving the quality of the surface layers of the mechanical seal ring ends, according to operating requirements.

It is proposed a system of directional choosing of technology of forming surface layers of the mechanical seal ring ends, which takes into account all stages of their lifecycle. Economic and environmental requirements are taken into account at the same time.

In order to achieve the required quality of rings surface, depending on its type and requirements to it, either separate methods or their combinations may be used.

It is proposed a mathematical model of abrasion which can measure the quantification of weight and linear abrasion when the work lost in friction is known for different quality of the surface layers of the mechanical seal ring.

It is developed the methodology for computing abrasion's constants E_A and constants (Δh_{max}, Δm) of equation of abrasion for various materials of sliding couples.

References

[1] Kondakov L A, Golubev A I, Ovander V B 1986 *Seals and sealing technics* (Moscow: Mechanical engineering) p 464
[2] Zaharov B S, Zaharov I B 2011 *Seals for oil centrifugal and piston pumps* (Moscow: Open Society) p 204
[3] Garkunov D N 1978 *Tribotehnika* (Moscow: Mechanical engineering) p 327
[4] Mayer E1978 *Mechanical seals* (Moscow: Mechanical engineering) p 288
[5] Golubev G A, Kukin G M, Lazarev G E, Chichinadze A V 1976 *Contact seals of rotating shafts* (Moscow: Mechanical engineering) p 264
[6] Golubev G A 1973 The friction and wear researches of seals at the sound speed *The rational using of friction materials in sliding couples* (Moscow: Science) pp 172–210
[7] Elkin A I, Nikolaev V V 1967 The frictional property rubbermaterials at heats in vacuum *Reports AS the USSR* (Moscow: Science) **3** pp 152-161
[8] Kolchenko A V, Silin A A 1957 About a mode and the friction reliance of rubber in oil *Reports AS the USSR* (Moscow: Science) **4** pp 112-134
[9] Lukjanow S 1972 Automatyczne sterowanie stanowiska do badania pier scieni uszczelniajacych *Biuletyn informacja techniczny motoryzacja* (Warszawa: Politecnica) **4** pp 28-33
[10] Martsynkovskyi V A 2012 *The dynamics of centrifugal machines rotors* (Sumy: Sumy state university) p 563
[11] Tarelnik V B, MartsinkovskyV S, Antoshevsky B 2012 Sliding surface formation in machine seals by various methods *Bulletin of the Samara State space university* (Samara: Samara state university) **3 (34)** chapter 1 pp 217-225
[12] Martsynkovskyy V, Zahorulko A, Gudkov S, Mischenko S 2012 Analysis of buffer impulse seal *Procedia Engineering* **39** pp 43-50
[13] Antoszewski B 2014 Influence of Laser Surface Texturing on Scuffing Resistance of Sliding Pairs *Advanced Materials Research* (Warszawa: Politecnica) **874** pp 51-55
[14] Błasiak S, Kundera Cz 2012 Numerical analysis of the grooved surface effects on the thermal behavior of a non-contacting face seal *Procedia Engineering* **39** pp 315-326
[15] Kirik G V 2011 *New composite materials* (Sumy: the University book) p 310
[16] Vorobev A A 1969 Base alterations of material's quality due to electrical discharge alloying *Material electronical treatment* (Moscow: Mechanical engineering) **6** pp 25-30

IOP Conf. Series: Materials Science and Engineering **233** (2017) 012038 doi:10.1088/1757-899X/233/1/012038

Express - method for determining the quality of a magnetic fluid for operation in the working gap of a magnetic fluid seal

A V Radionov[1], A D Podoltsev[2], A A Radionova[1]

[1] SMPE Ferrohydrodinamica, Bolshaja Morskaja Street, 45/5, Nikolaev 54030, Ukraine
[2] Institute of Electrodynamics NAS of Ukraine, 56, Pobedy av., Kiev-57 03680, Ukraine

Email: ferrohydrodynamica@gmail.com

Abstract. The reliability of the magnetic fluid seals operation is reduced to an assessment of the quality control of a magnetic fluid associated with its stability in the working gap of the seal in many cases. The previously developed relatively simple method for express analysis of the quality of a magnetic fluid is supplemented by studying the dependence of the dynamic characteristics of a nanofluid on the size of magnetite particles. This method can be easily implemented in an industrial enterprise using devices with a magnetic fluid.

1. Introduction

Recently, magnetic fluid seals (MFS) have found wide application for sealing systems of bearing units operating under severe operating conditions (abrasive, moisture, dust, etc.) [1]. This is due to the fact that the use of MFS instead of traditional seals leads to a significant increase in the reliability and operability of process equipment, as well as to reduce the risk of environmental and technological accidents [2].

In many cases the problem of reliable MFS operation leads to an assessment of magnetic fluid (MF) quality control, associated with its stability in the working gap of the seal. Magnetic colloids are complex objects with polydisperse particles capable of forming complex magnetostructural bonds under the influence of strong magnetic fields, which have a significant effect on the colloidal stability of MF.

The synthesis of any magnetic fluid is a compromise problem that should satisfy conflicting requirements: a relatively weak magnetophoresis of particles in a gradient magnetic field should be combined with a sufficient saturation magnetization and an acceptable value of viscosity.

Quality control of MF and its diagnostics as a complex physico-chemical system sets a number of technical problems. First of all, it is the task of selecting quality parameters for an adequate evaluation of the MF's performance in certain specific conditions.

The magnetic characteristics of MFs are the most often used ones in scientific research. However, reliable results are obtained using the expensive and narrow-profile instrumentation, which is unacceptable for arranging entrance control at enterprises operating equipment with installed MFS.

A relatively simple method for determining the sedimentation stability of MFs was proposed in [3,4], based on the measurement of the force acting on a sample with a MF in an inhomogeneous magnetic field, as well as measuring the dynamics of the time variation of this force using high-precision electronic weights [5]. Feature of the work [4] was the conduction of experimental studies on a specially constructed gear structure sample with a MF, similar to the one that exists in MFS. Eight types of industrial MFs on different bases manufactured in LLC "Ferrohydrodynamica" have been selected as samples. However, the effect of the dimensions of magnetite nanoparticles on the performance of MFs has not been studied in this paper. High-dispersion magnetite for all eight types of magnetic fluids was synthesized by the chemical condensation method, which is based on the process of precipitation of the salts of two- and trivalent iron by a concentrated aqueous solution of ammonia (Elmore reaction [6]). The average particle size obtained by the Elmore reaction is 7 ... 20 nm. At room temperature magnetite particles are in a superparamagnetic state and are characterized by a practically zero remanent magnetization in this range of sizes. It is the only range of particle sizes sedimentationally stable magnetic fluid can be synthesized.

The aim of this work is to study the influence of the dimensions of magnetite nanoparticles on the magnitude of the magnetic force acting on a sample filled with a magnetic fluid, and also on the character of the change of this force over time for the implementation of the express method for determining the MF sedimentation stability.

2. Research results

It is known that the physical and chemical properties of a magnetic fluid depend on the technology of MF preparation in a rather complicated manner. The MF synthesis features have an affect on this properties. They are the rate of inflow of a solution of salts of two- and trivalent iron to alkali, the intensity of mixing of the mixture, the choice of a precipitant, the removal of water, and the temperature regime. In addition, in a certain way, the properties of MF depend on the type and concentration of the stabilizer, the composition of the dispersion medium, the presence of various additives. The number of factors and features of synthesis technology that affect the properties of industrial MF is so great that it is not always possible to control them and, moreover, obtaining a fluid with given properties is often an unsolvable problem. Therefore, a magnetic fluid from one batch, but with different centrifugation times have been selected as reserch sample. Standard MF have been prepared with a centrifugation time of 1 hour, which corresponds to the technological regulations adopted in the "Ferrohydrodynamica". Also, the second part of the MF batch was centrifuged for 10 minutes, and the third batch was centrifuged for 4 minutes. Vacuum oil VM-3 (most widely used in industry), apiezon (a promising new material for a magnetic fluid), and kerosene (the properties of kerosene-based MF have been studied in detail by other authors and there is the possibility of our data comparison with the data of other researchers) have been chosen as the dispersion media.

The technical characteristics of the three samples of the magnetic fluid changed insignificantly, depending on the time of centrifugation. This is clearly seen from Table 1, where the characteristics of MFs based on kerosene are presented.

Table 1. Technical characteristics of a kerosene based magnetic fluid with different centrifugation times.

Time of centrifugation (min)	Volumetric concentration (%)	Saturation magnetization (kA / m)	Density (kg / m3)
60	16,8	60,75	1315,0
10	16,9	57,63	1312,8
4	18,1	59,84	1358,1

At the same time, the yield of the magnetic fluid at the centrifugation time for 1 hour decreased by approximately 10% compared to the reduced centrifugation time, and the sediment mass was approximately 2 times greater.

The precipitate when synthesizing standard MF was thick, black tar-colored and did not spread. And when obtaining a magnetic fluid with reduced centrifugation time, the sediment consisted of two fractions (solid and liquid). The solid fraction was more friable, the liquid fraction solidified in 24 hours. The observed processes were common to all three tested liquid bases. Figure 1 shows the precipitation after centrifugation of MF based on kerosene.

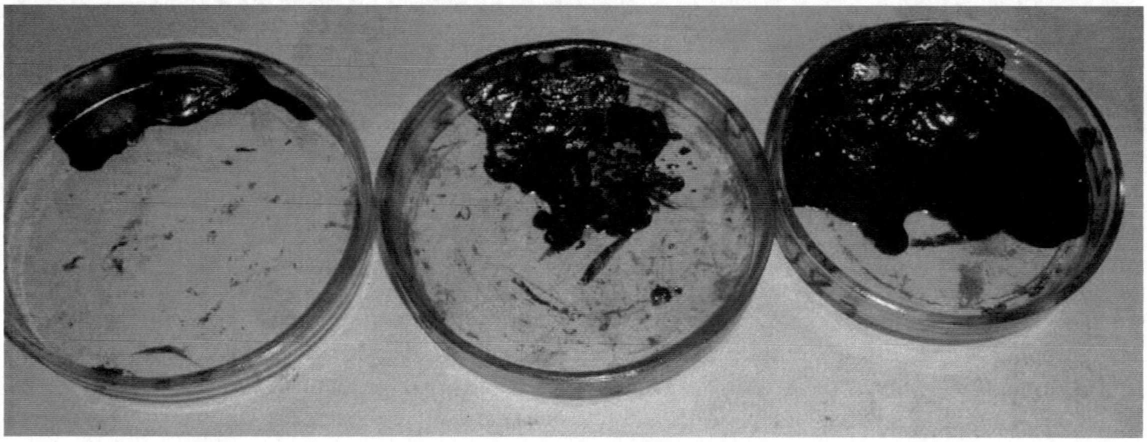

(a) (b) (c)

Figure 1. Precipitation after centrifugation of the kerosene-based magnetic fluid at the time of centrifugation: (a) 4min; (b) 10 min; (c) 60 min.

Changes in the formation of incompletely centrifuged MFs should be manifested in the nature of the frequency and temperature dependences of the magnetic parameters of the MF. In order to study the internal state of magnetic colloids, magnetic susceptibility is traditionally chosen as the studied magnetic characteristic. This is due to the fact that it is the most structurally sensitive parameter, in addition, the results of its study, can also provide information on the mechanism of relaxation of the magnetization of magnetic colloidal systems [7].

The magnetic susceptibility can be determined from the slope angle of the initial section of the magnetization curve. The linear initial sections for three types of MF based on kerosene with different centrifugation times, are shown in Figure 2. It follows that a decrease in the centrifugation time leads to an increase in the angle of inclination, and, accordingly, to an increase in the magnetic susceptibility. A lot of articles, detailed reviews in a number of monographs, for example, in [7-9], etc., have been devoted to the experimental determination of the frequency and temperature dependences of the magnetic susceptibility and the relaxation time for MFs based on kerosene. Their amount considerably exceeds the number of publications for MFs on more viscous bases. In order to be able to compare our results on the real and imaginary parts of the effective magnetic susceptibility with the work of other authors, they are presented for the MF based on kerosene.

The magnetic susceptibility was determined by the bridge method, described in detail in [7].

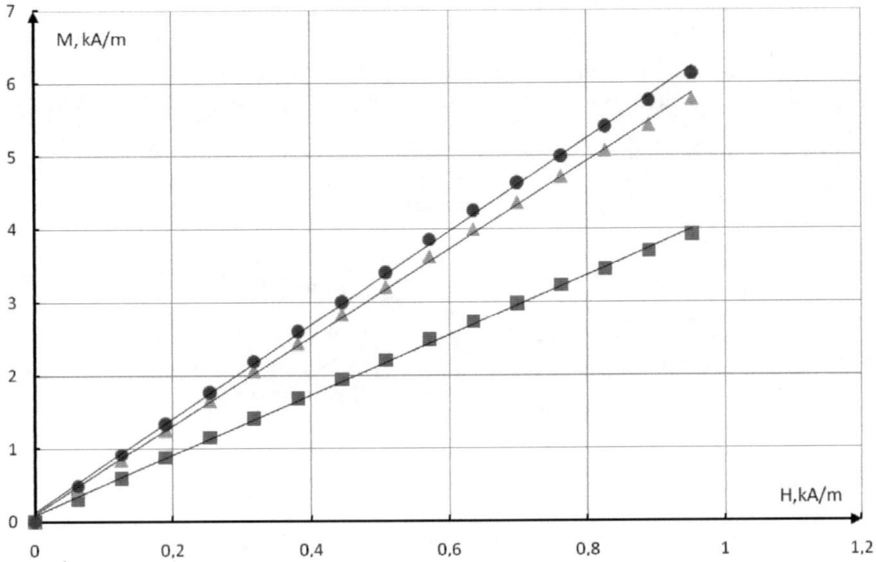

Figure 2. Linear portion of the magnetization curve for a kerosene-based magnetic fluid at centrifuge time: ● - 4 minutes, ▲ - 10 minutes, ■ - 60 minutes.

Figure 3 shows the dependence of the real part of the effective magnetic susceptibility on the temperature for two cases: in the absence of a magnetic field (Figure 3 (a)) and when a constant magnetic field is applied (Figure 3 (b)). The measurements were carried out at a frequency of 1000 Hz.

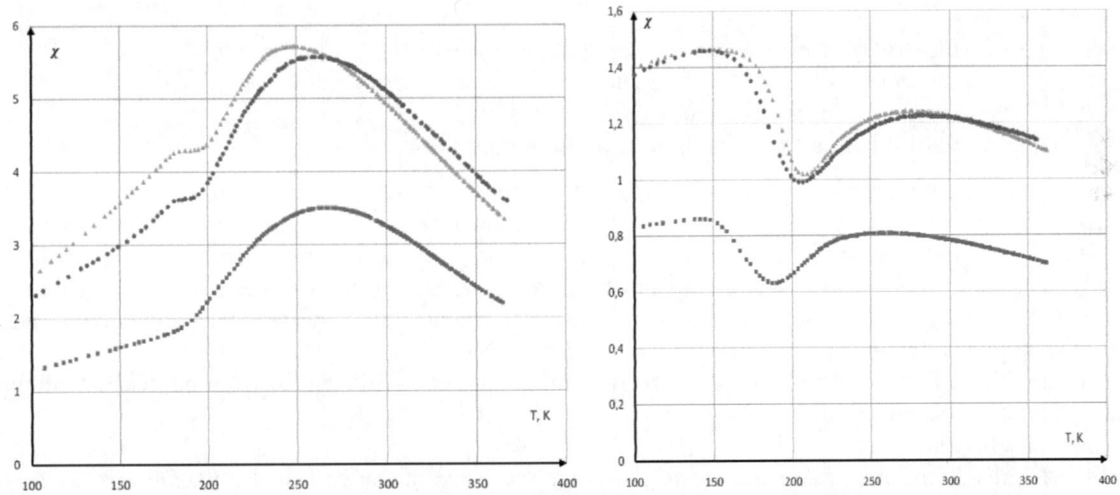

(a) (b)

Figure 3. Dependence of the magnetic susceptibility on the temperature without a magnetic field and when applying a field for a kerosene-based magnetic fluid at the time of centrifugation: ● - 4 minutes, ▲ - 10 minutes, ■ - 60 minutes.

From their consideration it follows that the maximum values of χ when the magnetic field is applied shift to the region of higher temperatures for the MF with a reduced centrifugation time. In view of this, according to [10], MF samples that have undergone partial centrifugation have larger particles, since the energy barrier for the particles of these samples has a higher value, and therefore their transition to the superparamagnetic state should occur at a higher temperature. The same conclusion can be drawn from an analysis of the frequency response of the imaginary part of the magnetic susceptibility presented in Figure 4, since in this case the extremum shifts toward low frequencies.

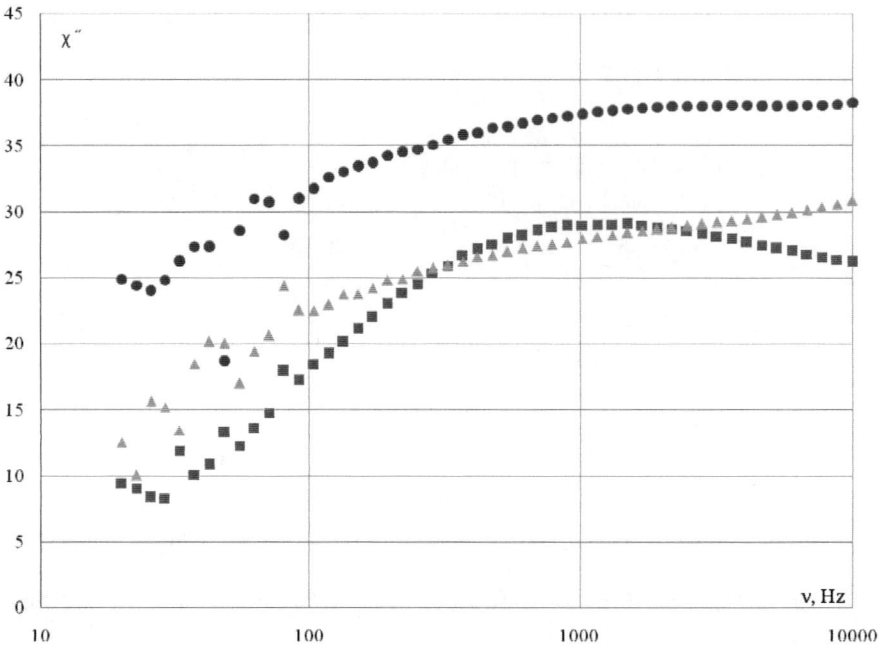

Figure 4. Dependence of the imaginary part of the magnetic susceptibility on the frequency of the magnetic field for a kerosene-based magnetic fluid at the time of centrifugation: ● - 4 minutes, ▲ - 10 minutes, ■ - 60 minutes.

So, for further research, we obtained samples of a magnetic fluid on three liquid bases (vacuum oil, kerosene and apiezon) with a centrifugation time of 60 min, 10 min, and 4 min. From the foregoing, it is clear that we can neglect the influence of random factors, depending on the production technology of MF. Therefore, we can study the character of the time variation of the magnetic force acting on the MF in the inhomogeneous field of a permanent magnet. The experiments were carried out according to the procedure detailed in [4].

To quantify the dynamic processes occurring in a magnetic fluid we use the coefficient of the magnetic force change as a diagnostic parameter, calculated as

$K_F = \Delta F_m / F_m(0)$,

Where ΔF_m is the maximum change in the magnetic force at the end of the magnetic relaxation process, and $F_m(0)$ is the magnitude of the force at the initial instant of time.

The results of measuring the time-varying relative value of the magnetic force acting on samples of a magnetic fluid with different liquid bases and the centrifugation time are shown in Figure 5.

IOP Conf. Series: Materials Science and Engineering **233** (2017) 012038 doi:10.1088/1757-899X/233/1/012038

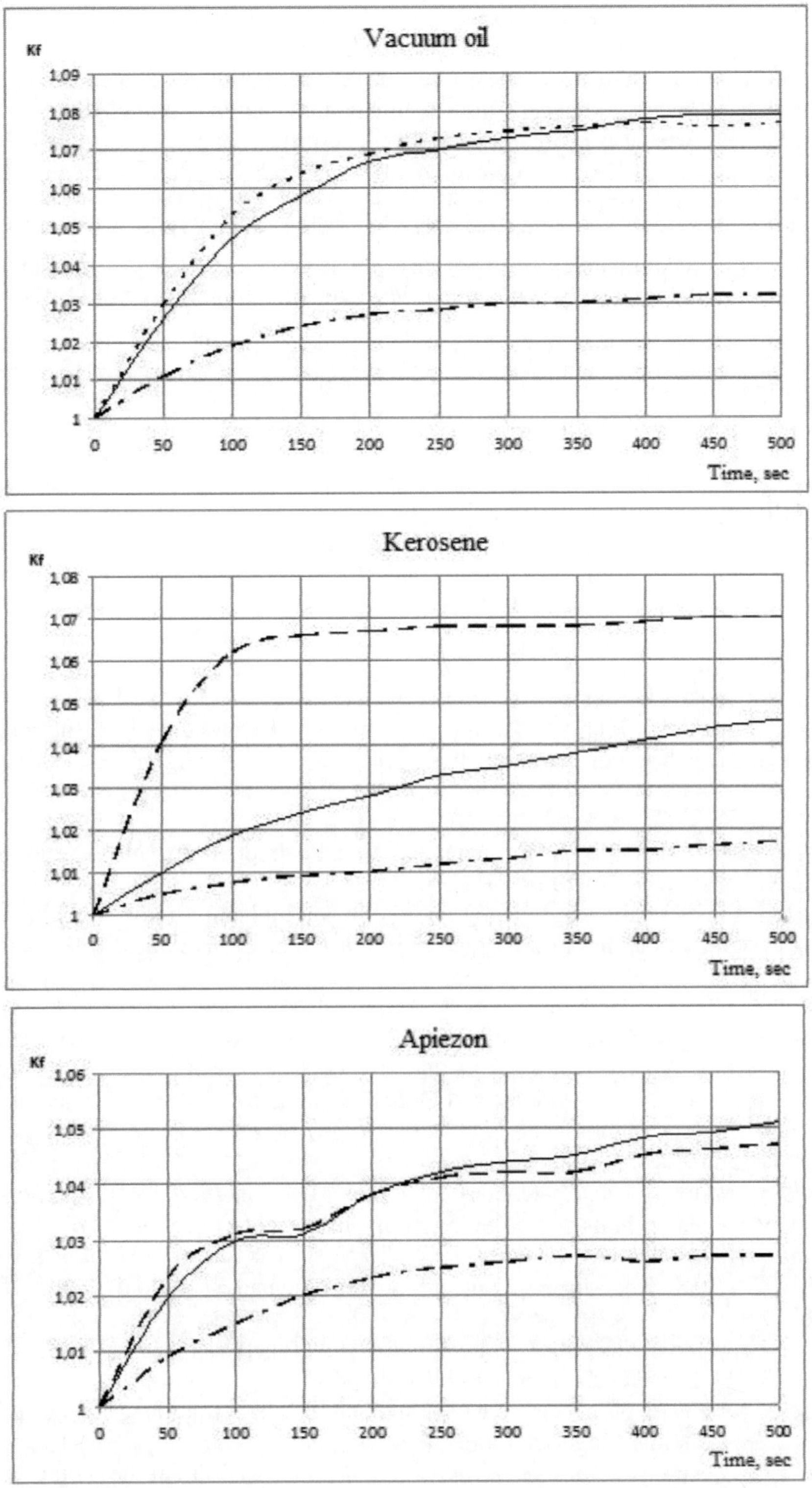

Figure 5. Dependence of the relative magnetic force on time for magnetic fluids on different bases and duration of centrifugation:_ _ _ - 4 min., ____ - 10 min., _ ._ ._ - 60 min.

From the graphs it follows that the nature of the change in the relative magnitude of the magnetic force with the passage of time is the same. For magnetic fluids that have undergone complete centrifugation, the change in the magnetic force during 10-15 minutes did not exceed 3%, which, according to the analysis in [4], indicates a high sedimentation stability of the MF.

The value of the K_F coefficient is virtually independent of the reduced centrifugation time (4 and 10 min) for magnetic liquids based on vacuum oil BM-3 and apiezon, and the greatest increase in the relative magnetic force at the beginning of the experiment was different for kerosene-based MFs during centrifugation for 4 minutes. In general, MF, which has not undergone standard centrifugation for one hour, can not be recommended for long-term use as part of MFS. The remaining large particles significantly reduce the stability of the MF, and, accordingly, the reliability of the MFS.

Conclusions

1. A relatively simple method of rapid analysis of the quality of a magnetic fluid proposed in Ref. [4] was supplemented by studying the dependence of the dynamic characteristics of MFs on the size of magnetite particles.

2. Experimental samples differing in the size of magnetic nanoparticles were obtained, which was achieved by different centrifugation times of the initial MF. It is shown that the performance characteristics of these MF samples differ by not more than 10%.

3. It has been confirmed that in spite of the fact that magnetorite particles are isolated with the Elmore reaction without residual magnetization, the centrifugation stage is mandatory for obtaining a stable MF. The average particle size in the MF should not exceed 8 nm, and the change in the magnetic force acting on the sample with a MF in a given inhomogeneous magnetic field should not exceed 3% in a time of 10-15 minutes.

4. The proposed method is relatively simple and can be implemented in an enterprise operating a magnetic fluid seals to analyze the characteristics of the MF before filling it into the seal.

References

[1] Radionov A V 2015 Experience in the magnetic fluid seals operation at oil refineries of the CIS countries *Chemical engineering* **10** pp 11-17

[2] Radionov A V 2016 Estimation of technogenic risk when replacing traditional seals with magnetic fluid sealing complexes taking climatic factors into account *Collection of scientific works NUK* **3** pp 120-127

[3] Bashtovoy V G Rex A G, Klimovich S V 2014 Experimental study of the influence of magnetophoresis and Brownian diffusion on the "magnetic weight" of a magnetic fluid *Power engineering. News of higher educational institutions and energy associations of the CIS* **1** pp 65-69

[4] Radionov A V Radionova A A, Podoltsev A D 2017 Experimental study of dynamic processes in a magnetic fluid in an inhomogeneous magnetic field of a rotary shaft seal *Tehnichna elektrodinamika* **2** pp 77-82

[5] Kyrylenko O V, Podol'tsev O D, Kondratenko I P 2013A method for measuring the magnetic susceptibility of magnetic fluids and a device for its implementation *Patent for Ukraine № 102749. Bull № 15 dated August 12, 2013*

[6] Elmore W C 1938 The magnetization of ferromagnetic colloid *Phys. Rev.* **54** (12) pp 1092–1095

[7] Kunikin S A 2011 Magnetic colloids. Features of the functional dependences of the magnetic susceptibility *LAP LAMBERT Academic Publishing* pp 152

[8] Polunin V M 212 Acoustic properties of nanodispersed magnetic liquids (M.: FIZMATLIT) p 384

[9] Odenbach S 2009 Colloidal Magnetic Fluids *Springer* p 438

[10] Yu I Dikansky, A G Ispiryan S A Kunikin, A V Radionov 2015 On the nature of the maximum of the temperature dependence of the magnetic dependence of magnetic *fluids Journal of Technical Physics* **85** (8) pp 100-104

IOP Conf. Series: Materials Science and Engineering **233** (2017) 012039 doi:10.1088/1757-899X/233/1/012039

Determination of mechanical characteristics of stuffing box packings

A Zahorulko[1], C Kundera[2], S Hudkov[1]
[1] Department of General Mechanics and Dynamics of Machines, Sumy State University, 2, Rimsky-Korsakov Street, Sumy 40007, Ukraine
[2] Faculty of Mechatronics and Machine Building, Kielce University of Technology, Aleja Tysiąclecia Państwa Polskiego 7, PL- 25314 Kielce, Poland

E-mail: a.zagorulko@omdm.sumdu.edu.ua

Abstract. *The resource of stuffing box seal is determined not only by the unit design and by the conditions of its operation, but also essentially depends on the correct choice of stuffing box packing. Not one of stuffing box packing manufacturers does not provide data on mechanical properties. Therefore, in this paper the results of experimental investigations on determining the mechanical properties of packings are presented. The obtaining mechanical characteristics allow to clarify existing methods of calculation and to increase sealing and durability of stuffing box packing seals.*

1. Introduction

The braided stuffing box packings impregnated with lubricants are most widely used for sealing of centrifugal pumps for general industrial usage. An important factor affecting the work of stuffing box seal are tribomechanical characteristics of the packing, the account of which affects the seal life. However, none of producers of the packing does not provide data on physical and mechanical properties (elastic modulus and Poisson's ratio), which are one of the main characteristics and are mandatory in calculating the packing seals.

2. Definition of physical and mechanical properties of the packings

The experimental studies of physical and mechanical properties of the packing were carried out on the rig that described in [1, 4].

In the experiment, a specimen of the stuffing box packing is pressurized three times to determine the effect of pressure testing on physical and mechanical properties of the packing. Before testing the specimen of packing is pressurized with a maximum load, in order to compress the packing by deleting its initial waviness and voids between fibers. After that we apply vertical load F_y to the test specimen of packing gradually increases to a maximum value, varies with a predetermined step. Axial deformation Δ_y is incremented at each step. The increment of axial deformation Δ_x can be observed while unloading spring pack by screw. Obtained data define elasticity modulus E and Poisson ratio ν as the relation of the respective magnitudes of increments [1-6].

The elastic modulus is determined from the ratio

$$E = \frac{\Delta\sigma_y}{\Delta\varepsilon_y} \quad (1)$$

where $\Delta\varepsilon_y = \Delta_y / b_0$, $\Delta\sigma_y = \Delta F_y / A_0$, A_0 – the cross sectional area of the packing.

Poisson's ratio is determined by the formula

$$\nu = \frac{\Delta_x}{\Delta_y} \quad (2)$$

Experiments were carried out for various stuffing box packings, such as GAMBIT 608 in size of 10×10 (made of a yarn of fibrous PTFE, filled with graphite and saturated with silicone oil), GAMBIT 6088 in size of 8×8 (made of a yarn of fibrous PTFE, filled with graphite) MS-360 POLMAN in size of 8×8 (interlacing of fibers made of PTFE with graphite and aramid fibers).

The experiments are presented in Figure 1-5. Figures 1-3 are shown dependencies of relative deformation on the axial load of several cyclic loading. On the first loading cycle, the packing compression occurs by filling the voids between the existing warp and partial extrusion of lubricant. It appears residual deformation. On subsequent loading cycles occur only compression of the material and extrusion of impregnation. With increase in load, the packing is strengthened, and its rigidity is increased. As can be seen from these figures the dependence of the relative deformation on applied load under uniaxial stress state is non-linear. Figures 4 and 5 show the dependence of the Poisson ratio on the axial load.

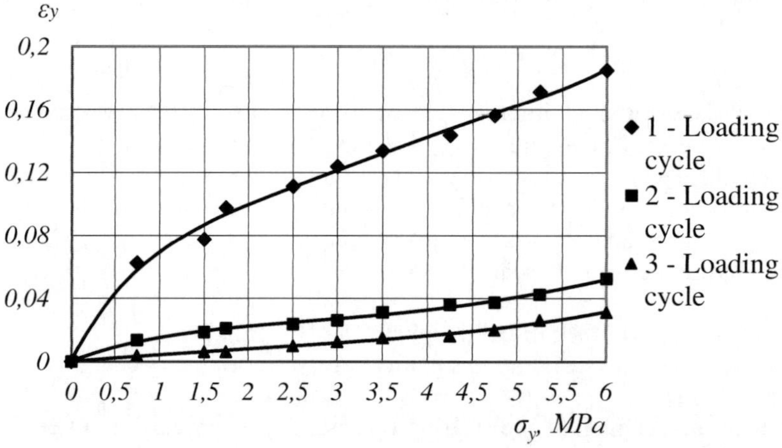

Figure 1. The dependence of the relative deformation on the axial load at several cyclic loadings (GAMBIT 6088 in size of 8×8).

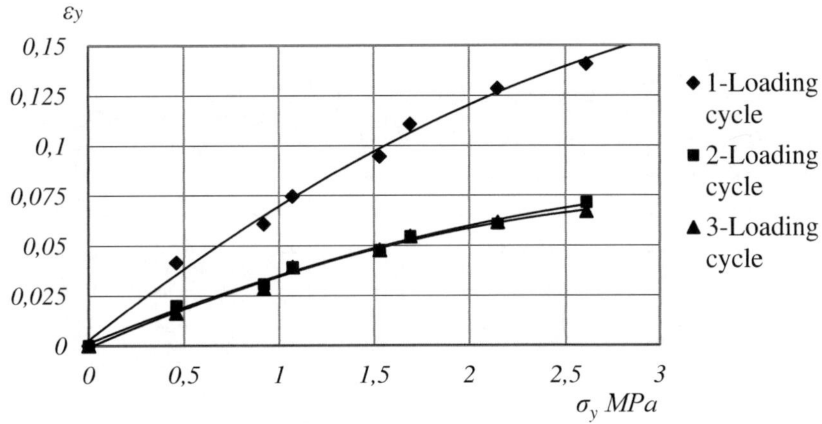

Figure 2. The dependence of the relative deformation on the axial load at several cyclic loadings (GAMBIT 608 in size of *10×10*).

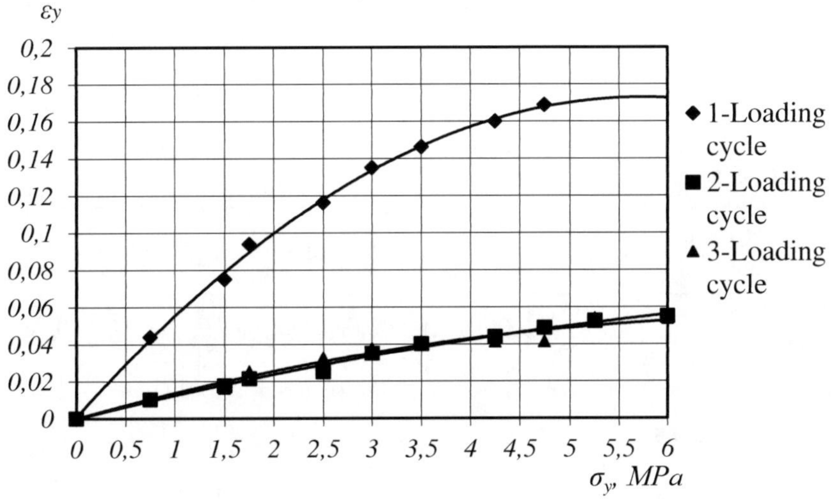

Figure 3. The dependence of the relative deformation on the axial load at several cyclic loadings (MS-360 POLMAN in size of *8×8*).

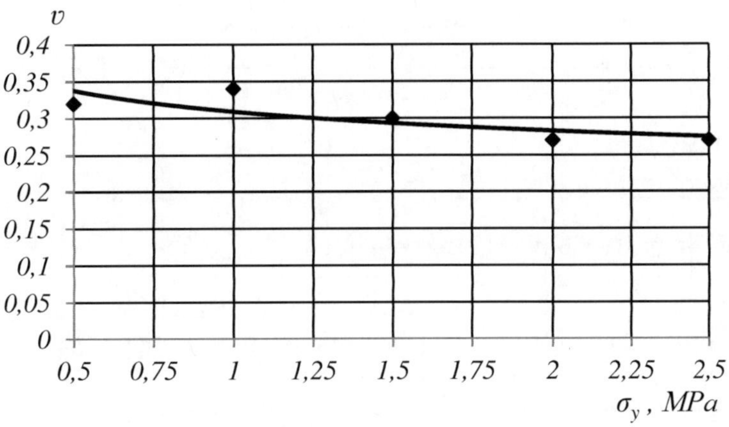

Figure 4. The dependence of the Poisson ratio on the axial load (GAMBIT 608 in size of *10×10*).

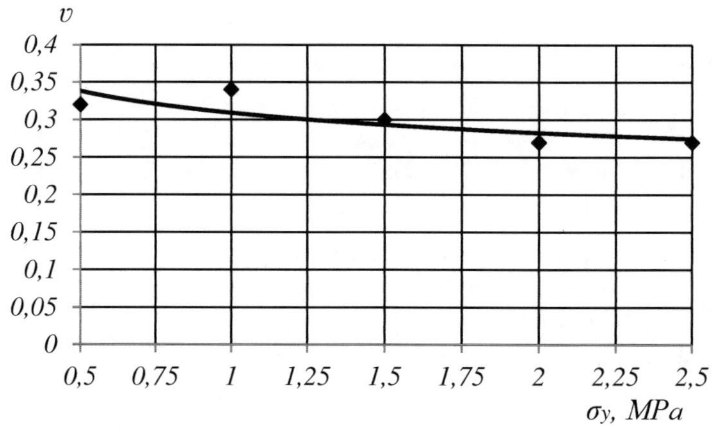

Figure 5. The dependence of the Poisson ratio on the axial load (MS-360 POLMAN in size of *8×8*).

Because the dependence of the elastic modulus and Poisson's ratio on the axial load is not linear, it is necessary to choose an analytic function with sufficient precision approximating the experimental data. Least-squares method (LSM) is used for this purpose.

Tables 1 and 2 show the regression equations derived LSM as well as indexes of correlation.

Table 1. The regression equations of relative deformation on the axial load.

Type of stuffing box packing	Equation of regression	Correlation Index
GAMBIT 6088 in size of *8×8*	$\varepsilon_y = 0{,}0007 \cdot \sigma_y^2 + 0{,}0011 \cdot \sigma_y + 0{,}0018$	0,9857
GAMBIT 608 in size of *10×10*	$\varepsilon_y = -0{,}0057 \cdot \sigma_y^2 + 0{,}0412 \cdot \sigma_y - 0{,}001$	0,9963
MS-360 POLMAN in size of *8x8*	$\varepsilon_y = -0{,}00 \cdot \sigma_y^2 + 0{,}0147 \cdot \sigma_y$	0,9838

Table 2. The regression equations of Poisson's ratio on the axial load.

Type of stuffing box packing	Equation of regression	Correlation Index
GAMBIT 608 in size of *10×10*	$\nu = 0{,}341 \cdot \sigma_y^{-0{,}1364}$	0,8186
MS-360 POLMAN in size of *8x8*	$\nu = 0{,}3121 \cdot \sigma_y^{-0{,}1317}$	0,8219

3. Determination of the stuffing box packing roughness

Stuffing box packing has a braided structure, so by the friction it will have a relief-textured work surface. At that leaks through the seals are defined by microchannels resulting from contact of the surface with support disk. Determination of roughness profile of stuffing box packing using traditional contact method is impossible, since packing has quite a small hardness. Therefore, it is necessary to use modern non-contact optical methods for determining of roughness.

IOP Conf. Series: Materials Science and Engineering **233** (2017) 012039 doi:10.1088/1757-899X/233/1/012039

Roughness profile of stuffing box packing (GAMBIT 608 in size of *10* × *10*) was defined by optical method. Before determining the roughness of a sample of the stuffing box packing was first pressed in order to reduce the impact of packing braided structure, then conducted packing running.

The roughness of the packing was measured using a three-dimensional non-contact profilograph-profilometer.

Figures 6-8 show the results of measuring the roughness profile of the stuffing box packing surface in the design of face packing seal with hydrodynamic unloading of friction pair [5]. Roughness profile of stuffing box packing has a uniform distribution of peaks and valleys. This shows the uniform distribution of the contact pressure over the width of the friction pair and wear of the friction surfaces of the stuffing box packing and the support metal disc.

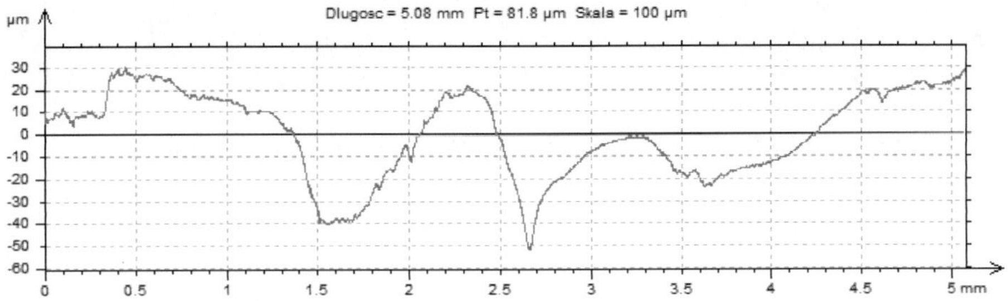

Figure 6. Two-dimensional roughness profile of the stuffing box packing.

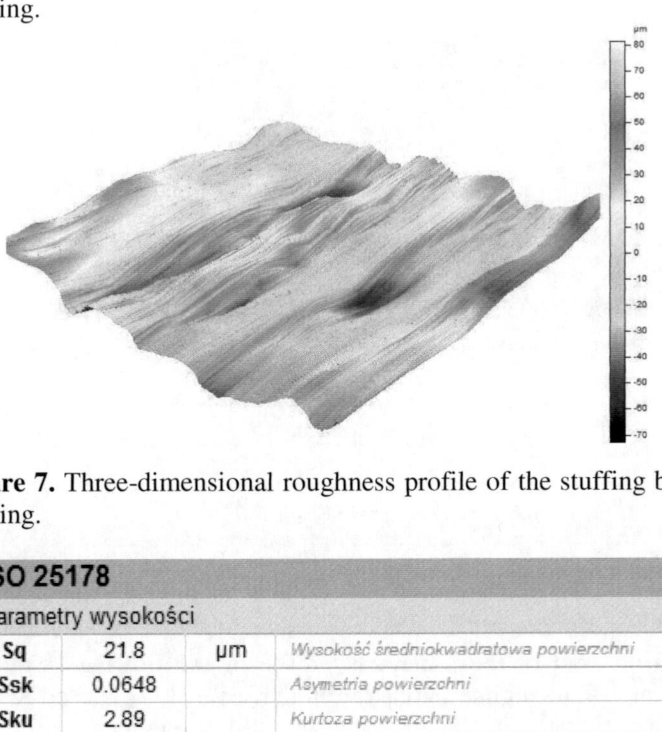

Figure 7. Three-dimensional roughness profile of the stuffing box packing.

ISO 25178			
Parametry wysokości			
Sq	21.8	µm	Wysokość średniokwadratowa powierzchni
Ssk	0.0648		Asymetria powierzchni
Sku	2.89		Kurtoza powierzchni
Sp	81.3	µm	Maksymalna wysokość piku powierzchni
Sv	72.6	µm	Maksymalna wysokość wgłębienia powierzchni
Sz	154	µm	Maksymalna wysokość powierzchni
Sa	17.9	µm	Średnia arytmetyczna wysokość powierzchni

Figure 8. Parameters of surface roughness of the stuffing box packing.

IOP Conf. Series: Materials Science and Engineering **233** (2017) 012039 doi:10.1088/1757-899X/233/1/012039

In Figures 9-11, the results of measurements of the roughness profile of the stuffing box packing in the traditional design of face packing seal are presented [6]. These figures show that the roughness profile is divided into two sections: section with greater roughness corresponding overloaded section of face packing seal and with a function of a sealer, and a section of a lower surface roughness, on which there is no contact between the stuffing box packing and support metal disk.

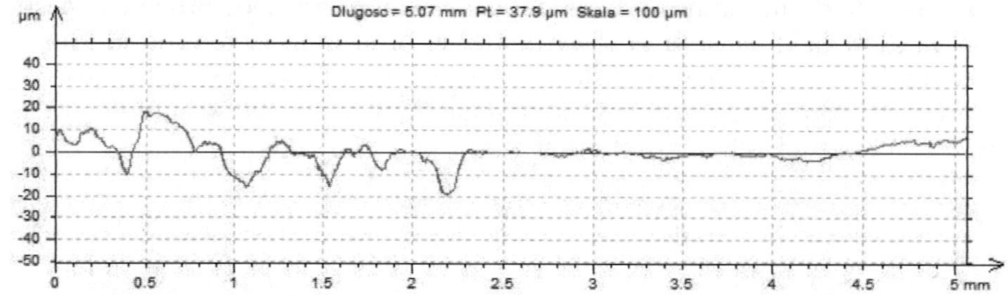

Figure 9. Two-dimensional roughness profile of the stuffing box packing.

Figure 10. Three-dimensional roughness profile of the stuffing box packing.

ISO 25178			
Parametry wysokości			
Sq	7.21	µm	Wysokość średniokwadratowa powierzchni
Ssk	0.283		Asymetria powierzchni
Sku	4.50		Kurtoza powierzchni
Sp	34.3	µm	Maksymalna wysokość piku powierzchni
Sv	23.9	µm	Maksymalna wysokość wgłębienia powierzchni
Sz	58.2	µm	Maksymalna wysokość powierzchni
Sa	5.41	µm	Średnia arytmetyczna wysokość powierzchni

Figure 11. Parameters of surface roughness of the stuffing box packing.

The results of measurement of the roughness profile of the stuffing box packing enable us to determine at the stage of calculating the initial gap in the seal (3) [7] needed to solve the problem of elastohydrodynamical lubrication of face packing seal [8]. Roughness value of the stuffing box packing depends on the method of braiding, packing material, impregnation and its physical and mechanical properties.

$$h = 0.5\left(\frac{R_{max1}^2}{R_{a1}} + \frac{R_{max2}^2}{R_{a2}}\right) \tag{3}$$

4. Conclusions

The results of experimental studies of physical and mechanical properties of the stuffing box packing indicate their non-linearity. The amount of the applied load and the number of loading cycles is significantly affected on the physical and mechanical properties of the stuffing box packing. With increasing load stuffing box packing is strengthened and becomes more rigid. According to the experimental studies physical and mechanical properties of the stuffing box packing stabilized only after the third loading cycle. Nonlinearity of physical-mechanical properties of the stuffing box packing may affect on the results of seal calculation, and consequently, its performance and service life. Obtained empirical relationships make it possible to determine the elastic modulus and Poisson's ratio of the stuffing box packing, depending on the applied load. Therefore, it is necessary that the stuffing box packing manufacturers cited data on modulus elasticity and Poisson's ratio for each type of stuffing box packing, depending on the method of braiding, material, impregnation, and also its size.

Measurement results of the roughness profile of the stuffing box packing provide an opportunity to identify the seal initial gap needed to solve the problem of elastohydrodynamical lubrication of face packing seal. Roughness value of the stuffing box packing depends on the method of braiding, packing material, impregnation and its physical and mechanical properties.

References

[1] Zahorulko A, Gudkov S, Gaft J 2008 Experimental investigations of physicomechanical properties of stuffing packings *Zeszyty naukowe "Nauki techniczne - budowa i eksploatacja maszyn* (Kielce: Politechnika Swietokrzyska) **2** pp 115-121

[2] Zahorulko A 2015 Experimental investigation of mechanical properties of stuffing box packings // *Sealing Technology* **8** pp 7-13

[3] Adamczak S, Bochnia J, Kundera C 2012 Stress and strain measurements in static tensile tests *Metrology and Measurement Systems* **XIX** (3) pp 531-540

[4] Timoshenko S P, Goodyear J 1979 *Theory of Elasticity* ed. G S Shapiro; transl. M I Reitman (M: Science) p 560

[5] Zahorulko A V, Gudkov S M 2007 Experimental investigations of new designs of face packing seals with hydrodynamic unloading of friction pair *Journal of East Ukrainian National University of Volodymyr Dahl* **3** pp 91-97

[6] Gaft Y, Zahorulko A, Martsynkovskyy V, Shevchenko S 2000 Face packing seals - new opportunities for pump rotor hermetic sealing *Proc. 16th Int. Conf. on Fluid Sealing* pp 335-349

[7] Mayer E 1978 *Mechanical seals* (M: Mechanical engineering) p 288

[8] Zahorulko A, Gudkov S 2010 Solution of problem concerning elastohydrodynamic lubrication for friction pair of face packing seal *Proc. XIII Int. Conference in Sealing Technology* pp 317-326

IOP Conf. Series: Materials Science and Engineering **233** (2017) 012040 doi:10.1088/1757-899X/233/1/012040

Prediction of conditions of flange joints depressurization during the assembly of pipelines using technological and standard fastening elements

O V Bondarenko[1], A P Dzyuba[1]

[1]Oles Gonchar Dnipro National University, Gagarina avenue 72, Dnipro city, 49010, Ukraine

E-mail: bov1977@i.ua

Abstract. This article shows the prediction method of conditions of flange joints depressurization during the assembly of pipelines using technological and standard fasteners.

1. Introduction. Problem statement

Efficiency of solving the problem of reducing the pipelines mass is directly related to the manufacturing process of their assembly, which ensure the load reduction on parts of pipelines. This is significant primarily for space-rocket hardware and aircraft engineering, less for chemical and energy equipment, river craft and sea-borne craft. One of the approaches to the load reduction on flanges, fastening elements and pipes is pressurization of pipeline flange joints using technological and standard fastening elements.

At present it is known three modes of pressurization of pipeline flange joints with using such approach [1-3]. All of them are designed to joints pressurization, which sealed by soft metal gaskets. The possibility of reducing the tightening torque of fastening elements in comparison with standard value [4] is created due to occurrence of flange seal face and gasket firm-and-impervious joint in contact zone [1,5,6]. Reducing of torque and axial forces of standard fastening element tightening allows, in turn, to reduce the flange sizes and use lighter and less durable materials for their manufacture [7,8].

For different modes of pressurization of flange joints correlation between tightening torque of standard and technological fastening elements, in which the joint remains the pressurization, is different. Conducted experiments showed that joints depressurization can primarily occur during the replacement of technological fastening elements with standard fastening elements [9]. In this case, the initial value of the leakage of the air-helium mixture (10% He) is 10^{-12}–10^{-10} W, after the depressurization the leakage was 10^{-1} W and more.

Calculated dependencies for determining the axial forces and the corresponding unscrewing torques which influence on standard fastening elements using pressurization flange joints methods are given in paper [10,11], and this article shows only formulas for determining the axial forces which influence on standard fastening elements.

The method [1] consists in tightening the technological and standard fastening elements, temporary aging and technological fastening elements unscrewing. For this method the axial force which acting on the standard fastening elements determines in following way:

$$Q_{12} = \frac{Q_1 \cdot \lambda_{CP} + Q_1 \cdot \lambda_F}{\lambda_F} \tag{1}$$

here λ_F – compliance of technological and standard fastening elements (technological and standard elements are the same, they are screwing up by the same torque and simultaneously) λ_{CP} – summary compliance of flanges and gaskets; Q_1 – tightening axial force of technological and standard fastening elements; Q_{12} – axial force which influence on standard fastening elements after unscrewing the technological fastening elements.

The method [2] consists in technological fastening elements tightening, temporary aging and consecutive replacement of technological fastening elements with standard fastening elements. For this method the axial force which influence on standard fastening elements determines in following way:

$$Q_{2F} = \frac{Q_1 \cdot \lambda_{CP} + Q_2 \cdot \lambda_{F2} - Q_2 \cdot \lambda_{CP}}{\lambda_{F2}} \tag{2}$$

where Q_1 – tightening axial force of technological fastening elements; Q_2 – tightening axial force of standard fastening; λ_{F2} – compliance of standard fastening elements; Q_{2F} – axial force which influence on standard fastening elements after the replacement of technological fastening elements.

The method [3]consists in technological fastening elements tightening, temporary aging, standard fastening elements tightening and technological fasteners unscrewing. For this method the value of axial force Q_{23}, which influence on the standard fastening elements after unscrewing the technological standard parts, determines by the formula:

$$Q_{23} = \frac{Q_1 \cdot \lambda_{CP} + Q_2 \cdot \lambda_{F2}}{\lambda_{F2}} \tag{3}$$

It should be noted that in papers [10,11] calculations are made with assumption that all technological fastening elements are unscrewing simultaneously and all standard fastening elements are screwing up simultaneously. In spite of such idealization this approach allows to describe accurately the final state of flange joints.

With the view of prediction the joint depressurization this article proposes the determination method of correlation of tightening torques (axial forces) of technological and standard fastening elements in which contact density of flanges seal faces and gaskets (for cases of the most common lock and spherical flange joints of pipelines) are breaking.

2. Basic research material

Papers [9-11] show that axial force which corresponding to unscrewing torque of standard fastening elements exceeds the axial force which corresponding to their tightening torque.

This is due to "extension" of the flanges as a result of decrease of deflecting forces acting on them. It is the "extension" of the flanges that can affect on firm-and-impervious joint with the force Q_R and lead to contact density violation of seal faces and joints depressurization. Obviously in this case the force of "extension" of the flanges comes to difference between the summary tightening axial force of technological and standard fastening elements.

In order to determine whether the force Q_R is sufficient to depressurize the joint it should be considered what kind of the firm-and-impervious joint is in the contact zone of flanges seal face and gaskets in spherical and lock joints.

Area over which Q_R acts, in contact zone of spherical surface comes to calculated area of spherical surface of gasket F_{SSP} .

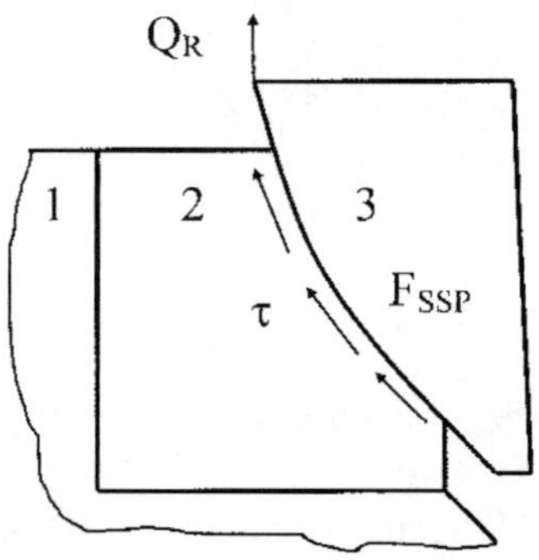

Figure 1. Direction of force Q_R and shearing stresses on contact spherical surface. Figure is showing 1 –flange with the seat for a gasket, 2 – gasket, 3 – flange with the tip, Q_R – force of flange "extension", τ – shearing stresses caused in gasket under the influence of force Q_R.

Calculated dependencies for determining the geometric dimensions including the gasket spherical parts are given in paper [12]. Experimental data and results of exploitation of real joints show that violation of density of firm-and-impervious joint occurs by gasket material at the level of the deepest ledge on flange spherical surface. Taking into account the fact that height of microroughnesses of flange spherical surface is usually Ra 1,6 μm (Rz 6…10 μm) «thickness» of firm-and-impervious joint zone is 6…10 μm [9]. Magnitude of a force Q_R, at which joint depressurization takes place can be determined from the following relation

$$Q_R = F_{SSP} \cdot \tau \tag{4}$$

where τ – shearing strength of gasket.

Beside that, it should be noted the microgeometry of flanges seal faces and gasket. Usually flanges manufacture by turning, therefore microroughnesses on their cylindrical, spherical and plane seal faces are of the form the helical line with pitch equal to the value of cutter feed [13]. According to the data [14] the value of the feed for turning operation of surfaces with roughness Ra 1,25…2,5 μm is from 0,04 mm/rev. to 0,12…0,13 mm/rev.

So far as gasket material fills the microroughnesses on the flange surface, for joint depressurization is necessary that during the unscrewing process of technological fastening elements, their replacement with standard fastening elements and flange "extension", the ledges on their surfaces were deformed the gasket material on the distance which equal to the size of microirregularity mounting base. Experimental data have shown that as a rule at the initial leakage value 10-12-10-10 W depressurization immediately occurs to a leakage value 10-5 W or more.

Microrelief of plain flange surfaces of lock joints has appearance of helical line, with microirregularity height about 10 μm (Ra 2,5 μm, Rz 10 μm) and pitch from 0,04 mm to 0,12…0,13 mm [13]. Moreover, on plain flange surfaces usually execute V-shaped pressurizing risks, which are filled with the gasket material; when the fastening elements are tightened [4]. Microrelief of cylindrical flange surfaces in which contact with gasket takes place, is similar to microrelief of plain surfaces and ledges are formed (in the process of gasket plastic deformation) with height h_1 (gasket external) and h_2 (gasket internal diameter). At the same time the height of ledges h_1 and h_2 are different along the perimeter of gasket [15]. This is due to hardening and increasing the elasticity of the gasket material under deformation caused by the action of axial forces on the fasteners tightening. In the process of successive fastening elements tightening in the total deformation of the gasket, the proportion of the plastic decreases and the fraction of the elastic component increases, which causes the different heights h_1 and h_2 along the perimeter. Cumulative value of strain remains constant and varies along the sinusoid [1,4]. While constructing the method and taking into account the assumptions made earlier, it is assumed that the average value of h_1 and h_2 is 170 ... 200 μm for pressure on a gasket made of aluminum 160 ... 170 MPa and of copper - 270 ... 320 MPa.

The greatest influence on the joint depressurization is the violation of the contact density along the cylindrical surfaces. It is the appearance of the ledges with heights h_1 and h_2 that testifies about filling the microroughness on the plain surfaces of the flanges by gasket material and about the intensive plastic flow of the gasket material. Subsequent filling by the gasket material of microirregularities already on the cylindrical surfaces of the seat for a gasket and the tip of the flanges leads to a phenomenon which similar to the punch jamming during the pressing. Such "jamming" can be one of the building mechanisms of firm-and-impervious joint in contact zone of flange seal faces and gasket.

Therefore, for total joint depressurization, there should be a violation of the contact density over plain surfaces, in the area of pressurize matchmarks along the cylindrical surfaces. Force Q_R which necessary to depressurize the joint is determined by the following formula:

$$Q_R = Q_{FS} + Q_{GS} + Q_{CS} = F_{FS} \cdot \sigma_B + F_{GS} \cdot \tau + F_{CS} \cdot \tau \tag{5}$$

where Q_{FS}, Q_{GS}, Q_{CS} – forces, which necessary for violation of contact density along the plain seal faces, in the area of pressurize matchmarks and в long the cylindrical surfaces respectively; σ_B – breaking stress of the gasket material, F_{FS} – area of plain surfaces, F_{GS} – area of pressurize matchmarks, F_{CS} – contact area along cylindrical surfaces, $F_{CS} = \pi \cdot d \cdot h_1$, where d – external diameter of gasket.

For the assessment of capability for violation of the contact density, it is considered the conditions for the displacement of the fastening elements (bolts, studs, screws), flanges and gaskets in the process of the flange "extension" . Thus, for the violation of the contact density, displacement of the of the seal surfaces points as a result of the flange "extension" should be sufficient to deform the gasket material on an amount equal or greater than the size micro irregularity mounting base on the flange surface.

If the gasket is out of the operating zone of bending moment on the flanges from the axial forces of tightening fastening elements, it is obvious that the effect of the flanges"extension" can not lead to joint depressurization. In this case the gasket has the appearance only linear movement in the direction of pipeline axis – «springing», causes by decreasing of the force which acting on it. However, the usage of such flange joints is constrained by large overall dimensions in comparison with currently used and insufficiently developed engineering methods for their design and manufacturing.

In a majority of modern flange joints, the gasket is located in the operating zone of bending moment from the axial force of the fastening elements tightening. In this case the value of points displacement of seal faces depends on the joints pressurization method, compliance of fastening elements, gasket, flanges, eccentricity from axis of working line of tightening axial force to gasket external diameter, gasket width. It is considered that under the influence of bending moment, flange cross section revolves around the center its center of mass, although in practice the center of rotation is more often assumed to be the center of the pipe wall at the point of its transition into the flange ring or the place of ring support of the free flange on the tip. In calculated dependencies which determine the the value of the axial force which influence on the standard fastening elements [10], the values of the flange replacements are determined on the axis of the fastening elements (on the action line of the tightening axial force). Accordingly, the replacement of the points of flanges seal surfaces can be defined as the correlation of the distance from the gasket external diameter to the rotation center to the distance from the rotation center to the axis of the fastening elements (action lines of tightening axial forces) multiplied by the displacement value which obtained by the formulas [10]. If this value is equal or greater than the size of the microirregularity mounting base on the flange surface, the joint is depressurized.

3. Conclusions

According to this article prediction method of conditions of pipelines flange joints depressurization method was obtained. Method consists of the following steps: testing of the joint design and find out whether the gasket is in the operation zone of bending moment from tightening axial force of fastening elements; establish data about roughness of flange seal face and gasket; determine the compliance of flanges, fastening elements and gasket; determine the value of tightening torque (axial force) of technological and standard fastening elements; determine the values of flange "extension" force which are necessary for violation of contact density of flange seal faces and gasket, displacement of flange points on seal faces and in such way establish the possibility of joint depressurization.

The assumption in this article about instantaneity and simultaneity of unscrewing of technological fastening elements and their replacement with standard fastening elements, in spite of idealization of process, lets use analytical dependencies for prediction of conditions of flange joints depressurization during the assembly (on initial stages of pipelines designing and technological process of their manufacture), without recourse to labour-intensive computer simulation by applying finite-element models.

References

[1] Danchenko V G 1993 Pressurization method of pipelines flange joints *USSR. – no.4917009/29; Reg. 11.01.91; Publ.07.05.93. Bul. no.17. Discoveries. Utility inventions*

[2] Dzhur E O, Danchenko V G, Zevako V S, Bondarenko O V 2003 Pressurization method of pipelines flange joints *Patent 60067 A Ukraine no. 003010507; Reg. 21.01.2003. Publ. 15.09.2003. Bul. no.2*

[3] Dzhur E O, Bondarenko O V 2009 Pressurization method of pipelines flange joints *Patent UA no.87100. Bul. no.12. – 25.06.2009*

[4] State Standard 19749-84 Fixed split hydraulic systems joints. Closed closures. Types and technical requirements]

[5] Starcev N I 1976 Gas turbine engine pipelines (Moscow. «Mashinostroenie») p 272

[6] Ajnbinder S B 1957 Cold welding of metals (Riga: Academy of Sciences of the Latvian SSR Publ.) p 165

[7] Tishhenko Yu S, Bondarenko O V 2007 Mass optimization of the of designs of detachable joints of aggregates and assemblies of pnemo- and hydro-mains of launch vehicles *Space systems. Military rocketry. Scientific and technical collection* (Dnipropetrovsk) **2** pp 147-158

[8] Tishhenko Yu S, Bondarenko O V 2007 Improving the design of detachable flange joints *Space systems. Military rocketry. Scientific and technical collection* (Dnipropetrovsk) **2** pp 159-162

[9] Bondarenko O V 2005 Tightness of the flange joints for piping assembly using technological and standard fasteners *Geotechnical mechanics. Interdepartmental collection of scientific papers* (Dnipropetrovsk) **57** pp 182-193

[10] Bondarenko O V 2005 Determination of the forces value which influence on the flanges, gasket and fasteners during the assembly of pipelines using technological and standard parts *System technology. Regional interuniversity collections of scientific papers* (Dnipropetrovsk) **2** (37) pp 28-38

[11] O Bondarenko, A Dziuba 2014 Sealing of Pipelines Flange Connections in Conditions of Fasteners Tightening Torgue Reducing *Applied Mechanics and Materials* (Trans tech Publications: Switzerland) **630** pp 283-287

[12] Bondarenko O V, Zevako V S 2003 Engineering technique of designing pipeline flange-spherical joints of carrier rockets *Bulletin of Dnipropetrovsk University. Rocket and space technology. Dnipropetrovsk University Publ.* **7** pp 12-20

[13] Sanin F P, Dzhur E O, Kuchma L D, Najd'onov V A 1995 Tightness in rocket and space

technology *Textbook (*Dnipropetrovsk. Dnipropetrovsk University Publ.) 168 p
[14] Kosilov A G and Meshcheryakov R K 1985 *Directory of technologist-machine builder* In 2 vol., Vol. 2 (Moscow: Mashinostroenie) p 496
[15] Dzyuba A P, Bondarenko O V, Tishhenko Yu S 2011 Experimental determination of theplastic deformation value of the gasket in the lock flange joint of the pipeline *Vibrant reliability and tightness of centrifugal machines: monograph* Ed Marcinkovskyy V A, Zagorul'ko A V (Sumy: Sumy State University) pp 46-152

IOP Conf. Series: Materials Science and Engineering **233** (2017) 012041 doi:10.1088/1757-899X/233/1/012041

Dynamic Equilibrium Surfaces for Conical Fluid-Film Bearings

A Y Koltsov[1], A Y Korneev[1], L A Savin[1], Shengbo Li[2]

[1]Orel State University

[2]School of Mechanical and Automotive Engineering, Xiamen University of Technology, 361024, 600 Ligong Road, Jimei District, Xiamen, China

E-mail: me@alexander-koltsov.ru

Abstract. The paper considers the procedure of constructing curves of dynamic equilibrium and problems of applicability of the procedure of linearization of lubricant film reactions around the equilibrium point for conical fluid-film bearings.

1. Introduction

An important role in the study of rotor-bearing units is played by the assumption that it is possible to linearize, generally speaking, essentially nonlinear reactions of the lubricating layer of a fluid-film bearing in the area of a certain equilibrium position. By the position of equilibrium in this case the position of the center of the journal's center is meant, where the external load is completely compensated by the reactions of the lubricating layer, and the set of such points at different rotational speeds of the rotor constitute a spatial curve called the dynamic equilibrium curve. These curves for the case of cylindrical supports constitute a function ϕ in the domain of determining the permissible rotational speeds ω in the plane of the radial cross-section. Such functions are constructed for a given radial load vector F . Equality means $\phi(\omega) = (X, Y)$ that there is equality $\vec{R}(X, Y) = -\vec{F}$ at the point (X, Y), where R is the reaction force of the bearing.

Studies of cylindrical full-coverage and thrust bearings confirm that if there is an asymptotic convergence of the trajectory of the motion of the rotor, then it converges to a point on the dynamic equilibrium curve. In the case when the trajectory of motion has the form of a limit cycle, its geometric center is close to the point of dynamic equilibrium. This gives a formal basis for the linearization of the lubricant layer reactions at these points.

The aim of this paper is to investigate the feasibility of using this method for the case of conical fluid-film bearings. A distinctive feature of this problem is that the dynamic equilibrium curve (if it exists) lies in three-dimensional space, can have discontinuities and other features that are not characteristic of either cylindrical or thrust bearings. At the same time, the solution of this problem is necessary to determine the degree of conformity of the description of the dynamic characteristics of a conical bearing by calculating the stiffness and damping coefficients.

IOP Conf. Series: Materials Science and Engineering **233** (2017) 012041 doi:10.1088/1757-899X/233/1/012041

2. Mathematical model of a 'rotor-conical bearing' system

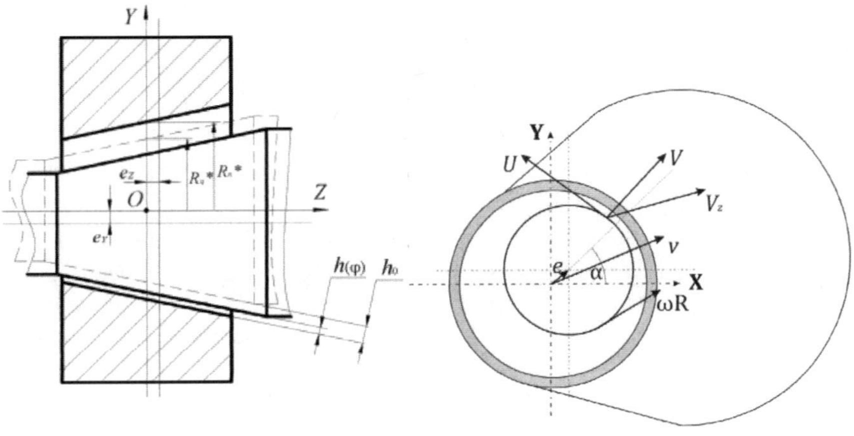

Figure 1. Calculation diagram of a conical fluid-film bearing.

To study the above issues, the mathematical model of the laminar flow of a thin layer of a viscous lubricant in the form of a two-dimensional Reynolds equation [1-3] will be used, which has the following form:

$$div\left(\frac{\rho h^3}{\mu}\, gradp\right) = 6\, div(\rho h V_t) + 12\rho V_n \tag{1}$$

$$V(\alpha) = \begin{pmatrix} V_x \\ V_y \\ V_z \end{pmatrix} = \begin{pmatrix} \cos\alpha & \sin\alpha & 0 \\ -\sin\alpha & \cos\alpha & 0 \\ 0 & 0 & 1 \end{pmatrix} \begin{pmatrix} \dot{X} \\ \dot{Y} \\ \dot{Z} \end{pmatrix} + \begin{pmatrix} 0 \\ \omega R(z) \\ 0 \end{pmatrix}$$

Here V_t, V_n - tangential and normal components of the projection of the velocity $V(\alpha)$ of the lubricant on the bearing surface; $V_x = u, V_z = v$ $V_x = u, V_z = v$ (fig. 1), V_z - speed of the lubricant in the axial direction, $\dot{X}$, $\dot{Y}$ and $\dot{Z}$ - speed of the center of the journal, ω - angular speed, rad/s, h - radial gap function, α - the angle of rotation corresponding to the motion along the axis OX along the supporting surface of the bearing, $\alpha = \frac{2\pi x}{R}$.

In the framework of this paper, isothermal formulation of the problem is taken, assuming, therefore, that the viscosity and density are constant throughout the lubricating layer.

The calculated area in this case is a "supporting surface", topologically equivalent to the surface of the truncated cone. Equation (1) in this case is conveniently written in polar coordinates, taking as the pole the point of convergence of a given cone [2]:

$$\frac{\partial}{\partial r}\left[\frac{r\rho h^3}{\mu}\frac{\partial p}{\partial r}\right] + \frac{\partial}{\partial \phi}\left[\frac{\rho h^3}{\mu}\frac{\partial p}{r\partial \phi}\right] = 6\frac{\partial}{\partial r}(r\rho h V_r) + 6\frac{\partial}{\partial \phi}(\rho h V_\phi) + 12 r\rho V_n$$

$$\begin{pmatrix} V_\phi \\ V_n \\ V_r \end{pmatrix} = \begin{pmatrix} 1 & 0 & 0 \\ 0 & \cos\gamma & -\sin\gamma \\ 0 & \sin\gamma & \cos\gamma \end{pmatrix} \begin{pmatrix} V_x \\ V_y \\ V_z \end{pmatrix} \tag{2}$$

The boundary conditions to equation (2) are the pressure values p_l and p_r, set on the boundaries and taken as equal to 1 bar within this study.

Determination of the gap function for the general case, which allows the rotation of the local coordinate reference frame of the rotor relative to the fixed bearing's frame, is given in [3]. In the present case of the problem, a simpler formulation of the problem will be considered in which only one conical fluid-film bearing is present, and the coordinate system associated with the rotor can be obtained from the coordinate system associated with the bearing only by the parallel transfer operation. This allows one to write the gap function in a simpler and more convenient form using the expression for the cylindrical bearing clearance function defined by three parameters: the radius of the bearing, the radius of the journal and the displacement vector of its geometric center in the radial plane:

$$\begin{aligned} &h_{conic}(R^b_{\max}, R^b_{\min}, R^r_{\max}, R^r_{\min}, X, Y, Z, \alpha, z) = \\ &= h_{cylindric}(\tau(0,z)R^b_{\max} + (1-\tau(0,z))R^b_{\min}, \tau(Z,z)R^r_{\max} + (1-\tau(Z,z))R^r_{\min}, X, Y, \alpha) \end{aligned} \tag{3}$$

$$\begin{aligned} &h_{cylindric}(R^b, R^r, X, Y, \alpha) = \\ &= R^b - (X\cos(\alpha) + Y\sin(\alpha) + \frac{\sqrt{(2X\cos(\alpha) + 2Y\sin(\alpha))^2 - 4\cdot(X^2 + Y^2 - R^2)}}{2}; \end{aligned}$$

$$\tau(Z,z) = \frac{z - Z}{L}$$

Here R^b and R^r - radiuses of the bearing and the rotor accordingly, L - length of the bearing. In practical constructions, the parameters α and z that determine the coordinates of a point on a reference supporting surface are considered free, and all others are considered connected.

One of the characteristics of a fluid-film bearing is the reaction force vector of the lubricating layer, the components of which can be obtained by integrating the projections of the vector $p\cdot\vec{n}$ along the surface, where p - pressure, $\vec{n}$ - the internal normal to the surface of the bearing:

$$\begin{aligned} R_X &= \int_{-\beta}^{\beta} \int_{R^b_{min}/\sin\beta}^{R^b_{max}/\sin\beta} -p(r,\phi)\cdot\cos\left(\frac{X(r,\phi)}{R(Z(r,\phi))}\right)\cos\alpha\, dr\, d\phi \\ R_Y &= \int_{-\beta}^{\beta} \int_{R^b_{min}/\sin\beta}^{R^b_{max}/\sin\beta} -p(r,\phi)\cdot\sin\left(\frac{X(r,\phi)}{R(Z(r,\phi))}\right)\cos\alpha\, dr\, d\phi \\ R_Z &= \int_{-\beta}^{\beta} \int_{R^b_{min}/\sin\beta}^{R^b_{max}/\sin\beta} p(r,\phi)\cdot\sin\alpha\, dr\, d\phi \end{aligned} \tag{4}$$

$$\alpha = arctg\left(\frac{R^b_{max} - R^b_{min}}{L}\right), \quad \beta = \pi \sin\alpha, \quad Z(r,\phi) = \left(r - \frac{R^b_{min}}{\sin\beta}\right)\cos\alpha$$

$$X(r,\varphi) = (\tau(0,Z(r,\varphi))R^b_{\max} + (1-\tau(0,Z(r,\varphi)))R^b_{\min})\cdot\frac{\pi(\varphi+\beta)}{\beta}$$

For the numerical integration of equation (1), the method of finite differences is used with an adaptive grid, described in [4]. This scheme does not imply the uniformity of the grid along the coordinate lines and has the $o(h^2)$ order of accuracy, where h - the average step of the grid. A typical pressure distribution in a fluid-film conical bearing in the steady-state motion is shown in the Figure 2

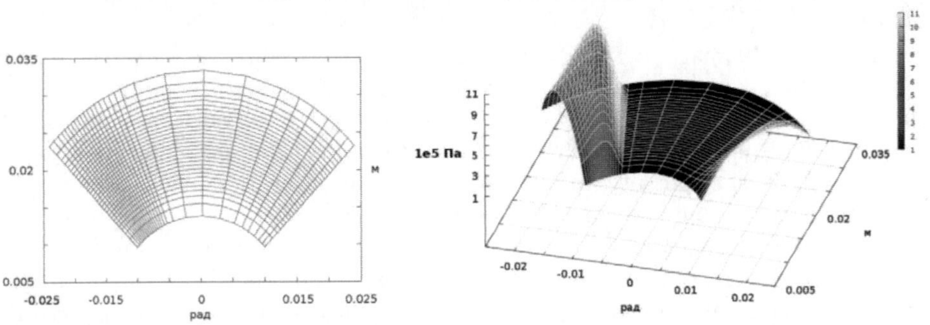

(a) (b)

Figure 2. To the solution of (2): (a) adaptive calculation grid; (b) pressure distribution on a conical bearing.

3. Dynamic equilibrium curves formation

Using the presented mathematical model, the question of determination of points of a dynamic equilibrium curve could be addressed, given the geometric parameters of the bearing, thermophysical properties of the lubricant and the given external force [2]. It is easy to show, that if for some ω such point exists, then it is the only one. To do this, it suffices to note that the norm of the bearing's reaction force vector is a monotonically increasing eccentricity function, that is, the amount of displacement of the rotor relative to its central position. Then, because of the symmetry of the bearing, the vector equation $\vec{R}(X,Y) = -\vec{F}$ can be divided into two equations:

$$\| R(X,0) \| = \| F \| \tag{5}$$

$$\frac{\langle A(\alpha)R(X,0), F\rangle}{\| F \|^2} = 1 \tag{6}$$

where $A(\alpha)$ is the matrix of the left-hand rotation by the angle α, $\langle\cdot,\cdot\rangle$ is the Euclidean scalar multiplication. Obviously, these equations can be solved in this order, that is, first to find the reaction force vector with a predetermined displacement direction averaged to the given load, and then determine the angle to which the given vector should be rotated in order to completely balance the applied force. Because of the monotonicity of the reaction, simple one-dimensional optimization methods can be used to solve the first equation, in particular, the half-division method.

For the case of a laminar isothermal flow of the lubricant in the bearing, it can also be shown that there is a single curve of dynamic equilibrium, determined to within an arbitrary factor in terms of

IOP Conf. Series: Materials Science and Engineering **233** (2017) 012041 doi:10.1088/1757-899X/233/1/012041

accuracy having the dimension of dynamic viscosity. Indeed, for the case of constant viscosity, one can make a $p \to p_0\mu$ substitution of variables and solve the Reynolds equation (1) in the form:

$$\text{div}\left(\rho h^3 \text{grad} p_0\right) = 6\,\text{div}(\rho h V_t) + 12\rho V_n$$

Then the reaction forces will be determined as

$$R_X = \int_\Omega -\mu p_0(\vec{x})\cdot\cos\alpha(\vec{x})dS = \mu\int_\Omega -p_0(\vec{x})\cdot\cos\alpha(\vec{x})dS = \mu R_X^0$$

$$R_Y = \int_\Omega -\mu p_0(\vec{x})\cdot\sin\alpha(\vec{x})dS = \mu\int_\Omega -p_0(\vec{x})\cdot\sin\alpha(\vec{x})dS = \mu R_Y^0$$

So, for a laminar isothermal flow, the shape of the dynamic equilibrium curves is determined only by the gap function and, therefore, is a only geometric characteristic of the bearing.

For the case of a conical bearing, it becomes necessary to consider a three-dimensional reaction vector, which makes it impossible to directly use the above-described decomposition technique. One can, however, get rid of this difficulty, if one observes that for each permissible axial displacement of the rotor there is a corresponding curve of mobile equilibrium in the cutting plane for a given projection of the external counterbalanced force F on it. So, for conical bearings, it is possible to define a surface that can reasonably be called a dynamic equilibrium surface, such that all radial curves of dynamic equilibrium for all permissible axial displacements of the rotor lie on it. On this surface, it is possible to determine the field of forces directed in the axial direction and which are the projection of reaction forces at the corresponding "equilibrium" points of reaction forces to the axis of the bearing. In this case, the problem of finding equilibrium points for a conical bearing can also be decomposed into the following steps:

1) form the surface of dynamic equilibrium Ω;

2) solve the equation $R(X,Y,Z) = -F$ with limitation $(X,Y,Z) \in \Omega$.

It is clear that every equilibrium point lies on the surface of mobile equilibrium. We note, however, that already at the stage of setting the problem, it becomes clear that it is impossible to speak in the general case of any continuous three-dimensional mobile equilibrium curve for a conical bearing. Moreover, the computational experiments carried out show that, in general, there is no reason for a conical bearing to assume the presence of *at least one* point in which both the radial and axial components of the applied force can be simultaneously balanced. A typical example of a mobile equilibrium surface and its corresponding field of axial reactions are shown in Figures 3-6.

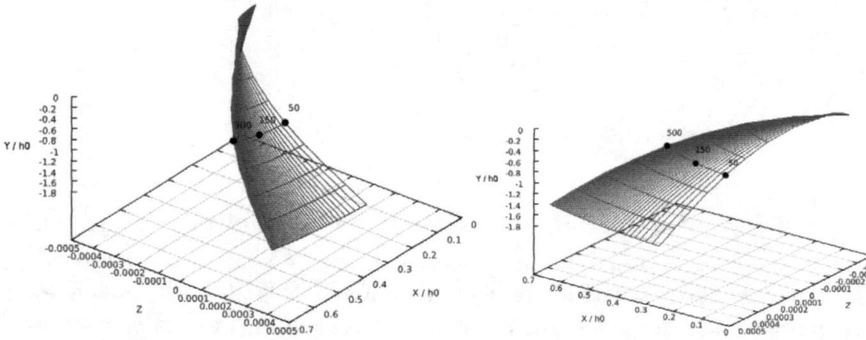

Figure 3. Surfaces of the dynamic equilibrium of the conical bearing; Lubricant: Water.

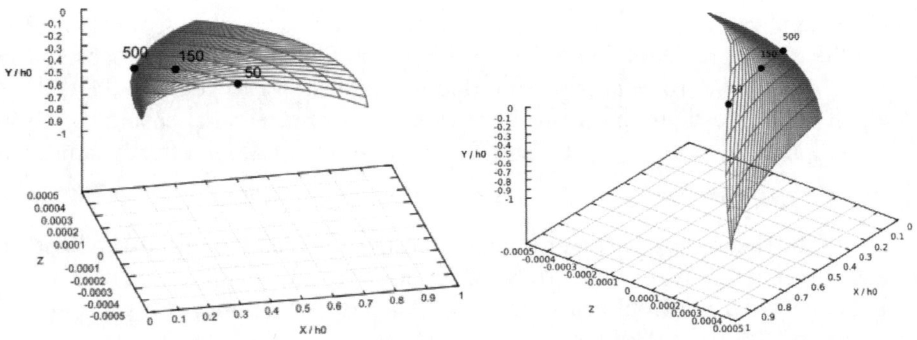

Figure 4. Surfaces of the dynamic equilibrium of the conical bearing; Lubricant: Oil TP-22.

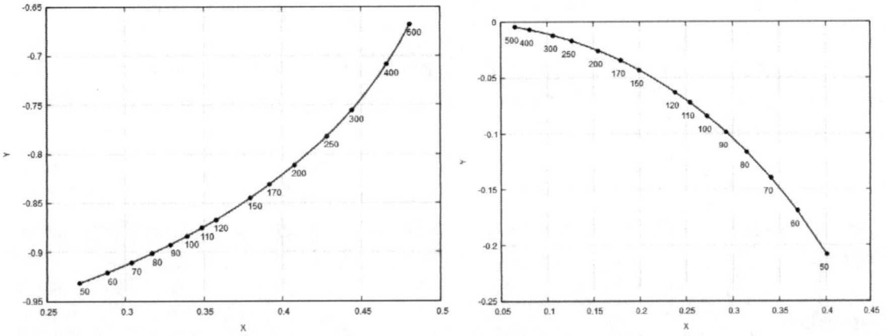

(a) (b)

Figure 5. The curves of dynamic equilibrium in the absence of axial displacement: (a) water; (b) turbine oil TP-22.

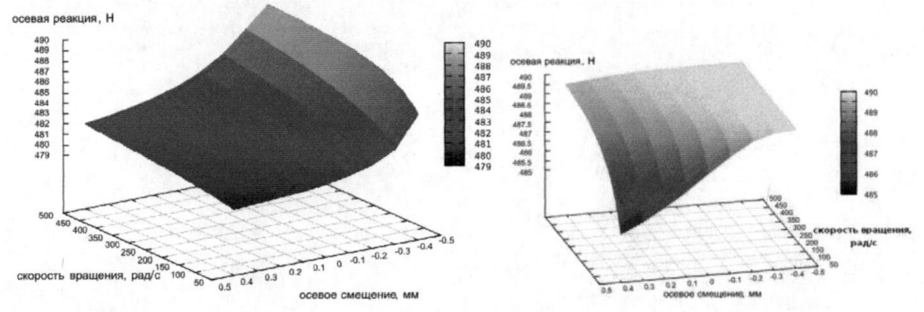

(a) (b)

Figure 6. Distributions of axial reactions forces of a conical bearing: (a) water; (b) turbine oil TP-22.

The figure shows that the field of axial reaction forces is practically uniform. From this the following conclusion could be drawn: The conical bearing, by its design at a given speed of rotation ω, is able to stably balance only a certain set of external forces, for which F_Z lies in some sufficiently close proximity $\phi(F_X, F_Y, \omega, 0)$.

First, for a conical bearing, to a lesser extent than for the rest, a procedure for linearizing the reaction in the area of a certain point is suitable, and a description of the effect of the lubricating layer on the rotor through the stiffness and damping coefficients. Secondly, a small dispersion of the axial component of the reaction of the lubricating layer at points lying on the surface of mobile equilibrium indicates that the application of a constant axial force in the general case must lead either to a displacement of the rotor in the direction of increasing gap and loss of bearing's load capacity or to the contact of the surfaces of the rotor-bearing unit due to the inability to compensate the applied axial load. The study of these questions and the determination of parameters that ensure the operability and stability of the motion of the rotor in conical supports require the solution of the dynamic problem.

References

[1] Savin L A 1998 Theoretical bases of calculation and dynamics of fluid-film with multiphase lubricants *Doct. Tech. Sciences thesis: 01.02.06* (Orel)

[2] Kirk R G, Gunter E J 1973 Nonlinear Transient Analysis of Multimass Flexible Rotors – Theory and Applications (NASA CR-2300)

[3] Lund J W 1965 Rotor-Bearing Design Technology *Part V Technical Report AFAPL-TR-65-45* (Aero Propulsion Lab, Wright-Patterson Air Force Base, Dayton, Ohio)

[4] Korneev A Y 2004 Dynamic and static characteristics of conical bearings *Cand. Techn. Sciences thesis: 01.02.06* (Orel)

[5] Koltsov A Y, Korneev A Y 2013 Determination of the generalized gap function in the case of a rotor misalignment in conical fluid-film bearings *Izvestia OrelGTU, series "Fundamental problems of technique and technology* **4** pp. 11-15

[6] Koltsov A Y 2015 Application of the method of statistical optimization of the computational grid for solving the equations of mathematical physics by the method of finite differences *Proceedings of the State University - UNPK, series "Fundamental Problems of Technology and Technology* **5** pp 171-177

Acknowledgments

This work was supported by the Russian Science Foundation under the project № 16-19-00186, by the National Natural Science Foundation of China (Grant No. 51405410) and the Natural Science Foundation of Fujian Province (Grant No. 2015J01228).

IOP Conf. Series: Materials Science and Engineering **233** (2017) 012042 doi:10.1088/1757-899X/233/1/012042

Optimum control of the active magnetic bearing for various parameters of the program movement

S V Majorov[1], N N Makhova[1]

[1]Department of Mechatronics and International Engineering, Orel State University, Naugorskoe sh. 29, Orel, Russia, 302020

E-mail: sergeyostu@yandex.ru

Abstract. We consider the mathematical model of stabilization of the situation of a rotor is given in an active magnetic bearing. Also various options of control system are considered Investigated influence of the program movement and free controller parameters on quality of control process.

1. Introduction

One of the main elements of many machines is the rotor supported system that causes need of creation of these which would meet various requirements related by operating conditions and at the same time in reliability and quality. In cases when emphasis is placed on long term of work and high reliability at high speeds and heavy dynamic loadings, especially there is a use of active magnetic bearings. Their work is based on use of the magnetic attraction operating on a ferromagnetic body from electromagnets (Figure 1).

During the operation of active magnetic bearings resources by means of which we can exercise control are limited that does urgent use of optimum control with such parameters of the program movement which would provide the greatest energy efficiency.

Support of rotors with an electromagnetic way of creation of the bearing ability have all reasons for expansion of scope of application in different types of transport and technological machines, including main pump units. It is in many respects connected with improvement of electronic devices, manufacturing techniques of electrotechnical elements, and also with the advent of new methods and control algorithms in nonlinear systems [1 – 3].

However at operation of cars with rotors on an active magnetic suspension there are problems connected with stability of the movement of a rotor [4 – 9]. At the same time it is necessary to consider as the general questions of dynamics of rotor systems so [10 – 13], and the aspects connected with questions of calculation of forces in magnetic support [4, 7, 8, 14] and synthesis and the analysis of a control system [4, 7]. Thus, the problem of design of rotor system on an active magnetic suspension needs to be solved together with creation of a control system which would provide steady a rotor with providing operational indicators and decrease in power losses of the unit in general.

2. Mathematical model

The linear model of «rotor – active magnetic bearing» system can be presented in the form of ordinary differential equations system as follows [4].

$$\begin{cases} m\mathrm{d}^2 y/\mathrm{d}t^2 - c_y y - h_i i = Q; \\ L\mathrm{d}i/\mathrm{d}t + (h_i/2)\mathrm{d}y/\mathrm{d}t + ry = u, \end{cases} \tag{1}$$

where
m – mass of ferromagnetic body (rotor);
y – rotor displacement reference about equilibrium position (control coordinate);
i – coil current;
u – voltage control;
Q – external force (gravity);
L – индуктивность;
r – coil resistance ;
c_y – «negative» position stiffness of bearing;
h_i – current stiffness of bearing;
t – time.

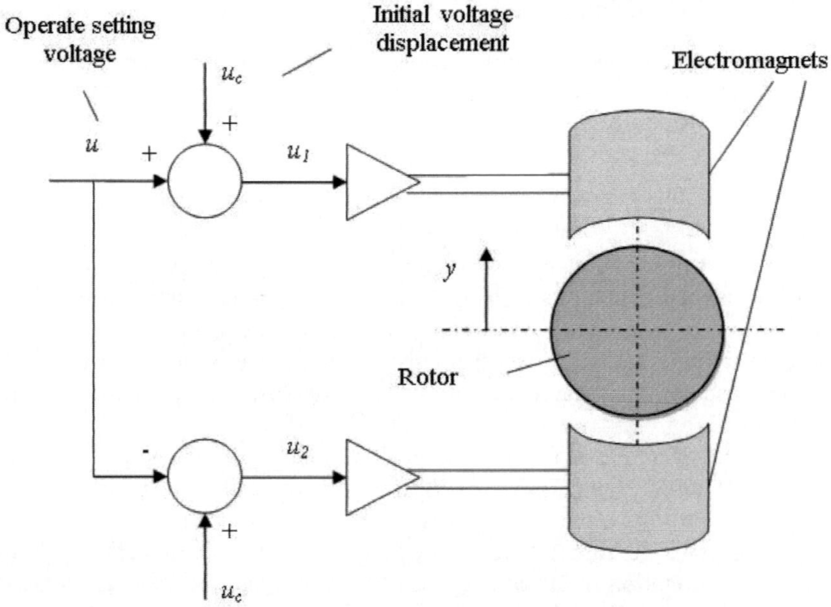

Figure 1. Model of a two-sided «rotor – active magnetic bearing» system.

Position and current stiffness can be determined by the following relations:

$$c_y = \frac{2Li_c^2}{\delta^2} \tag{2}$$

$$h_i = \frac{2Li_c}{\delta} \tag{3}$$

where
i_c – initial displacement current,
δ – bearing radial clearance.

We will note that use of linear model of the active magnetic bearing it isn't always justified in some cases it is necessary to use other approaches to determination of forces in a magnetic contour, such as in work [14].

At optimum control it is required to find such $u=u^o$ which when translating system from any initial state in a zero state minimizes some functionality of quality [15]:

$$\int_0^\infty \left[y^T(t)y(t) + \rho u^T(t)u(t) \right] dt \tag{4}$$

where ρ – a positive weight scalar. The size of value of management u and accuracy of positioning of system depends on this parameter. The more ρ, the is less than value u. Control at $\rho \to 0$ is called «cheap», at $\rho \to \infty$ – «expensive».

3. Optimal current control

For the optimum PD-regulator the control law looks as follows [15]:

$$i^o = -(k_1^o y + k_2^o \dot{y}) \tag{5}$$

where

$$k_1^o = m\frac{\omega_0^2 + k^2}{h_i}, \quad k_2^o = m\frac{\sqrt{\omega_0^2 + k^2}}{h_i} \tag{6}$$

Here k – module of poles of an control object,

$$k = \sqrt{\frac{c_y}{m}} \tag{7}$$

ω_0 – natural frequency undamped system.

Coefficients in expression (6) are received by equating of a characteristic polynom of the optimum closed system with a polynom of a program system. At the same time control parameters ρ and ζ are from relationships:

$$\rho = \frac{1}{(\omega_0^4 - k^4)}; \quad \zeta = \frac{1}{2}\sqrt{2\left(1 + \frac{k^2}{\omega_0^2}\right)} \tag{8}$$

Equations (6) contain the varied ω_0 parameter (natural frequency undamped system). As ρ shall be the positive, this parameter shall meet the following condition:

$$\omega_0 \geq k \tag{9}$$

Therefore we will set ω_0 on formula:

$$\omega_0 = k_\omega k, \tag{10}$$

where k_ω – free varied parameter.

We receive the most «expensive» control at k_ω=1 (Figure 2 (a)) when ω_0=k.

In Figures 2 and 3 dependence of rotor displacement and control current on time is shown respectively at various k_ω and ω_0 values.

Apparently from schedules, we receive the smallest expenses on current (Figure 3 (a)) at expensive management where k_ω=1 and at the set parameters of the ω_0=118.9 rad/s, however to the same case there corresponds the longest transient process (Figure 3(a)). With increase in ω_0 time of transient process is reduced, but at the same time the size of the current necessary for bearing control increases. In work [4] it is specified that in practice of ω_0 value 20 – 500 Hz usually make, however It should be noted that this size substantially depends on such parameters as inductance of L, current of shift of i_c

and size of a gap between rotor and electromagnets poles δ. In particular, reduction of δ leads to increase in the minimum ω_0 value which is necessary to provide stability of system.

4. Optimal voltage control

Optimum control of u^o is set as follows [15]:

$$u^o = -(g_1^o y + g_2^o \dot{y} + g_3^o i + g_4^o \sigma) \tag{11}$$

where

σ – integral of rotor displacement by time;

g_1^o, g_2^o, g_3^o, g_4^o – optimal values of feedback strengthening coefficients respectively on rotor displacement, speed, control current and time domain integral from rotor displacement.

We will consider a problem of optimum voltage control at ρ→0 («cheap» control).

For definition of optimum values of feedback strengthening coefficients of it is necessary to solve the characteristic equation of system:

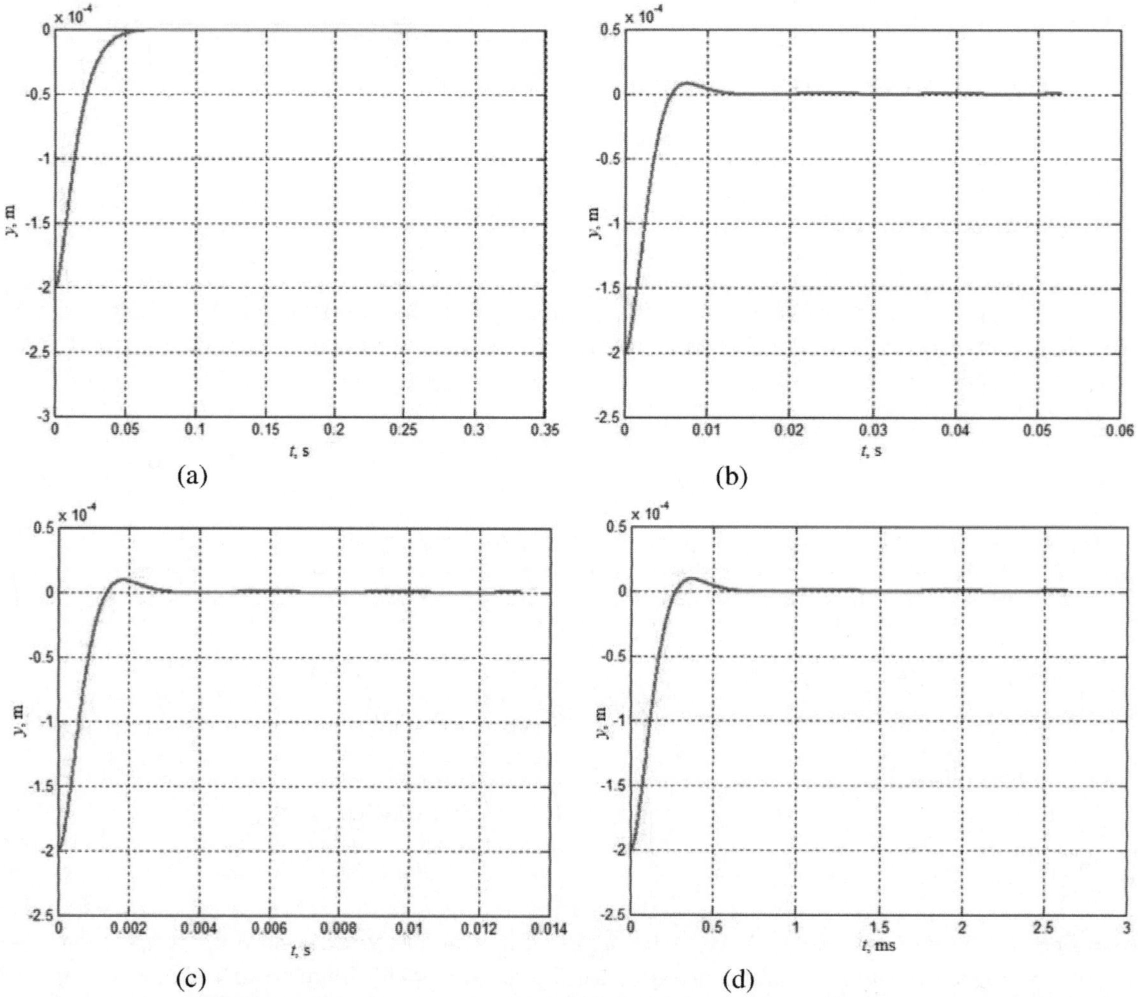

Figure 2. Dependence of rotor displacement on control time at (a) k_ω=1; (b) k_ω=5; (c) k_ω=20; (d) k_ω=100.

$$s^3 + \frac{r}{L}s^2 - \frac{c_y r}{mL} = 0 \tag{12}$$

Therefore we find poles of an control object of π_1, π_2 и π_3 which can be expressed:

$$\pi_1 = \alpha_1;\ \pi_2 = -\alpha_2 + \mathrm{j}\beta_2;\ \pi_3 = -\alpha_2 - \mathrm{j}\beta_2 \tag{13}$$

where

α_0, α_1, α_2, β_2 – positive numeric values,

j – imaginary unit.

At $\rho \to 0$ $\pi_1 = s_1$, $\pi_2 = s_2$, $\pi_3 = s_3$. To consider an integrated component we enter an additional pole of $s_4 = \alpha_0$, small on the module.

We receive the following coefficients of optimum voltage control:

$$\begin{cases} g_1^o = \left\{ c_y L(\alpha_0 + \alpha_1 + 2\alpha_2) + mL\left[(\alpha_0 + \alpha_1)(\alpha_2^2 + \beta_2^2) + 2\alpha_0\alpha_1\alpha_2 \right] \right\} / h_i; \\ g_2^o = mL\left[(\alpha_2^2 + 2\alpha_1\alpha_2 + \beta_2^2 + \alpha_0(\alpha_1 + 2\alpha_2) \right] / h_i; \\ g_3^o = L(\alpha_0 + \alpha_1 + 2\alpha_2) - r; \\ g_4^o = mL\alpha_0\alpha_1(\alpha_2^2 + \beta_2^2). \end{cases} \tag{14}$$

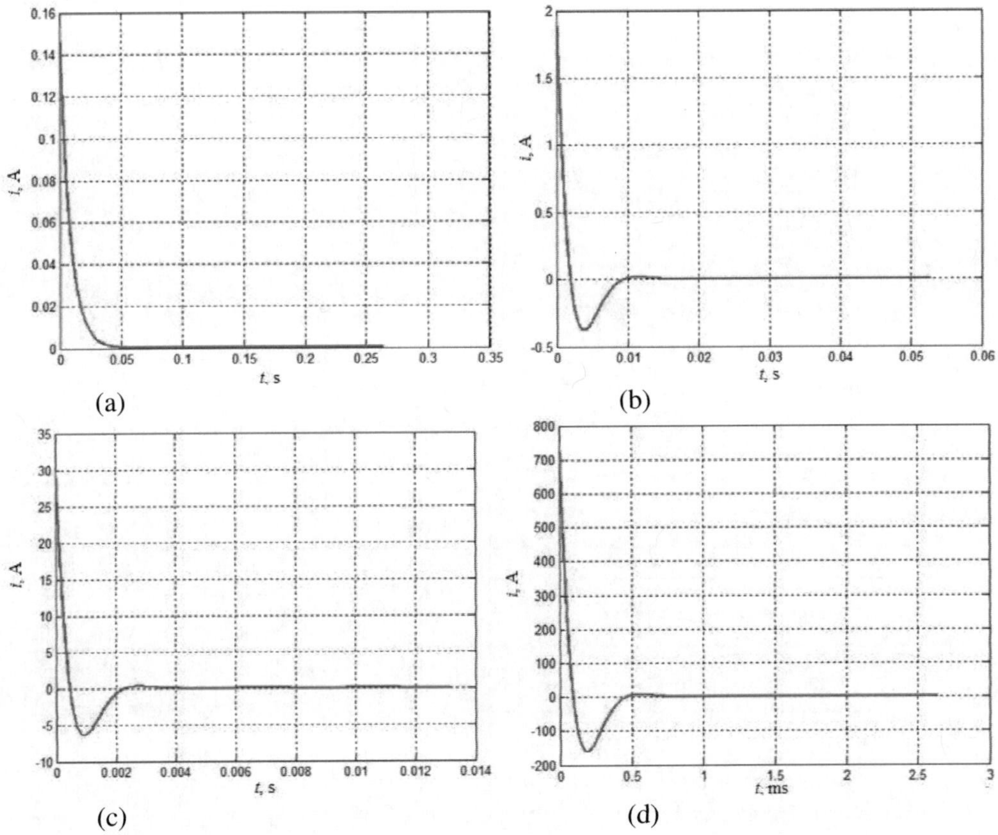

Figure 3. Dependence of control current on control time at (a) k_ω=1; (b) k_ω=5; (c) k_ω=20; (d) k_ω=100.

In this case the varied parameter is α_0 which we will set as the smallest of the available coefficients of α_1 and α_2 increased by some coefficient. Transient time depended characteristics at various α_0 values are provided on Figures 4 and 5.

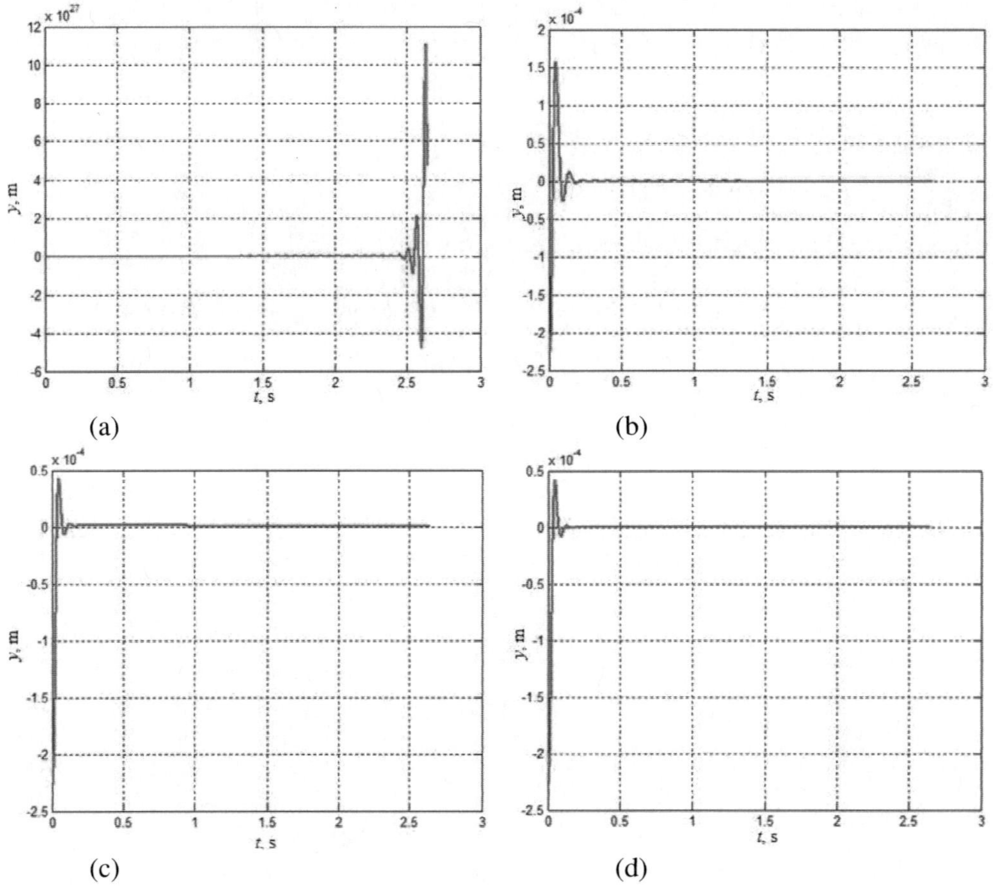

Figure 4. Dependence of rotor displacement on control time at (a) α_0=9.2; (b) α_0=0.9; (c) α_0=0.009; (d) α_0=0.00009.

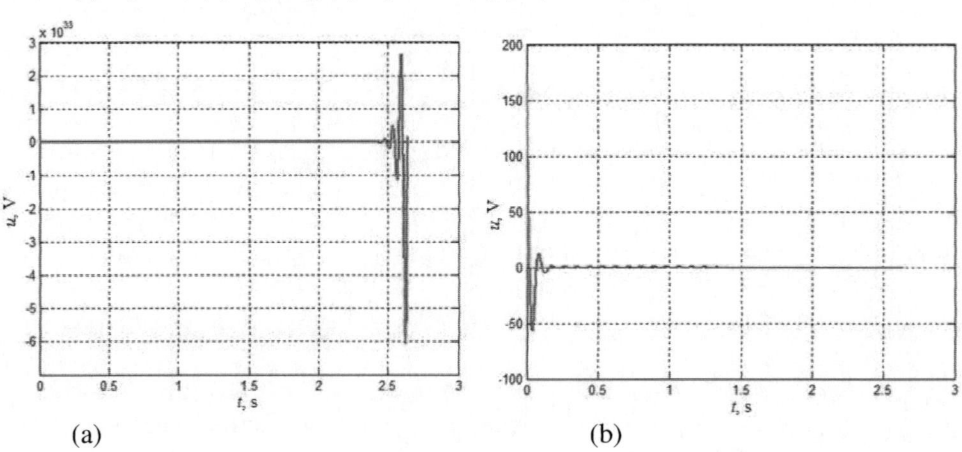

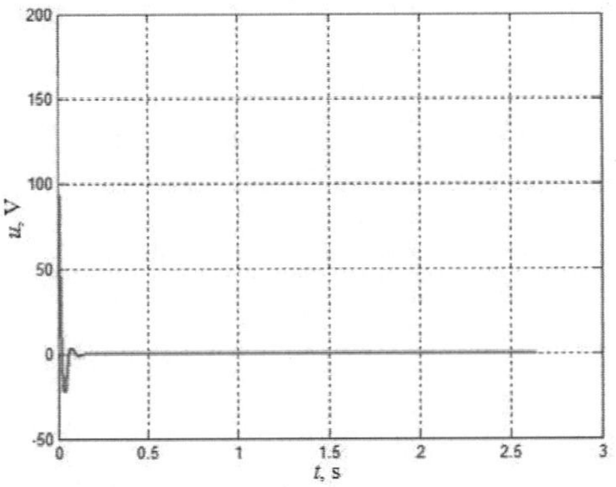

(c)

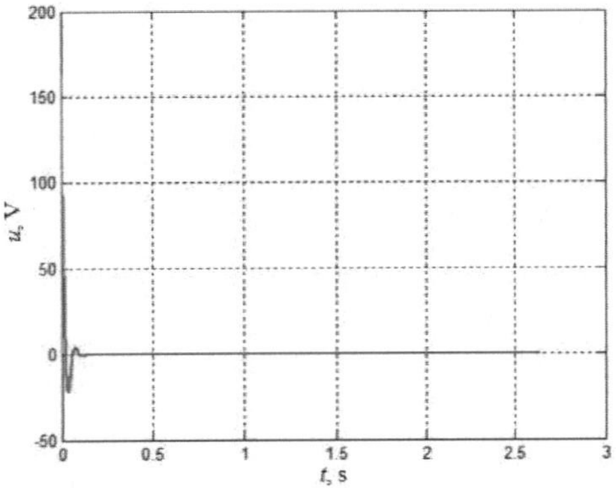

(d)

Figure 5. Dependence of control voltage on control time at a) α_0=9.2; b) α_0=0.9; c) α_0=0.009; d) α_0=0.00009.

At α_0 is 10 times less, than coefficients of α_1 and α_2, process it is unstable (Figure 4 (a) and 5(a)). At further reduction of α_0 we observe improvement of quality of transitient process (Figure 4 (b-d) and 5 (b-d)).

5. Conclusion

The model of an electromagnetic suspension with optimum current and voltage control is realized in the environment of Matlab, varying parameters of the program movement. Need of the correct selection of parameters for stability of system and its energy efficiency, and also for receiving transient processes of the required quality is obvious. Further it is planned to realize use of the studying systems for the purpose of simplification of selection of parameters of the program movement for each separate system.

References

[1] Abduragimov A S and Vereshchagin V P 2010 Features of the digital equipment of control of electromagnetic bearings of gas-distributing units *Electromechanics questions* (Moscow VNIIEM) **115**

[2] Vereshchagin V P and Rogoza A V 2008 Features of design of magnetic bearings for large machines *Electromechanics questions* (Moscow VNIIEM) **106**

[3] Sarychev A P 2009. П. Features and experience of creation of electromagnetic bearings for a series of compressors of gas-distributing units *Electromechanics questions* (Moscow VNIIEM) **106**

[4] Zhuravlev Y N 2003 *Active magnetic bearings: Theory, design, application* (St. Petersburg: Polytechnica) p 206

[5] Sarychev A P 2009 Development of electromagnetic bearings for a series of compressors of gas-distributing units *Electromechanics questions* (Moscow VNIIEM) **110**

[6] Sarychev A P 2008 Mathematical model of a rotor for the analysis of control of magnetic bearings *Electromechanics questions* (Moscow VNIIEM) **107**

[7] Schweitzer G and Maslen E H 2009 *Magnetic bearings. Theory, design, and application to rotating machinery* (Springer-Verlag Berlin Heidelberg) p 523

[8] Maslen E H 2000 *Magnetic bearings* (University of Virginia) p 245

[9] Schweitzer G 2002 Active Magnetic bearings – chances and limitations *Proc. 6th Internat. IFToMM Conf. on Rotor Dynamics* (Sydney, Sept. 30 – Oct. 3)

[10] Lee C W 1993 *Rotating Vibration analysis of rotors* (Springer) p 312

[11] Genta G 2005 *Dynamics of rotating systems* (NY: Springer) p 660

[12] Lalanne M and Ferraris G 1998 *Rotordynamics prediction in engineering* (J Wiley&Sons) p 266

[13] Friswell M I 2010 *Dynamics of rotating machines* (Cambridge University Press) p 512

[14] Solomin O V Dorofeev L V and Majorov S V 2008 Calculation of magnetic field in the radial active magnetic bearing / *Proc. 6th Internat. Conf. Vibration machines and technologies* (Kursk: KSTU) pp 779 –784
[15] Nogin V D 2008 Introduction to optimum control (St. Petersburg: UTAS) p 92

Acknowledgment
This work was partly supported by an RSF grant, project 16-19-00186 "Planning of optimal energy efficient rotor trajectories in mechatronic modules with complex rheology medium".

IOP Conf. Series: Materials Science and Engineering **233** (2017) 012043 doi:10.1088/1757-899X/233/1/012043

Actuators of active tribotechnical systems of the rotor-bearing type

L Savin[1], D Shutin[1], A Kuzavka[1]

[1] Department of Mechatronics and International Engineering, Orel State University, Naugorskoe sh. 29, Orel, Russia, 302020

E-mail: rover.ru@gmail.com

Abstract. The article describes the perspectives of using active bearings in rotor-bearing systems. The principal scheme of a mechatronic tribotechnical system anв classification of actuators used in such system are shown. Piezo actuators are considered from the point of view of use as actuators in active bearings. The comparative characteristics of different types of actuators

1. Introduction

Bearings in rotating machinery are a large class of tribotechnical systems. Fluid-film bearing are a very complex system from the point of view of tribology because different types of friction can appear during their operation. In addition to wear caused by friction, vibrations also appear in rotor-bearing systems due to different reasons including nonlinear properties of the lubricant film. Rotor oscillations increase friction and wear, lead to lead to the failure of the important parts of the machine. So, improving dynamic characteristics of rotor-bearing systems, including reducing vibrations level, is actual engineering problem. This problem is being solved in different ways. The classic way is investing in constructive and technological solutions in fluid-film bearings. However, the process of improving bearings this way often meets some physical limits, and even small improvements may take a lot of resources. The other way is integration of informational and control systems in bearings. This way is becoming more popular due to the development of electronic, computing and other technical means necessary for the implementation of such systems. So, improving the hardware characteristics by lead to new results in developing modern rotor-bearing systems in recent decades [1].

Actuators are one of the most important elements of such systems. Actuators transform the input signal (electrical, mechanical, pneumatic, hydraulic, etc.) into the output signal acting on the controlled object. Active fluid-film bearings are rather complex systems so the characteristics of the used actuators must meet rather strict requirements.

2. The principle of operation of a typical tribotechnical system of the rotor-bearing type

The way how active control is implemented in fluid-film bearings is illustrated in [2]. The shown method is the simplest one to be implemented and is perspective for application in rotating machinery. In such systems control is implemented by changing the lubricant pressure in the feeding chambers of a hybrid (combined hydrostatic and hydrodynamic) bearing. Such active bearings usually are a closed-loop control system containing sensors, controller, intermediate devices and actuators. The control signal is generated in the control loop that is external with respect to the structural design of the bearing. The typical scheme of such mechatronic bearing is shown in the Figure 1.

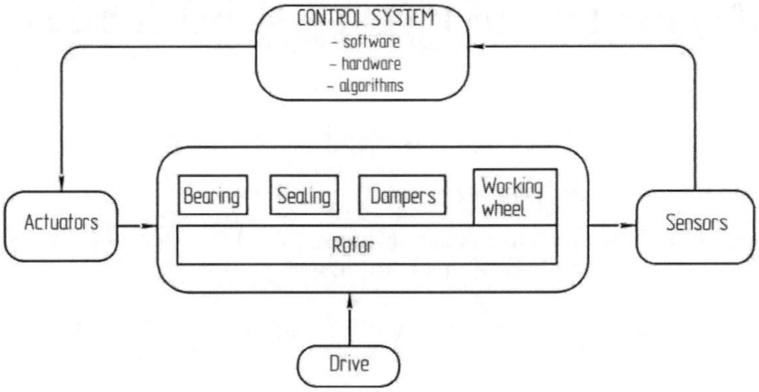

Figure 1. Common scheme of an active tribotechnical system of the rotor-bearing type.

3. Classification of actuators

In mechatronic bearings the rotor is usually a control object. The controlled coordinates describe the position of the rotor in the bearing. At the same time, the rotor's coordinates themselves are not as important for the machine's operation as its complex characteristics, such as oscillations amplitude, power losses due to friction. These characteristics are obtained using the data describing the position of the rotor in the bearing, size and shape of the rotor's trajectory.

The control signal produced by controller is transmitted in a control loop to the actuators and changes that way the parameters of operation of the rotor-bearing systems in the required manner.

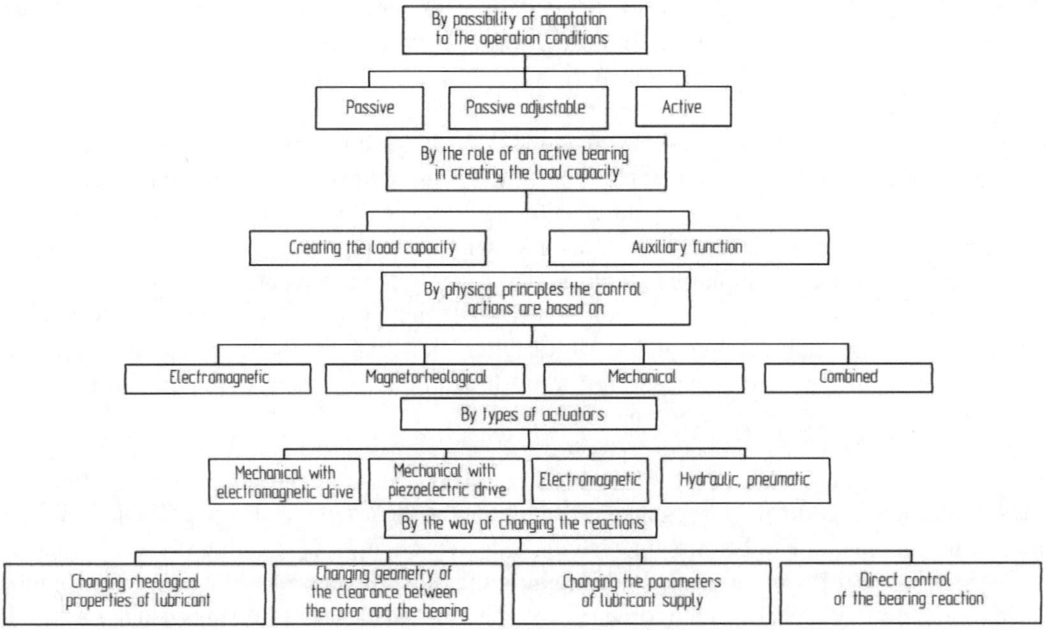

Figure 2. Classification of actuators of tribotechnical systems of the rotor-bearing type.

There are several approaches to creating the mechanical control action in fluid-film and gas bearings:

a) changing the geometry of the gap between the rotor and the bearing. The geometry of the gap is usually changed due to the movable elements of the bearing. Such elements are driven by actuators. The example of such bearing is shown in [12];

b) changing the parameters of supply of lubricant. This process is usually implemented by special controlling devices (throttles, pressure controllers) mounted in supply tracts [13];

c) combined actions obtained by joint using of the approaches mentioned above [14-16].

Active bearings in which the mentioned approaches are implemented are equipped by different types of actuators. The most commonly used types of actuators in fluid film bearings are described below.

1. The mechanical actuators with electromagnetic drive. Such actuators are rather cheap and simple. However, such actuators are often limited in performance due to high inertial properties of moving parts. It makes it more complicated to use these actuators in rotor-bearing systems with direct control of the rotor position at high rotation frequencies. Mechanical actuators with electromagnetic drive are more applicable in systems with indirect rotor position control, e.g. when the damping or stiffness of the whole bearing are controlled.

Mechanical actuators with electromagnetic drive can have different types of impact on the parts of the bearing. In some systems it is direct mechanical impact on the moveable parts of the bearing, e.g. in active tilting-pad bearings [17]. In other cases the considered actuators are a part of a device that regulates other parameters that are used for controlling the rotor-bearing systems. The example of such a device is a servovalve controlling the lubricant pressure [13].

2. The mechanical actuators with piezo drive. Such devices transform the input electrical signal into the output mechanical force or displacement due to the inverse piezoelectric effect [18]. There are different types of piezoactuators [19-21], some of them are shown in the Figure 3.

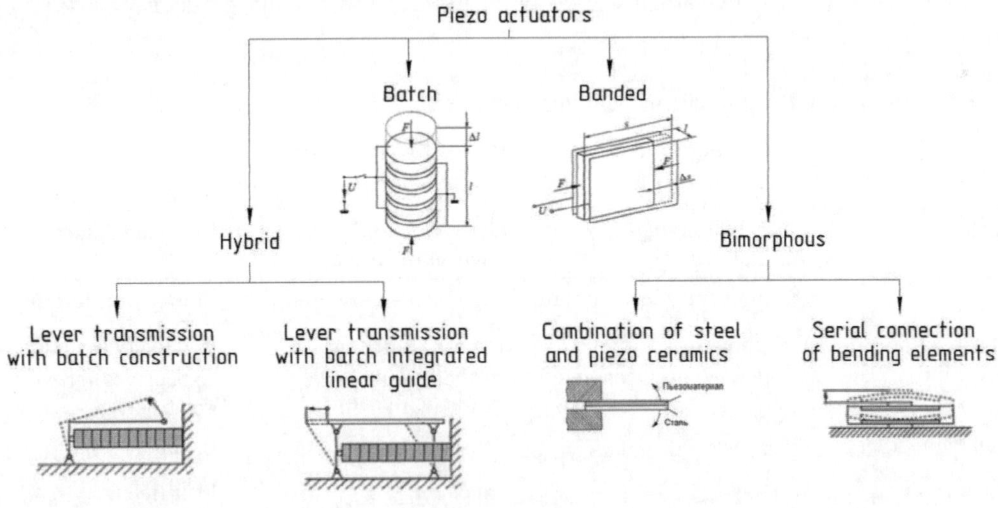

Figure 3. The most commonly used types of piezoactuators.

Piezoactuators have sufficiently better performance comparing to the category described above and very low inertial properties, so they can be used in high-speed rotating machinery with direct control of the rotor position in the bearing. However, piezoactuators are also limited in the possible range of mechanical displacement which is tens or hundreds of microns for simple batch piezoactuators. It limits application of piezoactuators in some cases.

When piezoactuators are used for having the direct impact on the moveable parts of the fluid-film bearing, the limited displacement range makes it impossible to use them in bearings with large geometric dimensions. In such bearings the required ranges of displacement of the moveable parts that provide the appropriate control action range exceed the possible actuator's displacement range. Its increase by using mechanical transmission are usually not justified due to the performance decrease caused by the inertial properties of the transmission elements and the appearance of concomitant negative effects such as backlashes and increased friction.

Bimorphous piezoactuators provide larger displacement range than batch piezoactuators but rather low generated mechanical force that is not enough in many cases for moving the parts of the bearing with the required dynamic characteristics.

3. The electromagnetic actuators. They are not actuators in the direct sense of this term. Electromagnetic actuators generate the electromagnetic field. In turn, it leads to appearance of the force action on the rotor-bearing parts due to different physical effects, such as magnetorheological effect. Such systems have almost unlimited performance but have a limited range of applications in tribotechnical systems.

A common unified evaluation of the various drives is not possible due to some reasons. First of all, actuators solve different technical problems in different types of bearings. Besides, different types of actuators are at different stages of development [20]. The table 1 presents the comparison of individual properties of the actuators in tribotechnical systems.

4. Hydraulic and pneumatic. Devices of this category mechanically change the parameters of a tribotechnical system by transforming the energy of liquid or gas flow. Pneumatic and hydraulic have rather simple design and good force-to-size ratio.

However, hydraulic and pneumatic systems have substantially limited frequency range, so using them in active bearings, especially with direct rotor position control, is also limited. Actuators can be based on hydraulic, pneumatic cylinders, hydraulic motors, compressors, as well as special types of drives in which mechanical movements are caused by a change in the pressure of a liquid or gas [10,11].

4. Comparison of actuators in rotor-bearing systems

Table 1. Comparative characteristics of the properties of the executive mechanisms of tribotechnical systems.

Properties	**Electromagnetic actuators**	**Hydraulic and pneumatic actuators**	**Piezo actuators**
Benefits	Simple transfer function	High force-to-size ratio	Simple transfer function
Drawbacks	High mass Relatively low generated force	Transfer function depends on pneumatic-hydraulic dynamics	Relatively long design Expensive high-voltage amplifier
Frequency range	Good	Satisfyingly	Excellent
Displacement range	Good	Excellent	Insufficient
Rigidity	Insufficient	Good	Excellent
Force-to-size ratio	Satisfyingly	Good	Good

Despite a number of drawbacks, piezo actuators are highly promising in terms of using them in active fluid-film bearings. This is facilitated, in particular, by high dynamic characteristics of piezo actuators that are very essential in view of the current tendency to increase the rotational speed of rotating machinery to increase its power-to-weight ratio. High-performance active bearings can provide long service life, reliability and safety such equipment. In addition, piezo actuators have a number of additional advantages that are essential for use in active bearings:

- almost unlimited resolution: piezo actuators convert electric power directly to mechanical, they provide movement in subnanometric ranges. This property is useful when correcting the clearance geometry between the rotor and the bearing housing;
- the response rate in microseconds;
- the absence of a magnetic field, the action of actuators is based on electric fields. They do not generate magnetic fields, nor are they affected by such. This is especially important when used in equipment that should be proofed from electromagnetic interference;
- low energy consumption in static state regardless of the external load magnitude , even under the influence of large loads, actuators do not consume energy. It allows developing energy-efficient rotor-bearing systems in which power losses are reduced by active control [18], and energy consumption of a control system is lower than energy savings due to reduced friction;
- there is no wear of piezo actuators because their operation is based on the dynamics of the solid;
- ability to work in extreme conditions, piezo actuators do not require lubrication, and piezoelectric effect is present even at low and cryogenic temperatures. It allows using piezo actuators in bearings with cryogenic lubrication, e.g. liquid oxygen, hydrogen, etc.

5. Application of piezo actuators in rotor-bearing systems

Let's consider the examples of using of piezo actuators in fluid-film bearings and gas bearings that operate mostly on the same principles as fluid-film bearings.

The paper [22] presents an active foil bearing where piezo actuators pull and push the foil and influence the rotor behavior that way. The bearing is used in small-sized gas turbine engines, centrifugal compressors, turbo expanders, aerospace auxiliary units, conventional vehicles. The scheme of the bearing is shown in the Figure 4.

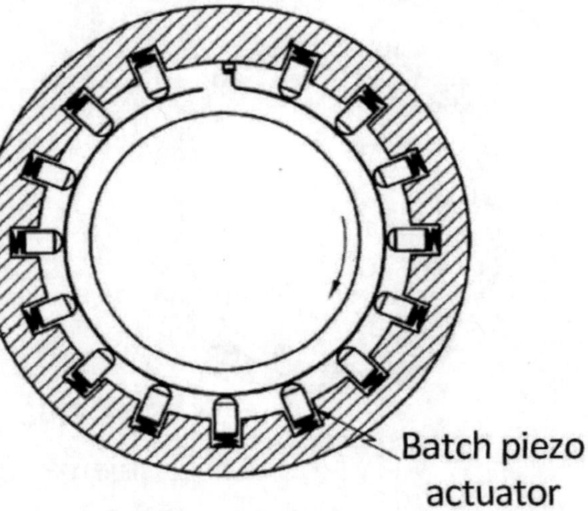

Figure 4. Foil gas bearing.

The devices described in [23, 24] also contain piezo actuators. In [23] the deflections of each individual corrugation of the circular corrugated element are directly proportional to the change in the magnitude of the current in the circuit. It allows monitoring the deflections and change the stiffness of the circular corrugated element (and the rigidity of the whole bearing as well) by increasing the current

(Figure 5 (a)). In [24] the control is implemented in a similar way, but the deflections are obtained by means of the foils (Figure 5 (b)).

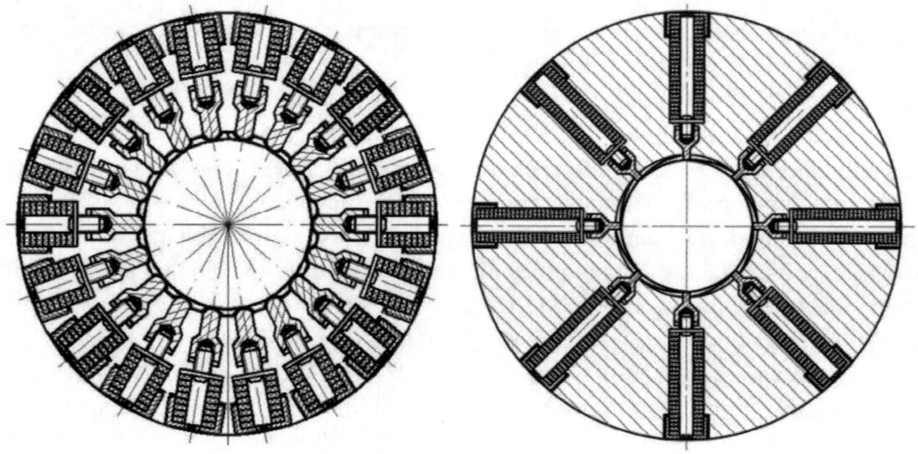

(a) (b)

Figure 5. Gas bearings with piezo actuators: (a) – single-foil bearing with the circular corrugated element; (b) – multi-foil bearing.

Piezo actuators are also used in the device describeв in [25]. The bearing differs from the foil bearings described above by usage of bimorphous piezo elements that consist of the foil and the piezo element (Figure 6).

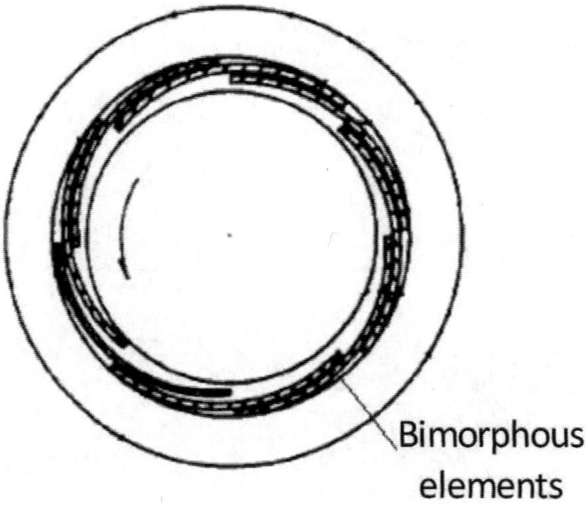

Figure 6. Multi-foil gas bearing with bimorphous elements.

Piezo actuators are used in active bearings in combination with other actuators as well. Such systems are described in [15, 16].

One of the main problems solved by active bearings is reducing the unwanted rotor movements during operation of the rotor-bearing system. The rotor position in the bearing is described by the coordinated of its center in the coordinate system based on the bearing. A certain position of the rotor is chosen as the setpoint of the control system. The control system minimizes the deflection of the rotor from the setpoint. The structure of a two-loop control system of a journal active hybrid bearing is described in [2] and shown in the Figure 7. SP is a device determining the setpoint, C is a converter, CT is a controller, A is an amplifier, RBU is a rotor-bearing unit, FB is a feedback loop, SV is a servovalve changing the pressure of the lubricant. The main drawback of the servovalve is its low dynamic properties. Using piezo actuators instead of the servovales with electromagnetic drive allows obtaining much better dynamics of the system.

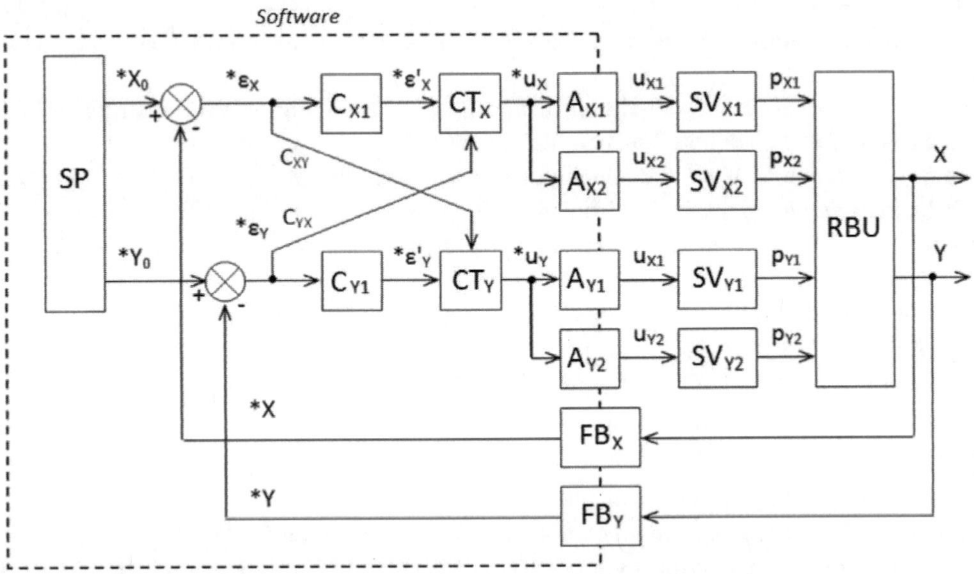

Figure 7. Structure of the control system of an active journal hybrid bearing.

6. Conclusion

Development of modern piezo actuators results in new developments in tribotechnical systems. The presented active mechatronic bearings use various ways of using piezo actuators in such systems. The combination of known physical phenomena (piezoelectric effect, electromagnetism, hydro- and pneumo-mechanics, etc.) into integrated system contributes to the appearance of modern devices which are actively introduced into mechatronic rotor-bearing systems.

References

[1] Maslen E H 2008 Smart machine advances in rotating machinery *9th International Conference on Vibrations in Rotating machinery* (IMechE, Exeter, UK) pp 3-14

[2] Shutin D V 2015 Improving the hybrid bearings by controlling the parameters of lubricant supply: *Dissertation of Cand. Techn. Sciences: 05.02.02* (Orel) p 192

[3] Alciatore, David G, Michael B *Introduction to mechatronics and measurement systems* p 507

[4] Revich Yu V 2016 *Entertaining Electronics.* – 3rd ed., Revised. and additional (SPb.: BHV-Petersburg) p 576

[5] Bul B K 1970 Basics of the Theory of Electric Devices *A manual for electrical engineering specialties of universities* ed. G V Butkevich. (M: "Higher School") p 600

[6] Chunikhin A A 1988 Electric Devices: General course *Textbook for high schools 3rd ed., Per. And additional* (Moscow: Energoatomizdat) p 720

[7] Belov M P, Novikov A, Rassudov L N 2007 *Automated electric drive of standard production mechanisms and technological complexes: a textbook for students of higher education* (Moscow: Publishing Center "Academy") p 576

IOP Conf. Series: Materials Science and Engineering **233** (2017) 012043 doi:10.1088/1757-899X/233/1/012043

[8] Zhuravlev Yu N 2003 *Active magnetic bearings. Theory, calculation, application* (SPb: Publishing house of the Polytechnic) p 208
[9] J Hesselbach, C *Abel-Keilhack* 2003 Active Hydrostatic Bearing With Magneto-Rheological Fluid *Journal of Applied Physics* **93** (10) pp 8441-8443
[10] Bashta T M 1972 *Hydraulic drive and hydropneumoautomatics* (Moscow: "Mechanical Engineering") p 480
[11] Robert H Bishop 2002 *The mechatronics handbook* (Bosa Roca, ISA, CRC Press Inc, 1st ed) p 507
[12] Wu A, Cai Z 2008 Model-Based Control of Active Tilting-Pad Bearings *IEEE/ASME Transactions on Mechatronics* (Louisiana) **12** pp 689-695
[13] Shutin D V, Savin L A 2012 Modeling of active journal hybrid bearings *Proceedings of SWSU. Series: Engineering and technology* **1** pp 54-60
[14] Patent US №20050275300. *Methods of controlling the instability in fluid film bearings* / Aly El-Shafei. Published 18.09.2008.
[15] R N Polyakov, M E Bondarenko Patent RU №2605228. *Hybrid bearing* (Published 20.12.2016. Bulletin №35)
[16] L A Savin, R N Polyakov, N V Komarov, D V Shutin Patent RU №258281. *Hybrid bearing with elastic elements* (Published 27.06.2016. Bulletin №18)
[17] Santos I F1994 Design and Evaluation of Two Types of Active Tilting Pad Journal Bearings *The Active Control of Vibration* (Mechanical Engineering Publications Limited) pp 79-87
[18] Panich A E 2008 *Piezoceramic actuators: Textbook* (Rostov-on-Don) p 159
[19] King T G, Preston M E, Murphy B J M, Cannell D S 1990 Piezoelectric ceramic actuators: A review of machinery applications *Precision Engineering* **12** (3) pp 131-136
[20] Haymann B, Gert V, Popp K, Repetsky O 2010 *Mechatronics: Components, methods, examples: Textbook* (Novosibirsk: SB RAS Publishing House) p 602
[21] Bardin V A, Vasiliev V A, Chernov P S 2014 Current state and development of nano and micro range actuators *Nika* pp 123-127
[22] Lee Yong Bok, Kim Chang Ho, Lee Nam Soo, Kim Tae Ho Patent US № 6527446 *Foil journal bearing utilizing semi-active dampers* (Published 26.06.2003)
[23] L A Savin, A V Sytin, D V Shutin, V O Tyurin Patent RU №2568005 *Foil gas bearing with active control* (Published 10.11.2015. Bulletin №31)
[24] L A Savin, A V Sytin, V O Tyurin, S D Antonov Patent RU №2581101. *Multi-foil gas bearing with active control* (Published 10.04.2016. Bulletin №10)
[25] Tsumaki Nobuo Patent JPN №0454309 *Gas bearing* (Published. 21.02.1992)

Acknowledgments
The present research was implemented as a part of the RSF project No. 16-19-00186 'Planning of optimal energy efficient rotor trajectories in mechatronic modules with complex rheology medium'.

Influence of technological factors on characteristics of hybrid fluid-film bearings

A Koltsov[1], A Prosekova[1], A Rodichev[1], L Savin[1]

[1] Department of Mechatronics and International Engineering, Orel State University, Naugorskoe sh. 29, Orel, Russia, 302020

E-mail: prosekova.anastasia@yandex.ru

Abstract. The influence of the parameters of micro- and macrounevenness on the characteristics of a hybrid bearing with slotted throttling is considered in the present paper. The quantitative assumptions of calculation of pressure distribution, load capacity, lubricant flow rate and power loss due to friction in a radial hybrid bearing with slotted throttling are taken into account, considering the shape, dimensions and roughness of the support surfaces inaccuracies. Numerical simulation of processes in the lubricating layer is based on the finite-difference solution of the Reynolds equation using an uneven orthogonal computational grid with adaptive condensation. The results of computational and physical experiments are presented.

1. Introduction

The use of fluid-film bearings in a number of cases is practically non-alternative. However, the selection, calculation and design is done by the designer for each case individually. The fluid-film bearing from the point of view of design is a multifactor complex model that allows to take into account the influence and interaction of design, technological and operational parameters. Accounting technology parameters associated with the study of the effect of macro and micro deviations on the characteristics of bearings [1], as the surface shape is formed during processing. The geometric parameters of the unevenness of the bearing surface depend on the combination of processing modes, rigidity and errors of the "machine-tool-detail" system. These errors can both improve the characteristics of the bearing, and adversely affect its performance. So, the study of the operation of a bearing with a different combination of parameters requires the formation of an individual approach, conceptual and mathematical models based on the joint numerical solution of the Reynolds equation and the radial gap function.

2. Mathematical model development

In some cases, the radial method of supplying lubricant to the fluid-film bearing is impossible or difficult for various reasons. If it is necessary to create a hybrid lift in the lubricating layer in a radial bearing, slit throttling from one side of the bearing can be used. The creation of this type of lifting force during the operation of hydrostatic dynamic bearing with slotted throttling is explained by the Lamakin-Etinger effect [2]. The bearing capacity of the lubricant layer is formed due to the pressure flows formed as a result of a specially created pressure drop at the sides and hydrodynamic effects when the spindle rotates. If the pressure at the sides of the developed bearing becomes equal to atmospheric pressure, the bearing becomes hydrodynamic. Bearings with this method of creating of load capacity are used in machines for pumping liquids, lubrication is possible with a processed medium under pressure in the working area. For this bearing design, it is necessary to provide a device for discharging the used lubricating fluid from the opposite side inlet. To determine the type of mathematical dependencies describing the processes in the lubricating layer of a plain bearing, the design diagram of which is shown in the Figure 1, we introduce a number of assumptions [3]:

1) the lubricant is assumed to be a Newtonian fluid;

2) the flow of the lubricant under steady-state operation is assumed to be laminar, which will be provided by such a combination of operating parameters of the rotor-support system that $Re \leq 41.3\sqrt{\dfrac{D_{ном}}{2h_0}}$,

3) the lubricant completely adheres to the support surfaces and during the operation of the system;

4) the lubricant is considered incompressible;

5) the bearing is completely filled with lubricant, in the loaded area the rupture of the lubricating film is not allowed;

6) there is no misalignment of shaft and bearing axes;

7) the motion along the longitudinal axis of the bearing is absent;

8) the isothermal flow of the lubricant in the gap is considered, as a result of which the viscosity of the lubricant is constant for the constant temperature.

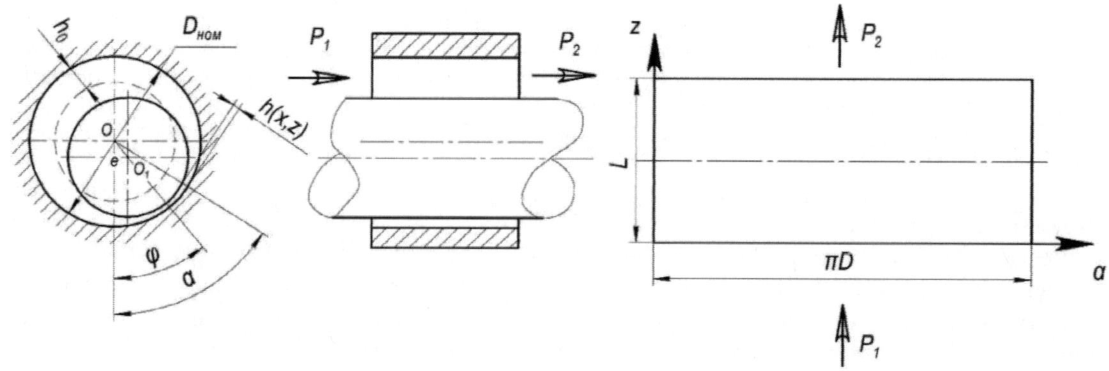

Figure 1. Calculation diagram of a fluid-film bearing with slotted throttling.

On the basis of the assumptions above and the principle of operation of the fluid-film bearing, the determination of the pressure distribution in the lubricating layer is carried out by numerically solving the Reynolds equation in an isothermal formulation, which has the form:

$$\frac{\partial}{\partial x}\left[h^3\frac{\partial p}{\partial x}\right]+\frac{\partial}{\partial z}\left[h^3\frac{\partial p}{\partial z}\right]=6\,\mu\omega R\frac{\partial h}{\partial x} \tag{1}$$

To obtain the correct solution of (1), let us set the following boundary conditions, starting from the calculation scheme and the experience of numerical determination of pressures in the lubricating layer [2, 3]:

boundary conditions of the sides of the bearing:

$$p(x,0)=p_1; \quad p(x,L)=p_2 \tag{2}$$

condition along the coordinate x = α·R shall be written based on the Sommerfeld hypothesis, according to which the bearing is covered by the lubricating layer. Then we can write the conjugation conditions:

$$p(0, z) = p(2\pi R, z); \quad \frac{\partial p}{\partial x}(0, z) = \frac{\partial p}{\partial x}(2\pi R, z) \tag{3}$$

The radial clearance between the journal and the bearing sleeve exerts a significant influence on the operation of the fluid-film bearing. When considering fluid-film bearings with rough surfaces, the function of the radial gap becomes more complicated. In the general case, all errors can be divided into dimensional errors, deviations in the shape of the surface, and roughness. These errors appear in the manufacturing process, their magnitude depends directly on the parameters of the process. To model the gap function, taking into account the real topology of the operating surfaces of the bearing assemblies, measurements were made and the surface of the experimental samples was studied.

An important issue in modeling the radial clearance function is the consideration of the effect of roughness, when its magnitude is commensurable with the thickness of the film. So, in [4-6] it was revealed that the maximum pressure and side leakage are reduced, and the coefficient of friction is increased. Further study of microroughness showed that for the longitudinal and transverse direction of the relief, the integral characteristics take different values. This observation became the basis for the development of the direction of creating bearings with textured surfaces [7]. The presence of a special texture allows to achieve optimum performance of the bearing unit.

To establish the mathematical relationships between the parameters of the roughness and its description, a measurement was made of the altitude parameter Ra and the pitch of microroughness, and the study of the portion of the supporting surface on a metallographic microscope. As a result, it was found that the basis of microroughness is the risks left as a result of the passage of the cutting tool over the surface. To describe microroughnesses, one can use the following function:

$$h_{micro}(x, z) = A_1 \sin(2\pi / k_1 z - sx) + A_2 \sin(k_2 x) \tag{4}$$

where

A_i - amplitude, which is equal to half the height of microroughness and is measured from a certain midline;

k_i - the coefficient that determines the step of the microroughnesses;

s - the coefficient taking into account the displacement of the direction of microroughness, which corresponds to the feed of the tool during the processing of the supporting surface.

The shape of the microroughnesses described with the help of this function has the form shown in the Figure 2.

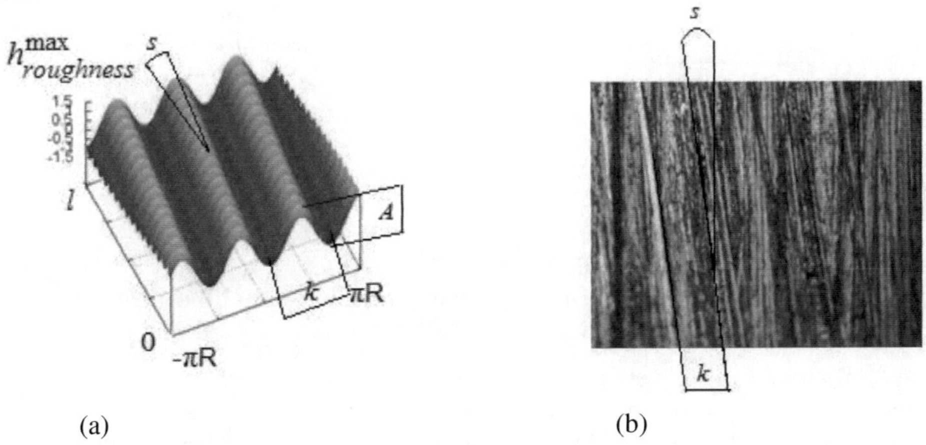

Figure 2. Microroughness of a bearing's surface.

The obtained mathematical description corresponds to a sufficient degree to the shape of the roughness of the bearing's surfaces of the fluid-film bearing obtained as a result of machining and implements a combined approach to the description of surface's microroughness.

To develop a mathematical model of the real surface, measurements were also made of the deviation of the profile of the support surface from roundness in different sections of the experimental samples, and also the circular log data were analyzed. The results of the analysis showed that the support surface in general is a superimposed macro- and micro-deviations. Moreover, the real surface profile can be obtained as a consequence of imposing several types of deviations. Conditionally this overlay can be represented as a superposition of surface irregularities, since unevenness reduces the radial clearance by an amount equal to the sum of the heights of the irregularities [8].

$$h(\varphi, z) = h_{ideal}(\varphi, z) + h_{skewness}(\varphi, z) + h_{macro}(\varphi, z) + h_{micro}(\varphi, z) \quad (5)$$

From this equation, the component of the misalignment should be eliminated, since earlier in the system of assumptions it was assumed that the axes of the shaft and the bearing are parallel. For the case of an arbitrary eccentric shaft position in the bearing, during operation, the gap function is most conveniently represented as:

$$h(\varphi, z) = R(\varphi, z) - r(\varphi, z) \quad (6)$$

where

$R(\varphi, z)$ - bearing's radius function;

$r(\varphi, z)$ - rotor's radius function.

To describe the macro deviations of the support surface of a radial hybrid bearing with slotted throttling, it is advisable to use the second-order surface equation, which is affected by some affine transformation operator, which establishes the corresponding rotation and displacement relative to the canonical coordinate system. Microroughness is imposed with respect to some given macroroughness [9]. Then the cross-section of the support surface will have the form shown in the Figure 3.

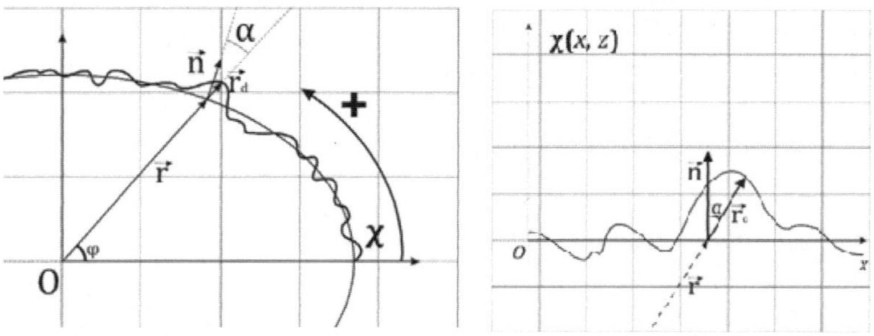

(a) (b)

Figure 3. Determination of the gap function with assumptions of the microroughness of the surface: (a) the cross section in the coordinate system associated with the carrier surface; (b) studied cross-section in the coordinate system of the scan.

Let us perform numerical simulation of the processes taking place in the lubricating layer with a fluid-film bearing with slotted throttling on a finite-difference model on an uneven-dimensional orthogonal computational grid with adaptive condensation according to the procedure described in [10, 11]. The boundary conditions for finite-difference approximation take the following form:

$$p_{i,j} = p_{N,j}; \quad p_{i,1} = p_{i,N} = p_2 \quad (7)$$

A typical solution of the Reynolds equation for a slotted throttling bearings is shown in the Figure 4.

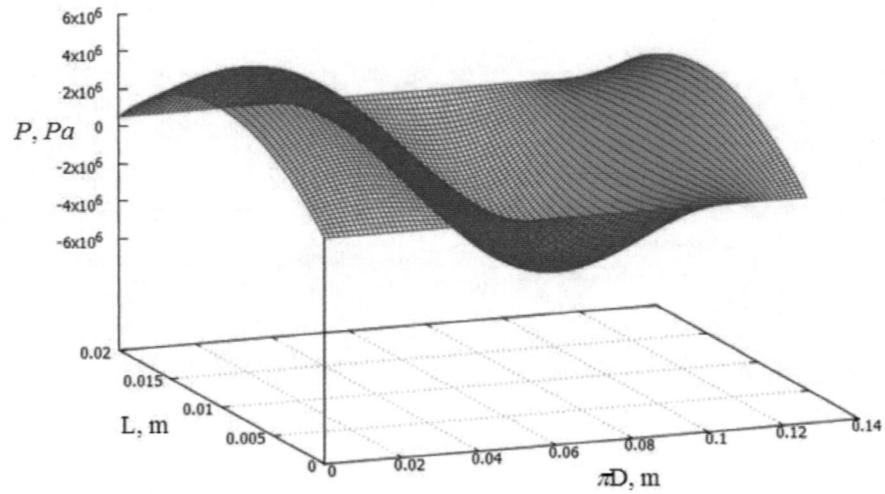

Figure 4. Pressure distribution of a bearing with slotted throttling.

For the tasks set, we consider the basic integral static characteristics of a fluid-film bearing: the bearing capacity, the flow rate a lubricant, the power loss due to friction. These characteristics are calculated through the following relationship [8]:

$$W = \sqrt{R_X^2 + R_Y^2} \tag{8}$$

where R_X, R_Y - hydrodynamic force in the projection on the X and Y axes, which are determined by numerical integration of the pressure distribution using the formulas [2]:

$$R_X = \int_0^L \int_0^{\pi D} p \cdot sin(2x/D) \cdot dxdz; \quad R_Y = \int_0^L \int_0^{\pi D} p \cdot cos(2x/D) \cdot dxdz \tag{9}$$

$$Q_V = \int_0^{\pi D} -\frac{h^3}{12\mu K_x} \cdot \frac{\partial p}{\partial x} dx \tag{10}$$

$$N_{fr} = \omega_0 \frac{D}{2} \cdot \int_0^L \int_0^{\pi D} \left[\frac{h}{2} \cdot \frac{\partial p}{\partial x} + \frac{U \cdot \mu \cdot K_x}{h} \right] \cdot dxdz \tag{11}$$

A computer experiment was performed to investigate the characteristics of a fluid-film bearing with slotted throttling, taking into account the geometric parameters of the unevenness of the real surface. As geometric parameters of the experiment, the geometric parameters of the irregularities are taken: the height and step of the microroughness, the magnitude of the displacement relative to the longitudinal axis of the bearing, and the magnitude of macroroughness. For carrying out the computational experiment, the value of the macro-unevenness is taken as a percentage of the tolerance value for the bearing size. The main results of the computational experiment given in the Figure 5

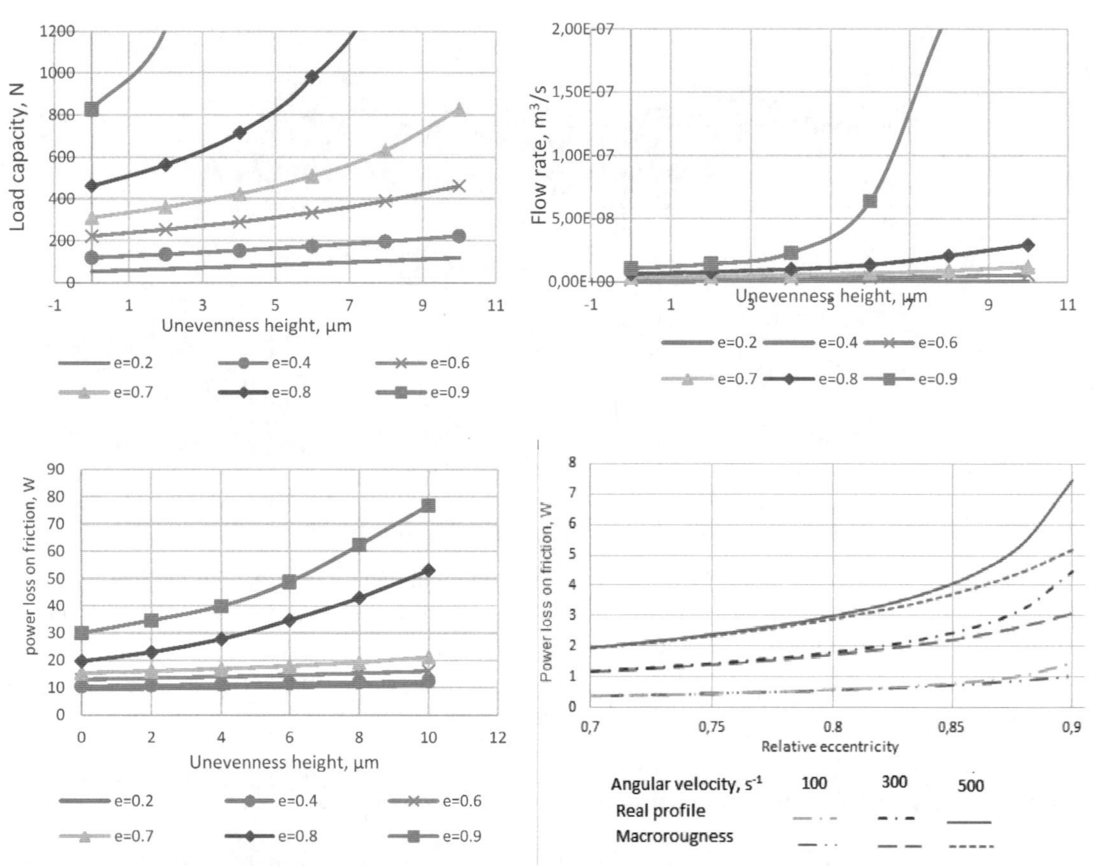

Figure 5. Integral characteristics of a hybrid fluid-film bearing with slotted throttling versus unevenness's height.

According to the findings of the research, inequalities of bearing area may not be considered if the ratio of the height of surface roughness to radial clearance does not exceed 0.015. For small and average radial clearance, a change in roughness by 1 μm leads to the change in bearing capacity 19-32%, flow rate - 7-13%, power loss on friction - 8-15%. An increase in roughness leads to an increase in lubricant consumption, lower bearing capacity and power loss on friction within 9%. An increase in the rate of micro roughness displacement in the longitudinal direction of the plain bearing leads to an increase bearing capacity within 5%, increase in flow rate by 7-8%. When taking into account the deviation of the shape when describing the surface profile, the bearing capacity decreases by up to 50% compared to perfect bearing. The affect of by the lateral shape and corsetry when using the real surface profile differs by 1.5 ... 2%, which is insignificant. If you look at the slider bearing with the real surface profile, you can see from the graph on Figure 5 that with an increase in the relative eccentricity, a degree of roughness appears, which is superimposed on macro-unevenness. It was found that the influence of roughness as part of a real profile at large relative eccentricities leads to the increase in bearing capacity and frictional power loss by 23-48%, depending on the parameters of surface roughness.

3. Confirmation of the adequacy of the developed model

To confirm the adequacy of the previously presented mathematical model for the operation of a bearing with slotted throttling and nonsmooth support surfaces, experimental studies were carried out. For this, a set of bearings of different lengths with different parameters of micro and macroroughness was developed. Each experimental sample was subjected to measurement of the parameters of irregularities according to the requirements of state standards with the use of a profilometer and a circular meter.

In describing the real surface profile, it was assumed that the support surface is the result of superposition of macro and micro deviations. To confirm this assumption, a study was made of the deviation of the shape of the support from roundness on the basis of the obtained circular logs. The results of the measurement of the cross section were obtained taking into account the altitude parameter of the roughness. To confirm the hypothesis put forward, a comparison is made between the experimental and theoretical curves (Figure 6) describing the profile of the cross section.

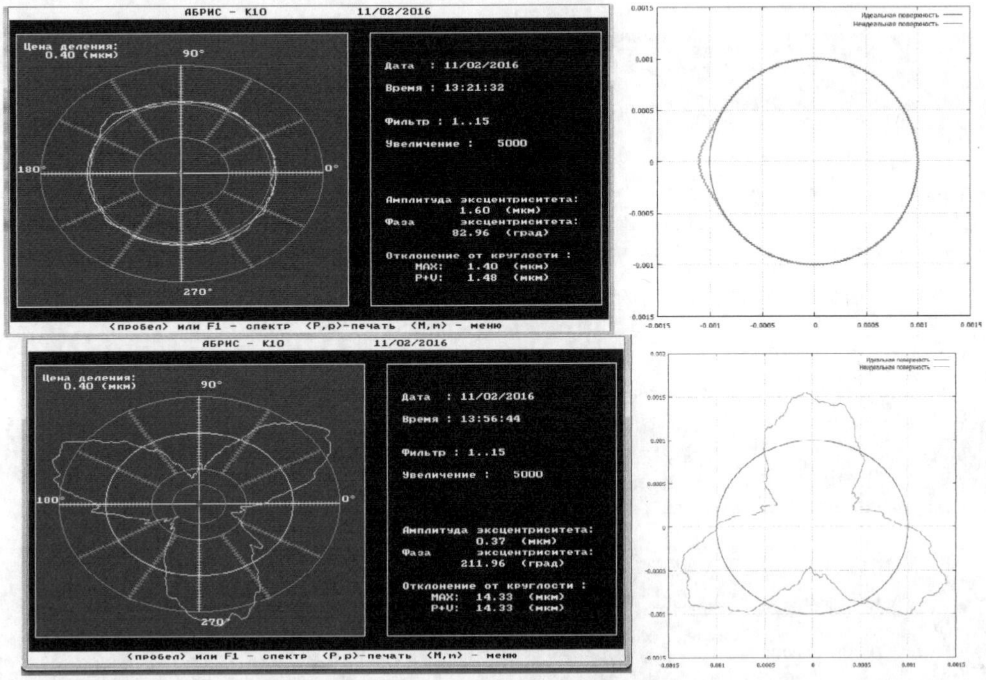

Figure 6. Comparison of real and simulated curves.

As it can be seen from the comparison of the presented results of modeling the real surface profile and measuring the deviation of the profile from roundness, the theoretical and experimental data agree satisfactorily. When specifying the parameters of the unevenness obtained in the course of measurements of the profile of the experimental sample in the developed mathematical model, the discrepancy between the experimental and theoretical profiles did not exceed 5-7%.

The results of measurements of microroughness of prototypes at 12 points, located through equal segments, with the help of a profilomer (Figure 7) showed that in most cases the shape of the distribution of irregularities on the reference length is close to a harmonic function over which a certain random variable is superimposed. In addition, the values of the height parameters and the pitch step of the irregularities assumed during the simulation were confirmed.

Thus, it can be concluded that the proposed assumptions in the mathematical description of the surface roughness were fully confirmed during the experimental study of the profile of the supporting surface. When a roughness measurement parameter was substituted for a mathematical model, the discrepancy between the theoretical and real profile was not more than 4%. As a result, it can be concluded that the mathematical description of the microroughnesses obtained to a sufficient degree corresponds to the shape of the roughness of the reference surfaces of the fluid-film bearing obtained as a result of machining.

The parameters of unevenness of the experimental samples of sliding bearings obtained as a result of measurements are recorded in the developed mathematical model in order to obtain numerical

IOP Conf. Series: Materials Science and Engineering **233** (2017) 012044 doi:10.1088/1757-899X/233/1/012044

values of the basic integral characteristics for each of the samples. A physical experiment was carried out to confirm the adequacy of numerical modeling of processes in the lubricating layer of a plain bearing with a slit throttling. For this purpose, the installation Figure 8), the methods of conducting and processing the results, which are described in detail in [12], were developed. The experimental module includes a rotary installation (with a dynamic loading system) with fluid-film bearings, one of which is a radial hydrostatic bearing with a radial lubricant supply, and the second support is a cylindrical bearing with axial throttling of the lubricant, a device for supplying a lubricant, in addition, the module has places for placing the primary converters of the information-measuring system.

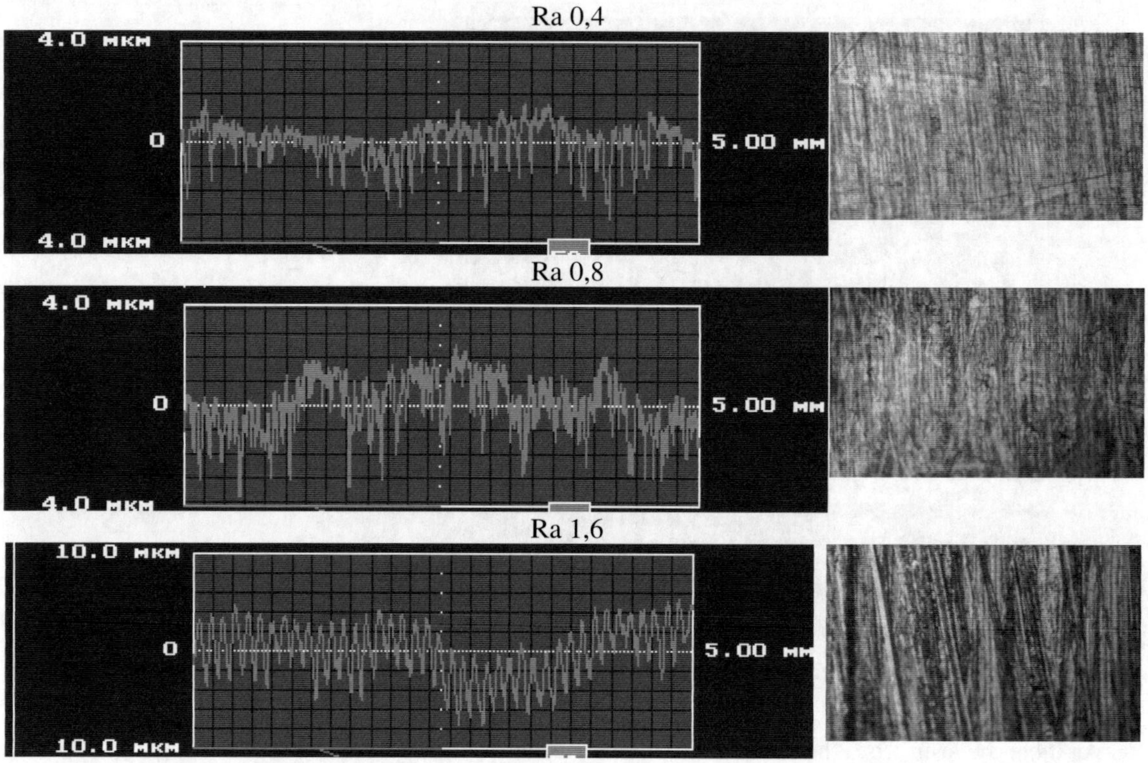

Figure 7. Representation of roughness measurement results using the profilometer model 296.

Bearing nodes are installed in the body of the experimental installation from two sides. To ensure the accuracy of the assembly of the housing with the bearing assemblies and to ensure their alignment, as well as to measure the real radial clearance between the bearing and the shaft, a laser centering device Kvant-LM was used. Water is used as the lubricant in the test rig.

Figure 8. Photo of the test rig.

The basis of the information-measuring system is the multifunction module NI6052E, which features multichannel digital and analog input-output and counters-timers. As a tool for managing the test rig, acquiring and processing data from primary converters, the software developed by the authors in the LabView was used (Figure 9).

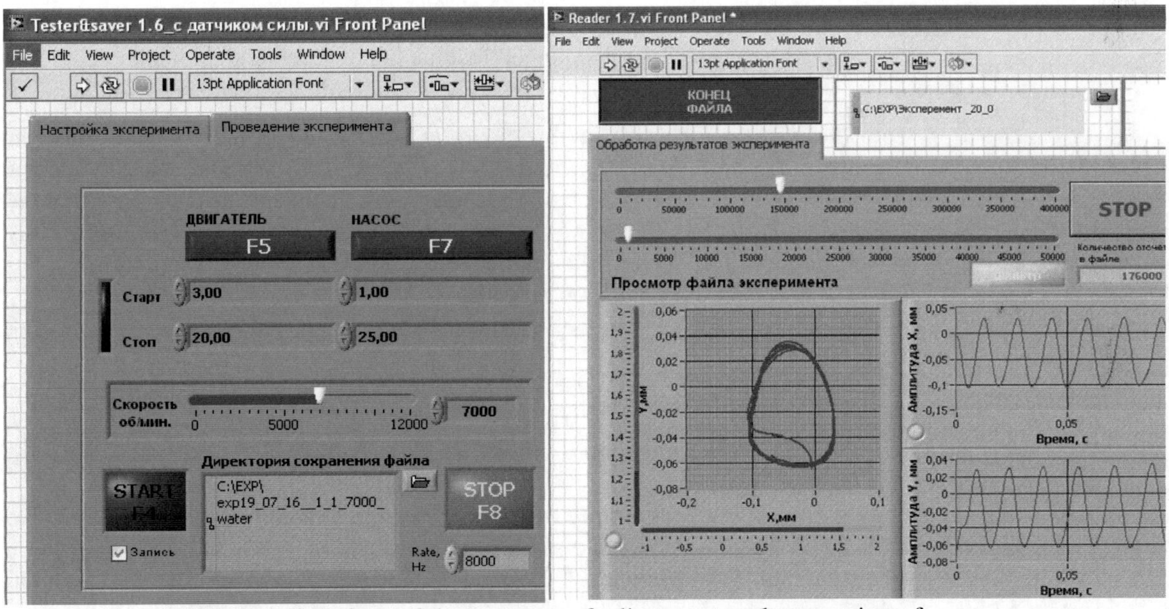

Figure 9. Interface of the program of adjustment and processing of experimental data.

Based on the readings of the sensor system of the rig, the motion of the rotor was measured in the bearing during steady-state regime, as well as the angular velocity with a free run-out of the rotor. We process the results of measurements on the basis of the technique given in [8,12]. Figure 10 shows the results of a comparison of the calculated bearing capacity and the friction coefficient of the fluid-film bearing and the results of the experiment.

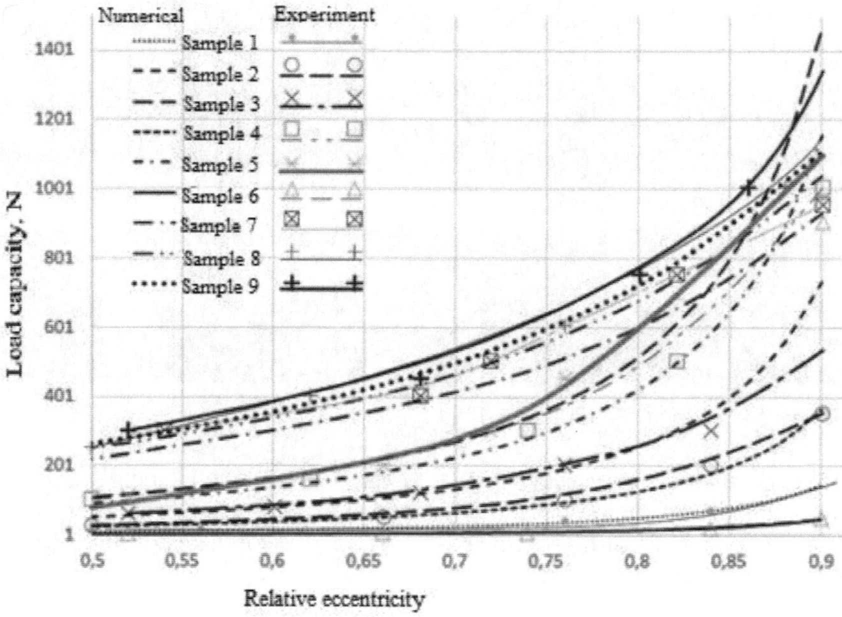

Figure 10. Comparison of theoretical and experimental studies of the load capacity of the lubricant layer in channels of complex geometry.

As can be seen from the graphs, the form of the theoretical curves corresponds to a sufficient degree to the experimental form. The discrepancy between the results of theoretical calculations and the physical experiment is within 12%. As a result, it can be concluded that the hypotheses, assumptions and methods of numerical modeling of processes in the lubricating layer proposed in the modeling process can be adopted. The developed mathematical model of the flow of lubricant in circular channels of complex geometry is considered adequate.

References

[1] L A Savin, O V Solomin 2006 *Modeling of rotor systems with fluid-film bearings: monograph* (M: Engineering 1) p 444

[2] V A Martsinkovskiy 2002 *Centrifugal machines vibrations. In 2 books. Book 1st. Hydrodynamics of throttling channels* (Sumy: SumSU) p 337

[3] M V Korovchinsky 1959*Theoretical foundations of sliding bearings* (M: Mashgiz) p 404

[4] Yaser S Jazayeri-M 2010 Effect of surface quality of a turbine rotor on performance of hydrodynamic journal bearing *Materials of 25th international power system conference* (Teheran, Iran) p 52-59

[5] S U Bhaskar, M M Hussain and Md. Yousuf Ali 2013 Stability analysis on plain journal bearing with effect of surface roughness *International Journal of Scientific & Engineering Research* **4** (3) p 1-7

[6] R Turaga, A S Sekhar and B C Majumdar 1999 The effect of roughness parameter on the performance of hydrodynamic journal bearings with rough surfaces *Tribology International* **32** pp 231-236

[7] R K Sharma, D Singh and S Kango 2012 Numerical investigation on the influence of surface texture on the performance of hydrodynamic journal bearing *Meccanica* **47** (2) pp 469-482

[8] I A Danchin 2007 Influence of deviations of the shape of support surfaces of hydrostatodynamic bearings on the dynamic characteristics of rotor systems *Dissertation for the degree of candidate of technical sciences* (Orel) p 160

[9] A Yu Koltsov, A V Prosekova 2014 Modelling surface of the rotor-bearing unit with macro and micro devations of form *Fundamental and applied problems of technology and technology* **5** (307). pp 39-43
[10] A Yu Koltsov 2015 Application of the statistical method of design grid for approximate PDE solution by finite differences method *Fundamental and applied problems of technology and technology* **5** (313) pp 180-185
[11] A Yu Koltsov 2015 Research of questions of applicability and computational efficiency of the statistical parametrical adaptation method in constructing the discrete analogue of mathematical models *Fundamental and applied problems of technology and technology* **6** (314) pp 17-23
[12] A V Prosekova, A Yu Rodichev and L A Savin 2016 Technological support of performance characteristics of bimetallic sliding bearings *Scientific and methodical electronic journal "Concept"* **11** pp 3426–30

Acknowledgments
The present research was implemented as a part of the RSF project No. 16-19-00186 'Planning of optimal energy efficient rotor trajectories in mechatronic modules with complex rheology medium'.

IOP Conf. Series: Materials Science and Engineering **233** (2017) 012045 doi:10.1088/1757-899X/233/1/012045

Complex double-mass dynamic model of rotor on thrust foil gas dynamic bearings

A Sytin[1], A Babin[1], S Vasin[2]

[1] Department of Mechatronics and International Engineering, Orel State University, Naugorskoe sh. 29, Orel, Russia, 302020
[2] Department of Design, Tula State University, pr. Lenina 92-9, Tula, Russia, 300600

E-mail: alex.mech.osu@gmail.com

Abstract. The present paper considers simulation of a rotor's dynamics behaviour on thrust foil gas dynamic bearings based on simultaneous solution of gas dynamics differential equations, equations of theory of elasticity, motion equations and some additional equations. A double-mass dynamic system was considered during the rotor's motion simulation which allows not only evaluation of rotor's dynamic behaviour, but also to evaluate the influence of operational and load parameters on the dynamics of the rotor-bearing system.

1. Introduction

Rotor systems are key elements of the majority of the military, research and industrial machines that are being developed nowadays. Bearings in them are used to maintain rotation axis steady relative to housing, for compensation of external forces, to decrease power loss due to friction and to damp vibrations. Various types of journal and thrust bearings are used, both roller-element and fluid-film, magnetic bearings and possible combinations of these, - they occupy a particular area of research because they determine and encapsulate functional peculiarities of a rotor machine. Due to a number of reasons, first off, interchangeability, the roller-element bearings are the most used, however are hardly applicable to highly loaded and/or high-speed rotors, where fluid-film bearings are virtually uncontested.

One of the possible ways of enhancing the performance of a rotor machine is increasing the rotation speed of the machine, and here of the most effective way of increasing the possible speed range is decreasing the viscosity of a lubricant. Bearings lubricated with gas have practically unlimited fleetness, high level of damping, operate with low power loss and heating and maintain a steady position of a rotor [1]. Here, the bearings with elastic elements, i.e. foil gas dynamic bearings, are the most promising. In the past 50 years a great variety of designs of foil bearings has been presented, however, the basic structure and basic elements generally remain intact. Foil gas dynamic bearings (FGDB) have a number of characteristics, complex of which provides these bearings with certain advantages in comparison to other types of bearings.

First of all, ambient air could be used as a lubricant for foil bearings, this simplifies maintenance and decreases dimensions of the lubrication systems quite significantly; it also provides possibility of operation under cryogenic conditions. This fact substantiates the use of FGDB in turbo-refrigerators, compressors, micro generators, and expanders in cryogenic and aircraft machinery. Secondly, the FGDB have dynamic properties that manifest themselves in stability of rotor's motion in a wide range of rotation speeds given an acceptable level of balancing. Combination of damping properties of the lubricant film and elastic elements allows avoiding biharmonic vibrations due to vortex instabilities. It should be noted, that design of the FGDBs is implemented presently based on empiric data. The absence of specialized tools for calculation considering these bearings could also be noted – this objectively justifies fundamental and practical, theoretical and experimental research of FGDB. A few areas of research could be distinguished in this field: 1) application of low viscosity media (hydrogen, oxygen, neon, methane, etc.) which are used as lubricants in fuel components of aircrafts' driver nodes, in cooling systems, etc. [2,3]; 2) development of foil bearings with self-diagnostics functions and active control of load capacity and damping properties [4,5].

Dynamic behavior of a rotor machine, which is characterized by a specter of natural frequencies, vibration level and amplitude of vibrations, and external forces acting on the machine's elements, is determined by stiffness and damping of bearings [1]. Main reasons of internal axial forces' occurrence are design factors and lubricant's properties during transient regimes.

2. Simulation model of a FGDB

The present paper considers a thrust FGDB, calculation diagram of which is presented in the Figure 1.

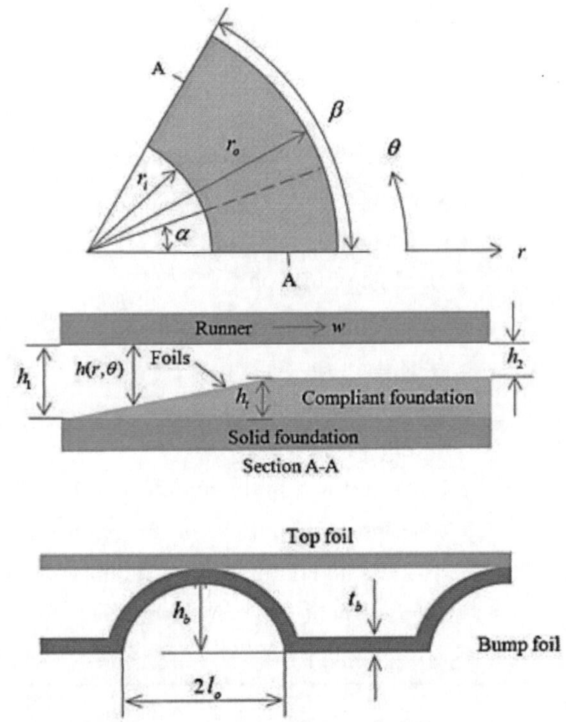

Figure 1. Calculation diagram of a thrust FGDB.

Simulation of rotor's dynamic behavior is based on the following steps: 1) simulation of gas film considering the compliant surface of a bearing with certain properties, 2) estimation of dynamic coefficients of the film, i.e. stiffness and damping, 3) derivation of motion equations for a double-mass dynamic system to study the influence of top foil's mass on the calculation of rotor's displacement. Generally, simulation of the dynamic behavior of a rotor on thrust FGDB is based on numerical solution of equations of gas dynamics, theory of elasticity and differential equations of motion with some additional relations. The operational and geometric parameters of the simulated bearing are presented in the table 1 along with the notation used in the paper.

Table 1. Notation used in the paper.

r_i	Inner radius
r_o	Outer radius
α	Tapered region angular size
β	Single pad's angular size
z	Number of pads
h_1	Maximum film thickness
h_2	Minimum film thickness
h_b	Height of the pad, $= h_1 - h_2$
l_o	Half bump length
t_b	Thickness of the bump foil
t_f	Thickness of the top foil
s	Bump spacing
r, θ	Polar coordinates, radial and circumferential axes accordingly
ω	Angular velocity
$p(r,\theta)$	Pressure distribution over a single pad
p_a	Ambient pressure
μ	Viscosity
$h(r,\theta)$	Axial gap function
V_θ, V_r, V_y	Velocity components
ρ	Density of the gas
v	Poisson's ratio
E	Young's modulus
R_y	Reaction force in the direction of film thickness
R_{y0}	Reaction force in the equilibrium position, $= mg$
c_y, b_y	Stiffness and damping coefficients
$\Delta Y,\ \Delta \dot{Y}$	Displacement and velocity increments
$F(t)$	External load

2.1. *Simulation of gas film on a compliant foundation*

Simulation of the gas film is based on numerical solution of the steady-state Reynolds equation. Here a compressible medium is considered in a non-isothermal formulation. The axial gap is simulated using the Bezier curves, the procedure is presented in more detail in [6]. This allows one to avoid calculation error in the region where the tapered land turns to plain surface, on the contrary to the known and assumed correct procedure in [7]. The governing Reynolds equation is as follows:

$$\frac{\partial}{\partial r}\left[\frac{\rho h^3}{\mu}\frac{\partial p}{\partial r}\right]+\frac{\partial}{r\partial\theta}\left[\frac{\rho h^3}{\mu}\frac{\partial p}{r\partial\theta}\right]=6V_\theta\frac{\partial(\rho h)}{\partial\theta}+12\rho V_y \tag{1}$$

Numerical solution of (1) has been implemented using the finite difference method. In order to simulate the deformation of the elastic elements, the following additional relations have to be introduced. The axial gap considering local deformation of the elastic element:

$$h(r,\theta) = h_0 + w \tag{2}$$

where h_0 – initial local film thickness, w – deformation of the foil.

Stiffness of the bump foil is, according to [7], governed by the following relation:

$$c_b = \frac{E t_b^{\ 3}}{2(1-\nu^2) l_o^{\ 3}}$$

so, the compliance rate of the bump foil is

$$k = \frac{2s}{E}\left(\frac{l_o}{t_b}\right)^3 (1-\nu^2)$$

Finally, to calculate the deformed top foil, the following transformation of (2) takes place:

$$\begin{aligned} h(r,\theta) &= h_0 + K1(p - p_a) \\ K1 &= \frac{\delta \cdot h_2}{p_a} \\ \delta &= \frac{2 p_a s}{h_2 \cdot E}\left(\frac{l_o}{t_b}\right)^3 (1-\nu^2) \end{aligned} \tag{3}$$

Equations (1) and (3) are solved iteratively using the finite difference method given an accuracy criterion ε expressed in terms of load capacity on the previous and current iterations. The load capacity of a thrust bearing could be determined using the following relation:

$$R = z \iint_{\theta r} p(r,\theta) r d\theta dr$$

Once the reaction force of the gas film given the deformation of the foil is determined, the dynamic coefficients of stiffness and damping could be obtained. This is based on the premise, that the reaction force, non-linear in reality, could be linearized using Tailor series in close proximity to the equilibrium position [8]. Equilibrium position in the present case is the position where all elastic and gas dynamic forces are equal to the rotor's weight *mg*.

$$R_y = R_{y0} + K_y \Delta Y + B_y \Delta \dot{Y}$$

In the Figure 2 and Figure 3, the calculation results of stiffness and damping coefficients are presented for a following setup: inner radius – 0.025 m, outer radius – 0.05 m, angular velocity – 8000 rad/s, minimum film thickness – 10 μm, viscosity of ambient air – $1 \cdot 10^{-5}$ Pa·s, bump foil's height – 1.75 mm, bump foil's thickness – 0.125 mm, bump spacing – 4.75 mm, half bump length – 1.75 mm, top foil's thickness – 0.125 mm, Poisson's ratio – 0.29, Young's modulus - $214 \cdot 10^9$ Pa. Angular size of the tapered land region is 0.6β.

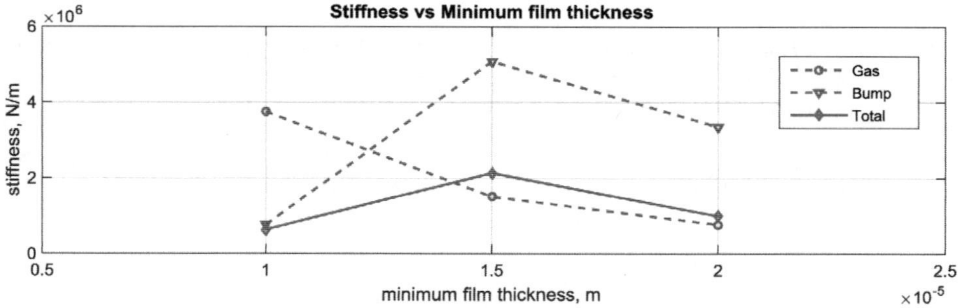

Figure 2. Stiffness of gas film, bump foil and total stiffness vs minimum film thickness.

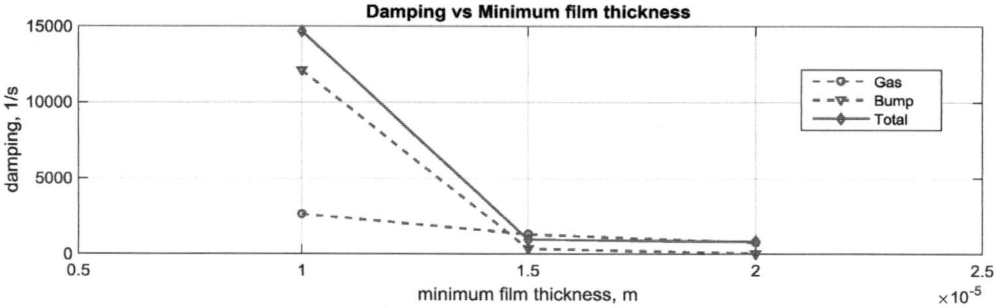

Figure 3. Stiffness of gas film, bump foil and total stiffness vs minimum film thickness.

It could be seen, that dynamic coefficients of a thrust FGDB depend significantly on the position of a rotor, i.e. the influence of presence of elastic elements is stronger at smaller values of minimum film thickness. Thus, problems of design of thrust FGDB always take specific operational and geometric parameters of a rotor machine into account.

2.2. Single- and double-mass dynamic system simulation

To study the influence of the top foil's mass on the dynamic behaviour of a rotor two models have been developed, the diagrams are presented in the Figure 4. Indexes of *c* and *b* correspond to gas film (index '1') and bump foil (index '2').

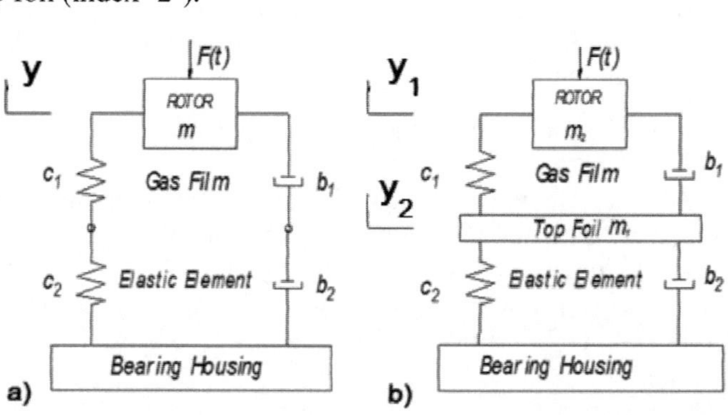

Figure 4. Single-mass (a) and double-mass (b) dynamic systems.

The stiffness and damping coefficients according to the corresponding calculation diagram in the Figure 4 are calculated as follows:

$$b_{total} = b_1 + b_2$$

$$\frac{1}{c_{total}} = \frac{1}{c_1} + \frac{1}{c_2}$$

The corresponding equations of motion are as follows: 1) single-mass dynamic system:

$$m\ddot{Y} = R_{y0} + c_y(\Delta Y) + b_y(\Delta\dot{Y}) - F(t) \tag{4}$$

where
c_y and b_y are total stiffness and damping accordingly,
m – rotor's mass.
2) double-mass dynamic system:

$$\begin{cases} m_1\ddot{Y}_1 = c_1(\Delta Y_1 - \Delta Y_2) + b_1(\Delta\dot{Y}_1 - \Delta\dot{Y}_2) - F(t) \\ m_2\ddot{Y}_2 = c_2(\Delta Y_2) + c_1(\Delta Y_2 - \Delta Y_1) + b_2(\Delta\dot{Y}_2) + b_1(\Delta\dot{Y}_2 - \Delta\dot{Y}_1) \end{cases} \tag{5}$$

where indexes '1' and '2' correspond to the rotor and the foil accordingly.

Equations (4) and (5) are solved using the ode23 solver in Matlab, with the following initial conditions and assumptions: equilibrium position of the rotor is defined by the equilibrium of forces, the equilibrium position of the top foil is the maximum deformation of the foil which is balanced by the gas dynamic forces and elastic forces in the bump foil. Thus, the following initial conditions have been set: initial position of the rotor is 50 μm, where the action of forces in the lubricant is virtually not deforming the foil, initial position of the foil is defined by the equilibrium position of the foil, which is taken -0.5 μm relative to 0 μm displacement at the initial moment in time. Initial velocities of both the rotor and the foil are zero.

The bearing with the same parameters as in par. 2.1 is considered. The rotor's mass is 1.75 kg, free fall acceleration is 9.8 m/s, which results in the steady-state minimum film thickness of 10 μm. The mass of a single top foil is 0.001 kg.

To study the dynamic behavior of the system, two types of external forces were considered – impulse of 8 N (Figure 5) and periodic force of 4 N at approx. 60 Hz (Figure 6).

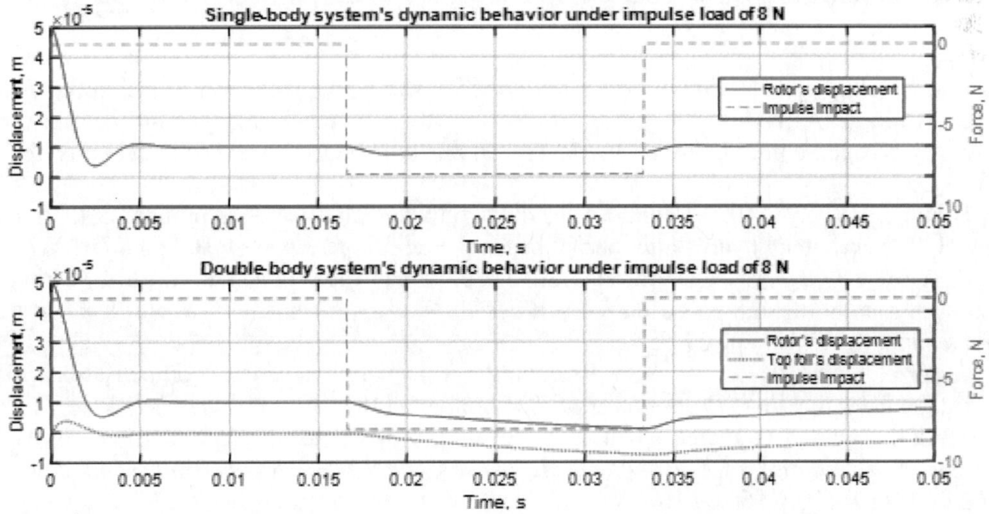

Figure 5. Dynamic behavior of a thrust FGDB under an impulse load.

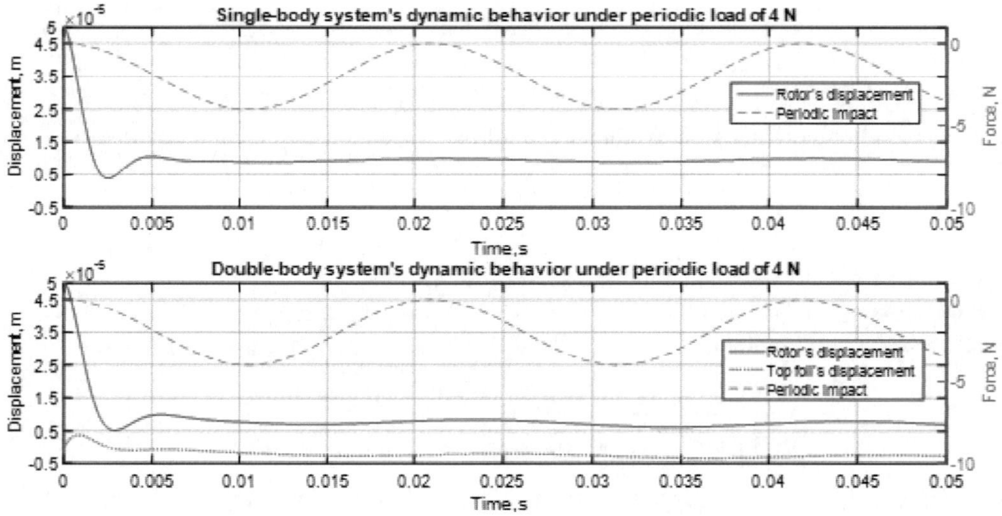

Figure 6. Dynamic behavior of a thrust FGDB under a periodic load.

3. Conclusions

It could be seen from the Figure 5 and the Figure 6, that introduction of the top foil's mass to the dynamic system does not influence the displacement of the rotor significantly. In comparison to the rotor's weight, the top foil usually weights several orders of magnitude less. As for the computational performance, in case of the single-mass dynamic system, the ode23 function returned results in less than a second, whereas in the case of the double-mass dynamic system the time it took for the program to return results was more than 15 min depending on various parameters. So, a conclusion could be drawn, that when performing the dynamic analysis of a rotor-bearing system with thrust FGDBs, top foil's mass could be neglected. The developed single-mass model could be used to study the dynamic response of a thrust FGDB to various types of external impact. Further research involves a series of experiments to verify the obtained numerical results.

References

[1] M V Korovchinsky 1959 *Theoretical foundations of sliding bearings* (M .: Mashgiz) p 404
[2] V O Tyurin, L A Savin, A V Sytin 2016 Lubrication of petal gas-dynamic bearings with low-viscosity cryogenic liquids *Bulletin of BSTU* (Bryansk: Bryansk State Technical University) **2** (50), pp 13-20
[3] M Braun, R Wheeler III, R Hendricks 1987A fully coupled variable properties thermohydraulic model for a cryogenic hydrostatic journal bearing *Journal of Tribology* **109** pp 405 1987
[4] A V Sytin, V O Tyurin, S D Antonov 2015 Active control of mechatronic foil bearing of rotor-supporting units of high-speed turbomachines *Fundamental and applied problems of technology and technology* **3** (311) pp 59-65
[5] G Aguirre, F Al-Bender, H Van Brussel 2008 Dynamic stiffness compensation with active aerostatic thrust bearings *Proceedings of ISMA* pp 105-118
[6] A Babin, L Savin 2016 Increase of Energy Efficiency by Means of Application of Active Bearings *International Journal of Mechanical, Aerospace, Industrial, Mechatronic and Manufacturing Engineering* **10, 9** pp 1667 – 1671
[7] K Feng, Z Guo, Z Zhao 2015 Parametric study on static and dynamic characteristics of bump-type gas foil thrust bearing for oil-free turbomachinery *Proceedings of the Institution of Mechanical Engineers Part J Journal of Engineering Tribology 1994-1996 (vols 208-210)* **229** (10)

[8] H Heshmat, J A Walowit, O Pinkus 1983 Analysis of Gas Lubricated Foil Journal Bearings *Journal of Lubrication Technology* **105** pp 647-655
[9] L A Savin, O V Solomin 2006 *Modeling of rotor systems with fluid-film bearings: monograph* (M.: Engineering 1) p 444

Acknowledgments
The present research was implemented as a part of the RSF project No. 16-19-00186 'Planning of optimal energy efficient rotor trajectories in mechatronic modules with complex rheology medium'.

IOP Conf. Series: Materials Science and Engineering **233** (2017) 012046 doi:10.1088/1757-899X/233/1/012046

Bi-metal foil gas dynamic bearings with bimorph piezoelectric foils

A Sytin[1], A Rodichev[1], A Kulkov[1]

[1] Department of Mechatronics and International Engineering, Orel State University, Naugorskoe sh. 29, Orel, Russia, 302020

E-mail: alex.mech.osu@gmail.com

Abstract. The present paper considers application of bi-metal materials and coatings to provide necessary strength and wear resistance of the surfaces of rigid and elastic gas dynamic bearings. Authors suggest using multi-layer foils with bimorph piezoelectric elements that operate in the generator regime to determine the deformation of elastic elements, and in the actuator regime to form an optimal shape of the surface of the bearing.

1. Introduction

Nowadays, rotor systems are key elements in machinery systems. These systems are subject to a number of mandatory requirements, namely: reduction in size, weight, increased requirements for reliability and energy efficiency indicators, high vibration resistance, etc. The most vulnerable and at the same time, the most important element of the rotary system is the bearing assembly, which is the most loaded during the operation of the device. Rolling-element bearings are standard elements with a developed methodology for selecting the required structural and geometric parameters. Fluid-film bearings in each particular case involve significant preliminary calculations for the choice of bearing type, geometric parameters and lubricant. However, there is a problem associated with the occurrence of situations not envisaged in the design process, which lead to disruptions in the operation of the rotor support assembly and the turbomachinery as a whole. One of the most modern and promising ways to protect parts and assemblies of various mechanisms from the resulting intense wear is the use of antifriction coatings – lubricants similar to paints, but containing fine particles of solid lubricants uniformly distributed in the mixture of binder elements and solvents instead of a coloring pigment. Modern technologies allow the creation of various thin-film coatings to improve the tribological characteristics of friction pairs. Examples of such coatings are wear-resistant and antifriction thin films, which increase the lifetime expectancy of machine parts. Providing durability and wear resistance of thin-film coatings is a complex technical task.

2. Application of antifriction coatings

Modern trends in the development of industrial production are characterized by increased requirements for quality and performance. The development of technology necessitates the creation of materials possessing a complex of properties that provide high strength, corrosion resistance, thermal conductivity, heat resistance, wear resistance, etc. Often, individual metals and alloys cannot provide the required range of properties. Therefore, layered metal compositions have been widely used. Such materials can be manufactured by combining dissimilar metals in a monolithic composition that preserves the reliable coupling of the components during further processing and under operating conditions. Such materials include bimetal-layer materials consisting of two or more metals and alloys. The use of layered metal compositions makes it possible to increase the reliability and durability of a large class of parts and equipment. As a result of the cost savings of non-ferrous metals (Ni, Cr, Cu, Mo, Ti, etc.), the cost of manufacturing them is reduced. The use of layered compositions allows for the development of more sophisticated design solutions for the creation of modern machines, instruments and apparatus. Today, bimetals are the most important group of industrial materials with a wide range of properties. The use of bimetals allows significantly increasing the efficiency of production of a wide class of parts and equipment for chemical, oil, agricultural, transport, energy and other machine-building industries. By application, all currently produced bimetals can be subdivided into the following types: corrosion-resistant, antifriction, electrotechnical (conductor and contact), instrumental, wear-resistant, thermobi-metal.

Values such as: roughness, friction coefficient of of rest / slip, hardness - could be determined experimentally using the characteristic equipment (hardness tester, tribometer, profilometer). In a more complicated situation, the definition of the wear resistance value is found. This process is extremely complicated, because it combines a large number of parameters. It is based on the friction force, which in turn is divided into adhesion and deformation components. To evaluate each of the components it is necessary to conduct a large number of different experiments, which in turn makes it difficult to select the coefficients. Simulation of the process is complicated by the influence of varying roughness, temperature, etc. In the process of friction, these quantities change unpredictably, which makes the process of evaluating the wear of thin-film coating difficult. At the moment, studying the wear resistance of thin-film coatings is an important technical task. There are many different domestic and foreign methods of investigation of thin antifriction coatings, which differ significantly from one another (the method of "friction difference", the method of measuring the main momentum and forces, Archard equation). Analysis of the studies showed that there are two ways to determine the wear resistance of thin films (coatings). In the first case, it is necessary to create a mathematical model for the destruction of the coating, by selecting the coefficients with a step-by-step stop of the wear process. In the second case, it is necessary to conduct a direct experiment until complete wear of the coating. The mechanism of action of antifriction coatings is as follows. After polymerization, the binder elements together with the dry lubricants fill the cavities of the microroughness of the surface and increase its supporting area (Figure 1).

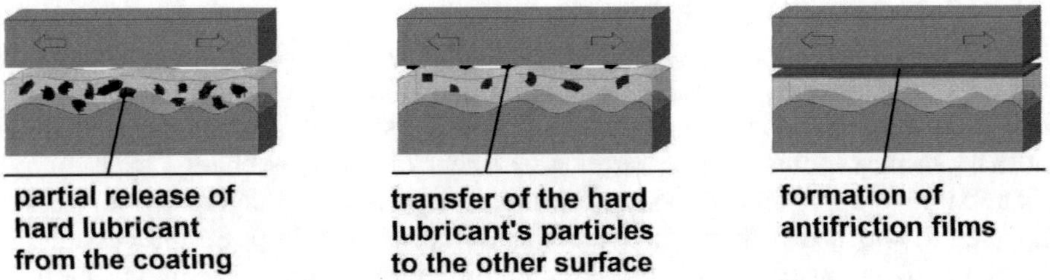

Figure 1. Antifriction coating mechanism.

In the process of friction, some of the lubricating particles are transferred to the conjugate surface. In this case, they are oriented parallel to the direction of motion, forming a smooth and slip protective anti-friction film on both surfaces. Such antifriction coatings are made on the basis of molybdenum disulfide, graphite, polytetrafluoroethylene, as well as compositions of various solids, the content of which, depending on the type of coating, can reach 70%. Dry lubricants in coatings provide the necessary complex of properties. Binders determine its protective properties, chemical resistance and type of curing.

The technological process of preparation depends on the material and condition of the workpiece. Adhesion and lifetime expectancy of antifriction coatings depend largely on the applied technology of preliminary preparation of parts' surfaces. In the Figure 2, a diagram of the dependence of the lifetime expectancy of the coatings on the methods of preliminary preparation of the surface is presented.

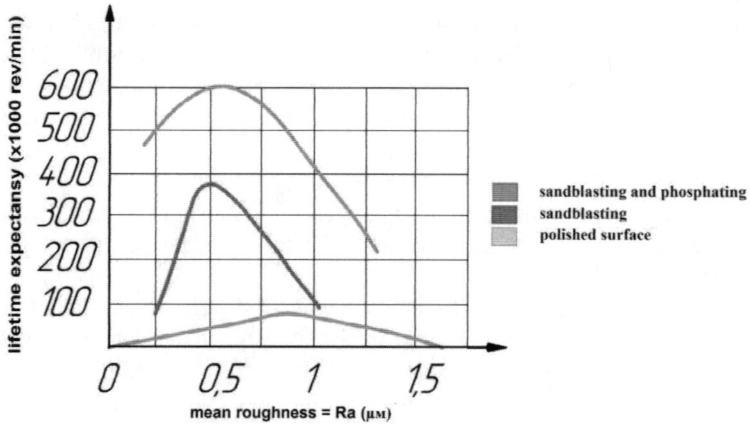

Figure 2. Lifetime expectancy versus methods of preliminary treatment of the surfaces.

Preparation for the application of antifriction coating begins with degreasing. This technological operation can be carried out with organic low-fat solvents, manually, and also with the help of ultrasonic cleaners. The washing operation is performed several times until all traces of fat are completely removed from the surface. In this case, a new solvent must be used for the next wash. After degreasing the parts are dried in air. Do not touch the treated surfaces with hands. The oxides present on the surface being treated are removed by chemical (etching, anodizing and phosphating) or by mechanical methods (sandblasting, polishing). For pre-treatment of metallic materials, phosphating and sandblasting are particularly well established. A rough, porous surface after preliminary treatment promotes mechanical adhesion to the antifriction coating, thereby achieving a much better adhesion. As a mechanical method for steel, titanium, aluminum, copper, magnesium and their alloys, sandblasting with alumina or cast steel (grain size about 55 microns). This type of preliminary surface preparation is also required for chrome and nickel-plated parts. In addition to removing corrosion, sandblasting makes the surface of the part more rough (Ra from 0.5 to 1.0 μm) and provides better adhesion of the antifriction coating. According to the technology, after completion of work, it is necessary to remove adherent sand particles with dry compressed air, not containing oil. To avoid the risk of corrosion on the surfaces treated in this way, antifriction coating is required in the shortest possible time. It should be taken into account that during sandblasting of parts, their linear dimensions change to 1.3 μm. After sandblasting, it is possible to apply an antifriction coating, however, for a longer service life of the coating, it is recommended to phosphate the part (subject the surface to pickling). For processing parts made of copper and copper alloys, instead of sandblasting, etching with a mixture of two or more acids (sulfuric, phosphoric, nitric, chromic, and hydrochloric) can be used. Etching of stainless steel is made by special solutions of oxalic acid. The immersion time, the concentration of the solution varies depending on the alloy and the state of the surface. After etching, thoroughly clean the parts to remove acid residues. Aluminum and its alloys are processed by electrolytic oxidation (anodizing). Aluminum alloys with a copper content of 0.5% or more, as well as alloys with a total content of dopants in excess of 7.5%, are treated in a solution of sulfuric acid. After treatment with sulfuric acid, the parts must be washed in water, then the surface film must be fixed by immersion in a 5% solution of sodium or potassium dichromate, rinsed and dried. Drying is performed at a temperature of no higher than 102 ° C. Other aluminum alloys and aluminum can be processed in a solution of chromic acid. Chromic acid forms a thin film, which provides protection from corrosion. After treatment with chromic acid, the part must be rinsed in hot water and allowed to air dry. To form a good surface film in acid solutions, it is necessary to use high purity water (with a low content of chlorides and sulfates). After anodizing before applying to the part of the antifriction coating, they must not be touched with bare hands.

Phosphating (Figure 3) is used for iron and steel, but is not suitable for stainless steel, galvanized cast iron and parts with cadmium and galvanic coating. Treatment with manganese phosphate increases the bearing's load capacity and lubricating properties of the coating. Zinc phosphate treatment improves its corrosion resistance. Iron phosphate improves adhesion with antifriction coatings. For phosphating, only those solutions are used that create layers of small crystals. Most of the particles deposited on the surface must be in the range of 3 to 8 μm. This is equivalent to the weight of the applied layer from 5 to 15 g / m^2. After treatment, the phosphate layer should have a smooth, uniform structure and color from gray to black. Spots of phosphating solution and corrosion marks on the surface of parts are not allowed. To avoid corrosion under the influence of moisture contained in the ambient air, antifriction coatings should be applied as soon as possible after phosphating.

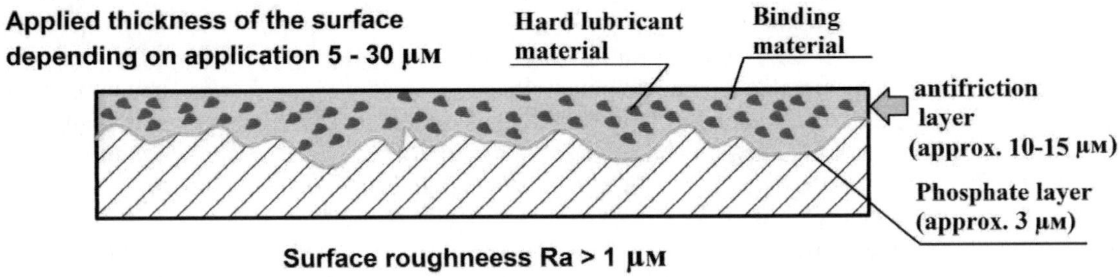

Figure 3. Model of antifriction coating formation.

The most important process is the application of the antifriction layer. Antifriction coatings (Figure 2) are dispersions of selected solid lubricants in solutions of organic or non-organic compounds in solvents or water. After applying and drying, antifriction coatings form a strong bond between the binding agent and solid lubricants. In the case of tribological requirements, the transferred solid materials are transferred to the mating component, a transfer film is formed, which leads to a reduction in the so-called shearing forces and, therefore, leads to a decrease in the friction coefficient. To date, there are a large number of different binders and hard materials with different properties, also based on nanoparticles.

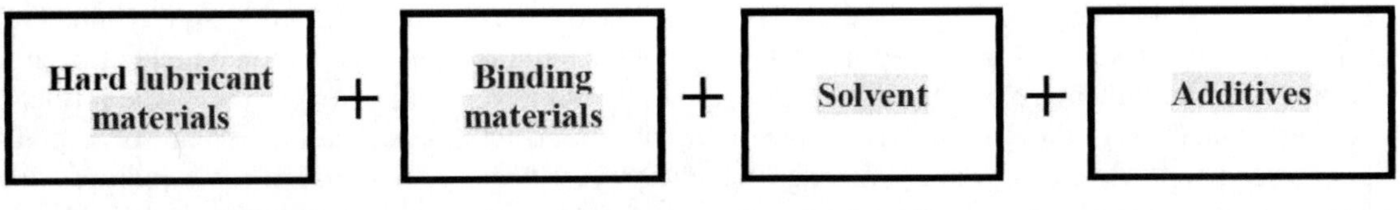

Figure 4. Contents of an antifriction coating.

Depending on the shape, size, weight, quantity and material of the work pieces, as well as surface properties, various methods of applying antifriction coatings are used. When choosing the method of applying antifriction coatings, it is necessary to take into account the requirements for the film, the configuration and location of the coated sliding surfaces. The quantity of the prepared composition is calculated taking into account the viability indicated in the technical documentation for the material and the area that can be covered during this time. Antifriction coatings can be applied by spraying, immersion, brushing, roller coating and screen printing. The final step is to control the thickness of the coating. The thickness of the film has a significant effect on the service life, friction coefficient and anticorrosion properties of antifriction coatings. It should exceed the size of the surface irregularities of the contacting surfaces, making, as a rule, from 5 to 20 microns. The application of a coating on both interfaced surfaces with a relatively thin layer provides a greater load-capacity than a thicker layer on only one surface.

3. Application of bimorph piezoceramic elements in foil gas dynamic bearings

The introduction of monitoring and active control in the rotor-bearing nodes will solve the problem of occurrence of situations unforeseen during the design s of by varying the shape of the radial and axial clearance, as well as by changing the rigidity of the elastic support surface. In addition, active control will expand the scope of sliding bearings. At the same time, active control is realized by various actuating mechanisms, which act on the regulated body by means of an electromagnetic field, mechanical forces, piezoelectric effect. The latter acquires the most popularity today and is realized due to devices called piezoactuators or piezoelectric drives. Such devices have unlimited resolution capabilities, piezoelectric converters convert electrical energy directly to mechanical, they are able to move in the subnanometric range (this property is useful when correcting the gap's geometry between the rotor and the bearing housing), the speed of operation of such actuators is in the microseconds range, existing to date powerful packet actuators are capable of moving 100 tons or more by 250-500 microns with a minimum step of 0.05 - 0.1 μm; and other properties [1].

The developing direction in the field of piezoactuators is the use of combined flexible lamellar elements containing metal layers and piezoceramics. A combination of such elements is called plate-like flexural actuators [2] or bi-morphs. There are two types of bimorphs: parallel and sequential. In the parallel bimorph, the direction of polarization of the plates is the same. This determines the high sensitivity and the ability to use for control the bias voltage, which generates an electric field parallel to the direction of polarization, thus reducing the risk of depolarization. In the sequential bimorph, the polarization direction of the plates is opposite. To avoid depolarization, the maximum electric field is limited to several hundred volts per millimeter of thickness. Bimorph can work in two modes (Figure 5). When using the direct piezoelectric effect, the mechanical energy applied to the bimorph is transformed into electrical energy. The mechanical force causes bending of the bimorph, at the same time compression of one of the constituent plates and expansion of the other occurs, which causes the formation of a charge in the bimorph. Bimorph works in a generator mode. Bimorph, operating in this mode, is used as a flexible sensor. The sensor of the generator type does not require an external power source to operate. Such a sensor is designed to convert dynamic deformations into electrical signals, followed by processing and recording by various devices. The sensor can be used as an independent converter of mechanical deformation into an electrical signal, or be an integral part of a more complex device. The sensor can be connected to a monitoring and control system in two main ways: according to the voltage acquisition scheme or the current acquisition scheme. Bimorph, using the inverse piezoelectric effect - conversion of electrical signals into mechanical forces – is a piezoelectric actuator. The bimorph piezoelectric actuator is a passive non-linear four-terminal network and requires the presence of a power source for excitation. In the sequential bimorph the plates are connected sequentially in relation to the power supply. In this case, the plate with positive polarity is under the influence of positive voltage and, due to the inverse piezoelectric effect, lengthens, then as a plate of negative polarity does not work and prevents this length change, which leads to bending of the bimorph. Since the piezoceramic element has the properties of a capacitor, the energy absorbed by it is released gradually after the voltage is cut off. In order for the bimorph to quickly return to its original state, there are two resistors between the electrodes.

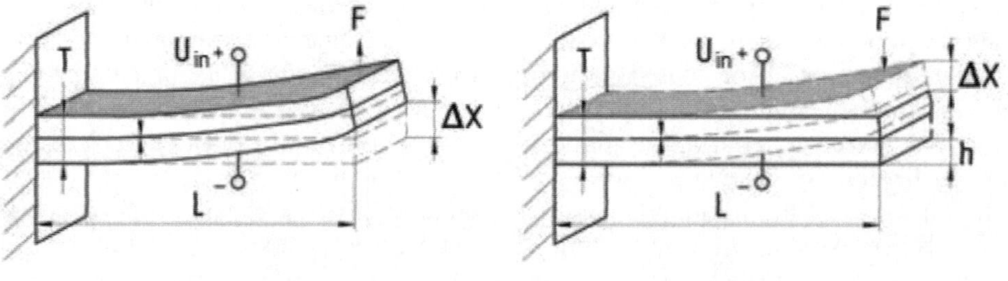

Figure 5. Operating modes of a piezoelectric bimorph.

Based on the piezoelectric effect in 1992, Tsumaki Nobuo proposed the construction of a foil gas dynamic bearing with bimorph piezoelectric foils (Figure 6), which operate in the generator mode, based on direct piezoelectric effect. The presented design is a foil gas dynamic bearing with a control system and allows recording of signals generated from deformation of piezoelectric bimorph elements. Determination of permissible deformations allows adjusting the structure to prevent an emergency situation, as well as continuously tracking the state of the elastic surface.

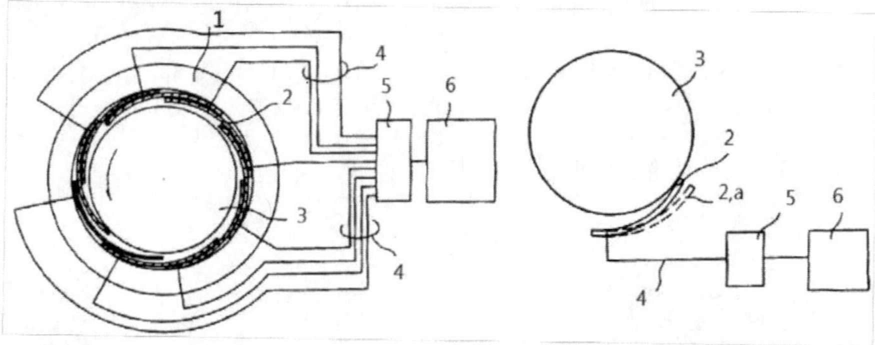

Figure 6. Foil gas dynamic bearing with bimorphs, patent JPH № 0454309: 1 – bearing's housing, 2 – bimorph piezoelement in the initial state, 2,a - bimorph piezoelement in a generator mode under the influence of gas dynamic pressure, 3 – rotor, 4 – wire connections, 5 – ADC, 6 – controller.

As is known from NASA reports, as well as scientific works of scientists in the field of petal gas-dynamic bearings, the most dangerous for this type of bearings are the periods of starting and stopping, when the friction interaction of the shaft surface and elastic elements occurs. To ensure the strength and durability of the surface of elastic piezoelectric elements, it is necessary to cover with wear-resistant antifriction coatings. In general, the separately taken elastic element will represent a multilayered structure (Figure 7), in the middle of which it is necessary to lay an elastic base, to increase the rigidity and fatigue strength of the bimorph piezoelectric element. The piezoceramic plates placed on opposite surfaces of the elastic base must have a different sign connection, as shown in the Figure 5. On top of the piezoceramic elements are covered with a layer of antifriction wear-resistant material.

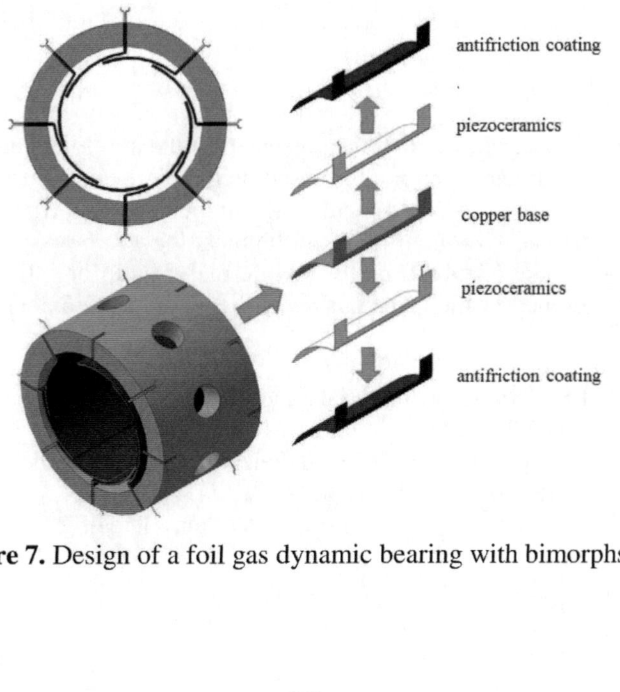

Figure 7. Design of a foil gas dynamic bearing with bimorphs.

According to the multi-layered bearing concept presented above, the technological process of manufacturing was developed, the results of which are shown in Figure 8. The central base of the petal can be made of various materials (copper, beryllium bronze, stainless steel). The base increases the mechanical strength of the bimorph, but reduces the amount of movement. In our case, the petals were made of copper (a). To protect the copper surface of the petal from corrosion and external exposure to the environment, a nickel-based coating was used, then the surface was coated with a special lacquer which is an electrical insulator, and also protects against humidity and other influences, which increases the lifetime expectancy b). Using the gluing method, apply a layer of piezoceramics on each side of the petal (c). The thickness of the adhesive layer is 10 - 15 μm. For gluing, epoxy or acrylic glue is used. The elasticity of the adhesive compound usually eliminates the appearance of vibration fatigue during prolonged operation. Electrodes made of lead or nickel are applied to the piezoceramic plate, and the thickness of the layer is 6-10 μm [3].

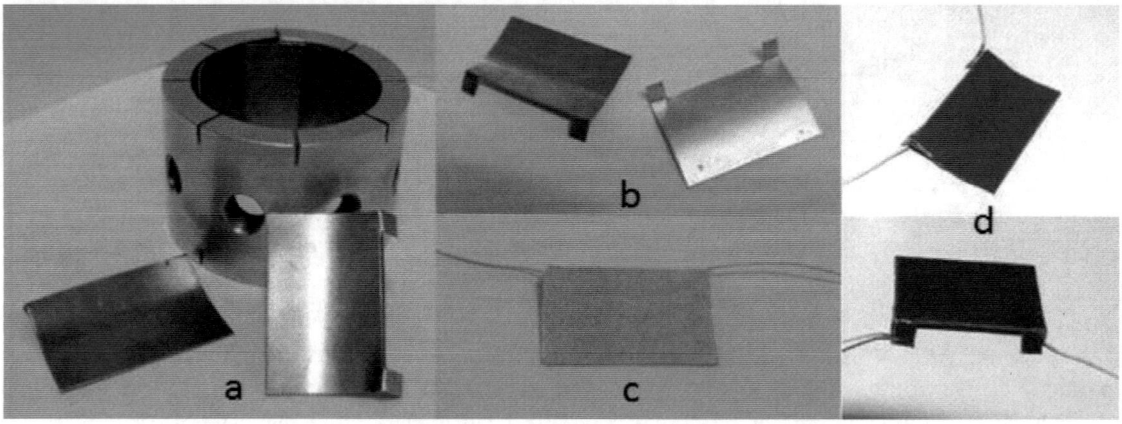

Figure 8. Foil gas dynamic bearing with bimorphs.

After applying the electrodes, the plate is polarized in a strong permanent electric field and acquires piezoelectric properties. Layered technology increases flexibility and allows increasing the degree of deflection of the actuator. In the final stage, antifriction coatings are applied to the surface of the petals using conventional staining techniques, such as spraying, screen printing, dipping, brushing. After coating and drying, the solvent evaporates and the binders polymerize and ensure reliable adhesion to the substrate (d). The choice of a particular method of applying antifriction coatings depends on the geometry of the parts to be coated and the desired result in terms of uniformity and durability of the coatings.

Conclusions

The ways of prevention of emergency situations presented in the article are based on the use of bimetallic coatings, as well as intelligent technologies based on the piezoelectric effect. This approach leads to the emergence of a new class of mechatronic elastic-damper supports with an increased service life and a minimum accident rate. Active monitoring in the most loaded rotor-support nodes will allow to constantly monitor the signature of vibration, to provide real-time information about the deformation of each individual elastic element, which also has a significant scientific interest.

References

[1] Spillman W B, Sirkis J S 1996 Smart materials and stuctures: What are they *International journal of solids and structures* **5** pp 247–254

[2] Edward F Crawley Javier de Luis 2002 Use of Peizoelectric Actuators as Element of Intelligent Structures *AIAA Journal* **25** (10) pp 1373–1388

[3] Sharapov V 2011 Piezoceramic sensors (Springer Verlag, Heidelberg, Dordrecht, London, New York) p 498

[4] Dzhagupov R G and Yerofeyev A A 1994 Piezoelectronic devices of computer facilities, control and management systems *Handbook* (St. Peterburg, Politekhnika) p 608
[5] Adelman N T, Stavsky Y 1980 Flexural extensional behavior of composite piezoelectric circular plates *J. Acoust. Soc. Am.* **67** (3) pp 819–822
[6] Lee C K 1990 Theory of laminated piezoelectric plates for the design of distributed sensors/actuators. Part I: Governing equations and reciprocal relationships *J. Acoust. Soc.Am.* **87** (3) pp 1144–1158
[7] Tsumaki Nobuo Patent JPH № 0454309. GASBEARING 21.02.1992.
[8] Babaev A E, Babaev A A, Yanchevskiy I V 2010 Influence of an oscillating circuit on the radiation of transient acoustic waves by an electroe- lastic cylinder *J. Acoust. Soc. of Am.* **127** (4) pp 2282–2289
[9] Hu J, Qiu Zh., Su Ts-Ch. 2011 Axisymmetric vibrations of a viscous-fluid-filled piezoelectric spherical shell and the associated radiation of sound *J. Sound and Vibration* **330** (24) pp 5982–6005
[10] Jhung M J, Choi Y H, Ryu Y H 2009 Free vibration analysis of circular plate with eccentric hole submerged in fluid *J. Nuclear Eng. and Technology* **41** (3) pp 355–364
[11] Avdiaj S, Šetina J, Syla N 2009 Modeling of the piezoelectric effect using the finite-element method *Materials and Technology* **43** (6) pp 283–291
[12] Chung G, Phan D 2010 Finite element modeling of surface acoustic waves in piezoelectric thin films *J. Korean Phys. Soc.* **57** (3) pp 446–450
[13] Kamath H, Willatzen M, Melnik R V N 2006 Vibration of piezoelectric elements surrounded by fluid media *Ultrasonics* **44** (1) pp 64–72

Acknowledgments
The present research was implemented as a part of the RSF project No. 16-19-00186 'Planning of optimal energy efficient rotor trajectories in mechatronic modules with complex rheology medium'.

IOP Conf. Series: Materials Science and Engineering **233** (2017) 012047 doi:10.1088/1757-899X/233/1/012047

Influence of technological deviations on the basic operational characteristics of hydrodynamic bearings

Y Tarasevych[1], I Savchenko[1], N Sovenko[1]

[1]General Mechanics and Machine Dynamics Department, Sumy State University, 2 Rymskogo-Korsakova st., 40007 Sumy, Ukraine

E-mail: tyulia@ukr.net

Abstract. The main purpose of this paper is to define a methodology to determine the analytical approximate closed-form expression of probabilistic characteristics of load capacity of liquid-lubricated journal bearings in the case of a steady flow regime. Influence of random change of basic geometrical parameters of journal bearings (gap value and eccentricity) on the load capacity of bearing is considered. It is shown that the actual value of hydrodynamic force in bearing can substantially differ from a calculation one. The got results confirm the necessity of application of such approach at consideration both geometrical and operating parameters of bearing, for example, axes misalignment, deviation of surface form, roughness and others, which have casual nature too.

1. Introduction and statement of the problem

The operational characteristics of rotary machines are largely determined by the efficiency of the supporting and sealing units. A number of specific requirements are imposed on the work of the supporting units. This requirements different from those that should be satisfied by the sealing units, namely: sufficient bearing capacity with small dimensions, high vibration stability under all operating modes, minimal friction and wear of the working surfaces for a given life-time, low rate of lubricating - cooling material, the possibility of using a working medium as a lubricant, easiness of installation and operation.

At present, complex tribotechnical devices with various physical principles of creating bearing capacity, including rolling bearings, magnetic bearings, fluid friction bearings and various combinations thereof, are used as rotor supports. The using of rolling bearings as supports for high-speed rotors is limited by their extreme speed and durability. Therefore, the use of sliding bearings makes it possible to provide reliable operation of the rotary machine in a wide range of speeds and loads. A variety of types of sliding bearings can be conditionally divided according to the following criteria:

- by the principle of operation (hydrodynamic, hydrostatic, combined);
- by the direction and nature of the load (radial, axial and radial-axial);
- by type of lubricant (liquid, gas). Moreover, under operation in liquid lubrication conditions, in some cases, it is possible to transfer part of the volume of the lubricant from the liquid to the gaseous phase and vice versa;
- by design.

The variety of these categories of sliding bearings is united by one feature: they are all non-standard elements of machines, in the design and calculation of which in each specific case additional experimental and theoretical studies are needed. Despite the fact that a significant number of works have been published in the area of calculating and designing of sliding bearings, created mathematical models and proposed methods of calculating characteristics of bearings [1-6] mainly do not take into account the fact that the majority of both the geometric and operational parameters of these complex device are random. There are only a few studies considering stochastic nature of some parameters. For example, the effect of a spatially randomly varying component of roughness has been studied in [7] using the stochastic finite element method. The potential inaccuracies in calculating the dynamic coefficients using the ratio of pressure change due to a film thickness change are reported in [8]

The thickness of the lubricating layer, which is one of the main operating parameters, is determined by the appropriate tolerances for the manufacture of parts and assembly of the machine, the quality of the surface treatment (the presence of macro and microroughness), which are random variables. In addition, possible misalignment of the main bearing surfaces and their variation under operation are also random. Therefore, the study of the probability characteristics of sliding bearing is not only an actual scientific, but also an important practical problem.

2. Main equations and the solution of the problem

A mathematical model of a full journal sliding bearing (Figure 1) is considered.

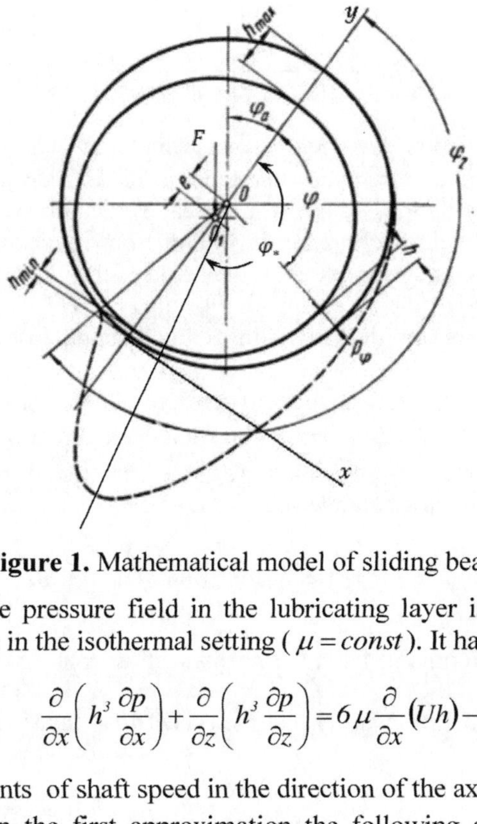

Figure 1. Mathematical model of sliding bearing.

The determination of the pressure field in the lubricating layer is based on the solution of the modified Reynolds equation in the isothermal setting ($\mu = const$). It has the form:

$$\frac{\partial}{\partial x}\left(h^3\frac{\partial p}{\partial x}\right)+\frac{\partial}{\partial z}\left(h^3\frac{\partial p}{\partial z}\right)=6\mu\frac{\partial}{\partial x}(Uh)-12\mu V \tag{1}$$

where U, V – components of shaft speed in the direction of the axes x and y, respectively.

To solve the problem in the first approximation the following assumptions are assumed. The lubricating fluid is incompressible, i.e. $\rho = const$, $\frac{\partial p}{\partial t} = 0$. The flow of the lubricant is steady, $U = const$. As the changes in the characteristics of the fluid flow in the axial direction (along the Oz axis) are significantly less than their changes in the circumferential direction, the second term in Eq.

(1) can also be neglected. Taking into account the assumed assumptions, the Reynolds equation for determining the pressure distribution in the bearing gap takes the form:

$$\frac{\partial}{\partial x}\left(h^3\frac{\partial p}{\partial x}\right)=6\mu U\frac{\partial h}{\partial x} \tag{2}$$

After integrating equation (2) once it is obtained:

$$h^3\frac{\partial p}{\partial x}=6\mu Uh+C$$

The constant C can be determined from the condition: under $h=h_*$, $p=p_{max}$ and $\left.\frac{\partial p}{\partial x}\right|_{h*}=0$. So, $C=-6\mu Uh_*$, and equation (2) is transformed to the form: $h^3\frac{\partial p}{\partial x}=6\mu U(h-h_*)$.

According to Figure 1, the gap in an arbitrary bearing cross-section can be presented by the formula $h=R-r+e\cos\varphi=H(1+\varepsilon\cos\varphi)$. The speed of the shaft wall is $U=\omega r$.

Father solution of the problem is conducted in the movable system of coordinates, one axes of that passes through the line of centers OO_1. Passing to the cylindrical coordinate $x=r\varphi$ and after integration, it is obtained

$$\frac{\partial p}{\partial x}=\frac{6\mu\omega r^2\varepsilon}{H^2}\frac{(\cos\varphi-\cos\varphi_*)}{(1+\varepsilon\cos\varphi)^3}$$

where H is a random variable, the measurement limits of which depend on the tolerances accepted. For values with two-sided tolerances, the normal distribution law is applied, so H can be assumed to have the normal distribution. As a mathematical expectation of this parameter can be taken its calculated value. The actual value of the gap deviation from its design value, due to the landing chosen for the bearing, can be commensurable with the value of the gap itself. For example, for a shaft diameter of 140 mm and for one of the recommended landings of H7 / c8, its tolerance is 0.103 mm, and the radial clearance caused by the deviations in the dimensions of the shaft and sleeve can vary from 0.1 to 0.15 mm.

The random variables H, ε are independent. Therefore, the complex probability density can be written in the form $f_3(H,\varepsilon)=f_2(H)f_1(\varepsilon)$. For eccentricity, all directions are equally possible, and proceeding from physical properties, this random variable can take only positive values, so for its description the truncated Rayleigh law can be used:

$$f_1(\varepsilon)=\frac{C\varepsilon}{\sigma_\varepsilon}\exp\left[-\frac{\varepsilon^2}{2\sigma_\varepsilon^2}\right] \tag{3}$$

where the constant C is determined from the normalization condition for the probability density: $C=\left(1-\exp\left[-\frac{1}{2\sigma_\varepsilon^2}\right]\right)^{-1}$. Since the probability that the eccentricity takes values close to zero and close to unity is equally small, it can be assumed: $\sigma_e=0{,}288$.

Under steady mode the external loading on bearing is balanced by hydrodynamic force, acting in a bearing gap:

$$F=\int_{\varphi_1}^{\varphi_2}p(\varphi_t)\frac{ld}{2}\cos[\pi-(\varphi_a-\varphi_*)]d\varphi_t=$$

$$= \frac{3l\mu\omega r^3 \varepsilon}{H^2} \int_{\varphi_1}^{\varphi_2} cos[\pi - (\varphi_a + \varphi_*)] \int_{\varphi_1}^{\varphi_t} \frac{cos\varphi - cos\varphi_*}{(1 + \varepsilon cos\varphi)^3} d\varphi d\varphi_t \quad (4)$$

Where φ_t is a current angle, $\pi - \varphi_*$ is angle under which the pressure distribution in a bearing gap is maximum, φ_1, φ_2 are angles which determining beginning and the end of bearing load zone. Under assumed problem statement (Fig.1): $\varphi_1 = 0$, $\varphi_2 = \pi + \varphi_*$. The angle φ_* is determined numerically, depending on the bearing geometry and the required bearing capacity.

The range of variation in pressure force of the bearing gap can be defined by the inequality:

$$p < \frac{6\mu\omega\varepsilon r^2}{H^2} \int_{\varphi_1}^{\varphi_2} f(\varepsilon, \varphi, \varphi_*) d\varphi$$

Then the density of the probability of pressure in the bearing clearance can be represented as:

$$g(P) = \int_0^1 f_1(\varepsilon) f_2((\gamma(P,\varepsilon)) \left| \frac{\partial \gamma(P,\varepsilon)}{\partial P} \right| d\varepsilon \quad (5)$$

where $\gamma(F,\varepsilon)$ is the inverse function by variable H, $f_2(H) = \frac{1}{\sigma_H \sqrt{2\pi}} exp\left(\frac{-(H - mH)^2}{2\sigma_H^2} \right)$, m_H is the mean value of the gap, σ_H is the standard deviation of the gap.

The basic characteristics of the bearing operation are: volume flow of the lubricant, the power loss due to friction, load capacity.

The effect of the random variation of H and ε on the load capacity of bearing is considered. The standard deviation of the gap value in the bearing can be represented as: $\sigma_H = \Delta m_H$. Using the obtained probability density (5), the basic moment characteristics of the hydrodynamic pressure force is determined: mean value and standard deviation.

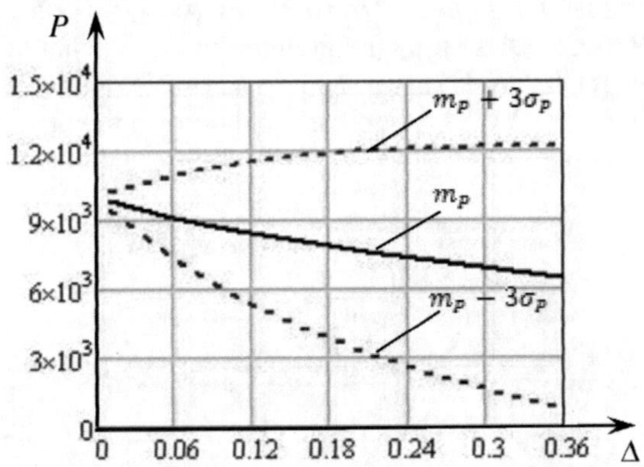

Figure 2. Dependence of mean value of pressure forces in the bearing gap and interval of possible values of radial force on the relative value of the standard deviation of the gap.

The change in the mean value as a function of the change in the root-mean-square deviation of the value of H and interval of possible values of hydrodynamic force (without taking into account the effect of a random change in the eccentricity) is shown in Figure 2.

As undertaken studies show with increasing of standard deviation value of geometric gap of bearing the actual value of radial force can be substantially differ from the calculation one.

IOP Conf. Series: Materials Science and Engineering **233** (2017) 012047 doi:10.1088/1757-899X/233/1/012047

3. Conclusions

As a rule the deterministic methods of basic characteristics calculation of sliding bearing are used. These methods do not take into account stochastic nature of geometric parameters of bearings. Lately a lot of different studies are devoted to the determination the effect of micro and macro roughnesses of bearing surfaces on its loading characteristic. However, taking into account admitted in engineering tolerances, the effect of these parameters can be negligible as compared to indetermination (stochastic) of calculated geometrical parameters. For example, under value of root-mean-square deviation of gap $0{,}1m_H$ (initial value of gap is $1{,}4\cdot10^{-4}$) the interval of possible gap values is ($9{,}8\cdot10^{-5}$; $1{,}82\cdot10^{-4}$)and deviation of calculation value from its mean value of bearing load capacity is 14%. And under value of root-mean-square deviation of gap $0{,}3m_H$ difference in calculation and possible value is 31%. As undertaken studies show with increasing of standard deviation value of geometric gap of bearing the interval of possible values of radial force is substantially increases. And as the expected value of bearing capacity decreases, the probability of bearing failure raises.

References

[1] Goenka P K 1984 Dynamically Loaded Journal Bearings: FiniteElement Method Analysis *Trans. ASME J. Tribol. Series F* pp 429-439.

[2] Kozma M 1995 Load Carrying Capacity of Dynamically Loaded Journal Bearing, *Periodica Polytechnica* **40** pp 39–44.

[3] Lund J W 1987 Review of the Concept of Dynamic Coefficients for Fluid Film Journal Bearings Trans. ASME J. Tribol. **109**, pp 37-41

[4] Voskresenskiy V A 1983 *Calculation and Desing of Fluid Film Bearings* (Moscow: mechanical engineering) p 232

[5] Chernavski S A 1963 *Sliding Bearings* (Moscow:Mashgiz) p 244

[6] Maksimov V A 1998 *Tribology of bearings and fluid film seals of high-speed turbo-machines* (Kazan: Fan) p 430

[7] Turaga R Sekhar A S Majumdar B C 1997 Stochastic FEM Analysis of Finite Hydrodynamic Bearings with Rough Surfaces *J.Tribology Transactions* **40** (4) pp 605-612

[8] Klit P and Lund J W 1986 Calculation of the Dynamic Coefficients of a Journal Bearing Using a Variational Approach *Trans. ASME J. Tribol.* **108** pp 421-425

[9] Tsybenko A.S. Tarasevich Y.Y. 2013 Probabilistic methods in mechanics (Sumy: SumDU) p 240

New method for strengthening surfaces of heat treated steel parts

V Tarelnyk[1], V Martsynkovskyy[1], O Gaponova[2], Ie Konoplianchenko[1], A Belous[1], V Gerasimenko[1], M Zakharov[3]

[1] Sumy National Agrarian University, 160 G. Kondrateva Str., 40021, Sumy, Ukraine
[2] Sumy State University, Sumy, 2 Rymskogo-Korsakova st., 40007, Sumy, Ukraine
[3] Regional Institute of Postgraduate Education, 5 Rymskogo-Korsakova st., 40007, Sumy, Ukraine

E-mail: tarelnik@i.ua

Abstract. There is described a method for strengthening surfaces of heat-treated steel parts, which includes technological procedures of carburizing by electroerosive alloying process (CEEA) and also nitriding processes. The nitriding process is simultaneously carried out with the CEEA operation for a period of time sufficient to saturate the surface layer of a part with nitrogen to the depth of a heat-affected zone. Thus, the process of so-cold nitro-carburizing process performed by the EEL method (NCEEA) takes place. To reduce a surface roughness, the NCEEA process is carried out in at least two stages, with a decreasing a discharge energy value at each subsequent stage.

1. Introduction

One of the main indicators of machines quality is their reliability. The most common cause of machine failures is recognized not a breakdown, but wear and damage of working surfaces of their parts and operating units. As a rule, destruction of a part begins with destruction of its surfaces.

The use of strengthening and protective coatings significantly improves the quality of products in machine building, provides reliable operation of units and parts on severe operating conditions for equipment, allows reducing material and energy costs for a machine operation, and also lowering consumption of expensive construction materials. Therefore, studies aimed at creating new and improving of the existing protective technologies are of current interest and well-timed.

From the technical point of view, one of the simplest methods for strengthening parts is a surface electro-erosive alloying (EEA) method. It is characterized by such advantages as a local impact, low energy consumption, lack of volumetric heating of material, etc. While using the EEA method, it is possible to increase hardness of a metal surface by applying a material of higher hardness onto the same or providing a diffusion introduction of necessary chemical elements into a surface layer from the environment or from an anode material [1].

However, the electro-erosive alloying (EEA) of heat-treated parts subjected to high specific loads, for example the parts of dies, rolling mill shafts and other similar items does not always bring a desired result. The cause of some items failures is that after processing with the use of the EEA method, under a layer of increased hardness, there is appeared a tempering zone, that is, a zone of reduced hardness. This results in a so-called breakdown of the strengthened layer and, as a consequence, in an intensive wear of the item. In this case, processing by the EEA method might be harmful, especially if the permissible wear of the alloyed surface exceeds the thickness of the layer of increased hardness.

According to method [2], carrying out the process with the use of ionic nitriding (IN) method (either before the EEA method or after the same) provides eliminating of the zones of reduced hardness when using electrodes made of pure, hard, and wear resistant metals. In addition, a smooth change in the hardness of the strengthened layer and also increasing the total depth of the zone of the increased hardness are observed.

The disadvantage of this method is the low productivity of the process, since even with increasing productivity up to 0.4 cm^2/min, undesirable roughness and continuity of the surface at the EEA process by chromium, tungsten and T15K6 hard alloy are, respectively, 4. 6; 7. 8 and 5. 4 μm; and 90; 55 and 80%, which fact significantly narrows the scope of the method application for strengthening machine parts.

There is known a method of carburizing steel parts with the use of the electro-erosive alloying (CEEA) [3], which has a number of advantages, the main of which are: providing 100% continuity for the strengthened surface layer, increasing the hardness of the surface layer of the part due to diffusion-quenching processes, alloying can be carried out in strictly specified areas with no protection of the rest of the surface of the part, lack of volumetric heating of the part and warpage associated therewith; simplicity of the technology application, flexible binding to existing equipment; strengthening process does not require special preparation and high qualification of the worker.

In this method, there is used the discharge energy of 0.036 to 6.8 J and the productivity of 1.0 to 0.2 cm^2 / min.

At carburizing a steel part with the use of the EEA method, the thickness of the strengthened layer depends on the discharge energy and alloying period of time (the process productivity). Increasing the values of discharge energy and alloying period of time results in increasing the thickness of the strengthened layer. And in doing so, there is increased the value of a surface roughness. Thus, in the case of processing an item made of medium-carbon alloy steel 40 Ch (Ra = 0.5 mkm) with the use of the EEA method performed by a carbon electrode on the condition of the productivity of 0.2 cm^2/min and the discharge energy of 6.8 J, the thickness of the increased hardness layer is more than 1.15 mm. The surface roughness corresponds to Ra = 11.7-14.0 μm.

There is also known the CEEA method, which is used to reduce the surface roughness of machine parts and, therefore, to expand the field of application thereof. This method is characterized by the fact that the CEEA process is carried out in stages with reducing the value of discharge energy at each subsequent stage [4].

The known method uses the values of discharge energy of 0.036 to 6.8 J and productivity of 14.0 to 2.0 cm^2 / min.

Despite above mentioned obvious advantages, the main one of which is a decrease in the surface roughness level of machine parts while maintaining the proper quality of the surface layer (the absence of microcracks, the availability of a layer of increased hardness, 100% continuity, etc.), this method has a number of drawbacks. First of all, this is a decrease in the microhardness of the surface layer as a result of tempering at a repeated (stage-by-stage) surface treatment by a graphite electrode, but with the use of a lower value of the discharge energy. In addition, along with the decrease in the microhardness of the surface, there is also decreased the depth of the layer having increased hardness.

In order to eliminate the above mentioned drawbacks, there has been developed a method for strengthening surfaces of heat-treated steel parts [5], which, like the methods known from the prior art, comprises both the operation performed by the EEA method and the operation performed by the IN method, wherein the IN operation has been being performed either before or after the EEA operation for a period of time that is sufficient for saturating the surface layer of the part with nitrogen to the depth of the thermal influence zone. The EEA operation is performed with a graphite electrode under condition characterized by the values of discharge energy of 0.1 to 6.8 J and productivity of 0.2 to 4.0 cm^2/min. At the same time, the CEEA process is performed in at least two stages with decreasing

IOP Conf. Series: Materials Science and Engineering **233** (2017) 012048 doi:10.1088/1757-899X/233/1/012048

the value of discharge energy at each subsequent stage, the first stage of alloying by a graphite electrode being performed at the values of discharge energy of 0.1 to 6.4 J and productivity of 0.2 to 4.0 cm^2/min, and the second one - at the values of discharge energy of 0.1 to 2.83 J and productivity of 0.2 to 2.0 cm^2/min.

In this case, the CEEA stage by stage process performed before the IN process results in decreasing the microhardness in the heat-affected zone (i.e., a zone of reduced hardness can be formed under the layer of increased hardness), and the CEEA process performed after the IN process for a period of time sufficient to saturate the surface layer of the part with nitrogen to the depth of the thermal zone influence results in eliminating the hardness failure.

The same results can be achieved if the IN process has been being conducted before the CEEA process for a period of time sufficient to saturate the surface layer of the part with nitrogen to the depth of the zone of thermal influence. And to reduce the surface roughness, the CEEA process should be carried out in stages, while reducing the value of discharge energy at each subsequent stage. In this case, the hardness in the zone of a thermal influence does not decrease, because the properties of the nitrided surface do not practically change at repeated heating up to 500-600 °C, while at heating a carburized and strengthened surface to 225-275 °C, its hardness has been decreasing.

At the use of the CEEA method for processing a steel nitrided surface, there is occurred a process similar to the process of so-called nitrocarburization, only in this case, the processes of saturations of the surface with nitrogen and carbon proceed alternately, in comparison to the traditional nitrocarburization method whereat the above mentioned saturation processes have been taking place simultaneously.

Despite a number of positive results presented above, the present method is not devoid of shortcomings. They are as follows: first of all, it is a long duration of the IN process (up to 24 hours), high costs of both electricity and required reagents, manufacturing of a witness sample to monitor the results of the process of strengthening, and the need to protect separate areas of a part surface, for example threads, from strengthening. Consequently, the technical problem of improving the quality of heat-treated parts has not lost its relevance.

To overcome the above said drawbacks, there has been developed a method for strengthening the surfaces of heat-treated steel parts, which, like the methods known from the prior art, includes the CEEA process, but which is significantly different in that nitrogen is supplied to the alloying zone. In this case, two processes, namely, the CEEA process and nitriding one have been taking place simultaneously. Such a method is essentially a method of nitrocarburizing by the EEA method (NCEEA).

Thus, the purpose of this work is to improve the technology of strengthening heat-treated parts at the expense of combining the methods of CEEA and nitriding by carrying out the EEA process with a graphite electrode in the environment of nitrogen.

2. Methodology and consideration of results

To study the processes of carburization and nitrocarburization by the EEA method, there were used special samples made of 40 Ch steel, which were heat-treated similarly to the method disclosed in publication [5] to obtain the value of hardness of 3900-4000 MPa. The samples were made in the form of coils consisting of two disks, 50 mm in diameter and 10 mm in width, connected by a spacer of 15 mm in diameter, which has two technological sections of the same diameter, Figure 1. The surface of the disks was ground to obtain Ra = 0.5 μm.

IOP Conf. Series: Materials Science and Engineering **233** (2017) 012048 doi:10.1088/1757-899X/233/1/012048

Figure 1. A sample made of the 40Ch grade steel for performing the CEEA and NCEEA processes.

The CEEA and NCEEA processes were automatically carried out using the unit of EIL-8A model. Table 1 shows the main modes of its operation, as well as the recommended period of time for alloying 1 cm^2 of a part surface (the EEA process productivity). For a storage capacitor capacitances: C = 20 μF and C = 300 μF, the unit had 8 operating modes.

Table 1. Operating Modes for the Unit of EIL-8A model.

Mode Number	EEA Process Productivity cm^2/min		Discharge Energy W, J	
	C=20 μF	C=300 μF	C=20 μF	C=300μF
1	0.14-0.13	0.25-0.20	0.01	0.14
2	0.14-0.13	0.33-0.25	0.014	0.22
3	0.17-0.14	0.33-0.25	0.02	0.28
4	0.20-0.17	0.50-0.33	0.024	0.35
5	0.25-0.20	1.0-0.50	0.03	0.42
6	0.25-0.20	1.0-0.67	0.034	0.49
7	0.33-0.25	2.0-1.0	0.038	0.56
8	0.33-0.25	2.0-1.0	0.043	0.63

The samples were fixed in the lathe chuck, and thereafter they were processed applying the following methods:

The CEEA method by stage-by-stage alloying with the graphite electrode of EG-4 (OST 229-83) grade at the discharge energy of 0.42 J (stage 1) and 0.14 J (stage 2), and at the productivity of 0.4 and 0.2 cm^2/min, respectively;

The NCEEA method by stage-by-stage alloying with the graphite electrode of EG-4 (OST 229-83) grade at the discharge energy of 0.42 J (stage 1) and 0.14 J (stage 2), and at the productivity of 0.4 and 0.2 cm^2/min, respectively;

The method of nitrocarburization is carried out at the expense of using a device that is attached to the vibrator of the EEA unit (Figure 2).

Due to the fact that at the NCEEA process, there is occurred cooling of the part portion being alloyed with a nitrogen stream, the productivity at the NCEEA process is twice reduced at the both stages, which fact doubles the process time for nitrocarburization.

The strengthened samples were cut to obtain the segments intended for preparing thin sections. The thin sections were examined on an optical microscope Neofot-2, whereon, there was estimated the quality of the layer, its continuity, thickness and structure of the under layer zones, namely, the diffusion zone and the thermally influenced zone. Simultaneously, there was carried out a durametric analysis to evaluate the distribution of microhardness in the surface layer and along the depth of a thin section starting from the surface.

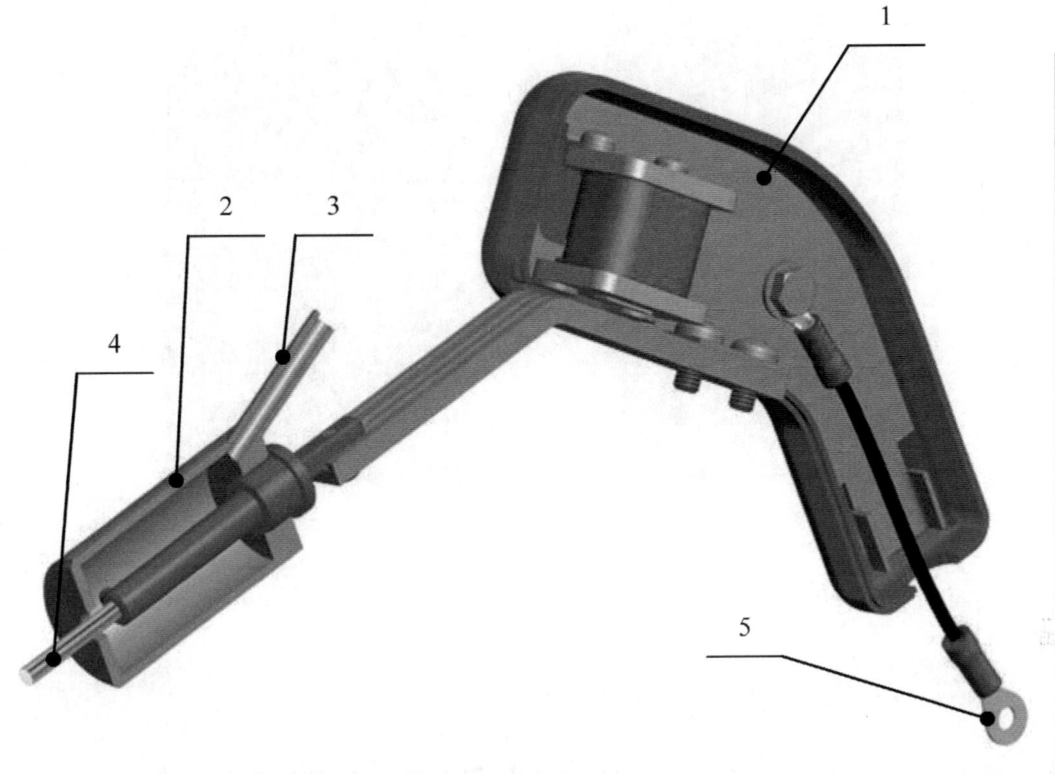

Figure 2. Device for supplying gas to the alloying zone: 1 – vibrator; 2 - mandrel for gas supply; 3 - gas inlet fitting; 4 – electrode; 5 – joint for connecting the vibrator to the EEA generator.

The microhardness measurements were performed with the use of the RMT-3 microhardnessmeter by pressing a diamond pyramid with a load of 0.05 N.

At all the stages of the process, the surface roughness was measured on the profilograph-profilometer of mod. 201 and of Caliber plant production.

Table 1 shows the microhardness distribution in the surface layer of the 40Ch steel samples having been heat-treated for hardness of 3900-4000 MPa and strengthened by various methods, and also the results of the influence of those strengthening methods on the roughness of the strengthened surface layer being formed.

Table 2. Results of processing the 40Ch Steel Samples surface layers by the CEEA and NCEEA methods.

Method of Strengthening	Microhardness, MPa (Measuring Step ~ 30 μm)					Ra, μm
	1	2	3	4	5	
CEEA	7010	3800	4300	4100	3900	0,8
NCEEA	10500	6200	5300	4300	4000	0,7

Figure 3 shows the microstructures of the surface layer of 40Ch steel and the microhardness distribution on the depth of the layer during carburization and nitrocarburization by the EEL method.

IOP Conf. Series: Materials Science and Engineering **233** (2017) 012048 doi:10.1088/1757-899X/233/1/012048

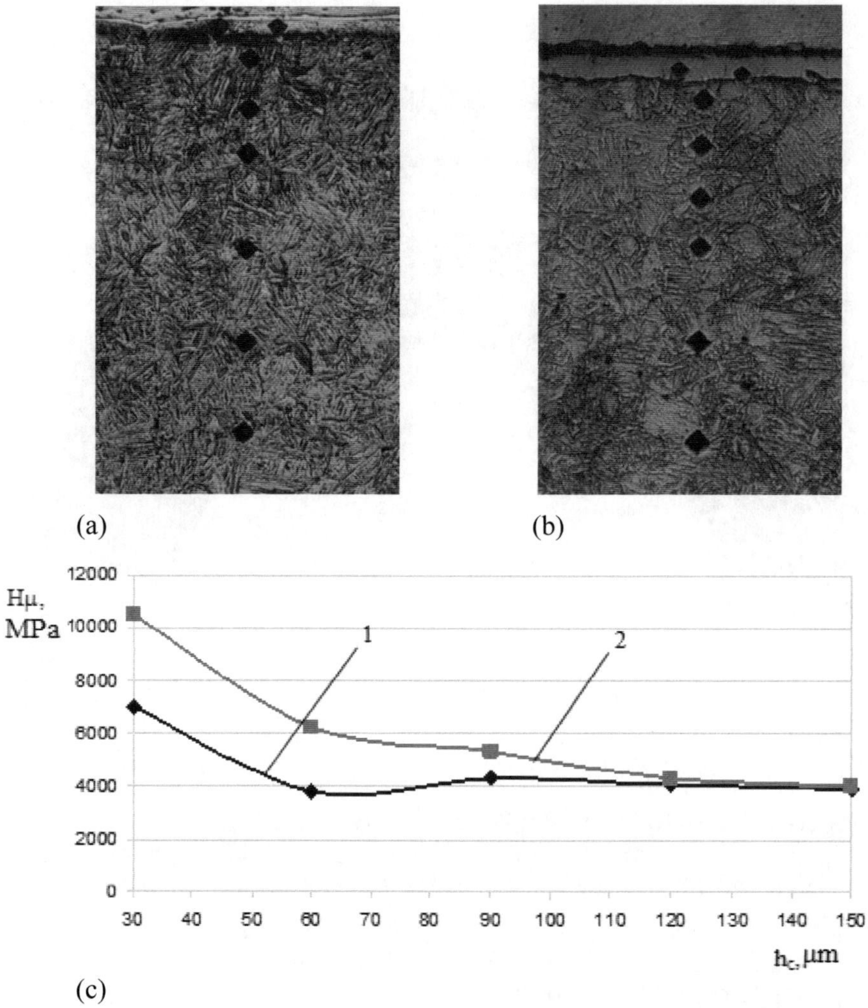

Figure 3. Microstructure of 40Ch steel samples after: a - CEEA; b – NCEEA; and c - microhardness distribution along the depth of the layer at: 1 - CEEA; 2 - NCEEA.

Analysis of Table 2 and Figure 3 shows that at the CEEL process of the 40Ch steel samples, under a layer of increased hardness, a tempering zone is located ("hardness dip"). In this case, this zone is located at the depth of ≈ 60 - 70 μm and is 3800 MPa.

Having been carried out on the above mentioned conditions, the NCEEA process eliminates the characteristic "hardness dip", with a general increase and a gradual decrease of the values of hardness in the transition zone. The surface roughness reduction at the NCEEA process is explained by the protection of the alloying zone with a flow of nitrogen from the ambient air (oxidizing) environment.

3. Conclusions

1. It has been stated that on strengthening heat-treated parts, the EEA process of the surface layer by carbon should be combined with nitriding, and in order to reduce the surface roughness, the NCEEA should be carried out in stages with reducing the discharge energy at each stage.

2. It was determined that the greatest microhardness (10500 MPa), the depth of the zone of increased hardness (120 μm), and the lower roughness (Ra = 0.7 μm) are recorded at the NCESEA process.

References

[1] Lazarenko N I 1976 *Electric Spark Alloying of Metal Surfaces*, (Moscow: Mechanical engineering) p 46
[2] Martsinkovskyy V S Tarelnyk V B (2013) A method for strengthening heat-treated steel parts surfaces *Patent for Invention UA 103701* Bul 21 p 3
[3] Martsinkovskyy V S , Tarelnyk V B, Belous A V (2008) A method for carburizing steel parts by electro-erosive alloying thereof *Patent for Invention RF 2337796* Bul 31 p 3
[4] Martsinkovskyy V S Tarelnyk V B Bratuschak M P (2013) A method for carburizing steel parts by electro-erosive alloying thereof *Patent for Invention UA 101715* Bul 8 p 3
[5] Tarelnyk V B, Martsinkovskyy V S , Kosenko P V, Volosko T P, Antoszewski B (2016) A method for strengthening surfaces of heat-treated steel parts Patent for Invention RF 2603932 Bul 34 p 3

IOP Conf. Series: Materials Science and Engineering **233** (2017) 012049 doi:10.1088/1757-899X/233/1/012049

New sulphiding method for steel and cast iron parts

V Tarelnyk[1], V Martsynkovskyy[1], O Gaponova[2], Ie Konoplianchenko[1], M Dovzyk[1], N Tarelnyk[1], S Gorovoy[1]

[1] Sumy National Agrarian University, 160 G. Kondrateva Str., 40021, Sumy, Ukraine
[2] Sumy State University, Sumy, 2 Rymskogo-Korsakova st., 40007, Sumy, Ukraine

E-mail: tarelnik@i.ua

Abstract. A new method for sulphiding steel and cast iron part surfaces by electroerosion alloying (EEA) with the use of a special electrode is proposed, which method is characterized in that while manufacturing the electrode, on its surface, in any known manner (punching, threading, pulling, etc.), there is formed at least a recess to be filled with sulfur as a consistent material, and then there is produced EEA by the obtained electrode without waiting for the consistent material to become dried.

1. Introduction

Improving reliability and durability of machine parts is inseparable from a surface layer quality. Manufacturing the parts entirely made of wear-resistant materials is impractical due to the difficulty and high cost of processing such materials. Therefore, to improve operational performance and increase the service life of the machine parts, there are used wear-resistant coatings, which have found wide application in manufacture of various products, a lot of various methods for surface hardening, and also other ways to prevent seizing and galling friction surfaces, which are particularly important for the parts working under conditions of dry friction.

2. Analysis of the main achievements and publications

Improvements of protective and tribology properties of the machine part surfaces can be achieved both through creating a special topography of the surface layer [1, 2], and applying special coatings [3, 4] thereto. The coatings, which are applied to the friction surfaces, can significantly improve reliability of machines and devices. On applying the coatings, there could be provided running-in ability of the friction surfaces, prevention of scuffing; lubrication at the initial time before oil supply to the friction surfaces, minimal wear of the friction surfaces, the lowest coefficient of friction and minimal changes of the same during operation, increasing contact friction fatigue of friction unit materials, and others [5].

At present in industry, there is an acute problem of creating multifunctional wear-resistant coatings for heavy duty friction pairs operating at high line speeds and applied pressure with insufficient lubrication or complete lack thereof [6].

Recently, there have been conducted works on producing products, which could be used without external lubrication. These products are specified as "self-lubricating" ones because they use no external lubrication in the course of operation, and there are mainly used composite materials having molybdenum disulphide (MoS_2) in their composition. Formation and research of coatings based on molybdenum disulfide is one of the most promising directions in the field of creating solid lubricating coatings [7].

The ability to demonstrate a "self-lubricating" effect, i.e. provide for operability of a sliding friction assembly unit without introducing external lubrication, is the main feature of the most amount of polymers used in friction units. The nature of the "self-lubricating" effect of those polymers lies in the features of a so-called "third body" that is formed on the polymer and metal surfaces. In this case, the structure and composition of this "third body" is a kind of a "lubricating" layer, which provides an opportunity of redeformation without any abrasive wear and galling [8].

There is known **a sulphiding method** that prevents seizing, and provides creating a film of sulfides on a part surface [9]. The sulfides increase the surface activity of metals and alloys, as well as the process of wetting the surfaces with active agents, and they also increase seizing resistance. An iron sulfide (FeS) film increases wearing resistance of friction surfaces and improves running-in ability thereof. A ferro - sulfide coating has relatively high porosity and it absorbs a large amount of grease providing the material with the self-lubricating property.

Traditionally, a **sulphiding method** is a process of saturating a metal surface layer (of steel, cast iron, titanium alloys, etc.) with sulfur in appropriate salt baths. A depth of a sulphided layer is up to 50 microns. Besides increasing a surface roughness and dimensions of a part being processed, the disadvantages of the conventional sulphiding method includes the necessity to heat the entire part, and accordingly, the structural changes of metal, distortions and warpage, large processing time, high consumption of electricity, negative impact on the environment, and others. [10].

There is known a method wherein sulfur is introduced into the part surface by electric spark alloying (ESA), the same as electroerosion alloying (EEA). To introduce sulfur, it is deposited on a part surface as a powdered sulfur layer, and then there is carried out the process of alloying the part surface through the above said layer by the materials of the part or a coating. In the course of processing, applying of the sulfur powder is repeated 2-3 times [11].

It should be noted that this method does not result in a desired result.

The special features of forming properties and surface structure are provided by EEA entity that is as follows. Having been thrown out of the anode, the material in the liquid-drop and vapor states as an expanding beam enters the inter-electrode space heated by the passed electrical pulse to a temperature, at which the substance can exist in ionized state. The particles of the material collide with the cathode surface (a part) also locally heated by electric pulse, interact with it to form alloys, solid solutions or mechanical mixture. The micro-bath materials formed on the cathode and the anode incoming particles interact with each other and with the elements surrounding the electrodes.

Sulfur is a non-electrically conductive powder and to be used in the EEL process, that is to provide the alloying electrode contact with the surface being alloyed, the sulfur should be pushed aside by the alloying electrode. In this case, entering of the sulfur into the liquid phase zone of the electrode materials and forming of chemical compounds becomes significantly difficult, and if it occurs, it only happens at some isolated places. Continuity of the sulphide film surface is very low.

There is known an ESA method performed by iron sulfide electrode for processing steels 45, H12F, gray and pearlitic malleable cast iron, which method made it possible to form a 40-50 microns thick layer enriched with sulfur on the surface being treated. As a result, there has been significantly reduced the gray cast iron tendency to scuffing at dry friction, its wear has 8-10 times decreased, drawing dies durability has also 3.5 times increased [12]. The electrodes made of iron sulfide have been being produced according to Author's Certificate No. 139 336 published in Bulletin of Inventions of the Former Soviet Union, Number 13, 1961.

To manufacture the electrodes, the iron powder of AM mark previously deoxidized in a hydrogen atmosphere at 1100-1150^{0}C has been mixing with technical sulfur (iron powder/sulfur in a ratio percent of 61/ 39 weight/weight) in a ball mill for an hour. After discharging, the mixture will be heated until the reaction of iron sulfide formation has been completed. The obtained iron sulfide is crushed and then melted in a selit or gas furnace. The electrodes are made of the molten iron sulfide.

The casting process is performed at 1100-1150^{0}C into carefully dried earthen form having a proper permeability. The electrodes are ready for operation after cooling and removing metal burnt-on [13].

Despite the positive results reported above, the use of iron sulfide as EEL electrodes for sulphiding process has several drawbacks: the limited use of steel (iron) electrodes made of unalloyed steels at EEL processing. The electrodes made of unalloyed steels are heated, oxidized, and destroyed in the course of EEL processes, which negatively affects the quality of a coating being formed (high roughness, low continuity, burnt-on, welding separate smaller particles of a damaged anode). The above said electrode manufacturing method is expensive, time consuming and environmentally harmful.

Thus, the further research aimed at eliminating the above mentioned disadvantages of sulphiding process of steel and cast iron parts by the EEL method should be considered relevant and timely.

The purpose of the paper is to develop a new low energy-consuming and ecologically clean method for producing and applying electrodes for sulphiding steel and cast iron parts by the EEL method.

3. Research Technique

To study the sulphiding process by EEL method, there were prepared samples (15 x 15 x 8 mm) made of steel 20 and ductile iron VCH60 which served as the cathodes. The EEL process of the samples was carried out on various conditions of the EEL device of Elitron-22A model. The electrode-tool, namely, the electrode made of 12H18N10T stainless steel wire ∅ 3mm served as an anode. On the surface of the wire, there were provided special recesses performed by thread M3 cutting (Figure 1 (a)).

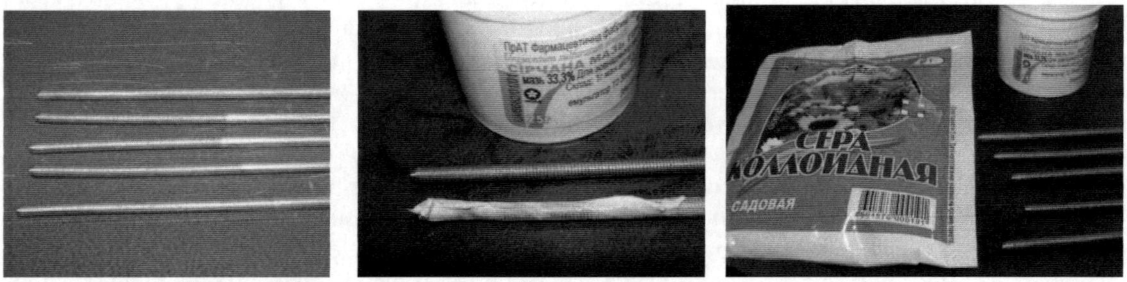

(a) (b) (c)

Figure 1. The electrodes are made of 12H18N10T stainless steel ∅ 3mm intended for sulphiding the parts by EEL method.

It should be noted that to manufacture a special electrode-tool, there may be used such a material as a corrosion-resistant metal, i.e. chromium, nickel, molybdenum, stainless steel 12H18N10T, 07H16N6, 30H13, nickel alloys HN58MBYUD, HN60VT, HN50VMTYUB etc., and grooves on their surface, besides threading, can be formed by any known method, e.g. punching, pulling through a die, and others.

Sulfur in the form of a consistent material was deposited into the grooves of the threaded surface. In this case, there was applied a sulfuric ointment wherein sulfur content was 33.3% (Figure 1 (b)). An amount of sulfur can be increased by adding the same, for example in the form of colloidal sulfur (Figure 1 (c)). Then, there was produced a process of sulphiding the samples by EEL method without waiting for drying the consistent material on the surface of the electrode-tool.

The qualitative analysis and quantitative composition of the coatings on the surfaces of the samples were carried out using the REMMA-102 scanning electron microscope of the Sumy "SELMI" production, which was equipped with X-ray spectrometer produced on the basis of lithium-silicon semiconductor detector.

The micrographs of the surface areas of the examined samples were obtained at various zoom levels under a mode of an image forming by secondary electron current at the microscope electron gun accelerating voltage of 20 kilovolts and the probe (beam) current of 200 pico amps. At the same time, on the surface of the coating, there were chosen three characteristic points (the smooth coating, the rough coating and the pore), and at each one of these points, there was determined an elementwise composition of the applied layer. In addition, the elementwise composition was determined along the depth of the layer with varying a scanning pitch.

As being under condition of X-ray microanalysis operation, the accelerating voltage was 20 kilovolt, and the beam current was one (1) nanoampere. Having been excited by electron beam, the characteristic X-rays were detected by a semiconductor detector.

The calculation of weight (mass) percent concentration was carried out by comparison of the examined samples with the standards. At the same time, there were taken into account the physical corrections for atomic number, fluorescence and absorption of radiation (three corrections method).

At all the research stages, the surface roughness of a sample was measured by the Profilograph-Profilometer (model 201) of the "Caliber" plant production. The device was modernized in such a way that the results of measurements were transferred to a computer through a special device.

4. Results of Researches

Figure 2 shows the topography of the surface portion of the coating formed by EEL method on a part made of steel 20 with the help of an electrode produced of stainless steel 12H18N10T at discharge energy Wp = 0.55 J, wherein there were selected three characteristic areas (1 - the smooth surface, 2 – the rough surface, 3 – the pore).

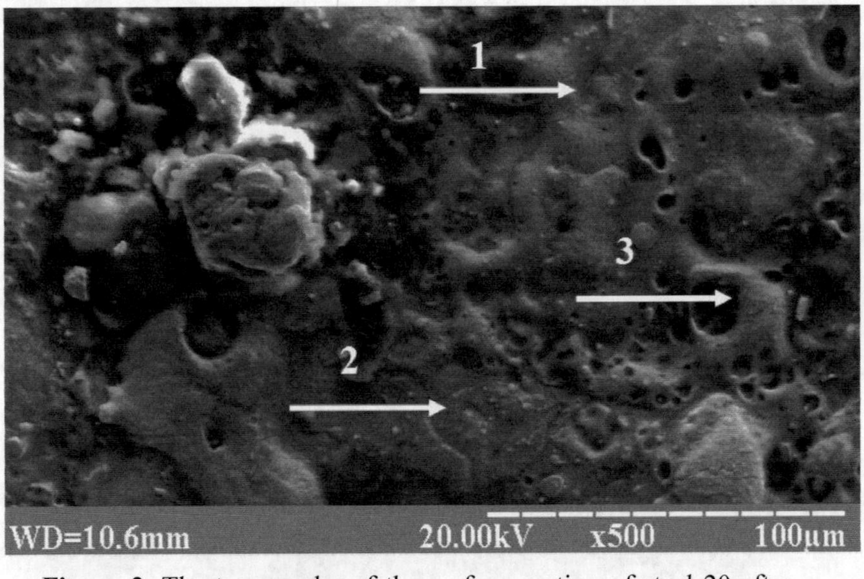

Figure 2. The topography of the surface portion of steel 20 after sulphiding by EEL method.

The spectrum of the surface and the elementwise composition both at the characteristic points, and on all the examined surface are, accordingly, shown in Figure 3 and Table 1.

IOP Conf. Series: Materials Science and Engineering **233** (2017) 012049 doi:10.1088/1757-899X/233/1/012049

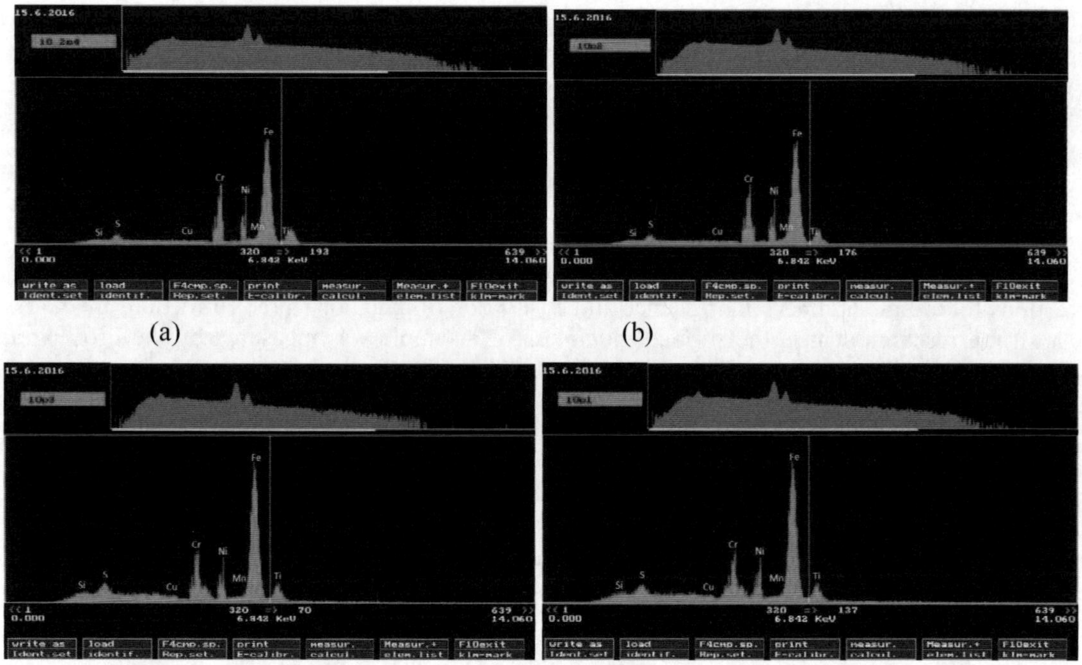

(a) (b)

(c) (d)

Figure 3. The spectra of the surface at the characteristic points (a – the smooth surface, b - the rough surface, c - pore), and on the entire surface (d).

Table 1. The elementwise coating composition at the characteristic points and on the entire examined surface.

The examined point and the surface area	Elements, %							
	S	Cu	Si	Mn	Cr	Ni	Ti	Fe
1	0.56	0.20	0.24	0.60	16.20	9.30	0.71	72.16
2	0.60	0.21	0.22	0.58	17.21	8.89	0.69	71.60
3	0.53	0.23	0.18	0.62	15.23	9.19	0.83	73.9
Σ	0.55	0.87	0.21	0.60	16.87	9.07	0.89	70.94

Figure 4. The points of scanning the coating along the layer depth.

Table 2. The elementwise coating composition with deepening from the surface.

The examined point of the surface	Elements, %							
	S	Cu	Si	Mn	Cr	Ni	Ti	Fe
1	0.41	0.23	0.24	0.58	9.21	5.21	0.61	83.51
2	0.31	0.19	0.22	0.51	5.23	3.19	0.34	90.01
3	0.12	0.20	0.21	0.62	3.19	2.23	0.21	93.22
4	0.09	0.21	0.23	0.57	1.16	1.21	0.17	96.36
5	0.06	0.16	0.16	0.54	0.53	0.76	-	97.79
6	0.03	0.24	0.17	0.53	0.21	0.25	-	98.57

In Table 3, there are represented the results of sulphiding the samples made of steel 20 and ductile cast iron VCH60 by a special electrode-tool made of steel 12H18N10T at various discharge energies.

Table 3. The results of sulphiding the samples made of steel 20 and ductile cast iron VCH60.

Sample material	Discharge energy, Wp, J	Sulphided layer depth, microns	Sulfur content, %	Surface roughness, Ra, microns
Steel 20	0.02	10	0.05*-0.85**	0.7
	0.05	25	0.05-0.80	1.7
	0.11	35	0.05-0.70	2.9
	0.36	50	0.05-0.65	5.2
	0.55	75	0.06-0.60	6.2
Ductile cast iron VCH60	0.02	10	0.05-0.85	0.8
	0.05	30	0.04-0.70	1.9
	0.11	40	0.05-0.65	3.1
	0.36	50	0.05-0.60	5.8
	0.55	75	0.05-0.50	6.7

* - The sulfur content in the lower portion of the sulfided layer;
** - The sulfur content on the surface portion of the sulfided layer.

The result of research aimed at improving the sulphiding process for the steel and iron cast substrates by the EEL method, presented in conclusions.

5. Conclusions

1. In the course of the sulfur saturation (sulphiding) process of the surfaces of steel 20 and ductile cast iron by EEL method with the use of a special electrode-tool made of stainless steel 12H18N10T at the device of "Elitron -22A" model, with increasing discharge energy from 0.02 to 0.55 J, the sulphided layer depth increases from 10 to 75 microns, and the surface roughness Ra increases from 0.7-0.8 to 6.2-6.7 microns.

2. With deepening, the sulfur concentration is gradually reduced from 0.60 - 0.85 and 0.50 - 0.85% on the surface, respectively, for steel 20 and ductile cast iron, and also respectively, from about 0.05-0.06 and 0.04- 0.05% at the depth of 75 microns.

3. An amount of alloying elements of the electrode-tool made of steel 12X18H10T, i.e. chromium, nickel and titanium, gradually decreases with deepening the coating for steel 20, respectively, from 16, 9; 9; and 0.9% on the surface to 1.2 and 0.2% at the depth of 60 microns.

References

[1] Antoszewski, B. 2014 Influence of laser surface texturing on scuffing resistance of sliding pairs. *In Advanced Materials Research* (Trans Tech Publications) **874** pp 51-55

[2] Błasiak S, Kundera C 2012 A numerical analysis of the grooved surface effects on the thermal behavior of a non-contacting face seal *Procedia Engineering* **39** pp 315-326

[3] Martsynkovskyy V, Yurko V 2014 Solutions for Increasing the Bearing Capacity of Thrust Bearings *In Applied Mechanics and Materials* (Trans Tech Publications) **630** pp 208-219

[4] Yurko V, Martsynkovskyy V 2014 Influence of Changing the End Floating Seal Dynamic Characteristics on the Centrifugal Compressor Vibration State *In Applied Mechanics and Materials* (Trans Tech Publications) **630** pp 356-364

[5] Hinber A M, Ivanova A F 1982 Wear-Resistant and Anti-Friction Coatings (Moscow: Mechanical engineering) p 43

[6] Shipway P H, McCartney D G, Sudaprasert T 2005 Sliding wear behaviour of conventional and nanostructured HVOF sprayed WC–Co coatings *Wear* **259** (7) pp 820-827

[7] Andreev M A, Markova L V, Moyseychik A N, Kaleda V V. 2011 Proc. Int. Conf. on Polymer Composites and Tribology (Gomel: Belarus) pp 32-33

[8] Krasnov A P et al 2011 Proc. Int. Conf. on Polymer Composites and Tribology (Gomel: Belarus) pp 11-12

[9] Zozulya V D et al 1990 Reference Dictionary. Friction, Wear and Lubrication of Machine Parts ed I M Fedorchenko (Kiev: Naukova Dumka) p 203

[10] Denisov N E et al 2006 *Tribological Materials Science and Tribotehnology* (Penza: Publ. Penza State University) p 246.

[11] N I Lazarenko N I 1976 *Electric Spark Alloying of Metal Surfaces* (Moscow: Mechanical engineering) p 46

[12] Hitlevich A E et al 1985 *Electric Spark Alloying of Metal Surfaces* (Chisinau: Shtintsa) p196

[13] Scherbinsky A G 1961 Method for Sulfur Saturation of Metal Surfaces *AS139336 Reports the USSR Inventions* (Moscow: Science) **13** pp 2-3

Improvement of integrated technology for restoring surfaces of steel and iron parts

V Tarelnyk[1], V Martsynkovskyy[1], A Sarzhanov[1], A Pavlov[1], V Gerasimenko[1], B Sarzhanov[1]

[1] Sumy National Agrarian University, 160 G. Kondrateva Str., 40021, Sumy, Ukraine

E-mail: tarelnik@i.ua

Abstract. There have been proposed new technical solutions including operations of applying coatings onto worn surfaces of steel and cast iron parts by electro-erosive alloying method (EEA). In this case, the EEA coatings are applied in two stages. At the first stage, there is applied a layer on the conditions providing the greatest thickness and continuity of the surfaces obtained. At the second stage, the operation is performed by the same electrode and on the conditions of the discharge energy and productivity corresponding thereto, which provide for the formation of the surface having a roughness value being about 2-4 times higher than that at the previous stage. At least one layer of a metal polymer material (MPM) is applied on the surface formed by the above said EEA method, with this layer having been reinforced by at least one wire reinforcement layer before polymerization.

1. Introduction

At long-term operation of machines, wear and tear of their components is accompanied by reduction in operating characteristics thereof. The wear of the working surfaces of the machine parts often requires their complete replacement, which event results in increasing the cost of production. The most important problems of repair and technical maintenance are as follows: supporting of operability, restoration of the resource of machines and equipment, providing their high reliability and the possibility of efficient use. Increasing the wear resistance of repaired machine parts is one of the topical problems of maintenance and repair.

Solution of these problems includes improvement of repair quality by introducing modern methods for a repair process organization and also optimal technologies to provide for strengthening and restoring of the parts. Consequently, as a rule, the resource of the restored parts should be much higher, owing to the efficient recovery methods and improved properties of strengthened surfaces.

2. Analysis of major technical achievements and publications thereon

Modern repair technologies have been developed following numerous methods for restoring parts of machines and equipment. A significant number of technological methods for covering and a variety of application areas of coatings, a wide range of materials for those purposes make it impossible to take an easy way in finding out an objective solution for choosing the coating and the optimal technology for its application under condition of a competitive approach.

According to publication [1], the operability and resource of the restored parts make up, on the average, of 60 to 80% of the above mentioned indicators for the new ones. However, at present, there are known the technological methods (electromechanical, electrophysical ones, etc.), with the help of which it is possible not only to completely restore the primary resource of the parts but even to increase it.

The process of restoration of the parts allows to save a considerable amount of scarce materials, to extend their service life by of 2 to 3 times, to reduce the production of spare parts at manufacturing plants and to reduce the primary cost of repairing machines and equipment.

It is one of the processes for improving the quality of the surface layer and reducing the cost of repairing of a machine, which includes providing multiple restorations of the shapes of the parts with the use of metal coatings and ensuring their interchangeability. At present, there are many different technological methods to compensate for the worn-out layer of metal parts [1-4]. Those are the most common methods of them, as well as their advantages and disadvantages that are represented in Table 1.

As a result of the analysis of the table, it can be noted that the main shortcomings that negatively affect the final result or significantly increase the cost of repairs are: the presence of distortions and warpage; weak adhesion of the applied layer to the base; the availability of pores, cracks and slag inclusions; decreasing of fatigue strength; increasing of environmental hazard.

Table 1. Technologies for compensation of a worn-out metal layer of metal parts.

Method	Advantages	Disadvantages
Surface welding	Increasing hardness and wear resistance, the possibility of unlimitedly increasing the worn surface to restore the same.	The formation of cracks, high porosity, availability of slag inclusions, reduction of fatigue strength, warpage, increased environmental hazard.
Galvanic coating	Preserving the structure of the part, high wear resistance and hardness of the surface.	Low running-in ability and wetting with oil, reduction of fatigue strength, low adhesion, increased environmental hazard.
Metallizatio n	The part material mechanical properties are not changed, and the part is not subjected to warpage, the layer is characterized by its high wear resistance.	High porosity (up to 10%), reduced fatigue strength, low adhesion, and increased environmental hazard.
Plastic deformation	Increased hardness, reduced roughness, enhanced wear resistance.	Low productivity, possible deformation of the surface by 5-10 microns or more, and there can be uniform metal lapping of 0.03-0.3 mm thickness.
Electroerosive alloying.	Local surface treatment, namely, an alloying process can be carried out in separate areas sized of several millimeters and more without protecting the rest of the surface; there is provided a strong joint between a transferred and base materials; the absence of general heating of the whole part during processing thereof, the possibility of using as processed materials the following ones: pure metals, alloys, metal-ceramic compositions, refractory compounds; providing increased hardness, heat - , wear - and corrosion resistance; there is no need to specifically prepare a surface for conducting the above said alloying process.	Increase of roughness, appearance of tensile residual stresses in a surface layer, reduction of a fatigue strength value of a part.
Application of metal polymer materials	The possibility of unlimitedly increasing the worn surface to restore the same, the deformation characteristics being close to metal, high adhesion ability.	There is need in a special surface preparation, including the formation of surface roughness. Relatively low hardness.

Among the considered methods aimed at restoring pats, there is paid much attention to the method of electroerosive alloying (EEA) and the application of the metal polymer materials (MFM), which are environmentally safe and more and more often used in repair production.

While choosing a technological method of restoration, the maximum wear value, whereat the part becomes unsuitable for use, is of great importance. In general 85% of the machine parts become unsuitable for operation at wear that does not exceed $(0.2 \div 0.3) \times 10^{-3}$ m [5].

According to publication [6], the conducted studies have set that on aluminum alloys, there can be obtained a wear-resistant coating with the use of the EEA method performed by electrodes made of Al-Sn based alloy, which coating wear resistance is 5-6 times higher than that of tempered steel. A specific feature of this coating is the presence of tin oxide micro/nano fibers in its structure which fibers have high microhardness HV 1200 kg / mm^2.

To perform a coating of increased thickness that is up to 0.3×10^{-3} m, there has been used the technology of so-called barrier layers [7]. The essence of the technology is as follows: after applying of 3 to 4 layers of AO20-1 alloy, when the increment of the layer is suspended, there is applied a barrier layer, which insulates the applied coating of AO20-1 alloy from the next layer of the same composition and allows to go on increasing the thickness of the coating.

It should be noted that increasing such an EEA operating condition as a discharge energy value, as a rule, results in reducing the continuity of the formed coating because of an appearance of through pores [8].

To improve the quality of the surface layer restored by the EEA method, we have proposed a process wherein the coating performed by the EEA method is applied in stages, with the layer applied at the first stage under conditions to provide the greatest continuity and thickness of the coating. Then, using the same electrode, there is performed the EEA process under such conditions of discharge energy and productivity corresponding thereto under which, there is performed a surface having a roughness of approximately 2-4 times higher than that at the previous stage. In this case, the metal of the cathode (the part) is ejected at the impulse application points, i.e. there is occurred sputtering of the metal in the most protruding zones of the surface, and in their places, there are formed depressions of a newly created coating, the bottom portions of which are disposed at the level of the surface of the previous coating. As a result, there is occurred a minimum increase in the level of the surface roughness.

As a reserve for increasing the thickness of the restored layer, there can be the combined technologies such as the EEA method and the method for applying metal polymer materials (MPM). In this case, each of the above said technologies in no way diminishes the dignity of the other one, but as combined, they complement one another and their combination eliminates the shortcomings inherent in each technology separately.

The advantages of the integrated technology of EEA + MPM (EEA method and further applying of metal polymer materials) are obvious: surface continuity is 100%; roughness is much lower than at the EEA method, hardness is much higher than at the MPM method. Due to the opportunity to apply the EEA coatings using a large scale of electrode materials, it is possible to vary the mechanical, thermal, electrical and other properties of the working surfaces of the parts within a wide range. Penetrating of the metal polymer materials (MPM) into the depressions and the microroughnesses of the restorable parts excludes the possibility of formation of corrosion foci in these depressions. Wear resistance, reliability and durability of the restored parts are higher than when for those restored with the use of the individual technologies.

It should be noted that when applying the method of the EEA + MPM integrated technology, various variants of surface formation are possible. The EEA method can be used to vary the height of the microroughnesses, and their subsequent processing by a blade cutting method (treatment with a blade of a cutting tool) can provide different ratios between the areas made of the deposited metal and the metal polymer material [9].

Despite the undeniable advantages, the method of the EEA + MPM integrated technology has some drawbacks. First of all, this is a low hardness and strength of the formed surface layer, especially in those cases wherein the MPM layer is disposed above the layer of the coating applied by the EEA method. The main field of the method application is a restoration of parts in permanent joints (seats for bearings, half couplings, etc.).

Thus, the aim of the work is to improve the method of the EEA + MPM integrated technology for restoring the parts by improving the quality of the coating formed by the EEA method and increasing the hardness and strength of the formed surface layer by reinforcing the same.

3. Methods of the research

From the Technological Instructions of the EEA equipment suppliers, literature sources and work experience, it is known that pure metals (chromium, nickel, etc.), stainless steels, for example, 12H18N10T, bronze, metal-ceramic hard alloys of BK (VC) and TK (TC) groups are recommended as the electrode materials for restoring parts surfaces .

Using the unit of Elitron-52A model, when applying the EEA method, the samples of 20x10x8 and 15x15x8 mm sizes and made of steel 20 were coated with electrodes made of BrO10F1 tin bronze, T15K6 hard alloy and 12H18N10T stainless steel under different conditions of discharge energy. In addition, onto the surfaces of the samples made of high-strength cast iron of VCh-60 grade, there were applied the coatings of chromium, nickel and T15K6 hard alloy using the unit of Elitron-22A model.

In doing so, there were produced three series of samples: for series 1, the coating was applied in one stage, namely, one sample was coated on one condition; for series 2, the coatings were applied in two stages, namely, the first coating was applied on the condition providing the greatest continuity and thickness of the coating, and the second coating was applied on the coarser condition providing a roughness of 2-4 times greater than the previous one; for series 3, the coatings were applied in two stages in the sequence that is the opposite one to series 2.

The thickness of the coating was measured with a micrometer, and the roughness of the surface was indicated on the profilograph-profilometer of 201 model of Caliber plant production.

4. Research results

4.1. Applying the tin bronze coatings

The qualitative parameters of bronze coatings, having been formed in accordance with the three series, are summarized in Table 2. As can be seen from the table, at applying the coating of tin bronze according to series 1, the maximum thickness of the coating (0.1 mm), at 100% continuity, is achieved at the discharge energy of Wp = 0.20 J. The surface roughness is Rz = 21 μm. The subsequent increase in the discharge energy up to Wp = 0.35 J causes, together with the increase in the thickness of the layer up to 0.43 mm, the sharp increase in the surface roughness up to Rz = 59 μm and the reduction in its continuity up to 80%. The successive increase in the discharge energy is accompanied by an even greater increase in the surface roughness and a decrease in the continuity. The greatest thickness of the layer, which is 0.92 mm, is formed at Wp = 0.90 J. At the same time, the surface roughness and the coating continuity are Rz = 98 μm and 60%, respectively. The further increase in discharge energy results in a sharp decrease in the quality of the coating (low continuity, high roughness, burns) and in the electrode burnout.

According to series 2, the first layer was performed of tin bronze, had the continuity of 100% and the thickness of 0.1 mm. The coating was applied on the condition of Wp = 0.20 J. Then there was performed the second layer at Wp = 0.35 and 0.55 J, where the roughness was about 3 and 4 times, respectively, more (Rz = 59 and 82 μm). The total thickness of the layer was, respectively, 0.65 and 0.81 mm, roughness Rz = 47 and 58 μm at 100% continuity (Table 2). The further increase in the discharge energy at applying the second layer up to Wp = 0.90 J, with a slight increase in the thickness of the layer, results in a sharp increase in the surface roughness from Rz = 58 to Rz = 81 μm.

The results of the quality parameters of the coatings, which were made of BrO10F1 material on steel 20 in stages according to series 3, were significantly worse than the same for the coatings of series 2 (the continuity of 65-75%, the roughness of Rz = 85-92 μm) with slightly different thickness of the applied layer.

IOP Conf. Series: Materials Science and Engineering **233** (2017) 012050 doi:10.1088/1757-899X/233/1/012050

Table 2. Dependence of the qualitative parameters of the tin bronze coatings, applied by the EEA method onto steel 20, on the discharge energy value.

Discharge Energy, W_p	Productivity, cm^2/min	Layer Thickness, mm	Roughness, Rz, μm	Continuity, %	Surface Image
			Series 1		
0.04	0.4	0.05	10	100	
0.11	1.5	0.06	12	100	
0.20	1.6	0.10	21	100	
0.35	1.7	0.43	59	80	
0.55	2.5	0.70	82	70	
0.90	3.4	0.92	98	60	
			Series 2		
0.04 + 0.35	Stage 1 0.4 Stage 2 1.7	0.55	57	90	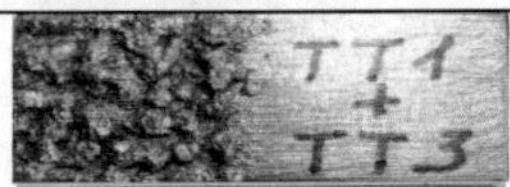
0.2 + 0. 35	Stage 1 1.6 Stage 2 1.7	0.65	47	100	
0.2 + 0.55	Stage 1 1.6 Stage 2 2.5	0.81	58	100	
0.2 + 0.90	Stage 1 1.6 Stage 2 3.4	0.95	81	100	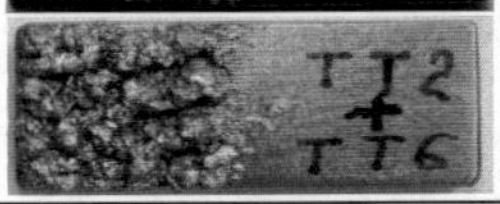
			Series 3		
0.90 + 0.04	Stage 1 3.4 Stage 2 0.4	0.87	92	65	
0.90 + 0.20	Stage 1 3.4 Stage 2 1.6	0.74	85	75	

Thus, the restoration of the steel parts using the EEA method performed with the electrode- tool made of tin bronze of BrO10F1 grade should be carried out in two stages, in this case, at stage 1, the process of forming the coating is carried out at the discharge energy of Wp = 0.20 J, and at stage 2 - at Wp = 0.55 J. As a result, there is obtained the coating having the thickness of 0.81 mm, the continuity of 100% and the roughness of Rz = 58 μm.

IOP Conf. Series: Materials Science and Engineering **233** (2017) 012050 doi:10.1088/1757-899X/233/1/012050

4.2. Applying the coating of T15K6 carbide (Sintered hard alloy)

As it can be seen from Table 3, there is obtained the coating of 100% continuity and having the layer thickness of 0.12 mm at the discharge energy of Wp = 0.55 J. The surface roughness is Rz = 21 µm. The further increase in the discharge energy results in increasing the thickness of the layer up to 0.19 mm and in significant decreasing the quality of the coating (the continuity of 85% and the roughness of Rz = 65 µm).

According to the proposed method, when forming the coating in stages with initially using the discharge energy of Wp = 0.55 J, and then Wp = 0.90 J, the thickness of the layer is 0.20 mm, the continuity is of 100% and the roughness is Rz = 37 µm. Applying the coating in the reverse order, that is first at Wp = 0.90 J and then at Wp = 0.55 J results in decreasing the continuity up to 90% and increasing the roughness up to Rz = 54 µm.

4.3. Applying the coating of 12H18N10T stainless steel

In Table 3, there are shown the results of the qualitative parameters of the coatings at EEA of steel 20 by the electrode- tool made of stainless steel 12H18N10T. As it can be seen from the table, the maximum coating continuity of 95% and the layer thickness of 0.08 mm are achieved at the discharge energy Wp = 0.35 J. The surface roughness is Rz = 11 µm. Further increasing in the discharge energy results in increasing the thickness of the layer up to 0.14 mm and significantly decreasing the quality of the coating (the continuity of 80% and the roughness of Rz = 31 µm).

According to the proposed method, when forming the coating in stages with initially using the discharge energy of Wp = 0.35 J, and then Wp = 0.55 J, the thickness of the layer is about 0.12 mm, the continuity is approaching 100% and the roughness is Rz = 17 µm. Applying the coating in the reverse order results in decreasing the continuity up to 80% and increasing the roughness up to Rz = 34 µm.

Table 3. Dependence of the quality parameters of the coatings made of T15K6 sintered hard alloy and 12Hh18N10T steel, applied by the EEA method on steel 20.

Discharge Energy, W_p	Productivity, cm^2/min	Layer Thickness, mm	Roughness, Rz, µm	Continuity, %	Surface Image
		T15K6 carbide (Sintered hard alloy)			
*0.55	2.5	0.12	21	100	
*0.90	3.4	0.19	65	85	
**0.55 + 0.90	Stage 1 2.5 Stage 2 3.4	0.20	37	100	
***0.90 + 0.55	Stage 1 3.4 Stage 2 2.5	0.14	54	90	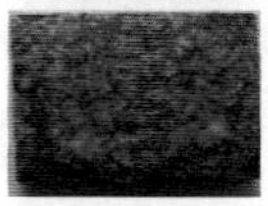
		12H18N10T stainless steel			
*0.35	1.7	0.08	11	95	
*0.55	2.5	0.14	31	80	

Discharge Energy, W_p	Productivity, cm^2/min	Layer Thickness, mm	Roughness, Rz, μm	Continuity, %	Surface Image
**0.35 + 0.55	Stage 1 1.7 Stage 2 2.5	0.12	17	100	
***0.55 + 0.35	Stage 1 2.5 Stage 2 1.7	0.8	34	80	

* - the coating was applied according to series 1;
** - the coating was applied according to series 2;
*** the coating was applied according to series 3.

Thus, according to the proposed method, the restoration of the steel parts by the EEA method with using the electrode tools made of T15K6 hard alloy and 12H18N10T stainless steel, is most expedient to carry out in two stages, (see Table 3).

It should be noted that the proposed method remains valid for the parts made of cast iron. In Table 4, there are presented the results of qualitative parameters of the surface layers being formed at EEA of the high-strength cast iron of VCh-60 grade with the electrodes made of chromium, nickel and T15K66 hard alloy at the unit of Elitron 22A model.

Table 4. Results of qualitative parameters of the formed surface layers at EEA of high-strength cast iron of the VCH-60 grade.

Electrode Material	Discharge Energy, W_p	Roughness, Rz, μm	Layer Thickness, mm	Continuity, %	Productivity, cm^2/min
EEA of the high-strength cast iron of VCh-60 grade at various discharge energy values					
Chromium	0.13	15.0	40	100	1.5
	0.27	17.5	50	100	1.7
	0.39	20.4	55	95	2.0
	0.55	35.0	60	90	2.5
Nickel	0.13	13.0	50	100	1.5
	0.27	15.5	55	100	1.7
	0.39	22.0	60	95	2.0
	0.55	33.5	70	85	2.5
T15K6	0.13	21.0	50	100	1.5
	0.27	26.0	60	95	1.7
	0.39	34.0	80	90	2.0
	0.55	42.5	100	80	2.5
EEA of the high-strength cast iron of VCh-60 grade according to the proposed method					
Chromium	Stage 1 0.27 Stage 2 0.55	30.2	80	100	Stage1 1.7 Stage 2 2.5
Chromium	Stage 1 0.55 Stage 2 0.27	24.0	55	90	Stage1 2.5 Stage 2 1.7
Nickel	Stage 1 0.27 Stage 2 0.55	29.5	95	100	Stage 1 1.7 Stage 2 2.5
Nickel	Stage 1 0.55 Stage 2 0.27	26.1	60	85	Stage 1 2.5 Stage 2 1.7
T15K6	Stage 1 0.13 Stage 2 0.55	43.3	110	100	Stage 1 1.5 Stage 2 2.5
T15K6	Stage 1 0.13 Stage 2 0.55	38.5	80	100	Stage 1 2.5 Stage 2 1.5

As it can be seen from the table, with increasing the discharge energy for all the electrode materials, the thickness values of the applied coatings and the surface roughness values increase, and the continuity indices decrease. Moreover, the continuity of 100% and the greatest thickness value of the layer are kept for chromium and nickel at $W_p = 0.27$ J, and for T15K6 hard alloy at $W_p = 0.13$ J.

Thus, when restoring the parts made of VHC-60 cast iron applying the EEL method with the use of the electrodes of chromium, nickel and hard alloy, the best quality parameters have been found out for the samples having the coatings formed in two stages, according to the above said proposed method.

Below, there is a description of the methods for reinforcing the MPM materials with wire. Depending on the required qualitative and geometric parameters of the surface to be restored, the process of reinforcement is carried out in various ways.

4.4. Restoring worn surfaces of parts of rotation body type

It is well known that the main cause of a machine part failure is not its breakage, but a wear of its surface layer. Sometimes there is appeared a need in restoring outer surfaces made of soft antifriction metals of the parts of rotation body type, for example, gear supporting fingers after destruction of a Babbitt layer.

In this case, layer 3 of a coating made of any soft antifriction metal (copper, tin, silver, tin bronze, etc.) is applied onto the worn-out surface of part 1 (Figure 1 (a)) by the EEL method. In doing so, between the applied metal and the part, there is formed transition layer 2, which is a mutual diffusion penetration of anode and cathode elements. Coatings can be applied with varying the discharge energy in the range of 0.036 to 6.8 J. Increasing the discharge energy results in increasing both the thickness of the applied coating and also the surface roughness obtained. Thus, the thickness of the layer can vary depending on the nature of the interaction between the anode and the cathode. There may be used either such units as of Elitron 52-A type equipped with a hand-held vibrator, or such units as mechanized devices of Elitron-347 or EIL-9 types equipped with multi-electrode heads. In the first case, the layer thickness may vary of 0.01 to 0.25 mm, and in the second one, it may vary of 0.05 to 2.0 mm, and also the height of the microroughness (Rz) may vary accordingly of 8.5 to 155.8 μm and of 20 to 200 μm and more. After that, metal-polymer material 4 is applied onto the surface processed by the EEA method.

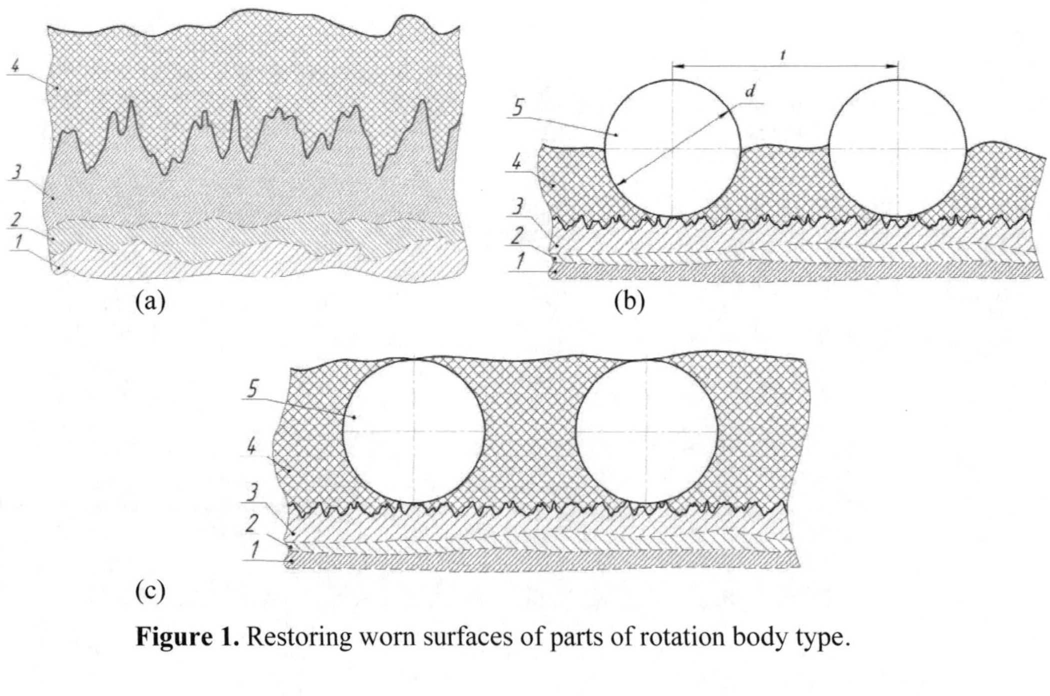

Figure 1. Restoring worn surfaces of parts of rotation body type.

Applying a material is one of the operations determining both the quality of the formed adhesion bonds and the service life of the restored part. The layer of the metal polymer material is carefully rubbed with a spatula or any other proper device into the surface of the part to be restored. On rubbing, the penetration of the metal polymer material (MPM) into the depressions and the microroughnesses of the restorable part, on the one hand, provide the adhesion improvement, and on the other hand, it excludes the possibility of the formation of corrosion foci in the depressions not filled with MPM.

Guided by the basic concept of applying MPM, which consists in the fact that its working layer should not be less than 1-1.5 mm, the pitch of the wire winding (t) on the shaft is calculated as follows:

$$t = d + 1 - 1.5\ mm$$

where: d is the diameter of the wire [3].

After winding the wire, it is necessary to continue applying the MPM until the applied layer has completely covered it (Figure 1 (c)).

In this case, to restore the worn surfaces, one can use a wire made of copper, tin, Babbitt, silver, etc.

In certain cases, there is a need to restore worn bearing necks and journal collars (the fitting surfaces), which have a hardness of the order of 35 to 40 units of HRC.

In this case, after applying a coating of a hard and wear-resistant metal by the EEA method followed by a layer of MPM on a worn surface, a heat-treated spring is put on the formed layer, with interference fit. In this case, the pitch of the spring winding should be at least of 1.0 to 1.5 mm. Having put the spring on, it is necessary to continue applying the MPM until it would completely cover the coils of the spring (Figure 1 (c)). As a material of the spring, there can be used the steel of 65G, 9HV2S grades, BrB2 beryllium bronze, etc.

The solidified metal polymer material can be treated by any of known methods, including grinding or processing with a blade cutting tool.

4.5. Restoring worn flat and curved surfaces of parts

At restoring flat and curved surfaces made of soft and antifriction metals, after applying a coating of a soft antifriction metal by the EEL method followed with applying a layer of MPM on the machined worn surface (1), in compliance with the technology described above (Figure 1 (a)), a grid is laid on the layer formed of the antifriction metal. The grid cell size is at least 1.0-1.5 X 1.0-1.5 mm (Figure 2).

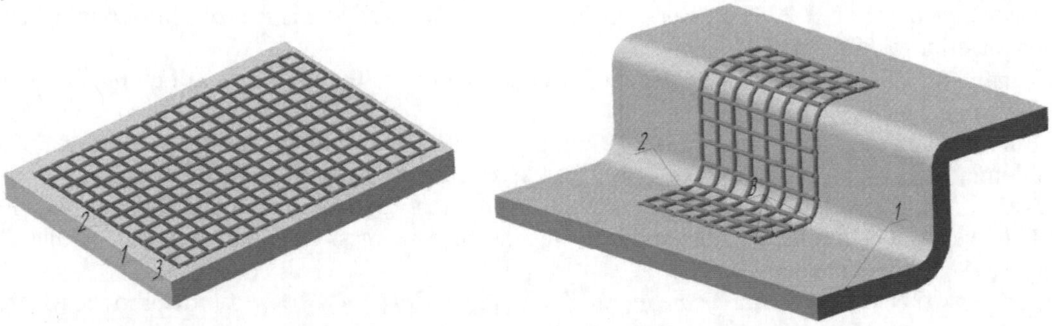

(a) (b)

Figure 2. Restoring of worn flat (a) and curved (b) surfaces.

The grid can be attached outside the surface to be restored by any known method, for example, by contact welding (3). Having set the grid, it is necessary to continue applying the MPM until it would completely cover it (Figure 1 (c)). In the case of considerable wear, overlapping of the grid can be repeated as many times as necessary.

The solidified metal polymer material can be treated by any of known methods, including grinding or processing with a blade cutting tool.

The worn flat and curved surfaces of the parts made of hard and wear-resistant metal are restored in compliance with the above described technology, except that the applied grid is made of heat-treated hard wire.

5. Conclusions

1. There is proposed a new method for restoring the worn surfaces of steel and cast iron parts by the EEA method. The method is characterized in that a coating is applied in two stages. At stage 1, there is applied a layer under conditions providing the largest surface thickness at the greatest continuity. Then at stage 2, onto the obtained surface, there is applied a layer of the coating using the same electrode and the EEL method at such a value of discharge energy and productivity corresponding thereto, at which there is formed a surface having a roughness of approximately 2-4 times higher than at the previous stage. As a result, at a relatively acceptable thickness of the restored layer, the most rational value of roughness is formed and the surface continuity rises up to 100%.

2. As a reserve for increasing the thickness of the restored layer, there can be the combined technologies such as the EEA method and the method for applying metal polymer materials (MPM). In this case, each of the above said technologies in no way diminishes the dignity of the other one, but as combined, they complement one another and their combination eliminates the shortcomings inherent in each separate technology.

3. It is proposed a new method for restoring machine metal parts, according to which onto the surface formed by the EEA method there is applied a layer of MPM, having been reinforced before its polymerization depending on the purpose and geometric features of the surface to be restored, with at least one layer of wire or wire connected to create a grid. As a result, there is formed a surface layer, with its quality, wear resistance, reliability and durability being higher than those obtained when using the EEL method and the method of applying MPM each separately for the worn surfaces restorations.

References

[1] Chabanny V Ya 2007 Repair of Automobiles: *Handbook* (Kirovograd: Kirovograd District Printery) p 720

[2] Ivanov V P 2007 Technology and Equipment for Restoring of Machine Parts: *Handbook* (Minsk: Tehnoperspektiva) p 458

[3] Ishchenko A A 2007 Technological Bases of Restoration of Industrial Equipment with Modern Polymeric Materials (Mariupol: PSTU) p 250

[4] Panteleenko F I et al 2003 Restoration of Machine Parts: *Reference Book* (Moscow: Mechanical engineering) p 672

[5] Kanarchuk V E et al 1995 Restoration of Automobile Parts. Technology and Equipment (Moscow: Transport) p 303

[6] Yurchenko E V, Yurchenko V I , Dikusar A I (2013) Nanostructuring of Surface Made of Aluminum Alloys on Conditions of Electroerosive Alloying *Nanoengineering* (2) pp 3-9

[7] Ivanov V I (2013) . Increasing Thickness of Electroerosive Coatings. Part 1. Reasons for Limiting Thickness of Coatings. Part 2. Methods for Increasing Thickness of Electroerosive Coatings *Papers of GOSNITI* pp.429-435, 150-456

[8] Lazarenko N I 1976 *Electric Spark Alloying of Metal Surfaces*, (Moscow: Mechanical engineering) p 46

[9] Martsinkovskyy V S Tarelnyk V B Pavlov O G Ischenko A O (2014) The method for Restoration of Worn Surfaces of Metal Parts (Variants) *Patent for Invention UA 104664* Bul 4 p 3

IOP Conf. Series: Materials Science and Engineering **233** (2017) 012051 doi:10.1088/1757-899X/233/1/012051

Definition of criteria for estimating alternative technologies of increasing quality of rotor shaft neck by electroerosive alloying and surface plastic deformation methods

V Martsynkovskyy[1], G Kirik[2], V Tarelnyk[3], P Zharkov[2], Ie Konoplianchenko[3], M Dovzhyk[3]

[1] TRIZ Ltd., Mashinostroiteley 1 St, 40020, Sumy, Ukraine
[2] Concern NICMAS, 6, Kurskiy ave., Sumy, 40020, Ukraine
[3] Sumy National Agrarian University, 160 G. Kondrateva Str., 40021, Sumy, Ukraine

E-mail: mbc@triz-ltd.com

Abstract. There are represented the results of influence of the surface plastic deformation (SPD) methods, namely, diamond smoothing (DS) and ball-rolling surface roughness generation (BSRG) ones on the qualitative parameters (residual stresses, fatigue strength and wear resistance values) of the steel substrate surface layers formed by the electroerosive alloying (EEA) method. There are proposed the most rational methods of deformation and also the composition for electroerosive coatings providing the presence of the favorable residual compressive stresses in the surface layer, increasing fatigue strength and wear resistance values. There are stated the criteria for estimating the alternative variants of the combined technologies and choosing the most rational ones thereof.

1. The problem statement

In the practice of mechanical engineering, Coatings, which can be formed in various ways, are of no small importance. Their main purpose is to improve the qualitative parameters of the surface layers of machine parts, namely, increasing microhardness, wear resistance, and also provide an opportunity for restoring a part surface area having been worn out, etc.

One of the promising technologies for improving a part surface quality is a method of electroerosive alloying (EEA). It is a universal method aimed at the followings:

- Increasing hardness, corrosion resistance, wears and heat resistance of a part surface;
- Decreasing the tendency of a part surface to frictional seizure;
- Restoring the dimensions of machine parts and mechanisms;
- Carrying out micrometallurgical processes on a part surface being treated to form necessary chemical compounds thereon;
- Creating a transition layer of certain roughness on a part working surface, etc. [1].

Despite the undeniable advantages, the EEA method has a number of drawbacks such as an increase in the surface roughness, an appearance of tensile residual stresses in a part surface layer, and a decrease in a part fatigue strength, which are often the limits for applying the above said method to a wider range of machine parts.

To increase the performance characteristics of the parts such as durability, wear resistance, the methods of surface plastic deformation (SPD) have been being increasingly used. Obviously, the use of the SPD methods combined with the EEA methods for processing a part surface layer is of particular interest since the above mentioned technologies complement each other [2-4].

In [5], there are represented the results of the investigations of the SPD methods influences on the values of roughness and microhardness for the coatings formed by an EEA method using the electrodes made of soft and hard metals and also their combinations, namely, combined electroerosive coatings (CEC).

In this paper, there are represented the results of the SPD influence on the electroerosive coating qualitative parameters such as residual stress, fatigue strength and wear resistance values, as well as the development of a methodology for determining the most rational alternative technologies to improve the quality of rotor shaft necks. The work carried out in this direction is of current interest.

The aim of the work is to improve the quality of the shaft surfaces of the rotors for dynamic equipment by determining the most rational deformation methods and the proper compositions for the EEA coatings in the course of manufacturing and repairing thereof, as well as setting up the criteria for estimating alternative variants of combined technologies and selecting the most rational one to achieve the above said aim.

2. Research Methodology

To study the SPD influence on the residual stresses of the surface layers having the coatings formed by the EEL method, there were made samples of steel 45 (∅ = 50 mm, L = 200-250 mm). The entire surface of the samples was processed with the use of the EEA method performed by applying hard wear resistant chromium metal, soft antifriction copper metal, and also the combination of the metals to provide the CEC of copper + chromium. Then each sample was subjected to the SPD process carried out using the methods realized by diamond smoothing (DS) and ball-rolling surface roughness generation (BSRG). Each operating condition was matched by a track on a sample of 10 to 15 mm wide.

The DS process was performed on a lathe with the help of a spring-hydraulic device equipped with ASPK-3 smoothers having the vertices radii of R = 3…4 mm depending on the value of hardness of the coating at speed V = 40 ... 80 m / min, smoothing force P = 60 ... 350 N and feed rate S = 0.02 ... 0.07 mm / rev. The BSRG process was carried out on a lathe with a spring-rod device equipped with the ball heads of ∅ 10 and ∅ 19 mm. The maximum breaking-in force for the ∅ 10 ball is 1200 N, and for the ∅ 19 mm ball is 3000 N. Feed rate S = 0,05 ... 0,21 mm / rev, speed V = 30 m / min. Lubrication was made with industrial oil.

The investigations of the axial residual stresses, which are the most dangerous stresses for shafts, were performed using the prismatic specimens made of steel 45 and having dimensions of 70 x 5 x 2 mm according to the method of I.A. Birger providing a stratified process of electropolishing stressed layers on the unit of the Pion type.

Based on the results of the investigation, there were constructed the graphs of the residual stresses depth distributions for the samples of various series.

At least three samples were examined within each series. In the course of the residual stress investigation, the state of a surface layer was determined for the following series:

EEA Cu; EEA Cu + EEA Cr; EEA Cu + SPD; EEA Cu + EEA Cr + SPD.

It should be noted that EEA processes performed by copper and chromium were produced on a unit equipped with a hand-held vibrator UILV-8, respectively, at the discharge energy of $W_u = 0.01$J and $W_u = 0.2$ J.

The copper coating SPD process was produced by the ball-rolling surface roughness generation process with the use of the ball head of ∅ 19 mm at the force of 500 N, and the SPD process of the CEC obtained by the Cu + Cr EEA method was produced by the BSRG process with the use of the ball head of ∅ 19 mm at the force of 1500 N.

The fatigue tests were carried out based on $1x10^6$ cycles on the UP-50 machine. To determine the fatigue strength, there were manufactured the 450 mm long samples made of steel 45 each and having the working diameters of 50 mm.

The fatigue strength was determined with an accuracy of 10 MPa for the same series of the samples as those having been used to determine the residual stresses.

The wear resistance values of the coatings were determined on the friction machine of SMC-2 type according to a disk-pad scheme. The disk was a round sample having the outer diameter ∅ 50 mm and the thickness of 10 mm made of 40HN steel. Babbitt B-83 was used as a material for a counterbody (the pad). With applying the EEA method, the surface of the disk was coated with chromium, tungsten, and hard alloy VK8 and CEC: EEA Cu + EEA W + BSRG; EEA Cu + EEA VK8 + BSRG, and EEA VK8 + BSRG + EEA Cu + EEA VK8 + BSRG. The lubrication of the friction surfaces was carried out under conditions when the lower part of the round sample was arranged in a tank filled with turbine oil T-22. The sliding speed was 0.78 m / s, the specific pressure was 5.0 MPa, the load on the pad was 1000 N, the test duration was 8 h, which corresponded to the friction path of 22.5 km. Before the tests, the counterbody made of Babbitt was bored on a lathe in a special device until the friction surface had coincided with a diameter tolerance corresponding to a sliding fit. The surface roughness of the counterbody was Ra = 1.6 μm. Then the samples were set on the friction machine and had been running-in for an hour under the specific load of 2.0 MPa. The process of running-in was deemed to be over when the portion of the treated surface was at least 95% of the nominal one.

To measure a linear wear, onto the working surface of the tested round sample on the Vickers hardness tester, there were made four grooves on diametrically opposite sides. To determine the integral wear of the entire surface, before the test, the sample was weighed with an accuracy of 0.0001 g on the analytical scales of the VLA-200 type.

The linear wear was determined with the use of the method of artificial basis by the depth difference of the prints measured before and after the experiment, and the integral (weight) wear was determined as the weight difference of the sample before and after testing.

In addition, the friction torque was measured in the course of the operation. At the same time, there was monitored the sliding speed and the load in the friction zone.

3. Results and discussion

The most dangerous indices for the fatigue strength of the alloyed shafts are the axial residual tensile stresses, which increase the amplitude of the actual stresses and contribute to the reduction of the durability of the parts.

Table 1 shows the results of measuring the axial residual stresses in the surface layers of steel 45 having the electroerosive coating of copper and CEC of copper + chromium, as well as the SPD effects on the magnitude and sign of these stresses.

Processing steel 45 with the use of the EEA method performed by copper (electrode) results in the formation of the tensile stress with the value of the depth of occurrence up to 0.1 mm in the surface layer and having its maximum value up to 170 MPa on the surface. Applying the CEC of copper + chromium onto steel 45 results in increasing the values both of the stress and the depth of occurrence, respectively, up to 210 MPa and 0.15 mm.

As a result of the BSRG strengthening of the samples after the EEA process by Cu and Cu + Cr, the deformation curves significantly change, because the deformation values have negative sign, which fact determines the presence of the favorable compressive stresses in the surface layer.

Table 1. The results of measuring the value, sign and depth of the residual stresses in the surface layer of steel 45 after the EEA and SPD processes.

Strengthening	Residual Stresses, σ, MPa	Depth Occurrence, h, mm	Roughness, Ra, μm
	EEA		
EEA by Cr	250	0.2	3.5 – 4.6
EEA by Cu	170	0.1	0.6 – 1.0
EEA by Cu + Cr	210	0.15	0.6 – 0.7
	EEA + SPD		
EEA by Cr + SPD	- 520	0.9	0.5
SPD + EEA by Cr + SPD	- 550	0.8	0.5
EEA by Cu + SPD	- 200	0.2	0.1
EEA by Cu + Cr + SPD	-350	0.4	0.1

The residual stresses occurring in the surface layers of the parts are inextricably linked to such an operational characteristic as fatigue strength.

When testing the field shafts with the CEC of Cu + Cr, it was found that, as a result of using the EEA method, the fatigue strength was 1.5 times decreased (from 395 to 255 MPa) in comparison with the uncoated samples, but it was 1.5 times higher than for the samples, which were only alloyed with chromium. The ball-running (BSRG) process of the CEC of Cu + Cr increased their fatigue strength by 16-20% in comparison with their basic version, namely, the uncoated samples (Figure 1)

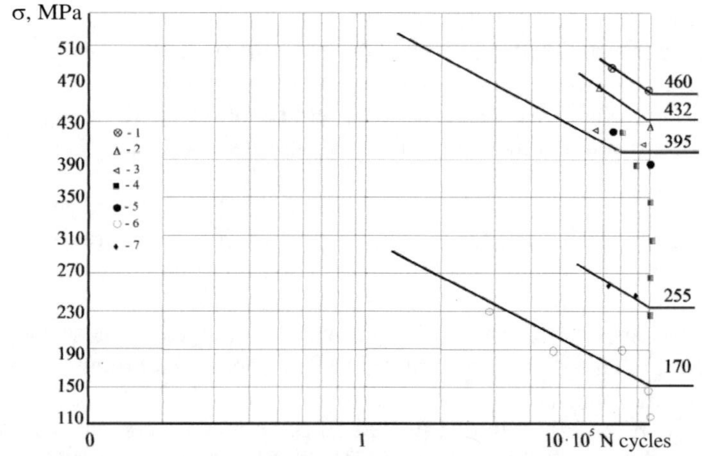

Figure 1. The fatigue strength values for the field samples made of steel 45 after EEA and SPD methods: 1. The EEA method by (Cu + Cr) + BSRG; 2. The EEA method by Cr + DS; 3. The EEA method by Cu + BSRG; 4. Without strengthening; 5. The EEA method by Cr + by Cr + DS; 3. The EEA method by Cu + BSRG; 6. The EEA method by Cr; 7 The EEA method by (Cu + Cr).

Increasing the fatigue strength values for the field shafts through the application of the CEC and the subsequent SPD processes is primarily due to decreasing the levels of the residual stresses thereof. In addition, increasing the fatigue strength is facilitated by the structure of the CEC coatings. Thus, having arisen on the surface of the shaft and growing down, a microcrack develops deeper and "bumps" into the CEC soft component and then it temporarily "damps" (relaxes), thereby increasing the fatigue strength of the shaft.

The analysis of Figure 1 shows that the fatigue strength of the field samples coated with a soft antifriction metal copper and subsequently processed using the SPD method is at the level of the uncoated samples, therefore, if necessary, the above said complex of strengthening technologies can be recommended for practical application.

Since the areas of the destructions of all the samples strengthened by the SPD method are located outside the coatings, increasing the fatigue strength values are even greater.

The wear results for the samples strengthened by the EEA + SPD methods and those without strengthening were compared with the results of the analysis of the samples strengthened by the method of pneumatic and centrifugal processing with the use of a ball ∅12, 7 mm and by the method of breaking – in with a roller ∅76 mm that were obtained at the Lubrication and Wear-Resistant Coating Laboratory, JSC "NPAO" VNIIcompressormash, Sumy, Ukraine [6].

It should be noted that after pneumatic and centrifugal processing with the use of a ball ∅12,7 mm and breaking – in with a roller ∅76 mm the microhardness of the surface layer was 270 and 305 HV, respectively.

In compliance with the obtained data, there were plotted the graphs demonstrating the friction coefficients changes, depending on the type of strengthening method and the test period of time (Figure 2). The results of measuring both the linear (Δh) and integral (Δm) wear rates of the samples are listed in Table 2.

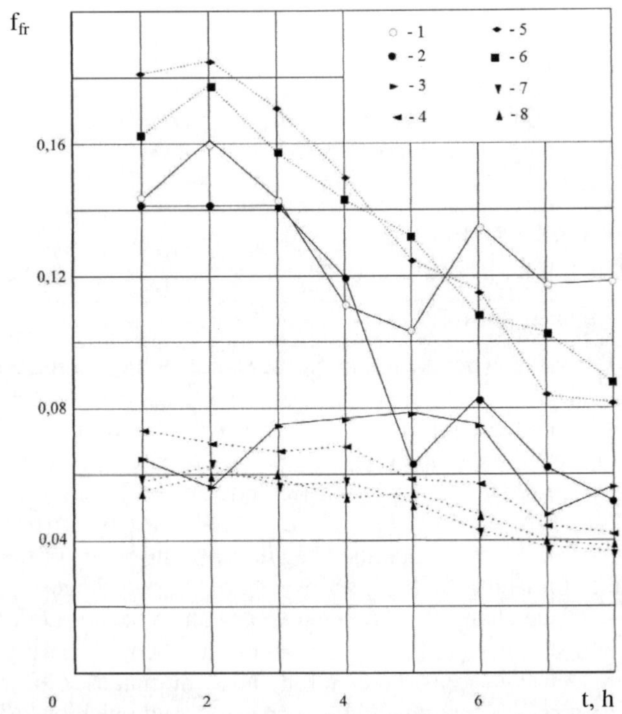

Figure 2. Changing the friction coefficient of steel 40XH on Babbitt depending on the type of strengthening and the interaction period time at: 1 - grinding; 2- pneumatic and centrifugal strengthening; 3 – breaking-in with a roller; 4 – The EEA method by Cr + Diamond Smoothing (DS); 5 – The EEA method by W + DS; 6 –The EEA method by VK8 + DS; 7 - Cu + the EEA method by VK8 + BSRG; 8 – The EEAA method by VK8 + BSRG + the EEA method by Cu + EEA method by VK8 + BSRG.

IOP Conf. Series: Materials Science and Engineering **233** (2017) 012051 doi:10.1088/1757-899X/233/1/012051

Table 2. Wear of strengthened steel 40HN paired with Babbitt B-83.

Strengthening	Wear	
	Weight wear x10^{3}, kg	Linear wear, μm
No strengthening	0.079	6.516
	0.081	6.501
	0.077	6.509
Pneumatic and centrifugal strengthening with the ball of ∅12,7	*	3.0**
Breaking-in with the roller of ∅76,0	*	2.0**
The EEA method by Cr + DS	0.023	2.043
	0.027	2.067
	0.022	2.003
The EEA method by W + DS	0.0039	2.507
	0.042	2.605
	0.040	2.489
The EEA method by VK8 + DS	0.035	2.322
	0.033	2.420
	0.034	2.303
The EEA method by (Cu + W) + BSRG	0.019	1.255
	0.017	1.204
	0.018	1.311
The EEA method by (Cu + VK8) + BSRG.	0.016	1.017
	0.015	1.104
	0.014	1.005
The EEA method by VK8 + BSRG + the EEA method by (Cu + VK8) + BSRG.	0.012	0.821
	0.010	0.826
	0.012	0.819

* - no data; ** - the data are taken from [6].

As a result of the SPD treatment (pneumatic and centrifugal strengthening, breaking-in by a roller), the coefficient of friction of the unstrengthened samples made of steel 40XH decreases by 1.5 - 2.0 times. With the combined strengthening process of EEA + SPD, the friction coefficient of the chrome coating is minimal, and for the EEA method by the hard alloy of VK8 it is maximal.

The use of copper as a pre-coating sharply reduces not only the roughness, but also the friction coefficient of the CEC being formed. Thus, with subsequent alloying by a hard alloy VK8, the copper, which was applied as a first layer, melts, fills the irregularities and pores of the coating, and covers it with a thin copper film that ultimately results in decreasing of the coefficient of friction.

Fig. 3 shows the nature of the change in the linear wear values of the samples made of 40HN steel strengthened by the method of EEA + SPD under condition when the hard alloy of VK8 with and without a sublayer of copper was used as the electrode alloying material. In the first approximation, the wear resistance curves resemble exponentially increasing relationships. All the samples initially expose more intensive wear, which after two hours of abrasion slows down and becomes more stable. The best wear resistance is possessed by the samples made of 40HN steel strengthened in the following sequence: the EEA method by VK8 + BSRG + the EEA method by Cu + the EEA method by VK8 + BSRG. In this case, the first and the last layers of the hard alloy of VK8 were applied at Wu = 0.2 J, and the intermediate (copper) layer was applied at Wu = 0.08 J. The BSRG process was produced after applying the first layer and the last one.

It should be noted that when a precoating of a soft antifriction material is used, it is most expedient to apply as the SPD method the BSRG or DS processes performed by the smoothing devices with R ≥

IOP Conf. Series: Materials Science and Engineering **233** (2017) 012051 doi:10.1088/1757-899X/233/1/012051

3 mm, otherwise the DS process results in "enveloping" of a soft material (in this case copper) on a diamond indenter, and, as a result, reducing the quality of the layer being formed.

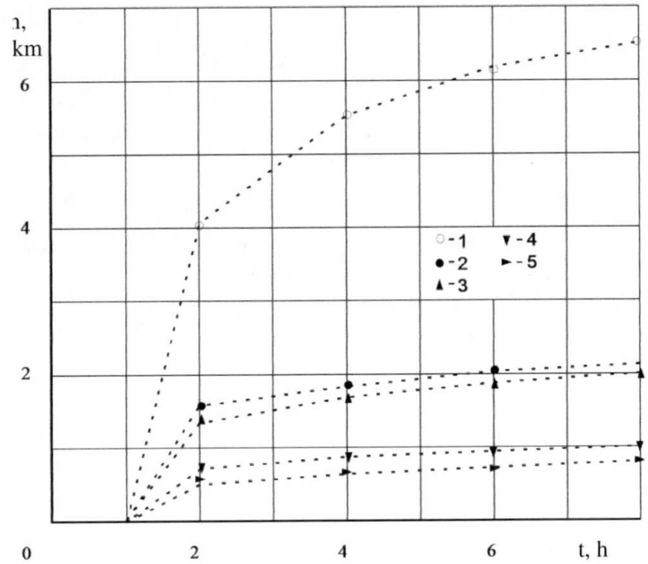

Figure 3. The nature of wear of steel 40HN working in tandem with Babbitt B-83: 1 - grinding; 2 - the EEA method by VK8 + DS; 3 - the EEA method by Cr + DS; 4 - the EEA method by (Cu + VK8) + BSRG; 5 - the EEA method by VK8 + BSRG + the EEA method by (Cu + VK8) + BSRG.

Earlier [7], it was established that to determine the wear equation constants for both the weight equation (1) and the linear one (2), it is necessary to approximate the experimental wear data for a given pair and the nature of the interaction of rubbing surfaces.

$$\Delta m = \Delta m_s \cdot e^{-E_A/A_{fr}} \quad (1)$$

where

Δm is weight wear;

Δm_s is saturation weight wear, i.e. the maximum permissible wear during the period of stable wear;

E_A is wear process activation energy;

A_{fr} is work done by friction.

$$\Delta h = \Delta h_s \cdot r_s / r \cdot e^{-E_A/A_{fr}} \quad (2)$$

where

Δh is linear wear;

Δh_s is a linear wear of saturation;

r, r_s are respectively, density of the surface layer during the period of stable wear and the same at the onset of the saturation wear, i.e., the wear for the greatest depth obtained during the period of stable wear Δh_s.

The results of the work done by friction at testing the samples made of steel 40HN, strengthened by the EEL and PPD methods and tested on the friction machine of CMC-2 according to the disk-pad scheme, wherein Babbitt B-83 was used as a pad material, are presented in Table 3. In addition, Table 3 shows the averaged values of the friction coefficients of f_{fr} for the samples of 40HN steel subjected to different types of strengthening operations (see Figure 3).

Table 3. Results of calculating a friction work value A_{fr} per 1 hour of the testing process.

Type of strengthening	Friction coefficient f_{fr}	A_{fr} per 1h, J	$1/A_{fr}$, $J^{-1} \times 10^6$
Without strengthening	0.12	336960	2.968
The EEA method by Cr + DS	0.06	168480	5.935
The EEA method by W + DS	0.14	393120	2.544
The EEA method by VK8 + DS	0.13	365040	2.739
The EEA method by Cu + the EEA method by VK8 + BSRG	0.05	140400	7.123
The EEA method by VK8 + BSRG + the EEA method by (Cu + VK8) + BSRG	0.05	140400	7.123

$$A_{fr} = f_{fr} \cdot P \cdot V \cdot t, \quad (3)$$

Where f_{fr} is a coefficient of friction for a sample of steel 40HN subjected to various types of strengthening; P is load acting onto the pad; V is a sliding velocity; T is a test time of 1 hour (3600 s).

To build a graph of relationship between $\ln\Delta m$ and A_{fr}^{-1}, it is necessary to determine the weight wear values of the samples at different values of friction work, i.e., at different periods of wear-out failure time (Table 4), and then to plot the curves of the relationships between $\ln\Delta m$ and A_{fr}^{-1} for the samples subjected to various types of strengthening (Figure 4). All the data necessary for this procedure are listed in Table 5.

Table 4. Relationships between the values of the weight wear Δm, as well as the values being inverse ones to the values of the friction work A_{fr}^{-1}, and test periods of time.

Type of strengthening	Test time t, h						
	1	2	3	4	5	6	7
	Weight wear Δm, kg x 10^5 / Value being inverse to value of the friction work $1/A_{fr} \times 10^6$, J^{-1}						
Without strengthening	2.8 / 2.968	5.1 / 1.484	6.2 / 0.989	7.0 / 0.742	7.5 / 0.594	7.7 / 0.495	7.9 / 0.424
The EEA method by Cr + DS	0.8 / 5.935	1.5 / 2.968	1.8 / 1.978	2.0 / 1.484	2.15 / 1.187	2.2 / 0.989	2.3 / 0.848
The EEA method by W + DS	1.4 / 2.544	2.5 / 1.272	3.0 / 0.848	3.4 / 0.636	3.65 / 0.509	3.8 / 0.424	3.9 / 0.363
The EEA method by VK8+ DS	1.3 / 2.739	2.4 / 1.370	3.0 / 0.913	3.3 / 0.685	3.5 / 0.548	3.55 / 0.457	3.6 / 0.391
The EEA method by Cu + the EEA method by VK8+ + BSRG	0.60 / 7.123	1.2 / 3.561	1.3 / 2.374	1.45 / 1.780	1.65 / 1.425	1.75 / 1.187	1.85 / 1.018
EEA method by VK8+ + BSRG +the EEA method by (Cu + VK8)+ BSRG	0.4 / 7.123	0.77 / 3.561	0.95 / 2.374	1.05 / 1.780	1.13 / 1.425	1.16 / 1.187	1.2 / 1.018

Table 5. Relationships between the logarithms of the weight wear Δm and the test time at abrading of the steel 40HN samples on Babbitt B-83.

Type of strengthening	Test time t, h						
	1	2	3	4	5	6	7
	Logarithm of weight wear Δm						
Without strengthening	-10.48	-9.48	-9.68	-9.57	-9.49	-9.47	-9.45
The EEA method by Cr + DS	-11.74	-11.10	-10.92	-10.82	-10.75	-10.72	-10.68
The EEA method by W + DS	-11.18	-10.60	-10.41	-10.30	-10.22	-10.18	-10.15
The EEA method by VK8+ DS	-11.25	-10.64	-10.41	-10.32	-10.26	-10.25	-10.23
The EEA method by Cu + the EEA method by VK8 + BSRG	-12.02	-11.33	-11.25	-11.14	-11.01	-10.95	-10.89
The EEA method by VK8+ BSRG +the EEA method by (Cu + VK8)+ BSRG	-12.43	-11.77	-11,56	-11.46	-11.39	-11.36	-11.33

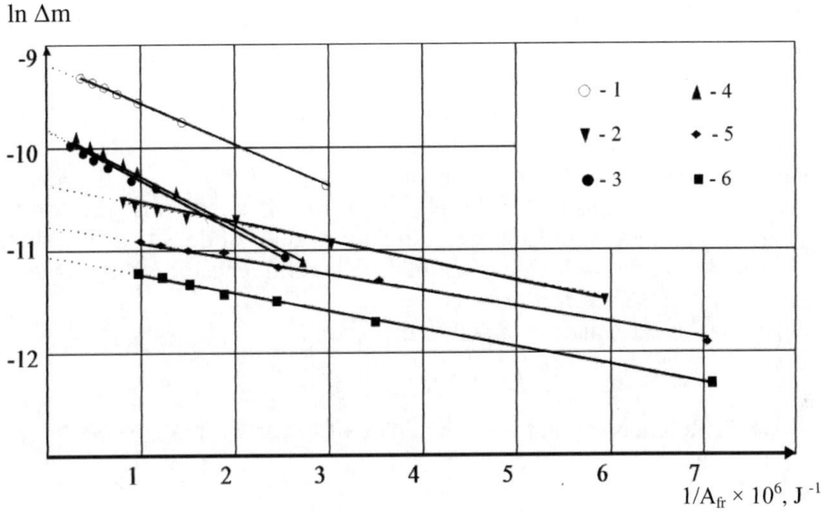

Figure 4. Relationships between the logarithm of the weight wear and the value being inverse to the value of the friction work for samples subjected various types of strengthening: 1 - without strengthening, 2 – the EEA method by Cr + DS, 3 – the EEA method by W +DS, 4 - the EEA method by VK8 + DS, 5 - the EEA method by (Cu + VK8) + BSRG, 6 - the EEA method by VK8 + BSRG + the EEA method by (Cu + VK8) + BSRG.

The tangent values of the slopes of the straight lines to the abscissa axis are listed in Table. 6. According to the wear equation $E_A = |\text{tg}\ \alpha|$.

The pre-exponential factor is obtained by the segment cut off on the ordinate axis of the straight line extended to the abscissa value of $A_{fr}^{-1} = 0$ ($\ln\Delta m = \ln\Delta m_s$ and at $A_{fr}^{-1} \to 0$).

The results of calculating the value of the wear process activation energy determined in two ways, that is, at $E_A = A_{fr}$ and at $E_A = |\text{tg}\alpha|$, correspondingly, and also the value of the saturation weight wear of Δm_s, are listed in Table 6.

Table 6. Calculation of the weight wear constant of E_A and the value of the weight wear saturation Δm_s.

Type of strengthening	$E_A = \lvert tg\,\alpha \rvert$, J	$E_A = A_{fr}$, J	%	$\Delta m_s x\ 10^5$, kg
Without strengthening	421160	434780	3.2	9.5
The EEA method by Cr + DS	205560	204080	0.7	2.7
The EEA method by W + DS	455970	443160	2.9	4.5
The EEA method by VK8+ DS	452720	443160	1.2	4.5
The EEA method by Cu + the EEA method by VK8 + BSRG	158641	160227	1.0	1.9
The EEA method by VK8+ BSRG +the EEA method by (Cu + VK8)+ BSRG	154429	156900	1.6	1.4

Some incompatibility (0.7 ... 3.2%) of the values of the wear-weight constants, that is, the values of activation energies, which were determined by different methods, can be explained as a result of various measurement errors. On the whole, the convergence of the results is satisfactory.

To build a relationship between the linear wear logarithm and the value being inverse to the friction work value, it is necessary to measure the linear wear values at the various values being inverse to the friction work values (Table 7).

The graphs of the relationships between the values of $\ln\Delta h$ and A_{fr}^{-1} for the samples made of 40HN steel treated with the various types of strengthening are shown in Figure 5. All the data that are necessary for building the graphs are given in Table 8. In addition, Table 8 shows a value of a density of an abradable layer as a function of a linear wear amount.

As it can be seen from Table 8, under the above said wear conditions, the surfaces of the test samples are subjected to a relatively uniform wear, and the values of density of the abradable surfaces insignificantly varies (2.5 ... 4.9%), which is most likely a consequence of errors in measuring both the weight wear and the linear one. Therefore, $r_s \sim r$.

It follows that the linear wear equation (2) takes the form:

$$\Delta h = \Delta h_s \cdot e^{-E_A / A_{fr}} \tag{4}$$

Table 7. Relationship between the value of linear wear Δh, as well as the value being inverse one to the value of friction work A_{fr}^{-1}, and the test time.

Type of strengthening	Test time t, h						
	1	2	3	4	5	6	7
	Weight wear Δh, μm Value being inverse to value of the friction work $1/A_{fr}$ x 10^6, J^{-1}						
Without strengthening	2.32 2.968	4.15 1.484	4.90 0.989	5.60 0.742	5.95 0.594	6.30 0.495	6.50 0.424
The EEA method by Cr + DS	0.71 5.935	1.3 2.968	1.6 1.978	1.77 1.484	1.87 1.187	1.94 0.989	2.0 0.848
The EEA method by W + DS	0.89 2.544	1.65 1.272	1.95 0.848	2.15 0.636	2.30 0.509	2.40 0.424	2.50 0.363
The EEA method by VK8+ DS	0.82 2.739	1.57 1.370	1.87 0.913	2.07 0.685	2.20 0.548	2.25 0.457	2.30 0.391
The EEA method by Cu + the EEA method by VK8 + BSRG	0.37 7.123	0.67 3.561	0.80 2.374	0.87 1.780	0.95 1.425	1.1 1.187	1.2 1.018
The EEA method by VK8+ BSRG +the EEA method by (Cu + VK8)+ BSRG	0.28 7.123	0.54 3.561	0.65 2.374	0.74 1.780	0.78 1.425	0.79 1.187	0.82 1.018

IOP Conf. Series: Materials Science and Engineering **233** (2017) 012051 doi:10.1088/1757-899X/233/1/012051

Table 8. Relationship between the logarithm of the value of linear wear Δh, as well as the value of the surface layer density, and the test time.

Type of strengthening	Test time t, h						
	1	2	3	4	5	6	7
	Logarithm of the value of linear wear Δh / Value of the surface layer density r x10^{-3}, kg/m^3						
Without strengthening	0.84 / 7.70	1.42 / 7.82	1.58 / 8.05	1.72 / 7.96	1.78 / 8.02	1.84 / 7.78	1.87 / 7.74
The EEA method by Cr + DS	-0.34 / 7.18	0.26 / 7.35	0.47 / 7.17	0.57 / 7.20	0.63 / 7.32	0.66 / 7.22	0.69 / 7.32
The EEA method by W + DS	-0.12 / 10.02	0.5 / 9.65	0.67 / 9.80	0.77 / 10.08	0.83 / 10.11	0.90 / 10.08	0.92 / 9.94
The EEA method by VK8+ DS	-0.20 / 10.09	0.45 / 9.74	0.62 / 10.22	0.72 / 10.15	0.78 / 10.10	0.81 / 10.04	0.83 / 9.97
The EEA method by Cu + the EEA method by VK8 + BSRG	-0.99 / 10.32	-0.50 / 10.4	-0.32 / 10.35	-0.24 / 10.24	-0.05 / 10.06	0.12 / 10.07	0.182 / 10.67
The EEA method by VK8+ BSRG +the EEA method by (Cu + VK8)+ BSRG	-1.27 / 9.10	-0.62 / 9.08	-0.43 / 9.31	-0.30 / 9.04	-0.25 / 9.23	-0.23 / 9.35	-0.20 / 9.32

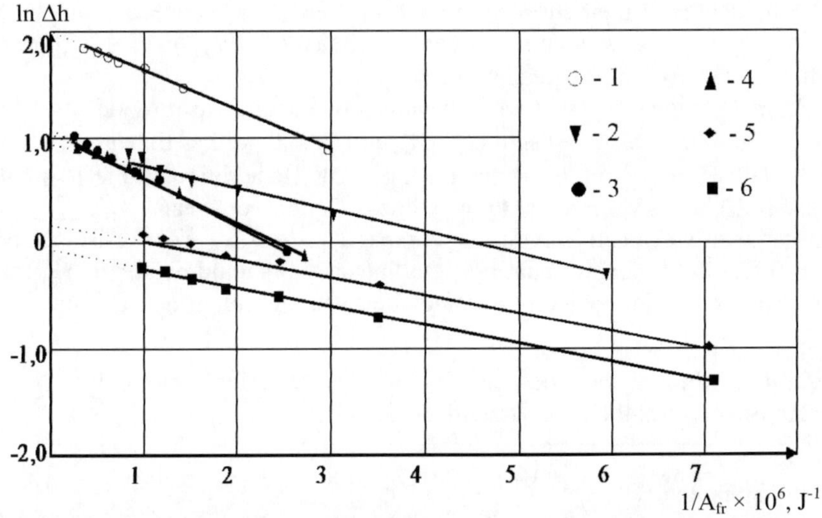

Figure 5. Relationships between the values of lnΔh and the values of A_{fr}^{-1} for samples subjected various types of strengthening: 1 - without strengthening, 2 – the EEA method by Cr + DS, 3 – the EEA method by W +DS, 4 - the EEA method by VK8 + DS, 5 - the EEA method by (Cu + VK8) + BSRG, 6 - the EEA method by VK8 + BSRG + the EEA method by (Cu + VK8) + BSRG.

The tangent values of the slopes of the straight lines to the abscissa axis are listed in Table 9. The pre-exponential factor is obtained by the segment cut off on the ordinate axis of the straight line extended (extrapolated) to the abscissa value of $A_{fr}^{-1} = 0$ (lnΔm = ln$Δm_s$ and at $A_{fr}^{-1} \rightarrow 0$).

The results of calculating the value of the wear process activation energy(the linear wear constant) E_A determined in two ways, that is, at $E_A = A_{fr}$ and at $E_A = |tg\alpha|$, correspondingly, and also the pre-exponential factor (saturation wear Δh_s) are listed in Table 9.

Table 9. Calculation of the wear constant of EA and the value of linear wear of saturation Δh_s).

Type of strengthening	$E_A = \|tg\ \alpha\|$, J	$E_A = A_{fr}$, J	%	Δh_s, μm
Without strengthening	407680	416670	2.2	7.8
The EEA method by Cr + DS	208930	203200	2.8	2.45
The EEA method by W + DS	467760	460950	1.5	2.9
The EEA method by VK8+ DS	456370	452160	0.9	2.85
The EEA method by Cu + the EEA method by VK8 + BSRG	164257	165407	0.7	1.16
The EEA method by VK8 + BSRG +the EEA method by (Cu + VK8)+ BSRG	150022	150969	0.5	0.98

Some incompatibility (0.5 ... 2.8%) of the values of the activation energies for the process of linear wear, which were determined by various methods, can be considered as results of various measurement errors. On the whole, the convergence of the results is satisfactory.

Since the processes of reducing the weights of the samples Δm (weight wear) and also the sizes of the samples along the depths of their surface layers Δh (linear wear) proceed simultaneously and are the derivatives of a common process, that is, a wear-out, then the activation energy (the wear constant) should be uniform, namely, $E_{\Delta m}$ must be equal to $E_{\Delta h}$.

In Table 10, there are shown the summarized values of energies to provide for activation of the weight and linear processes of wear, respectively, $E_{\Delta m}$ and $E_{\Delta h}$, as well as the values of the weight and linear processes of wear of saturation, respectively, Δm_s и Δh_s. In addition, Table 10 shows the data of the density of the surface layers worn down to the values of saturation wear.

It should be noted that, when carrying out the comparative tests for wear resistance of the samples with various kinds of the electroerosive coatings, the linear wear should be given the preference, since the weight wear cannot provide objective results because of the different values of density of abradable coatings.

Table 10. Summary of the wear equation constants for abrading the samples of steel 40HN on Babbitt B-83.

Type of strengthening	$E_{\Delta m}$, J	$E_{\Delta h}$, J	%	Δm_s x 10^5, kg	Δh_s, μm	R_s x10^{-3}, kg/m^3
Without strengthening	427970	412180	3.83	9.5	7.8	7.76
The EEA method by Cr + DS	204820	206070	0.61	2.7	2.45	7.02
The EEA method by W + DS	449570	464360	3.28	4.5	2.9	9.88
The EEA method by VK8+ DS	455460	454270	0.26	4.5	2.85	10.06
The EEA method by Cu + the EEA method by VK8 + BSRG	159434	164832	3.4	1.9	1.16	10.43
The EEA method by VK8 + BSRG +the EEA method by (Cu + VK8)+ BSRG	155665	150496	3.4	1.4	0.98	9.10

Using the above mentioned technique, it is possible to determine the constants of the wear equation, and therefore, envisage any wear for any material of a friction pair.

4. Conclusions

1. The use of combined electroerosive coatings (CEC), which are formed by alternately applying copper and chromium on steel 45, reduces the tensile stresses and the depth of their propagation in comparison with the coatings only made of chromium, respectively, of 250 to 210 MPa and of 0.2 to 0.15 mm. The use of the surface plastic deformation (ball-rolling) (SPD) results in changing the sign of deformation from the positive sign to the negative one, which fact witnesses about the availability of the favorable compressive stresses in the surface layer

2. The values of fatigue strength of the CEC, which are formed by alternately applying copper and chromium, are 50% higher than those of coatings only consisting of chromium. The SPD process increases the fatigue strength of the combined electroerosive coatings (CEC) made of Cu + Cr by 20% as compared to the base version, namely, the sample having no coating.

3. There has been developed and implemented a new combined technology to strengthen and repair the shaft-type parts. On combining the EEA and SPD methods, this technology consists in the formation of the surface layers having specified operational properties.

4. There have been set the criteria for evaluating alternative variants of the combined EEA technologies and for providing the choice of the most rational variants thereof. There has been developed the technique for determining wear constants, namely, the values of energy for activating wear processes for various materials of friction pairs, as well as for specifying the wear equation constants, namely, for the weight wear and / or the linear wear of saturation. All these approaches allow predicting the most reliable choice of the technology that is expedient to improve the quality of the surface layer of a product.

References

[1] Hitlevich A E et al 1985 *Electric Spark Alloying of Metal Surfaces* (Chisinau: Shtintsa) p 196
[2] Edigarov V R , Kilunin I Yu , Degtyar V V 2012 *Modern Science-Intensive Technologies* **3** pp 32-35
[3] Ivanov V I Burumkulov F H 2010 *Electronic Processing of Materials* **5** pp 27-36
[4] Romanenko D N 2007 Innov. Tech. and Equip. for a Mach.-Build. Complex Scient. Papers **10** pp 62-65
[5] Tarelnyk V, Martsynkovskyy V 2014 Upgrading of pump and compressor rotor shafts using combined technology of electroerosive alloying *Applied Mechanics and Materials* (Trans Tech Publications) **630** pp 397-412
[6] Scientific Research and Engineering Development Work on Increasing Wear Resistance of Crankshaft Necks of the Opposing Compressors of M 16, M 25, M 40 Bases (1990) *by Processing with SPD: Report on R & D* (Sumy: VNIIkompressormash) p 85
[7] Tarelnyk V B (2002) *Controlling the Quality of Part Surface Layers by Processing with the Use of Combined Electroerosive Alloying Method* (Sumy: MakDen Publishing House) p 323

IOP Conf. Series: Materials Science and Engineering **233** (2017) 012052 doi:10.1088/1757-899X/233/1/012052

Investigation of the torsional stiffness of flexible disc coupling

A Buryy[1,2], V Simonovsky[1] and V Obolonik[2]

[1] Department of General Mechanics and Dynamics of Machines, Sumy State University, 2 Rimsky-Korsakov Str., Sumy, 40007, Ukraine
[2] Special Design Bureau, PJSC "Sumy NPO", 58 Gorkogo Str., Sumy, 40004, Ukraine

E-mail: Andrey07ua@gmail.com

Abstract. Calculation of flexible coupling torsional stiffness is required when analyzing the torsional vibrations of the reciprocating machinery train. While having the lowest torsional stiffness of all the elements of the train, flexible coupling has a significant influence on the natural frequencies of torsional vibration. However, considering structural complexity of coupling, precise definition of its torsional stiffness is quite a difficult task. The paper presents a method for calculating the torsional stiffness of flexible disc coupling based on the study of its finite element model response under the action of torque. The analysis of the basic parameters that quantitatively and qualitatively affect the coupling torsional stiffness has been also provided. The results of the calculation as well as model adequacy, sufficient for practical application, have been confirmed at the experimental measurement of flexible disc coupling torsional stiffness. The obtained elastic characteristics (dependences of applied torque and torsional stiffness versus twist angle) are nonlinear in the initial stage of loading. This feature should be taken into account when creating reliable mathematical models of torsional vibrations of reciprocating machinery trains containing flexible disc couplings.

1. Introduction

When conducting the torsional vibration analysis of the reciprocating compressor train, there arose need for precise determination of the torsional stiffness of flexible disc coupling that connects engine and compressor rotors.

Flexible coupling often has the lowest torsional stiffness of all the elements of the train that results in a significant effect on the natural frequencies of torsional vibration. However, due to the complex design, it is difficult to accurately determine the torsional stiffness of the coupling [1]. This article is devoted to the description of this problem solution.

Torsional stiffness (c, Nm/rad) is an elastic property that is characterized by the ratio of the applied torque to the twist angle. In general, the coupling torsional stiffness is defined as a derivative of torque with respect to twist angle. The reciprocal of the torsional stiffness is the torsional compliance (e, rad/Nm).

$$c = \frac{1}{e} = \frac{dM}{d\varphi} \tag{1}$$

A distinction is made between linear and nonlinear couplings. For nonlinear couplings, the torsional stiffness is not constant but is a function of the torque magnitude. Typically, a torque vs. twist angle curve is obtained by fixing one end of the coupling and measuring the twist angle of another end with gradual torque increase.

The most accurate coupling torque vs. twist angle curve can be obtained only experimentally. However, this rather expensive process requires special equipment, as well as time for adjustment and measurements.

Flexible disc coupling whose design and operation are well described in [2] consists of different structural elements that deform in complex ways under load. Accordingly, the coupling torsional stiffness depends on a large number of parameters, so analytical calculation is not possible. A more promising approach is a numerical calculation based on the finite element method (FEM). The development of calculation technique and creation of adequate model will enable to obtain elastic characteristics of flexible disc couplings in a very short time.

The relevance of the work is also conditioned by the fact that disc couplings are the most common flexible element couplings due to their excellent performance, high misalignment capacity, compact design, minimal maintenance and relatively low cost. Furthermore, International Standard API 671 specifies metallic flexible-element couplings as the primary type permissible for special-purpose application. Such type of equipment includes high-torque and/or speed machines operating continuously for extended periods, as well as in adverse conditions [3].

2. Literature survey

There is a small number of articles devoted to modeling and numerical analysis of couplings with metallic flexible elements.

Dobre Daniel and colleagues investigated the stress-strain state of flexible membrane coupling under the influence of torque, centrifugal force, axial and angular misalignment [4]. JongHun Kang and his collaborators analyzed the effect of torsional stiffness of flexible disc coupling elements on the natural frequencies of the wind turbine shaft torsional vibrations [5]. In particular, the article gives the calculation results of torsional stiffness of flexible elements packs with different thickness and number of discs. However, both works do not cover the information on the parameters of finite element coupling models used for calculation.

The following two papers are directly dedicated to the modeling of flexible disc couplings and calculation of their torsional stiffness. However, the model proposed by Alex B. Francis in [6] does not take into account two important factors that significantly affect coupling torsional stiffness. They are the separation of flexible elements during coupling deformation from the applied load and bolted joints preload.

A group of scientists headed by Bin Zhao conducted the most complex numerical and experimental investigation of flexible disc coupling torsional stiffness [7]. But due to accounting for clearances between bolts and disc pack holes, their model is too complicated for practical application. In addition, clearances in the coupling must be kept to a minimum in order to avoid the increased imbalance and deterioration of the vibrational state of the unit.

The paper presents a method for calculating the torsional stiffness of flexible disc coupling based on the study of its finite element model response under the action of torque. The analysis of the basic parameters that quantitatively and qualitatively affect the coupling torsional stiffness has been also provided. The results of the calculation as well as model adequacy, sufficient for practical application, have been confirmed at the experimental measurement of flexible disc coupling torsional stiffness.

3. Numerical investigation of flexible disc coupling torsional stiffness

ANSYS Mechanical 15.0 simulation software suite was used as a finite element analysis tool.

The investigated flexible disc coupling consists of two hubs, two disc packs (26 circular discs with individual disc thickness of 0.5 mm), a spacer, as well as connecting bolts and nuts. The 3D model of the coupling is shown in Figure 1 and its basic design parameters are shown in Figure 2.

3.1. Finite element model of the coupling

When creating the finite element model, the coupling symmetry is used to reduce computational costs. Since the design is symmetrical to the middle of the spacer, first we define a twist angle of half a coupling. In addition, the coupling has 3-fold rotational symmetry, so it is enough to model only 1/3 of it (a sector with a 120° central angle) and apply cyclic symmetry boundary condition. The acquired finite element coupling model is shown in Figure 3. The mesh is generated in an automatic mode and additionally improved in contact areas.

Figure 1. 3D model of the flexible disc coupling.

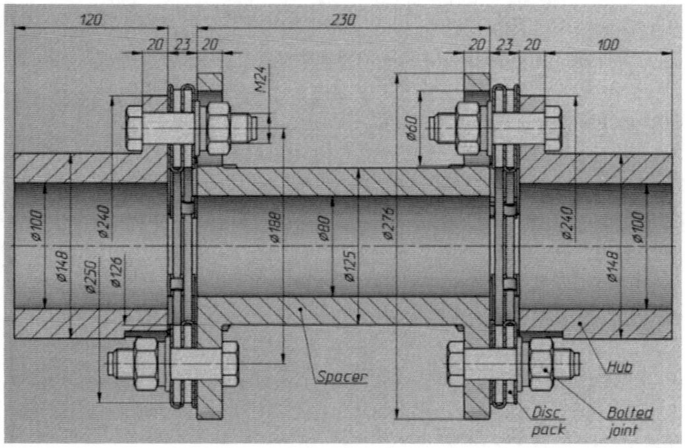

Figure 2. Basic design parameters of flexible disc coupling.

3.2. *Making contacts, fixing and loading*

Three numerical models were created to show the basic parameters that affect flexible disc coupling torsional stiffness. The models differ in contact interaction between the coupling elements and bolted joints preload. Table 1 shows the differences in coupling models.

Table 1. The parameters of coupling numerical models.

Model number	Primary contact type	Friction coefficient	Bolted joints preload F_p, N
1	"No Separation"	-	-
2	"Frictional"	0.15	10
3	"Frictional"	0.15	120830

Notes:

- in all models a bolt and a nut are rigidly connected (a contact type "Bonded");
- for model No. 2 a minimum bolt preload is specified to improve the calculation convergence;
- the initial clearances are absent in all frictional contacts.

Figure 3. Finite element coupling model.

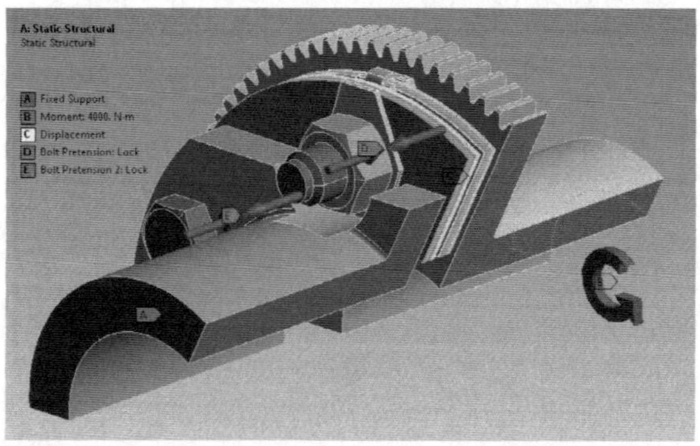

Figure 4. Constraints and loads arrangement.

IOP Conf. Series: Materials Science and Engineering **233** (2017) 012052 doi:10.1088/1757-899X/233/1/012052

All the models have the same constraints: the end surface of the hub is rigidly fixed, the radial motions of spacer cylindrical surface are ruled out. Figure 4 shows constraints and loads arrangement.

The main load in all models is a 4000 Nm torque (1/3 of the maximum torque) applied to the spacer end surface. The torque loading is divided into 20 substeps. The values of the coupling twist angle are calculated at each substep. The coupling torque vs. twist angle curve is plotted on the basis of the obtained values.

The bolt preload is calculated from the tightening torque on the formula [8]:

$$F_p = T / (Kd) \tag{2}$$

where

$T = 580$ Nm - tightening torque;

$K = 0.2$ - nut factor (when friction coefficient is 0.15);

$d = 0.024$ m - nominal diameter.

Thus,

$$F_p = 580 / (0.2 \cdot 0.024) = 120830 \text{ (N)}$$

3.3. *Obtaining results*

Figures 5 and 6 show the calculation results: coupling and disc pack deformation for coupling model No. 3 under maximum torque.

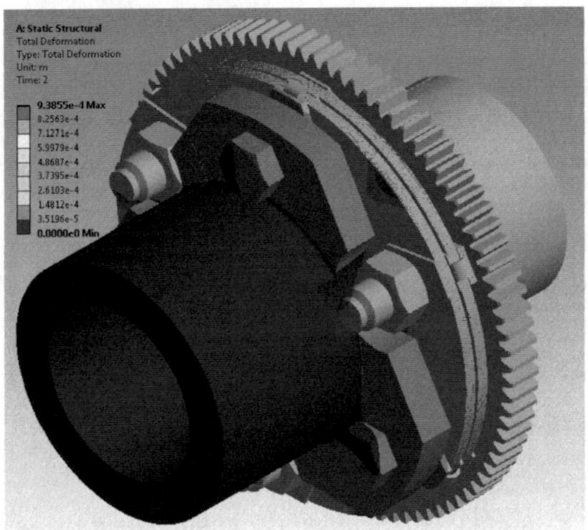

Figure 5. Deformation of the coupling (true scale).

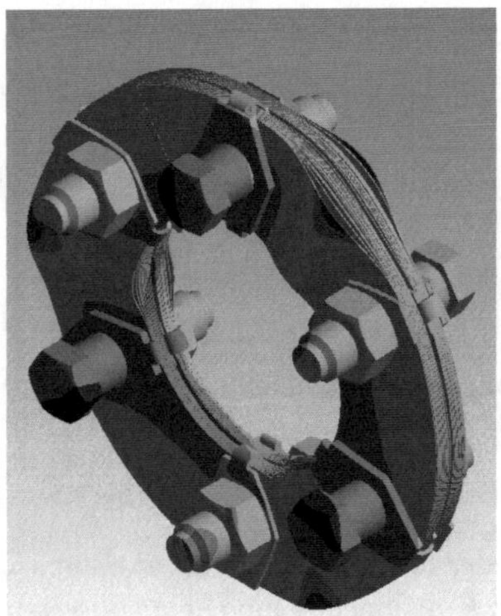

Figure 6. Deformation of the disc pack (10x deformation scale).

The tool "Deformation Probe" is used to obtain the values of the coupling twist angle. It determines the total displacement of the point located on the outer circumference of the spacer end face at each load substep (see Figure 7).

The twist angle of half a coupling is calculated by the total point displacement at each load substep:

$$\varphi_i^* = \frac{2\Delta_i}{D_s} \tag{3}$$

where

φ_i^* - the twist angle of half a coupling at the i-th load substep;

Δ_i - the total point displacement at the i-th load substep;

$D_s = 0.125$ m - outer diameter of the spacer cylindrical area.

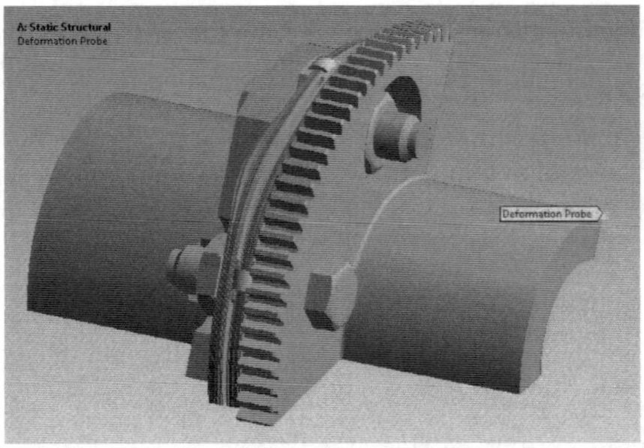

Figure 7. "Deformation Probe" point position.

To make the further use convenient and to compare with the experimental data, the total coupling torsional stiffness is determined excluding the torsional stiffness of hubs cylindrical areas.

The total coupling twist angle at each load substep is as follows:

$$\varphi_i = 2(\varphi_i^* - \varphi_c) = 2(\varphi_i^* - M_i \cdot e_c) \tag{4}$$

where

M_i - applied torque at the *i*-th load substep;

e_c - torsional compliance of the hub cylindrical area.

$$e_c = \frac{32}{\pi G} \cdot \frac{l_c}{D_c^4 - d_c^4} \tag{5}$$

where

$l_c = 0.1$ m, $D_c = 0.148$ m, $d_c = 0.1$ m - length, outer and inner diameters of the hub cylindrical area; $G = 7.8 \cdot 10^{10}$ Pa - shear modulus.

$$e_c = \frac{32}{3.14 \cdot 7.8 \cdot 10^{10}} \cdot \frac{l_c}{0.148^4 - 0.1^4} = 3.438 \cdot 10^{-8} \text{ (rad/Nm)}$$

In order to find the analytical expression for the coupling torque vs. twist angle curve, the values calculated by the equation (4) are approximated by a polynomial function. Then the coupling torsional stiffness vs. twist angle curve is calculated on the formula (1). The obtained curves for each numerical model will be presented in section 5.

4. Experimental determination of flexible disc coupling torsional stiffness

To determine which of the three numerical models is the most appropriate for torsional stiffness analysis of flexible disc coupling, experimental twist angle measurement of a real coupling was carried out. For this purpose, we have developed methodology that does not require coupling dismantling and allows to take measurements directly on the compressor unit.

The torque loading and measurement was conducted by using two specially manufactured levers with a half-flange mount (see Figure 8). The first lever, mounted on the coupling flange from engine side, was being gradually loaded by means of a jack-screw. Thus, the torque was created. The second lever was attached to the coupling hub flange on the compressor side and was rigidly connected to the 'ground' through the electronic dynamometer sensor. Thus, the second end of the coupling was fixed, while the sensor was measuring the applied torque.

The measurement of the coupling twist angle was taken by using two laser indicators mounted on the coupling hubs (see Figure 9). Laser beams were projected onto a screen with millimeter marking.

The screen was installed at a distance of 6 meters from the coupling axis. With increasing torque, there was a vertical movement of laser beam points on the screen. Movement of indicator points was recorded for each load step.

Before starting the measurements, the tentative loading (up to the value of the expected maximum torque) and unloading of the coupling were carried out to remove possible clearances between its elements.

Figure 8. General view of the experimental facility.

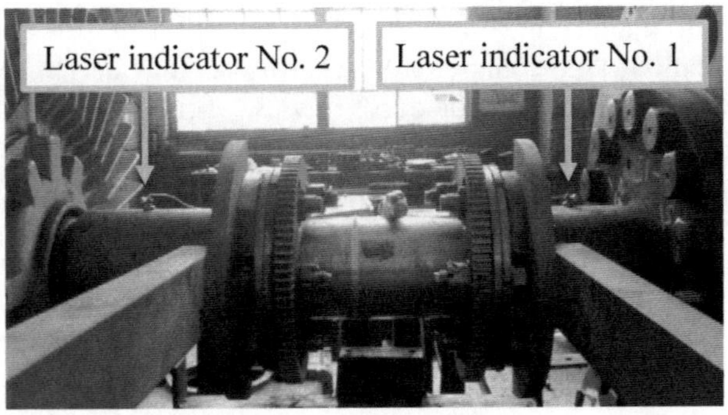

Figure 9. Laser indicators arrangement.

Obviously, laser beam No. 2 displacement is greater than laser beam No. 1 displacement by the coupling twist angle value (see Figure 10).

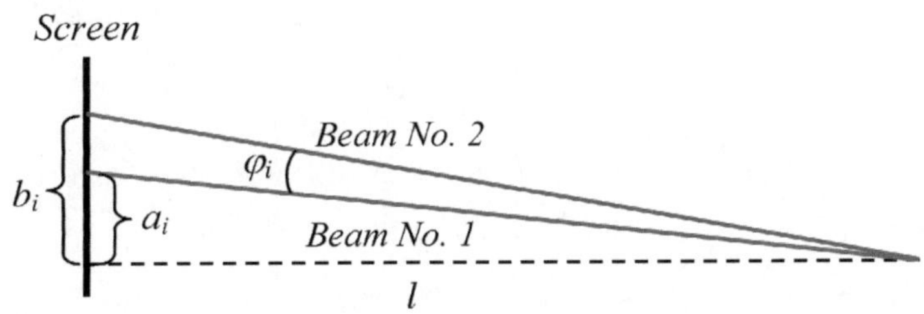

Figure 10. The scheme for defining the coupling twist angle.

The coupling twist angle on each load step is calculated by the formula:

$$\varphi_i = arctg(b_i / l) - arctg(a_i / l) \quad (6)$$

where

φ_i - the coupling twist angle at the i-th load step;

a_i, b_i - laser beam No. 1 and laser beam No. 2 displacements at the i-th load step;

l - the distance from the screen to coupling axis.

To find the analytical expression for the experimental coupling torque vs. twist angle curve, the values calculated by equation (6) are approximated by a polynomial function. Then the coupling torsional stiffness vs. twist angle curve is calculated according to formula (1).

5. A comparative analysis of the findings

The curves of the dependence of applied torque versus twist angle obtained for the numerical models, as well as experimentally, are presented in Figure 11. The curves of the dependence of coupling torsional stiffness versus twist angle obtained for the numerical models, as well as on the experimental data, are shown in Figure 12.

As seen from these graphs, the elastic characteristics of the coupling model No. 3 almost correspond to the real coupling characteristics. This proves to adequacy of this model and its acceptability for calculating torsional stiffness of flexible disc couplings with the other design parameters.

Coupling model No. 1 reproduces the model introduced in [6]. When specifying the contact interaction between the elements of this model, a linear contact type “No Separation” was employed. It assumes mutual sliding of model elements (without friction), but prohibits their separation. Thus, this model does not allow for separation of flexible elements when the coupling deforms from the applied load. This entails an artificial increase in torsional stiffness. Also, this model gives a linear torque vs. twist angle curve in view of the linearity of the contacts used.

When defining the contact interaction between the elements of coupling models No. 2 and No. 3, a more complex nonlinear contact type “Frictional” was used. It allows for friction between the model elements and, accordingly, assumes their separation. However, the second model does not consider the bolted joints preload. Comparing the curves obtained for these numerical models, we can conclude that it is very important to take into account the bolted joints preload. This option increases the torsional stiffness of flexible disc coupling, as well as gives the non-linear nature of the elastic characteristics at the initial stage of loading.

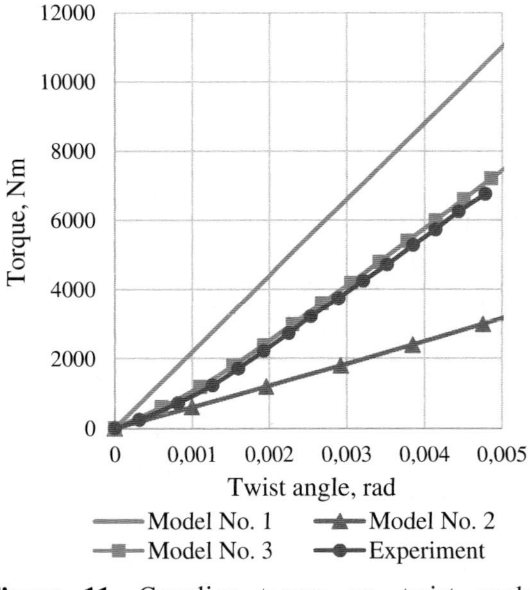

Figure 11. Coupling torque vs. twist angle curves.

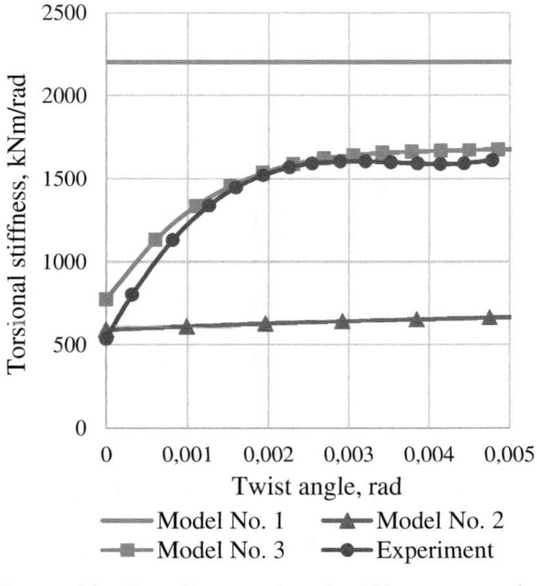

Figure 12. Coupling torsional stiffness vs. twist angle curves.

6. Conclusions

1. A method for calculating torsional stiffness of flexible disc coupling based on the study of its finite element model response under the action of torque has been developed.

2. When creating reliable numerical model of flexible disc coupling it is necessary to allow for friction between the coupling elements as well as the bolted joints preload.

3. The calculation results and sufficient for practical application adequacy of the numerical model are confirmed by the experimental measurement of flexible disc coupling torsional stiffness.

4. The obtained elastic characteristics (dependences of applied torque and torsional stiffness versus twist angle) are nonlinear in the initial loading stage. This feature should be taken into account when creating reliable mathematical models of torsional vibrations of reciprocating machinery trains that involve flexible disc couplings

IOP Conf. Series: Materials Science and Engineering **233** (2017) 012052 doi:10.1088/1757-899X/233/1/012052

References

[1] Wang Q, Feese T D and Pettinato B C 2012 *Proc. of the 41st Turbomachinery Symp.* (Texas A&M University)
[2] Mancuso J R, Gibbons C B and Munyon R E 1989 *Proc. of the 18th Turbomachinery Symp.* (Texas A&M University) p 141
[3] API Standard 671 2010 *Special Purpose Couplings for Petroleum, Chemical and Gas Industry Services*
[4] Dobre D, Simion I, Adir V G, Adir G M and Pascu N 2010 *Annals of DAAAM for 2010 & Proc. of the 21st Int. DAAAM Symp. (Zadar, Croatia)* vol 21 (Vienna: DAAAM International)
[5] Kang J, Bae J, On H and Lee H 2016 The Design of a Flexible Wind Turbine Coupling Structure considering Torsional Vibration International *IJAER* **11** pp 9093-98
[6] Francis A, Avdeev I, Hamann J and Ananthasivan S 2015 Accurate Characterization of Torsional Stiffness of Flexible Disk Couplings *J. Eng. Gas Turbines Power* **137** 082504
[7] Zhao B, Zhao Y, Feng J and Peng X 2016 Numerical and experimental investigation of the torsional stiffness of flexible disc couplings *IJMS* **114** pp 207–16
[8] Bickford J H 2007 *Introduction to the Design and Behavior of Bolted Joints, Fourth Edition Non-Gasketed Joints* (CRC Press) p 564

IOP Conf. Series: Materials Science and Engineering **233** (2017) 012053 doi:10.1088/1757-899X/233/1/012053

Dynamic analysis of centrifugal machines rotors supported on ball bearings by combined application of 3D and beam finite element models

I V Pavlenko[1], **V I Simonovskiy**[1] **and M M Demianenko**[1]

[1] Department of General Mechanics and Machine Dynamics, Sumy State University, 2 Rymskogo-Korsakova Street, Sumy, 40007, Ukraine

E-mail: i.pavlenko@omdm.sumdu.edu.ua

Abstract. This research paper is aimed to investigating rotor dynamics of multistage centrifugal machines with ball bearings by using the computer programs "Critical frequencies of the rotor" and "Forced oscillations of the rotor," which are implemented the mathematical model based on the use of beam finite elements. Free and forces oscillations of the rotor for the multistage centrifugal oil pump NPS 200-700 are observed by taking into account the analytical dependence of bearing stiffness on rotor speed, which is previously defined on the basis of results' approximation for the numerical simulation in ANSYS by applying 3D finite elements. The calculations found that characteristic and constrained oscillations of rotor and corresponded to them forms of vibrations, as well as the form of constrained oscillation on the actual frequency for acceptable residual unbalance are determined.

1. Introduction

The existing numerical methods for calculation of rotor dynamics, which are based on the use of modern programs on the basis of finite element method, use beam elements with taking into account the Bernoulli's Hypothesis or 3D finite elements [1]. In this connection, appliance of 3D models, for example in ANSYS, needs carrying out relatively long operations and computational time [2]. The appliance of beam models [3] is not a long-time process for the initial data preparation, as well as it makes the computational process faster. However, the problem of determination of critical frequencies of rotor systems by using existing programs for the advanced determined dependence of the bearing stiffness and gap seals on rotor speed leads to drawing the Campbell diagram [4] that complicates the computational process with additional graphic constructions. Besides, the determination of rotor oscillations by taking into account the system of imbalances and nonlinear bearing stiffness is the high-complicated scientific problem [5].

In accordance with the abovementioned, the aim of this research is a development of methods for determining eigenfrequencies and critical frequencies, as well as characteristics of rotor oscillations by applying a combined using 3D and beam finite elements on the example multistage centrifugal machines. To achieve this aim the following problems are set:

1) determination of the bearing stiffness with taking into account rotor speed by using ANSYS;

2) development of the mathematical model of rotor dynamics with the predetermined dependence of the bearing stiffness on rotor speed;

IOP Conf. Series: Materials Science and Engineering **233** (2017) 012053 doi:10.1088/1757-899X/233/1/012053

3) creation the files of numerical implementation of the purposed mathematical model with the use of beam finite elements;

4) implementation of the developed technique on the example of multistage centrifugal pump NPS 200-700.

2. Pump description and determination of the bearing stiffness

The multistage centrifugal oil pump NPS 200-700 (Figure 1) is designed to pump water, oil, petroleum product, and liquefied carbohydrate gas-oil products at temperature in the range -30… +200°C with particles less than 0,2 μm. The pump's rotor rotates with the operating speed 2950 rpm in two opposite ball bearings KF 7416AC.

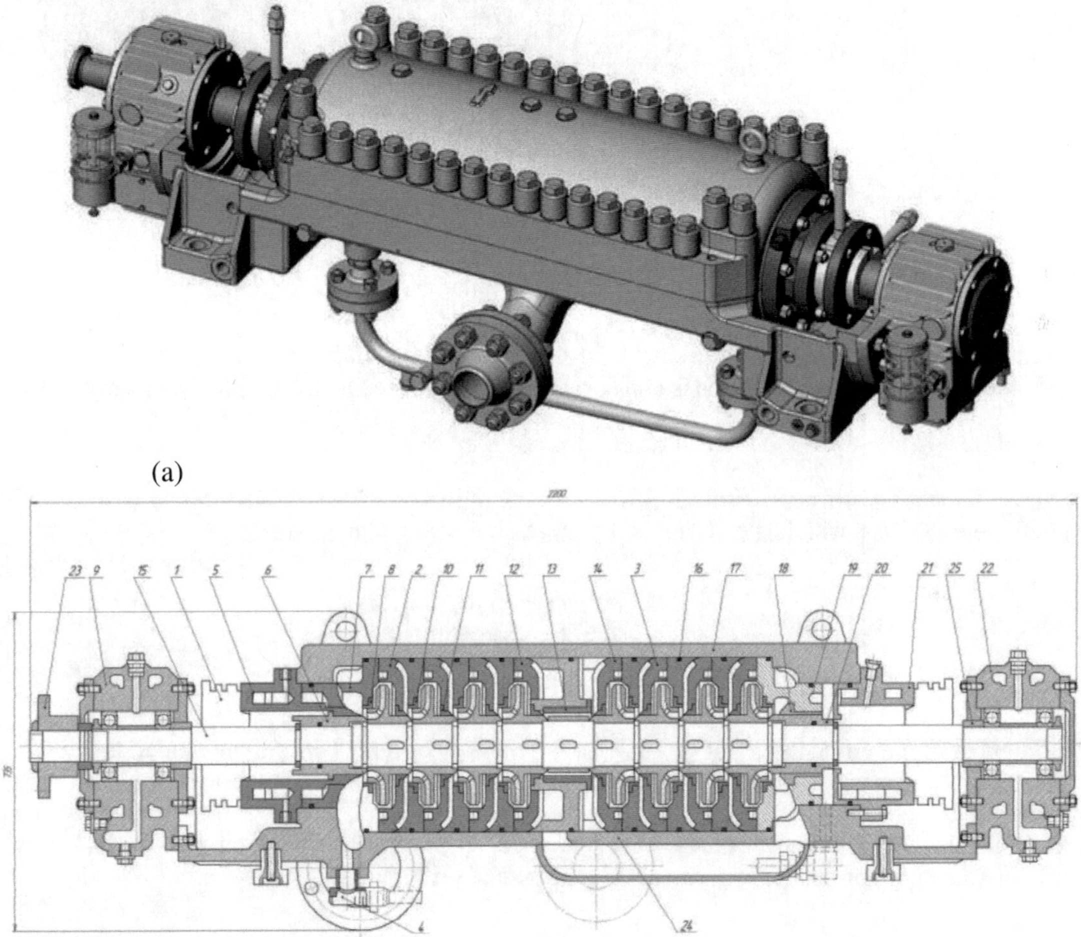

(a)

(b)

Figure 1. 3D model (a) and frontal section (b) of the centrifugal pump NPS 200-700.

The program ANSYS is used for determining the bearing stiffness with taking into account rotor speed for 2D beam elements. The related scheme of support loading is given on Figure 2 (a).

Considering the rotor speed and the housing stiffness is due to the fact, that the first factor determines the increasing quadratic dependence of bearing stiffness on rotor speed, and the second one leads to decreasing critical frequencies.

The geometry of pump housing and bearing support are chosen in accordance with the pump's engineering drawing. Rotation impact is carried out by applying the inertial forces to the inner bearing

IOP Conf. Series: Materials Science and Engineering **233** (2017) 012053 doi:10.1088/1757-899X/233/1/012053

race. Stiffness factor is defined by the analysis of loading of bearing assembly for different rotor speeds (Figure 2 (b)).

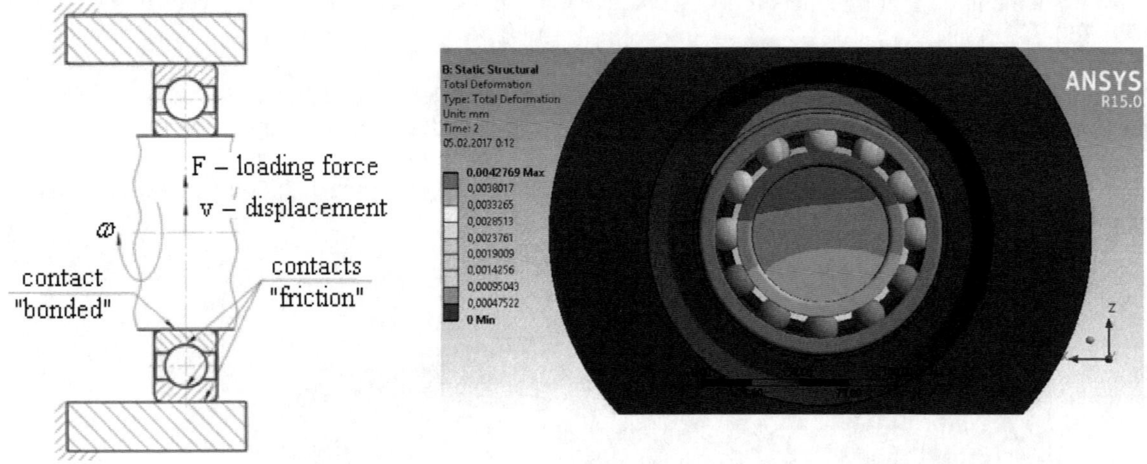

(a) (b)

Figure 2. Calculation scheme of support loading (a) and results visualization (b).

For developing the mathematical model of free and forced oscillations of the rotor with the impact of the rotor speed on bearing stiffness is considered by the proposed analytical dependence [6, 7]:

$$c(\omega) = c_0 + \alpha\,\omega^2 \tag{1}$$

where c – bearing stiffness, N/m; ω – rotor speed, rad/s; c_0 – bearing stiffness in case of $\omega = 0$; α – coefficient, N·s^2/m, which is determined by regression dependence [8-10]:

$$\alpha = \frac{\sum\limits_{k=2}^{n} (c_k - c_0)\,\omega_k^2}{\sum\limits_{k=2}^{n} \omega_k^4} \tag{2}$$

k – number of experimental point ($k = 2 \ldots n$); n – number of points within the numerical simulation by ANSYS; c_k – bearing stiffness, which is determined by the results of numerical simulation in case of frequency $\omega = \omega_k$ (Figure 3).

As a result, the following values were obtained: $c_0 = 2{,}4 \cdot 10^8$ N/m, and $\alpha = 46{,}6$ N·s^2/m. Approximating curve of the dependence (2) is represented on Figure 3.

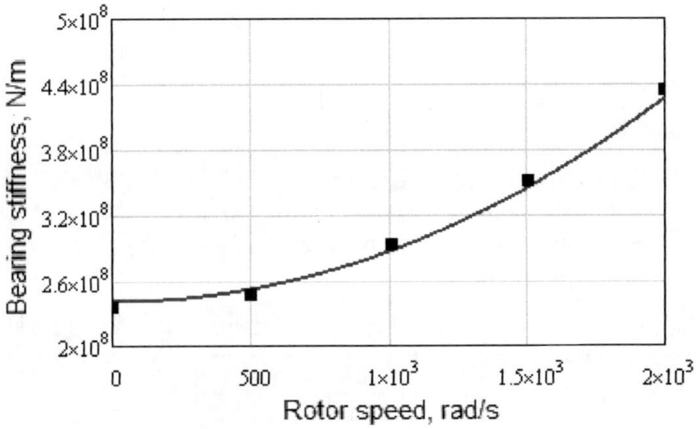

Figure 3. Dependence of the bearing stiffness on the rotor speed.

3. Mathematical model of rotor dynamics

Mathematical model of rotor dynamics system is based on the application of finite element method [3]. Local matrices of the stiffness $[C(\omega)]_e$ and inertia $[M]_e$ binodal finite elements with four degrees of freedom, which correspond to nodes cross motion of and rotation angles are determined by the following formulas:

$$[C(\omega)]_e = \frac{EI}{l^3}\begin{bmatrix} 12+c(\omega) & 6l & -12 & 6l \\ 6l & 4l^2 & -6l & 2l^2 \\ -12 & -6l & 12 & -6l \\ 6l & 2l^2 & -6l & 4l^2 \end{bmatrix}; \; [M]_e = \frac{m}{420}\begin{bmatrix} 156+m_0 & 22l & 54 & -13l \\ 22l & 4l^2-k_g & 13l & -3l^2 \\ 54 & 13l & 156 & 22l \\ -13l & -3l^2 & 22l & 4l^2 \end{bmatrix} \quad (3)$$

where E – Young's modulus of material, N/m^2; $I = \pi(D^4 - d^4)/64$– moment of inertia of cross section; D, d – outer and inner diameters of cross section, m; l – element length, m; $c(\omega)$ – bearing stiffness, N/m; $m = \pi\rho l(D^2 - d^2)/4$ – mass of the finite element; m_0 – mass of the bushing part or impeller; ρ – material density, kg/m^3; $k_g = 420\{(D^2 + d^2)/16 + l^2/3 + I_g/m\}$ – parameter, which considers gyroscopic moment of inertia for finite elements and moment of inertia Ig (kg·m^2) of impellers or bushing parts.

Global matrix of the stiffness [C(ω)] and inertia [M] are determined by summarizing the components of local matrices in corresponding nodes.

The numerical calculation of rotor's critical frequencies is to determine the roots of the frequency equation

$$det\left(\left[C(\omega)\right]-\omega^2\left[M\right]\right)=0 \quad (4)$$

But eigenfrequencies are determined for stiffness matrix $[C(\omega_0)]$, which corresponds to operating rotor speed ω_0.

Column vector $\{Y\}$ of the mode shape for rotor's forced oscillations is determined by the following formula:

$$\{Y\}=\left(\left[C(\omega_0)\right]-\omega_0^2\left[M\right]\right)^{-1}\omega_0^2\{D\} \quad (5)$$

where $\{D\}$ – column vector of imbalances (kg·m), which are applied in nodes of finite elements.

The mathematical model of free and forced oscillations of rotor systems is described and numerically implemented with the use of files "Critical frequencies of the rotor" [11] and "Forced

IOP Conf. Series: Materials Science and Engineering **233** (2017) 012053 doi:10.1088/1757-899X/233/1/012053

oscillations of the rotor" [12] within computer program MathCAD. The validation of the purposed methodology is proved in research works [13, 14] for the case of multistage centrifugal machines by comparing the corresponding results obtained in program complex ANSYS, as well as within research works [15-18].

4. Analysis of rotor dynamics for the centrifugal pump NPS 200-700

Figure 4 (a) demonstrates the fragment of the table, which contains the input data for calculation of rotor's critical frequencies for the multistage centrifugal oil pump NPS 200-700. The corresponding design scheme is presented on Figure 4 (b).

	L, m	D, m	d, m	m, kg	I_g, kg·m^2	c, N/m	α, N·s^2/m	Disb, kg·m
	1	2	3	4	5	6	7	8
1	0.014	0.065	0	0	0	0	0	0
2	0.034	0.065	0	0.64	0	0	0	0
3	0.069	0.065	0	15.75	0	0	0	$1.3 \cdot 10^{-4}$
4	0.025	0.07	0	0.79	0	0	0	0
5	0.026	0.07	0	0.7	0	0	0	0
6	0.057	0.07	0	0	0	$2.4 \cdot 10^{8}$	46.6	0
7	0.057	0.07	0	1.17	0	0	0	0
...								
46	0.057	0.07	0	0	0	0	0	0
47	0.026	0.07	0	0	0	$2.4 \cdot 10^{8}$	46.6	0
48	0.012	0.07	0	0	0	0	0	0

(a)

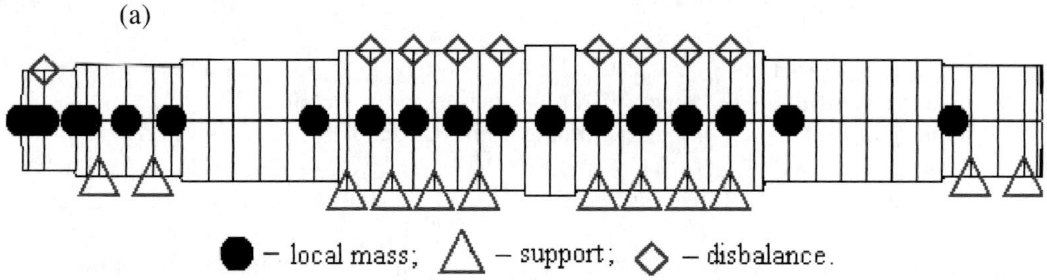

(b)

Figure 4. Fragments of the initial data table (a) and the corresponding design scheme (b) for calculating critical frequencies of the pump rotor.

The calculation results of rotor's eigenfrequencies and critical frequencies are summarized in Table 1, and corresponding forms of free oscillations are represented on Figure 5. Herewith, divergence of the results for eigenfrequencies and critical frequencies by using MathCAD and ANSYS is no higher than 0,7 % and 1,4 % respectively. The resonance safety factor is equal 23 %.

Table 1. Values of eigenfrequencies and critical frequencies, rad/s.

Frequency	Frequency number		
	1	2	3
Eigenfrequency	406 (406)*	890 (887)	1640 (1629)
Critical frequency	407 (407)	901 (896)	1805 (1780)

* The values in parentheses are defined by using ANSYS.

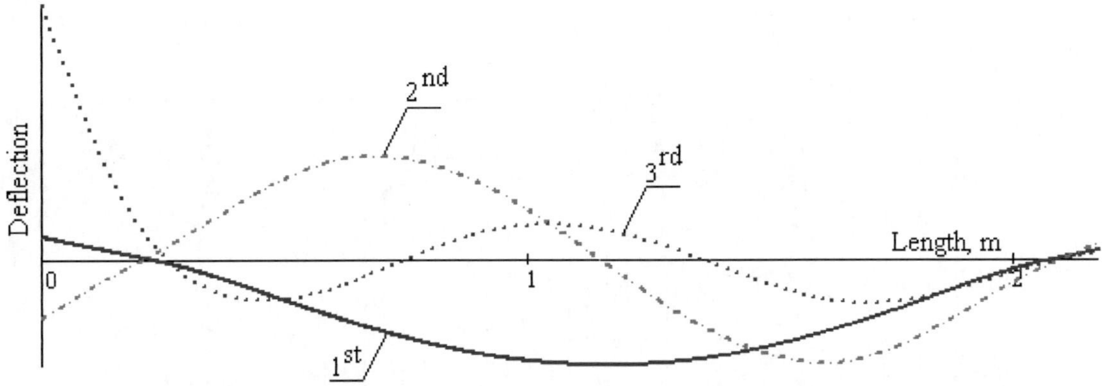

Figure 5. Forms of free oscillations of the rotor.

The results of numerical simulation (Table 1) correspond to the theorem about the spectrums of eigenfrequencies and critical frequencies [19]: since all critical frequencies are higher than operating rotor speed, they exceed the corresponding eigenfrequencies.

The additional benefit of applying the proposed methodology for determining eigenfrequencies and critical frequencies with the use of files [11, 12] of the computer program MathCAD is the lack of drawing Campbell diagram [20], which is represented on Figure 6 on the example of the rotor for the pump NPS 200-700 with the application of the program complex ANSYS.

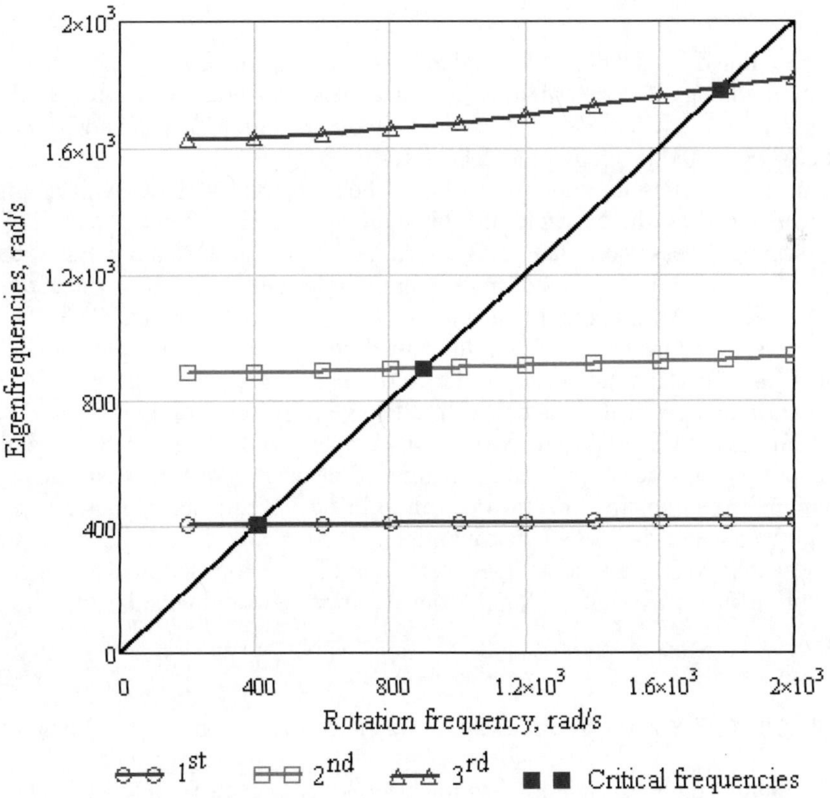

Figure 6. Campbell diagram obtained using ANSYS.

The form of the rotor forced oscillations on the operating frequency for the system of imbalances, which is determined in accordance with the standard [19], is presented on Figure 7. Maximum

IOP Conf. Series: Materials Science and Engineering **233** (2017) 012053 doi:10.1088/1757-899X/233/1/012053

deflection in the section of the central gap seal is equal 0,17 mm, which does not exceed the corresponding radial gap.

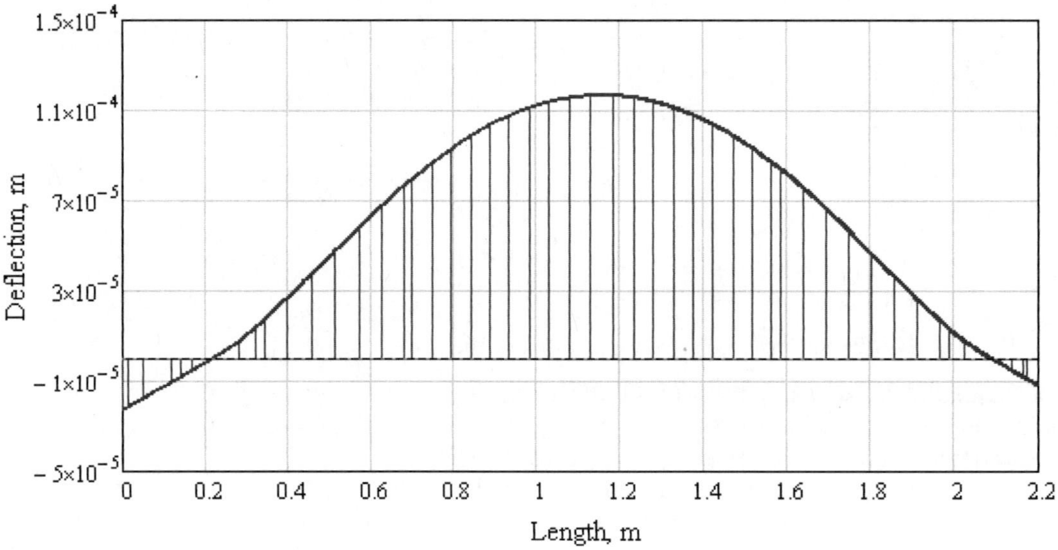

Figure 7. Form of forced oscillations of the rotor.

5. Conclusions

Thus, the work is aimed to develop the mathematical model of rotor dynamics systems and the corresponding methodology for determining eigenfrequencies and critical frequencies, as well as forms of free and forced oscillations by combined implementation of 3D and beam finite element models by using ANSYS and the computer algebra system MathCAD.

Bearing stiffness properties for supports with ball bearings KF 7416AC with taking into account rotor speed is determined by the example of multistage centrifugal oil pump NPS 200-700 by using ANSYS. The bearing stiffness coefficients and determined by the proposed regression dependence.

The represented mathematical model of rotor dynamics systems is based on the application of finite element method, which considers predefined analytical dependence of bearing stiffness on rotor speed, as well as gyroscopic moments of inertia for impellers and other bushing parts. The model is implemented in files “Critical frequencies of the rotor” and “Forced oscillations of the rotor” of the computer program MathCAD by the example of multistage centrifugal oil pump NPS 200-700.

The main advantages of the proposed approach in comparison on using ANSYS are the opportunity to take into account the advanced analytical dependence of bearing stiffness on rotor speed, and lack of drawing Campbell diagram in order to determine critical frequencies, as well as relatively fast preparation of input data and short computational time.

The further research will be aimed at implementation of the proposed methodology fir the case of nonlinear bearing stiffness, as well as for the development and numerical simulation of balancing methodologies.

References

[1] Vance J M, Zeidan F Y and Murphy B 2010 Machinery vibration and rotordynamics *John Wiley & Sons Inc.*
[2] Lee H 2015 Finite element simulations with ANSYS Workbench 16 *SDC Publictions*
[3] Pavlenko I V 2007 Finite element method for the problems of mechanical systems oscillations *Sumy State University*
[4] Genta G 2009 Vibration dynamics and control *Springer Science+Business Media*
[5] Ishida Y, Yamamoto T 2012 Linear and nonlinear rotordynamnics. A modern treatment with

applications *Willey-VCH*
[6] Pavlenko I V, Simonovskiy V I, Demianenko M M and Verbovyi A E 2017 The analysis of rotor dynamics for the multistage centrifugal pump using computer programs “Critical frequencies of the rotor” and “Forced odcillations of the rotor” *Applied scientific and technical research, Ivano-Frankivsk, Ukraine Academy of Technical Sciences*
[7] Pavlenko I V, Simonovskiy V I, Demianenko M M and Verbovyi A E 2017 Numerical simulation of rotor dynamics for centrifugal machines *Mechanics of machines – the main component of applied mechanics, Dnipro, The National Metallurgical Academy of Ukraine*
[8] Gadyaka V G, Symonovskiy V I 2005 Estimation of segment bearing stiffness while balancing flexible rotors for turbocharge units in the accelerating-balancing stand *Bulletin of Sumy National Agrarian University, Series “Mechanization and automation of industrial processes”* **11** 145–150
[9] Simonovskiy V I 2010 Refinement of the mathematical models for oscillatory systems by experimental data *Sumy State University*
[10] Simonovskiy V I 2015 Estimation of coefficients for mathematical models of the oscillatory systems *LAP LAMBERT AcademicPublishing*
[11] Pavlenko I V and Symonovskiy V I 2015 Computer program “Critical frequencies of the rotor” *Ukraine Certificate of Authorship 59855*
[12] Pavlenko I V and Symonovskiy V I 2015 Computer program “Forced odcillations of the rotor” *Ukraine Certificate of Authorship 61788*
[13] Nemykin D V, Pavlenko I V, Simonovskiy V I 2016 Investigation of turbocharger rotor dynamics *Modern technologies in industrial production*, *Sumy, Sumy State University* **1** 158
[14] Verbovyi A E, Pavlenko I V 2017 Computer simulation and numerical calculation of rotor dynamics for the centrifugal compressor *Modern technologies in industrial production*, *Sumy, Sumy State University* **1** 146
[15] Simonovskiy V I, Pavlenko I V et al. 2015 Development of new mathematical models of rotors for centrifugal machines and methods of their diagnosis *Scientific research 0113U001522, Sumy, Sumy State University*
[16] Simonovskiy V I, Pavlenko I V et al. 2016 Investigation of oscillations for centrifugal machines rotors related to nonlinearity of reactions in bearings and seals gaps, and their vibrational diagnosis *Scientific research 0115U000549, Sumy, Sumy State University*
[17] Martsynkovskyy V A, Zahorulko A V et al. 2016 Numerical simulation of rotor dynamics for liquid rocket engines with considering dynamic characteristics of bearings and seals *Scientific research 0115U000679, Sumy, Sumy State University*
[18] Symonovskiy V I, Pavlenko I V et al. Investigations of turbopump rotor dynamics for liquid rocket engines *Scientific research 51.24-01.15.SP, Sumy, Sumy State University*
[19] Simonovskiy V I 1986 Stability and nonlinear oscillations of the centrifugal machines rotors *Kharkiv, Higher School*
[20] GOST ISO 11342-95 Vibration. Methods and criteria for balancing flexible rotors *Moscow, Standards Publishing House*

IOP Conf. Series: Materials Science and Engineering **233** (2017) 012054 doi:10.1088/1757-899X/233/1/012054

Effect of Bearing Housings on Centrifugal Pump Rotor Dynamics

A S Yashchenko[1], A A Rudenko[1], V I Simonovskiy[2], O M Kozlov[3]

[1]PJSC "VNIIAEN", 2, 2nd Zheleznodorozhnaya str., Sumy, 40022, Ukraine
[2]General Mechanics and Dynamics of Machines Department, Sumy State University, 2, Rimsky-Korsakov str., Sumy, 40007, Ukraine
[3]JSC "Nasosenergomash Sumy",1, Pryvokzalna Sq., Sumy, 40011, Ukraine

E-mail: admin@vniiaen.sumy.ua

Abstract: The article deals with the effect of a bearing housing on rotor dynamics of a barrel casing centrifugal boiler feed pump rotor. The calculation of the rotor model including the bearing housing has been performed by the method of initial parameters. The calculation of a rotor solid model including the bearing housing has been performed by the finite element method. Results of both calculations highlight the need to add bearing housings into dynamic analyses of the pump rotor. The calculation performed by modern software packages is more a time-taking process, at the same time it is a preferred one due to a graphic editor that is employed for creating a numerical model. When it is necessary to view many variants of design parameters, programs for beam modeling should be used.

1. Introduction

The centrifugal pump rotor is one of the most complex and vital assembly units that determines reliability of the whole pump to a large extent.

Major components of a rotor are a shaft, impellers, shaft couplings, shaft sleeves, parts of mechanical seals, balancing devices, etc. Depending on a pump design the rotor is manufactured with different mutual arrangement of parts. A cantilever pump impeller is placed on the shaft end. As a rule, rotor in a single-stage double entry pump and ring-section type pumps is kept by symmetrically located outer bearings.

Bearings are placed in bearing housings. Bearing housings could have different shapes and arrangements. Figure 1 shows bearing housing designs of the barrel casing centrifugal boiler feed pump [1] intended to feed water supply from a deaerator into a plant unit steam generator.

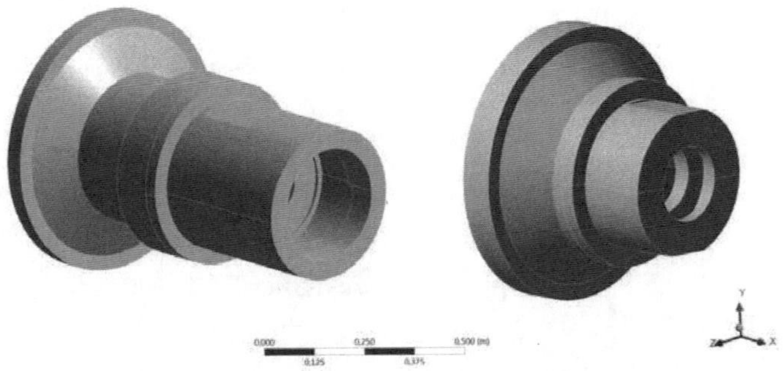

Figure 1. Design of Centrifugal Boiler Feed Pump Bearing Housings.

Rotor dynamics of the barrel casing centrifugal boiler feed pump rotor depends on various entities, the most important of which are hydrodynamic processes within plain bearings and annular seals, and the static stiffness of bearing housings.

The inclusion of the stiffness of bearing housing designs will allow to determine rotor dynamics of the barrel casing centrifugal boiler feed pump rotor more precisely.

When calculating rotor dynamics in conjunction with bearing housings it is possible to use both modern production software and programs for beam modeling. But when using such programs there are two key questions: execution time consumed by modern production software during a calculation, and accuracy of calculation being performed by beam modeling programs.

Following mathematical models of bearing elements of the design were used for research of system dynamics.

2. Plain Bearings

Plain journal bearings comprise the round cross-section shaft area (a shaft journal), rotating inside of a bearing shell which is round, as a rule. The diameter of the shaft journal is usually 99.8 – 99.9% of the bearing shell diameter, and resulting clearance is partially-filled with lubrication fluid.

A lubricating film in hydrodynamic plain bearings offers elastic property. Said another way, as a dynamic load is applied to the bearing, for example, as an unbalance consequence, the shaft journal has to move along round orbit about the static balance position, which it could be placed in case of only a static load presence. Effective stiffness and damping related to this elasticity of lubricating film have impact on critical vibration frequencies and system stability. Based on the above stated it is important to assess bearing dynamic characteristics during the design phase.

In study [2] Sommerfeld criterion (number) is used to determine the stiffness coefficient of hydrodynamic plain bearings.

$$S = \frac{\mu \cdot \omega \cdot D \cdot L}{W} \cdot \left(\frac{R}{h}\right)^2$$

where

μ - viscosity of the lubricating medium, $N \cdot f/m^2$;

ω - angular frequency of rotation, rad/s;

D – shaft diameter within the bearing;

L – bearing axial length, m;

W – bearing load, N;

R – shaft radius within the bearing, m;

h – radial clearance, m.

Depending on L/D and Sommerfeld number, stiffness and damping coefficients of the plain bearing are defined by the formula:

$$c_{xx}, c_{xy}, c_{yx}, c_{yy} = \frac{W \cdot \overline{K}_{xx}}{h}, \frac{W \cdot \overline{K}_{xy}}{h}, \frac{W \cdot \overline{K}_{yx}}{h}, \frac{W \cdot \overline{K}_{yy}}{h}$$

$$d_{xx}, d_{xy}, d_{yx}, d_{yy} = \frac{W \cdot \overline{B}_{xx}}{\omega \cdot h}, \frac{W \cdot \overline{B}_{xy}}{\omega \cdot h}, \frac{W \cdot \overline{B}_{yx}}{\omega \cdot h}, \frac{W \cdot \overline{B}_{yy}}{\omega \cdot h}$$

where $\overline{K}_{xx}, \overline{K}_{xy}, \overline{K}_{yx}, \overline{K}_{yy}, \overline{B}_{xx}, \overline{B}_{xy}, \overline{B}_{yx}, \overline{B}_{yy}$ are dimensionless coefficients varied with the Sommerfeld number.

3. The Bearing Housing

Stiffness of the bearing housing design C is defined in principle by the formula:

$$C = \frac{F}{\Delta x}$$

where F – force applied to the bearing shell, N;

Δx – displacement of the bearing induced by force F, m.

Software ANSYS Workbench 14.5 [3] was used to perform calculations. Table 1 lists stiffness values of bearing housings of the barrel casing centrifugal boiler feed pump.

Table 1. Results of the static analysis of the bearing housing displacement.

Location of the bearing	Loading	Stiffness of the design C, N/m
The pump bearing, drive end	in horizontal direction	$4.02 \cdot 10^9$
	in vertical direction	$4.50 \cdot 10^9$
The pump bearing, non-drive end	in horizontal direction	$2.50 \cdot 10^8$
	in vertical direction	$3.95 \cdot 10^8$

Mass of the drive end bearing housing is 127 kg, mass of the non-drive end bearing housing is 70 kg.

4. Analysis of Pump Rotor Vibration by the Method of Initial Parameters. Evaluation of the Bearing Housing Effect

When evaluating free bending vibration of the rotor subsystem, the pump rotor is simulated as a circular beam of a piecewise constant stiffness and its mass per meter with mass of shaft mounted parts (impellers, shaft sleeves, half-coupling, etc.) being discretely placed on the beam.

Bounds between the rotor and the pump casing are simulated by elastic supports at points of the bearing and annular seal location. Bearing housings are simulated by mass and stiffness of the bearing housing design.

Figure 2 shows such simulated model of the pump rotor. This is an unbranched frame structure comprising of the chain-arranged set sections of different types.

The calculation of rotor free vibration is performed by the method of initial parameters in a matrix form [4-6]. The calculation of the stiffness (Nm) of pump annular seals C is performed in accordance to the procedure [7].

Table 2 lists values of the first three natural frequencies of pump rotor vibration including the bearing housing design and without including it that have been obtained by the method of initial parameters.

Table 2. Values of first three natural frequencies of rotor vibration that have been obtained by the method of initial parameters.

Natural frequencies of vibration	f_1	f_2	f_3
Without including the bearing housing design, Hz	79.10	123.85	21.56
Including the bearing housing design, Hz	74.20	119.76	209.11
Relative impact of the bearing housing, %	-6.2	-3.3	-4.3

IOP Conf. Series: Materials Science and Engineering **233** (2017) 012054 doi:10.1088/1757-899X/233/1/012054

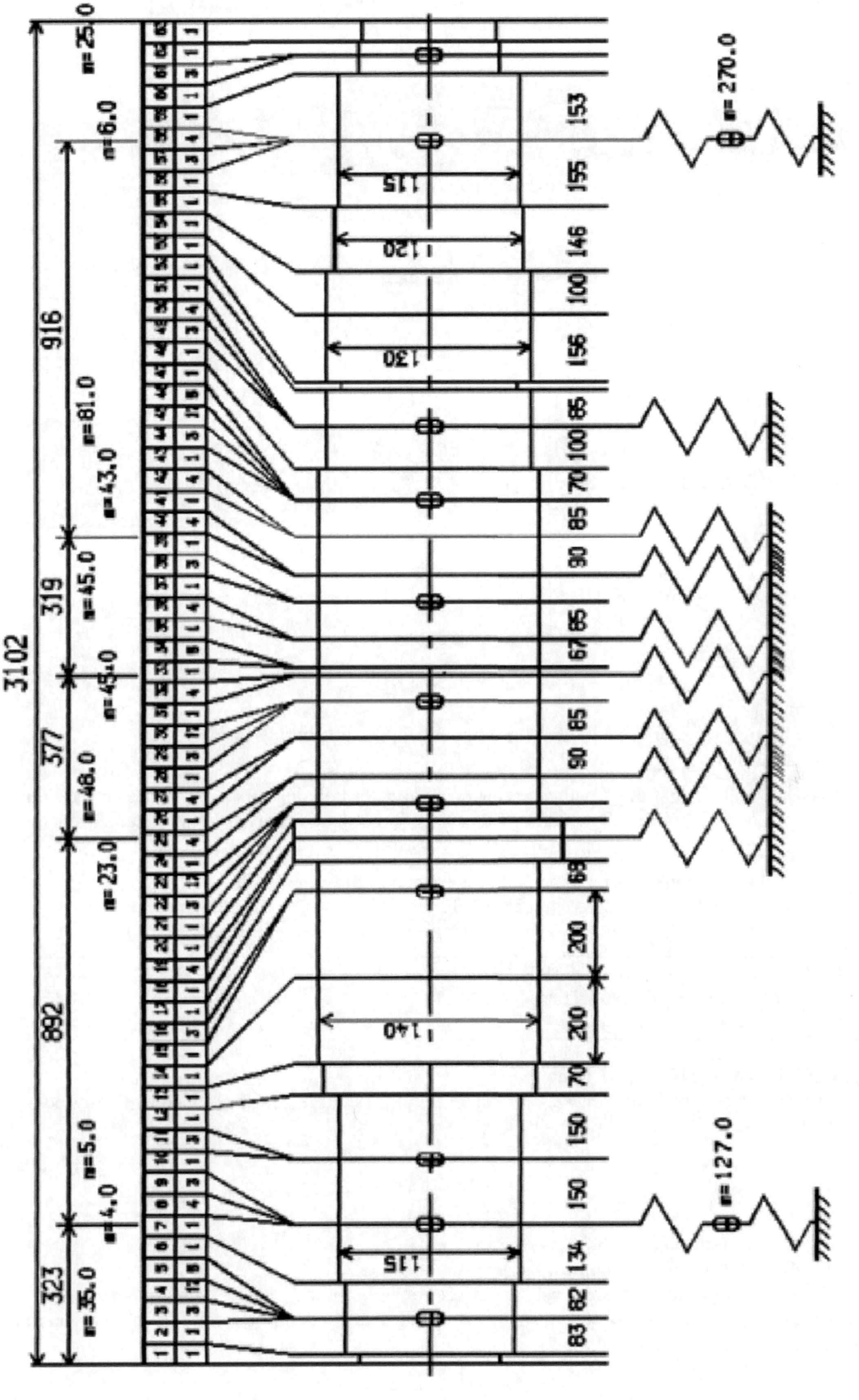

Figure 2. Simulation model of the barrel casing centrifugal boiler feed pump.

5. Analysis of Pump Rotor Vibration by the Finite Element Method

Figure 3 shows the rotor solid model of the barrel casing centrifugal boiler feed pump.

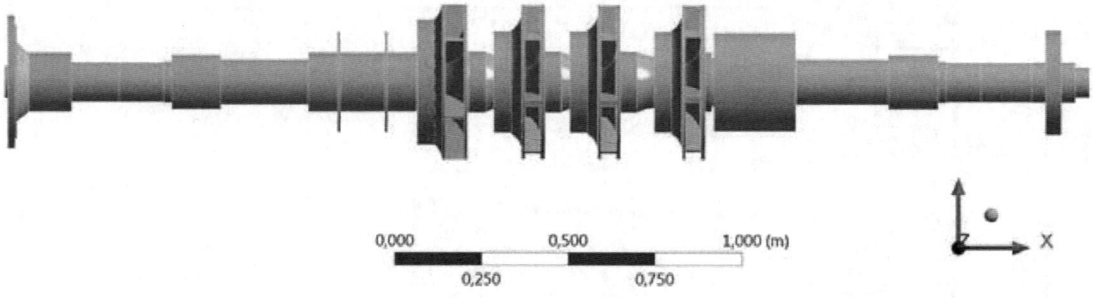

(a)

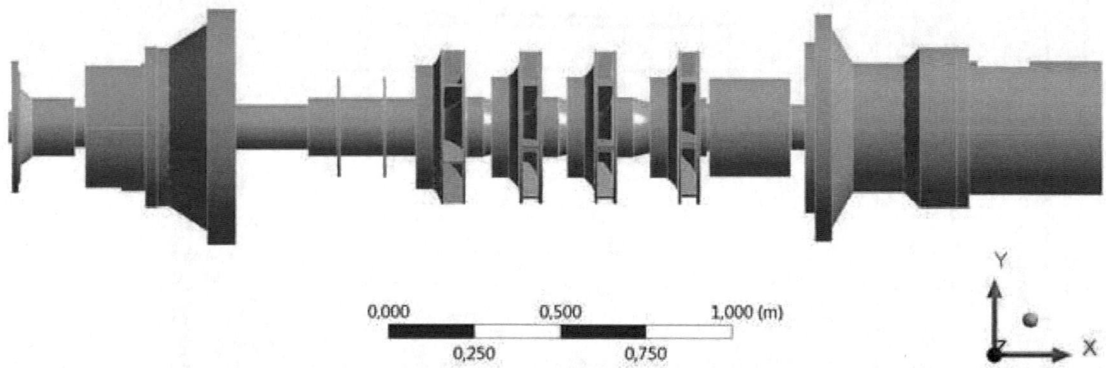

(b)

Figure 3. Rotor solid model: (a) without bearing housing; (b) with bearing housing.

Table 3 gives values of first three natural frequencies of rotor bending vibration including the pump bearing housing design and without including it that have been obtained by the finite element method (FEM).

Table 3. Values of first three natural frequencies of rotor bending vibration that have been obtained by FEM.

Natural frequencies of vibration	f_1	f_2	f_3
Without including the bearing housing design, Hz	84.47	132.83	207.89
Including the bearing housing design, Hz	80.47	129.26	202.33
Relative impact of the bearing housing, %	-4.7	-2.7	-2.7

6. Analysis of the Bearing Housing Effect

Equivalent stiffness of the "bearing housing - lubricating film" system $K_{eq.}$ is determined by the formula:

$$K_{eq} = \frac{K_{b.h.} \cdot K_{l.f.}}{K_{b.h.} + K_{l.f.}}$$

where

$K_{b.h.}$ is the stiffness of the bearing housing design;

$K_{l.f.}$ is the stiffness of the lubricating film.

High-stiffness bearing housing designs has less effect on pump rotor dynamics than low-stiffness ones.

When engineering calculation of rotor dynamics is performed, we can neglect the stiffness of bearing housings if following inequation is satisfied:

$$K_{b.h.} > 10 \cdot K_{l.f.}$$

IOP Conf. Series: Materials Science and Engineering **233** (2017) 012054 doi:10.1088/1757-899X/233/1/012054

7. Conclusions

According to performed numerical studies it was found that rotor dynamics of the barrel casing centrifugal pump depend on bearing housing designs.

Production software with the graphic editor for creating simulation models should be used to perform calculations of the bearing housing stiffness. Then obtained stiffness values should be used in calculations by means of beam models. Therewith numerical calculations of rotor dynamics are in good agreement with simulation results. Simulation analysis is a time-consuming process; this could be used as a final, validation method.

When it is necessary to view many variants of design parameters (in consideration of their variation range while in operation or for viewing design variants being developed) we could use programs for beam modeling.

References

[1] Martsykovskyy V A, Vorona P N 1987 *Pump for NPP* (Energoatomizdat, Moscow)
[2] J W Lund and K K Thomson 1987 A Calculation Method and Data for the Dynamic Coefficients of Oil Lubricated Journal Bearings *Proceedings of the ASME Design and Engineering Conference* (Minneapolis) p 1
[3] Programm Complex ANSYS 14.5, license agreement 673888
[4] Babakov I M 1965 *Vibration Theory* (Nauka)
[5] Ivovich V A 1981 *Transfer Matrix in Elastodynamics: Guide, 2nd enlarged edition* (Mashinostroenie, Moscow)
[6] Simonovskiy V I 2006 *Centrifugal Machine Rotordynamics* (Sumy State University, Sumy)
[7] Martsykovskyy V A 1980 *Noncontacting Seals of Rotary Machines* (Mashinostroenie Moscow)

IOP Conf. Series: Materials Science and Engineering **233** (2017) 012055 doi:10.1088/1757-899X/233/1/012055

Dynamics of the rotor on elastic-damping supports under action of kinematic effects

V Chernyshev[1], L Savin[1], O Fominova[1]

[1]Department of Mechatronics and International Engineering, Orel State University, Naugorskoe sh. 29, Orel, Russia, 302020

E-mail: chernyshev_46@mail.ru

Abstract. The article describes the elements of the theory of dynamic analysis of rotor systems. The mathematical model of a gyroscopic rotor as an elementary object on elastic-damping supports. The results of simulation of the trajectories of the rotor under kinematic loading with amplitude commensurate with the clearance in bearing assemblies of fluid friction.

1. Introduction

Since the rotor is considered an object of the gyroscopic type, and is a fast rotating symmetric body, the axis of rotation of which can change its direction in space, it is natural that this circumstance should be taken into account when designing mathematical models and choosing the appropriate inertial and elastic-damping parameters [1].

The observed gyroscopic effects due to the unusual reaction of the rotor to the "long" exposure to some influence when the force tends to rotate the rotor around one axis, and it rotates around another perpendicular axis, where precession occurs. When this precessionalmotion occurs during the entire time of action of external force and is terminated with a termination of the external forces' action, that identifies the property of inertia-free motion. Under the action of impulse impact the axis of the rotor practically does not change the original direction, but only performs fast oscillations with a small amplitude, that is the nutation motion [2].

When developing a mathematical model of the rotor on elastic-damping supports, as a rule, one has to overcome certain difficulties, which are mainly associated with the high dimensionality of the object being studied and the problem of adequate description of external force factors. In this case, during the preliminary stage of the study, it seems justified to consider a simplified mathematical model of the rotor, which reflects the main features of the dynamics of the rotor as a gyroscopic object and allows to evaluate the intensity of its vibrations [3]. It is obvious that the rate of vibration safety of the rotor is significantly affected by penetrating vibration propagating through the chain of "driver – housing – rotor". This is because the RMS value of the vibration is generally commensurate with the size of the gap in the fluid-film bearings, and their basic frequency can be taken equal to the natural frequency of the rotor.

The presence of a penetrating vibration determines the need to consider its effect on rotor's dynamic behavior. This can be done by reproducing the relevant kinematic effects directly in the process of integration of the mathematical model of the rotor. Note that the rotor on elastic-damping supports perceives the kinematic effects of using these supports and coupling it to the driver.

IOP Conf. Series: Materials Science and Engineering **233** (2017) 012055 doi:10.1088/1757-899X/233/1/012055

2. Model of an elementary gyroscopic object

The accepted calculation scheme of the elementary gyroscopic object in the form of a massive rapidly rotating disk in contact with the elastic-damping support shown in Figure 1.

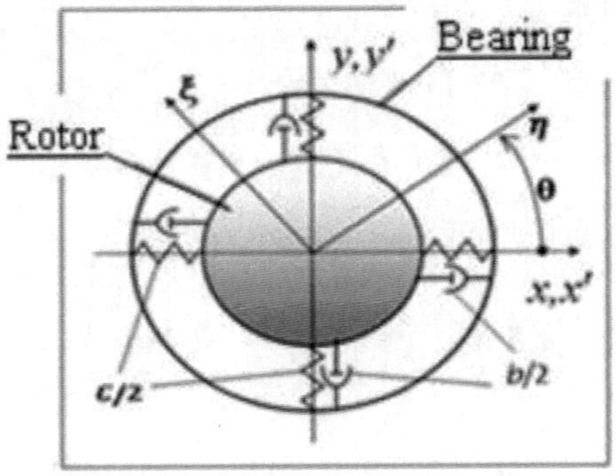

Figure 1. Diagram of the elementary gyroscopic object
Note: m – mass of the disk; J– axial momentum of inertia of the disc; c – stiffness of an elastic medium; b – viscous resistance of the medium.

It is assumed that the center of mass of the disc is at its geometric axis. The position of the disc is determined in fixed (absolute) coordinate system xy. In addition,additional mobile (non-inertial) coordinate systems $x'y'$ and $\eta\xi$ are introduced with a common origin. The coordinate system $x'y'$ moves forward together with the centre of mass of the disk.

The $\eta\xi$ system of coordinates is rigidly attached to the disk and rotates with it with angular velocity $\dot{\theta}=\omega$. It is believed that the resultant restoring and dissipative forces are directed from the center of mass of the rotor to the geometric axis of the support (bearing'ssleeve). In this case, the momentum of these forces about the center of mass of the disk is zero and in the steady-state regime, when $J\ddot{\theta}=0$ we have $\dot{\theta}=const.$

During the movement of the rapidly rotating disk, it is, apart from restoring and dissipative forces, influenced by the gyroscopic force. The influence of the latter will be taken into account by introducing of an additional component into the equations of motions in the form of the product of the mass of the disc m on some unknown subject and the definition of acceleration$w$. The projection of the additional component of the mw on the axis xy is equal to: mw_x and mw_y.

Let usassume that the bearing's sleeve is stationary. Then, the result of the application of the theorem of motion of center of mass,is the following interrelated differential equations:

$$\begin{aligned} m\ddot{x} + mw_x + b\dot{x} + cx = 0 \\ m\ddot{y} + mw_y + b\dot{y} + cy = 0 \end{aligned} \tag{1}$$

To obtain a closed system of differential equations we use the kinematicrelations that connect the projection of the velocity and acceleration of the center of mass of the disk on the axis of the stationary xy and rotating ηξ coordinate systems:

$$\begin{pmatrix} \dot{x} \\ \dot{y} \end{pmatrix} = R(\theta)\begin{pmatrix} \dot{\eta} \\ \dot{\xi} \end{pmatrix}, \qquad \begin{pmatrix} \ddot{x} + w_x \\ \ddot{y} + w_y \end{pmatrix} = R(\theta)\begin{pmatrix} \ddot{\eta} \\ \ddot{\xi} \end{pmatrix} \tag{2}$$

Here the rotation matrixis used:

$$R(\theta)=\begin{pmatrix} \cos\theta & -\sin\theta \\ \sin\theta & \cos\theta \end{pmatrix} \tag{3}$$

which has the following properties:

$$R(\theta)R^{-1}(\theta)=\begin{pmatrix} 1 & 0 \\ 0 & 1 \end{pmatrix}, \quad R(\theta)\dot{R}^{-1}(\theta)=\begin{pmatrix} 0 & +\dot{\theta} \\ -\dot{\theta} & 0 \end{pmatrix} \tag{4}$$

Performing the obvious sequence of matrix transformations:

$$\begin{pmatrix} \dot{\eta} \\ \dot{\xi} \end{pmatrix} = R^{-1}(\theta)\begin{pmatrix} \dot{x} \\ \dot{y} \end{pmatrix} \rightarrow \begin{pmatrix} \ddot{\eta} \\ \ddot{\xi} \end{pmatrix} = \dot{R}^{-1}(\theta)\begin{pmatrix} \dot{x} \\ \dot{y} \end{pmatrix} + R^{-1}(\theta)\begin{pmatrix} \ddot{x} \\ \ddot{y} \end{pmatrix}$$

$$\begin{pmatrix} \ddot{x}+w_x \\ \ddot{y}+w_y \end{pmatrix} = R(\theta)\left[\dot{R}^{-1}(\theta)\begin{pmatrix} \dot{x} \\ \dot{y} \end{pmatrix} + R^{-1}(\theta)\begin{pmatrix} \ddot{x} \\ \ddot{y} \end{pmatrix}\right] \tag{5}$$

$$\begin{pmatrix} \ddot{x}+w_x \\ \ddot{y}+w_y \end{pmatrix} = \begin{pmatrix} 0 & +\dot{\theta} \\ -\dot{\theta} & 0 \end{pmatrix}\begin{pmatrix} \dot{x} \\ \dot{y} \end{pmatrix} + \begin{pmatrix} 1 & 0 \\ 0 & 1 \end{pmatrix}\begin{pmatrix} \ddot{x} \\ \ddot{y} \end{pmatrix}$$

From the last matrix relationship (5) implies that acceleration:

$$w_x=\dot{\theta}\,\dot{y}, \; w_y=-\dot{\theta}\,\dot{x} \tag{6}$$

With (6) the system of differential equations of motion of the disc (1) can be written in the form:

$$\begin{aligned} m\ddot{x}+m\dot{\theta}\,\dot{y}+b\dot{x}+cx&=0 \\ m\ddot{y}-m\dot{\theta}\,\dot{x}+b\dot{y}+cy&=0 \end{aligned} \tag{7}$$

As it can be seen, the total force component of the gyroscopic force is zero.

Note that in the case of the kinematic effects, when the bearing's sleeve is not stationary, the system of differential equations of motion of the disk is transformed to the following:

$$\begin{aligned} m\ddot{x}+m\dot{\theta}\,\dot{y}+b(\dot{x}-\dot{g}_x)+c(x-g_x)&=0 \\ m\ddot{y}-m\dot{\theta}\,\dot{x}+b(\dot{x}-\dot{g}_y)+c(x-g_y)&=0 \end{aligned} \tag{8}$$

Here g_x, g_y and $\dot{g}_x$, $\dot{g}_y$ – projections of the displacement and velocity of the bearing'ssleeve on the xy axis.

3. Analysis of the equations of motion

Considering the system of differential equations (7). Using substitutions $c=mk^2$, $b=mn$ and $\dot{\theta}=\omega$, we get:

$$\begin{aligned} \ddot{x}+\omega\dot{y}+n\dot{x}+k^2x&=0 \\ \ddot{y}+\omega\dot{x}+n\dot{y}+k^2y&=0 \end{aligned} \tag{9}$$

Let us seek the solution to (9) as functions:

$$x=Ae^{\lambda t}, \; y=Be^{\lambda t} \tag{10}$$

Let us find the pairsλ, V of the system matrix

$$M = \begin{pmatrix} 0 & 1 & 0 & 0 \\ -k^2 & -n & 0 & -\omega \\ 0 & 0 & 0 & 1 \\ 0 & \omega & -k^2 & -n \end{pmatrix} \tag{11}$$

In result of calculationswe shall get the following:

$$\lambda = \begin{pmatrix} -\frac{n+i\omega}{2} + \sqrt{\left(\frac{n+i\omega}{2}\right)^2 - k^2} \\ -\frac{n-i\omega}{2} + \sqrt{\left(\frac{n-i\omega}{2}\right)^2 - k^2} \\ -\frac{n+i\omega}{2} - \sqrt{\left(\frac{n+i\omega}{2}\right)^2 - k^2} \\ -\frac{n-i\omega}{2} - \sqrt{\left(\frac{n-i\omega}{2}\right)^2 - k^2} \end{pmatrix} = \begin{pmatrix} -\alpha_1 + i\beta_1 \\ -\alpha_1 - i\beta_1 \\ -\alpha_2 + i\beta_2 \\ -\alpha_2 - i\beta_2 \end{pmatrix} \tag{12}$$

$$V = \begin{pmatrix} 1 & 1 & 1 & 1 \\ -\beta_1 & \beta_1 & -\beta_2 & \beta_2 \\ -1 & 1 & -1 & 1 \\ -\beta_1 & \beta_1 & \beta_2 & -\beta_2 \end{pmatrix} \tag{13}$$

In particular, when $n = 0$, the components of the vector λ are imaginary:

$$\lambda_{1,2,3,4} = \pm i\sqrt{-\frac{2k^2+\omega^2}{2} \pm \sqrt{\left(\frac{2k^2+\omega^2}{2}\right)^2 - k^4}} \rightarrow \lambda_{1,2} = \pm i\beta_1,\ \lambda_{3,4} = \pm i\beta_2 \tag{14}$$

Because $\alpha_1 \langle \alpha_2$ and $\beta_1 \langle \beta_2$, the precessionalmotion is determined respectively by the damping ratio α_1 and frequency β_1, nutation movement – by damping ratio α_2 and frequency β_2.

From the formula for determining λ, it follows that the greater the angular rotor speed ω, the lower the frequency of the precessional motion β_1 and, conversely, the more the frequency of nutation motion β_2.

Relation $(A, B)V = \lambda_i V_i$ allows determination that $B_1 = A_2,\ B_2 = -A_1,\ B_3 = -A_4,\ B_4 = A_3$.

So, the general solution of system (9) is written in the following form:

$$\begin{aligned} x_c(A,t) &= e^{-\alpha_1 t}(A_1 \cos\beta_1 t + A_2 \sin\beta_1 t) + e^{-\alpha_2 t}(A_3 \cos\beta_2 t + A_4 \sin\beta_2 t) \\ y_c(A,t) &= e^{-\alpha_1 t}(-A_1 \sin\beta_1 t + A_2 \cos\beta_1 t) + e^{-\alpha_2 t}(-A_3 \sin\beta_2 t + A_4 \cos\beta_2 t) \end{aligned} \tag{15}$$

From equations (15), it follows that when all initial conditions of the trajectory of precession and nutation motion are the torsion spirals (in the absence of damping it is a circle). This result displays the fact that the rotor has a pronounced error-correcting (instantaneous) properties.

IOP Conf. Series: Materials Science and Engineering **233** (2017) 012055 doi:10.1088/1757-899X/233/1/012055

Note that when $\alpha_1 \langle\langle \alpha_2$ transient (nutation) processes are quickly damped, and decrease in the radius of the spiral path will be slow.

Let us now consider the system of differential equations (8). Using substitutions $c = mk^2$, $b = mn$ and $\dot{\theta} = \omega$, we get:

$$\begin{aligned} \ddot{x} + \omega\dot{y} + n\dot{x} + k^2 x &= n\dot{g}_x + k^2 g_x \\ \ddot{y} - \omega\dot{x} + n\dot{y} + k^2 y &= n\dot{g}_y + k^2 g_y \end{aligned} \tag{16}$$

Taking $g_x = y_0 \sin \omega t \to -iy_0 e^{i\omega t}$ and $g_y = y_0 \cos \omega t \to y_0 e^{i\omega t}$, we seek particular solutions of system (16) as functions:

$$x(t) = x_h(t) \to K_x e^{i\omega t}, \quad y(t) = y_h(t) \to K_y e^{i\omega t} \tag{17}$$

Carrying out the appropriate substitutions and transformations, we shall find the complex amplitude K_x and K_y.

It is determined that the complex amplitudes K_x and K_y are equal in magnitude

$$|K_x| = |K_y| = K = y_0 \sqrt{\frac{\omega^2 n^2 + k^4}{4\omega^4 - 4\omega^2 k^2 + \omega^2 n^2 + k^4}} \tag{18}$$

and thus have the following phases

$$\begin{aligned} \varphi x &= a\,tan \frac{\omega^2 n^2 - 2\omega^2 k^2 + k^4}{2\omega^3 n} \\ \varphi y &= a\,tan \frac{-2\omega^3 n}{\omega^2 n^2 - 2\omega^2 k^2 + k^4} \end{aligned} \tag{19}$$

Hence, particular solutions of the system (16):

$$x_h(t) = K \sin(\omega t + \varphi x), \quad y_h(t) = K \cos(\omega t + \varphi y) \tag{20}$$

Uniting (15) and (20) finding that

$$x(t) = x_c(A,t) + x_h(t), \quad y(t) = y_c(A,t) + y_h(t) \tag{21}$$

Constants A_i are determined from the matrix equation:

$$A = \begin{pmatrix} 1 & 0 & 1 & 0 \\ 0 & 1 & 0 & 1 \\ -\alpha_1 & \beta_1 & -\alpha_2 & \beta_2 \\ -\beta_1 & -\alpha_1 & -\beta_2 & -\alpha_2 \end{pmatrix}^{-1} \begin{pmatrix} x(0) - Kx_h(0) \\ y(0) - Ky_h(0) \\ \dot{x}(0) - K\dot{x}_h(0) \\ \dot{y}(0) - K\dot{y}_h(0) \end{pmatrix} \tag{22}$$

4. The results of calculations and modeling

For the calculation and modeling of systems of equations (7), (8) the following main parameters are chosen: $\omega = 820\ s^{-1}$, $m = 1\ kg$, $c = 32000\ N/m$, $b = 5\ N\,s/m$. Variation of these parameters was carried out with the purpose of obtaining additional information and expansion of ideas about the properties of the system.

IOP Conf. Series: Materials Science and Engineering **233** (2017) 012055 doi:10.1088/1757-899X/233/1/012055

The influence of the angular velocity of the rotor, which is a gyroscopic object, on the character of manifestation of nutation processes is shown in the Figure 2.

For transients processesto be well observed, an impulse impact was simulated by setting the initial speed of the rotor $\dot{x}(0) \neq 0$. The angular speed of the rotor in each numerical experiment was different: $\omega = 500 s^{-1}$, $\omega = 1000\ s^{-1}$, $\omega = 1500\ s^{-1}$.

As expected, the amplitude and duration of high frequency oscillation of the rotor associated with the nutation induced processes decrease the faster, the more its angular velocity. The results of the numerical experiments confirm also that the rotor as a gyroscopic object is resistant to impact.

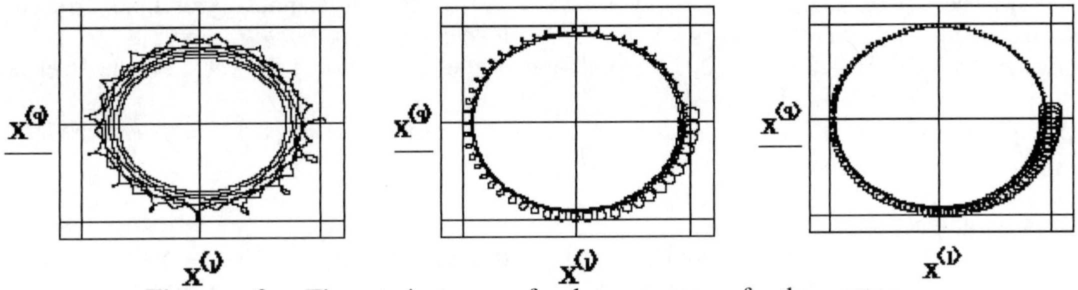

Figure 2. The trajectory of the center of the rotor
Note: $X^{\langle 1 \rangle} \leftrightarrow x(t)$, $X^{\langle 3 \rangle} \leftrightarrow y(t)$.

Figure 3 shows the characteristic oscillations of the center of the rotor in the absolute and relative motion under action of kinematic effects.

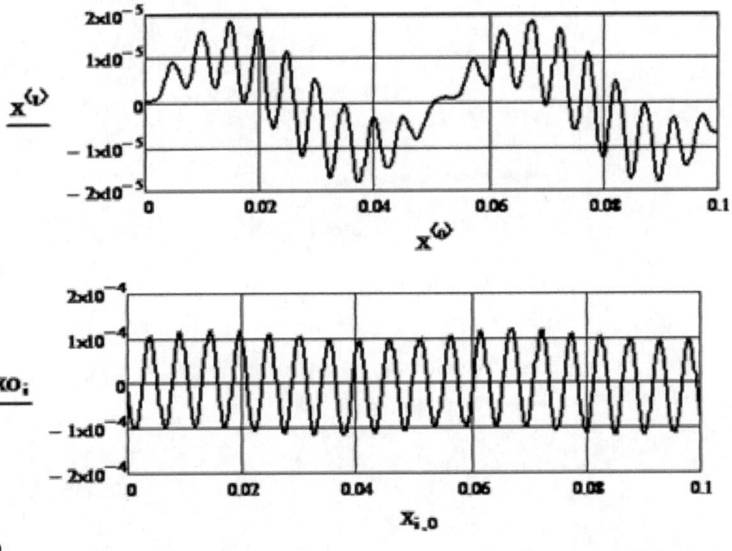

(a)

(b)

Figure 3.Oscillations of the center of the rotor in absolute (a) and relative (b) motion. Note:

$$X^{\langle 1 \rangle} \leftrightarrow x(t), XO_i = x(t) - y(t),\ X^{\langle 0 \rangle} = X_{i,0} \leftrightarrow t, \omega = 1200 s^{-1},$$

$$m = 1\ kg, c = 16 \cdot 10^4\ N/m, b = 0{,}1 N c/m,\ y(t) = 10^{-4}\ m.$$

In the absolute motion of the rotor, excited by a vibration and nutation processes with small amplitude and high frequency close to the frequency of the kinematic effects are superimposed on the precessional motion of the rotor, the frequency of which the smaller, the more its angular velocity. Since the maximum amplitude of the complex motion of the rotor is smaller than the amplitude of the kinematic effects, relative motion parameters of oscillograms of oscillation of the rotor practically do not differ from the parameters of the kinematic effects.

5. Conclusions

Proposed model of the rotor in the form of a massive rapidly rotating disk of elastic-damping support taking into account the accepted assumptions allows taking into account gyroscopic effects and evaluation of their influence on dynamic processes under vibrational kinematic effects.

References

[1] Chernyshev I V, Savin L A, Fominova O V 2016 Rotor dynamics-gyroscopic effects and management *Fundamental and applied problems of technics and technology* **5** pp 3-8
[2] Arenew S V, Fedyushkin A S 2013 Simulation of the vibrations of the rotor on the steel beam support *Mashinostroenie* **2** pp15-21
[3] Savin L A, Solomin O V 2006 *Modeling of rotary systems with bearings of fluid friction* (Moscow: Mashinostroenie-1) p 443

Acknowledgments

The present research was implemented as a part of the RSF project No. 16-19-00186 'Planning of optimal energy efficient rotor trajectories in mechatronic modules with complex rheology medium'.

IOP Conf. Series: Materials Science and Engineering **233** (2017) 012056 doi:10.1088/1757-899X/233/1/012056

Static calculation of the rotor unloading automatic machine for a high-pressure centrifugal pump

V A Martsynkovskyy[1], A Deineka[1], A Korczak[2] and G Peczkis[2]

[1]Sumy State University, Department of General Mechanics and Dynamics of Machines, 2, Rimsky-Korsakov Street, Sumy 40007, Ukraine
[2]Silesian University of Technology, Faculty of Energy and Environmental Engineering, St. Konarski 18 Street, 44-100 Gliwice, Poland

E-mail: a.dejneka@omdm.sumdu.edu.ua

Abstract. The article presents static calculation of an automatic machine for unloading of rotor axial forces in a high-pressure and high-speed pump. Analysis of the rotor displacement influence to values of mass flow rate and pressure in the unloading chamber of the unloading system is given. The main geometric parameters of the system are obtained.

1. Introduction

Most turbo pump units (TPU) are equipped with an automatic unloading machine (AUM) for the axial forces discharge, which consists of two tandem throttles (Figure 1) Throttle 5 is an annular gap which resistance is not influenced by the rotor axial displacement. The second throttle is a face gap, the resistance of which in turbulent flow regimes is inversely proportional to the axial gap value to the power of 1.5. In the first approximation, one can assume that radial vibrations of the impeller have no effect on the throttle's resistance. Rotor displacement to the left side reduces the face gap 4 and increases pressure in the chamber 3. It increases axial force pressure acting on the impeller, and rotor shifts to the right, approaching to its initial state. Thus a face throttle 4 provides a negative feedback between rotor position and the balancing axial force [1]. Gaps 4 and 5 located between fixed and rotating surfaces pose a threat of scoring, since small force and temperature deformations, as well as deformations from centrifugal loads, can cause local contact of rubbing surfaces [2].

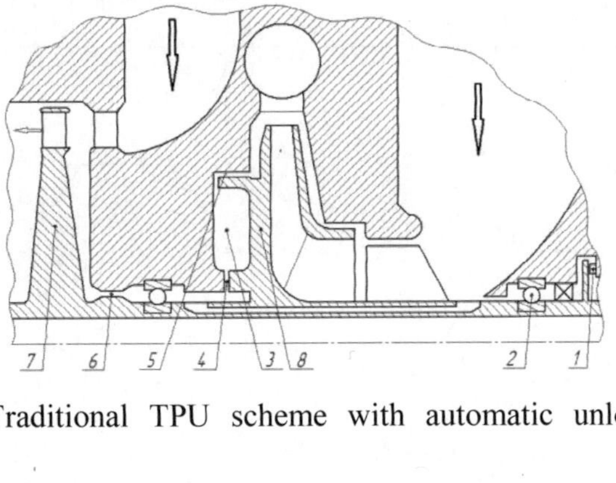

Figure 1. Traditional TPU scheme with automatic unloading machine.

IOP Conf. Series: Materials Science and Engineering **233** (2017) 012056 doi:10.1088/1757-899X/233/1/012056

Figure 2 shows a calculation scheme which differs from the traditional one, primarily by the feedback mechanism: unloading chamber is limited by two changing reciprocal slotted throttles. When rotor is moved to the right, the conductance g_1 of the upper throttle increases (hydraulic resistance decreases), and conductance of the lower one - g_2 decreases. Pressure p_2 in the unloading chamber increases, i.e. balancing force, returning rotor to its initial position increases. Scoring risk is much less due to reducing surface area forming the throttling groove.

High pressure values developed by the TPU cause turbulent flow regimes in the main hydraulic channel, so the relationship between mass flow rate and throttled pressure drop is characterized by a quadratic dependence

$$Q = g\sqrt{\Delta p}$$

where

Δp - throttled pressure drop in this section;

g - hydraulic conductance of the section.

Hydraulic system and conductance dependence of the rotor axial displacements significantly affect the AUM characteristics. Analytical representation of such dependences or consumption characteristics of the throttles are possible only for the channels having the simplest geometric form [3]. As a rule, it is necessary to use the results of experiments or perform numerical calculations with the help of appropriate computer programs.

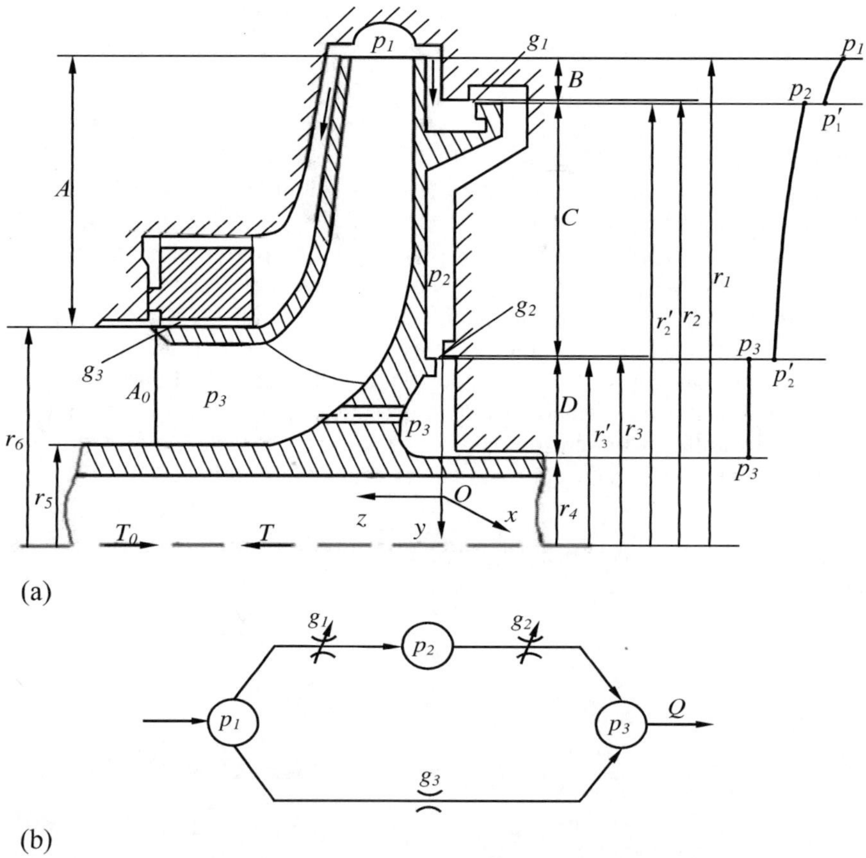

Figure 2. Calculation scheme of a centrifugal stage with an automatic unloading machine (a) and a hydraulic channel scheme (b).

Basic geometric parameters of the AUM are selected based on the static calculation results. It comes down to the fact that using the flow balance equation of the series-connected throttles and rotor axial balance equation one can obtain a dependence of steady axial displacement on the discharge pressure, unbalanced force T_0 and rotational speed, i.e. static characteristics are built. Static calculation also makes it possible to determine hydrostatic stiffness of the system - dependence of the regulatory impact (pressure p_2 in the unloading chamber) on the adjustable gap in the throttles. Negative value of the hydrostatic stiffness coefficient is a sign of the system equilibrium stability [4,5].

Unit rotor together with the automatic unloading machine is a complex dynamic system with distributed parameters that is a subject to periodic external perturbations. Each cross section of the rotor performs interrelated radial, angular and axial oscillations [6]. In regards to reliability, primarily axial vibrations can be dangerous, since their relatively large amplitudes limit the operating life of ball bearings wherein rotor is located. In this paper, as a first approximation, a simplified problem is considered: a rotor is studied as a solid body with an auto-loading system which performs one-dimensional axial oscillations along the support axis. Such a simplified model enables to obtain static and dynamic characteristics in an analytical way and accurately describe the main regularities of the system oscillations.

2. Calculation of the axial forces acting on the impeller

Purpose of the static calculation is to select such basic geometric parameters so that in the given variation range of the changing balancing force T_0 face gaps $z_{1,2}$ and flow rate Q values do not exceed the permissible limits. Analytical model of the automatic balancing system should take into account all the factors that have a significant effect on the system operation and at the same time it should not be cluttered with unnecessary details that complicate the analysis and give only minor quantitative corrections to the calculation results. When assessing the significance of certain factors, it must be borne in mind that an error in calculating equilibrated axial force, even in the nominal mode, can reach 50%, not to mention changing of this force during pump operation [7,8].

Balancing system will be considered as a hydro-mechanical automatic control system, for which the adjustable value is a size of the face gap. Regulatory impact F_z is a resultant axial force of pressure acting on the outer side of the main disk (balancing force). External influences are the discharge pressure p_1 and rotor speed ω. If a squeezing device is available the force of pre-compression of the springs $k\Delta$ acts as a control input. Block diagram of the balancing device as an automatic control system is shown in Figure 3.

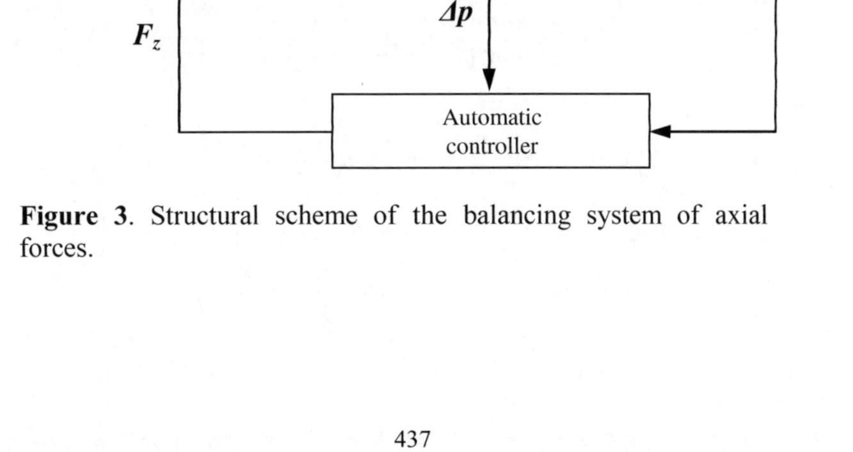

Figure 3. Structural scheme of the balancing system of axial forces.

Developing of static characteristics is based on the condition of equilibrium of the pressure axial forces acting on the impeller and the residual axial force T_0 on the side of the drive turbine [9]. The last force is considered to be given.

At present, an approximate estimate of the axial forces is obtained, as a rule, by taking the average liquid angular velocity across the chamber width to be stable: $\omega_c = 0{,}5\omega$ [10, 11]. A precise calculation of the angular velocity and pressure in the pockets of the impeller comes down to solving the problem of hydromechanics of a nonstationary three-dimensional turbulent flow with complex boundary conditions.

For preliminary estimation of axial forces we will use the recommendations of prof. V. Endrala [12], based on the numerical solution of the Reynolds equations of turbulent flow, both in the circumferential and radial directions. The recommendations' main message is that the average liquid angular velocity across the gap width ω_c is expressed with the swirl coefficient κ_0:

$$\omega_c = \kappa_0 \omega \tag{1}$$

Swirl coefficient of the flow depends primarily on the radial flow direction and for the relative widths of the chambers $b/r_2 = 0{,}02...0{,}04$ it has the following radial averaged values:

flow from the periphery to the center $\kappa_1 = 0{,}7...0{,}9$,

flow from the center to the periphery $\kappa_2 = 0{,}3...0{,}4$,

In the absence of radial flow $\kappa_0 = 0{,}45...0{,}5$.

In our case, in both pockets, from the covering disk and from the main disk, the radial (feed) flow is directed from the periphery to the center, so both pocket flows will be assumed: $\kappa \approx \kappa_1$

To calculate the axial forces acting on the impeller one should consider a section of the main disk $r_1 - r_2$ (chamber *B*, Figbre 2a). Sections of individual disk parts are indicated by capital letters *A, A₀, B, C, D*.

To find pressure distribution along the radii of the disks it is necessary to consider equilibrium of the elementary liquid volume dV (Figure 4), bounded by two cylindrical surfaces with radii r and $r + dr$ and meridional planes placed at an angle $d\varphi$ to each other: $dV = brdrd\varphi$ The average angular velocity is assumed to be constant for all chambers with a swirl coefficient of 0.7: $\omega_c = \kappa_1\omega = 0{,}7\omega$.

In the radial direction centrifugal force $dF_B = \omega_c^2 r\rho dV = \rho\omega_c^2 br^2 drd\varphi$ is balanced by the pressure force $dP = brd\varphi dp$. Comparing these forces, an equation $dp = \rho\omega_c^2 rdr$ is derived. Having integrated it considering the boundary condition $r = r_1, p = p_1$, it is obtained a parabolic law of pressure variation along the radius:

$$p_B = p_1 - 0{,}5\rho\omega_c^2\left(r_1^2 - r^2\right) = p_1 - \frac{1}{2}\rho\omega^2\kappa_1^2\left(r_1^2 - r^2\right) \tag{2}$$

By analogy for the section $r_2 - r_3$ (chamber *C*) and for the left chamber *A* the equations are the following:

$$p_C = p_2 - 0{,}5\rho\omega_c^2\left(r_2^2 - r^2\right) = p_2 - \frac{1}{2}\rho\omega^2\kappa_1^2\left(r_2^2 - r^2\right)$$

$$p_A = p_1 - 0{,}5\rho\omega_c^2\left(r_1^2 - r^2\right) = p_1 - \frac{1}{2}\rho\omega^2\kappa_1^2\left(r_1^2 - r^2\right) \tag{3}$$

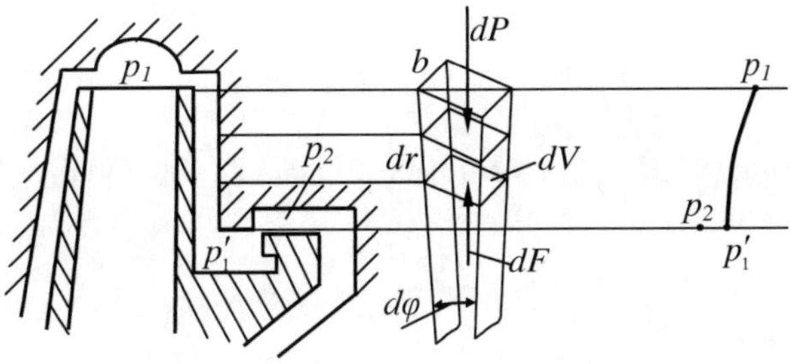

Figure 4. Evaluation of the axial force acting on the impeller.

Pressure in the chamber D and in the inlet orifice A_0 can be considered constant along the radius: $p_D = p_3 = const$. It is also assumed that pressure p_1 on the radius r_1 in both chambers is approximately the same.

Pressure values p_1', p_2' before the throttles differ from the pressure values p_1, p_2 at the inlet to the corresponding chamber by the amount of the inertial pressure:

$$p'_{A1} = p_1 - \delta p_{A1},\ \delta p_{A1} = \frac{1}{2}\rho\omega^2\kappa_1^2\left(r_1^2 - r_6^2\right) = \frac{1}{2\pi}\rho\omega^2\kappa_1^2 A,$$

$$p_1' = p_1 - \delta p_1,\ \delta p_1 = \frac{1}{2}\rho\omega_c^2\left(r_1^2 - r_2^2\right) = \frac{1}{2\pi}\rho\omega^2\kappa_1^2 B, \qquad (4)$$

$$p_2' = p_2 - \delta p_2,\ \delta p_2 = \frac{1}{2}\rho\omega_c^2\left(r_2^2 - r_3^2\right) = \frac{1}{2\pi}\rho\omega^2\kappa_1^2 C$$

Pressure values p_2, p_3 after the throttles differ from pressure values p_1', p_2' before throttles by the amount of pressure being lost to overcome resistance of the corresponding slotted throttles. The approximate pressure diagrams for the main disk are shown in Figure 2.

From the obtained equations, it is evident that the average pressure and, correspondingly, the pressure force on the wheel disks decreases if liquid rotation frequency in the pocket increases. As the ω_c decreases, the pressure diagram becomes more complete. This peculiarity is widely used in various designs: to reduce unbalanced axial force, all possible measures are taken to increase the average frequency of liquid rotation in the back pocket and to reduce it in the front one [6].

Pressure in pockets is axisymmetric, so it is possible to summarize axial forces acting on the elementary annular sections $2\pi r dr$. As a result, pressure forces on the cover and main disks can be determined:

$$T_A = 2\pi\int_{r_6}^{r_1} p_A r dr = \pi\left(r_1^2 - r_6^2\right)\left[p_1 - 0{,}5\rho\omega^2\kappa_1^2\left(r_1^2 - r_6^2\right)\right] = A\left(p_1 - \frac{\rho\omega^2}{2\pi}\kappa_1^2 A\right) \qquad (5)$$

$$T_{A_0} = \pi\left(r_6^2 - r_5^2\right)p_3 = A_0 p_3$$

$$T_B = 2\pi\int_{r_2}^{r_1} p_B r dr = \pi\left(r_1^2 - r_2^2\right)\left[p_1 - 0{,}5\rho\omega^2\kappa_1^2\left(r_1^2 - r_2^2\right)\right] = B\left(p_1 - \frac{\rho\omega^2}{2\pi}\kappa_1^2 B\right) \qquad (6)$$

$$T_C = 2\pi\int_{r_3}^{r_2} p_C r dr = \pi\left(r_2^2 - r_3^2\right)\left[p_2 - 0{,}5\rho\omega^2\kappa_1^2\left(r_2^2 - r_3^2\right)\right] = C\left(p_2 - \frac{\rho\omega^2}{2\pi}\kappa_1^2 C\right)$$

$$T_D = \pi\left(r_3^2 - r_4^2\right)p_3 = Dp_3,$$

IOP Conf. Series: Materials Science and Engineering **233** (2017) 012056 doi:10.1088/1757-899X/233/1/012056

where

$$A = \pi\left(r_1^2 - r_6^2\right), A_0 = \pi\left(r_6^2 - r_5^2\right), B = \pi\left(r_1^2 - r_2^2\right), C = \pi\left(r_2^2 - r_3^2\right), D = \pi\left(r_3^2 - r_4^2\right) \quad (7)$$

In the static equilibrium position the sum of all forces (5,6) is equal to the residual force T_0. This equation comes down to the form:

$$Cp_2 = T_0 + (A - B)p_1 + (A_0 - D)p_3 - \frac{\rho\omega^2\kappa_1^2}{2\pi}\left(A^2 - B^2 - C^2\right) \quad (8)$$

Pressure p_2 in equation (8) depends on the gaps in the slotted throttles, i.e. from the axial rotor position. Having derived p_2 from the flow balance equation one substitutes it in (8) and obtains an equation to find a static characteristic – a dependence z on the discharge and suction pressure, on the axial force T_0, and on the rotor speed.

3. Determination of the turbulent conductance of slotted throttles

The equation of flow rates through the upper and lower slotted throttles for the auto-modeling section of the turbulent flow has a form:

$$\sqrt{p_1 - p_2 - p_1'} = g_2\sqrt{p_2 - p_3 - p_2'} \quad (9)$$

g_1, g_2 - turbulent conductance values that represent a separate problem to be solved. To solve this problem, a liquid oxygen flow in a system of two serial slotted throttles at a different size of the axial gap are simulated. Turbulence model $k-\varepsilon$ was used in the calculation. As a result of each calculation, values of mass flow rate and averaged pressure p_2 in the chamber between the throttles were obtained, by means of which throttle turbulent conductance was calculated (Table 1). Figures 5 - 7 show the distribution of pressure p_2 for different axial gap values of the upper throttle. $h_1 = r_2 - r_2' = 0{,}18mm$, $h_2 = r_3 - r_3' = 0{,}23mm$ - radial gaps in the first and second throttles, respectively.

Axial displacement of a rotor z and face gaps of the upper and lower throttles are interrelated by the following dependences: $z_1 = z_n - z$, $z_2 = z$. $z_n = 0{,}65mm$ is a maximum possible value of the face gap protecting another throttle from being overlapped.

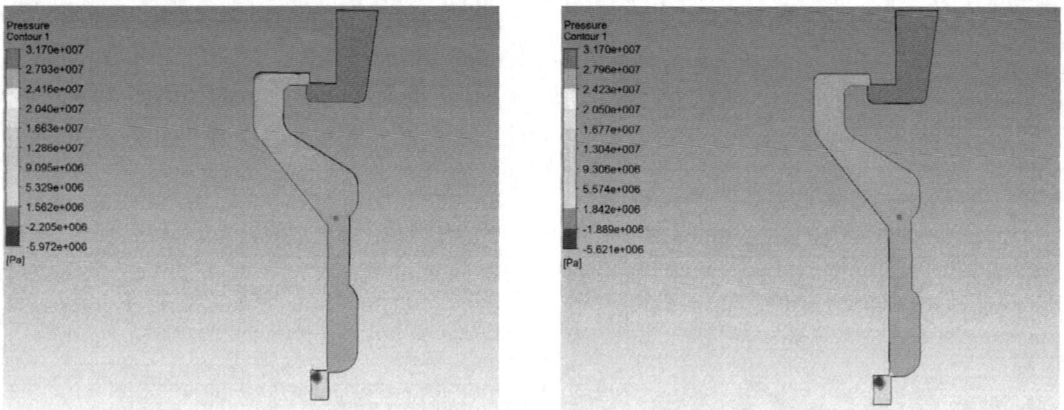

(a) (b)

Figure 5. Pressure distribution p_2: a-z = 0.2mm; b-z = 0.25 mm.

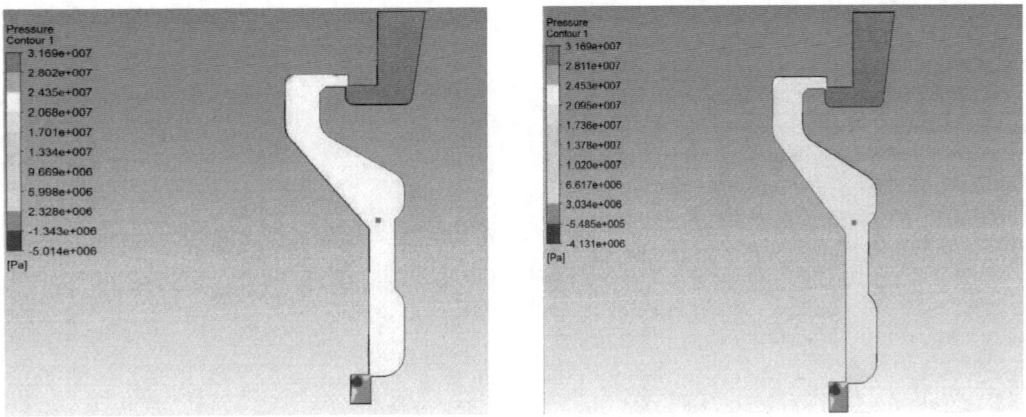

(a) (b)

Figure 6. Pressure distribution p_2 : z = 0.3 mm; Z = 0.35 mm.

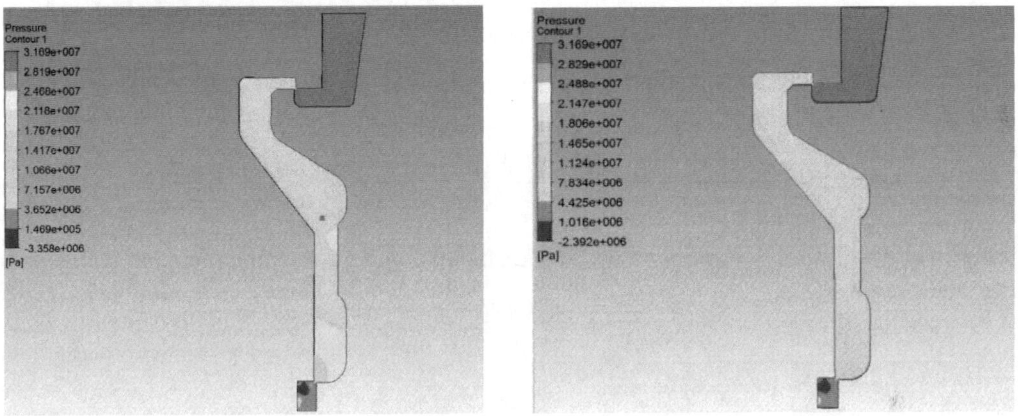

(a) (b)

Figure 7. Pressure distribution p_2 : z = 0.4 mm, z = 0.45 mm.

Based on the calculation results, summarized in Table 1, dependences of different throttles conductance on the rotor z displacement were obtained. Figure 8 shows that these dependences are close to linear, therefore it is adequate to take: $g_1 = a_1 u + b_1$, $g_2 = a_2 u + b_2$.

$a_1 = -10{,}81 \cdot 10^{-6}, b_1 = 14{,}26 \cdot 10^{-6}, a_2 = 8{,}14 \cdot 10^{-6}, b_2 = 2{,}49 \cdot 10^{-6}$

Table 1. Numerical calculation of conductance

Axial gap, mm 10^2	0	5	10	15	20	25	30	35	40	45	50	55	60	65
Mass flow rate, kg/s	18,6	19,9	21,7	23,8	26,0	28,0	29,6	30,6	30,8	30,0	28,4	26,4	25,5	23,9
Average pressure, MPa.	30,4	29,9	29,3	28,3	27,1	25,2	23,0	20,2	16,8	13,7	10,5	8,1	7,3	5,8
Conductance $g_1 \cdot 10^6$,$(m^7/kg)^{0,5}$	14,8	13,8	12,7	11,7	11,0	10,0	9,1	8,2	7,2	6, 4	4,9	4,6	4,2	4,0
Conductance $g_2 \cdot 10^6$,$(m^7/kg)^{0,5}$	3,2	3,4	3,8	4,2	4,7	5,3	5,9	6,5	7,3	8,0	8,8	9,7	10,1	11,1

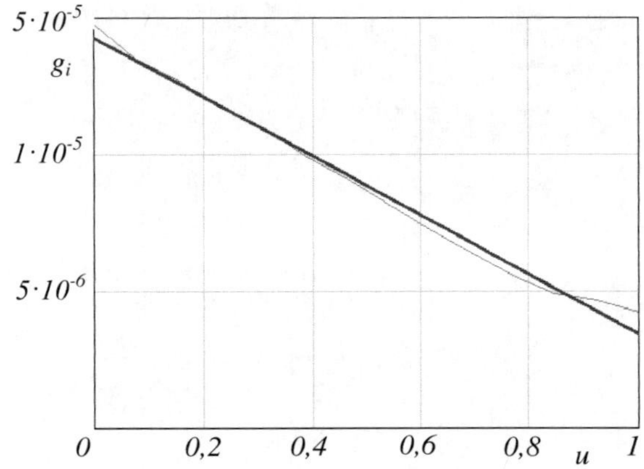

(a)

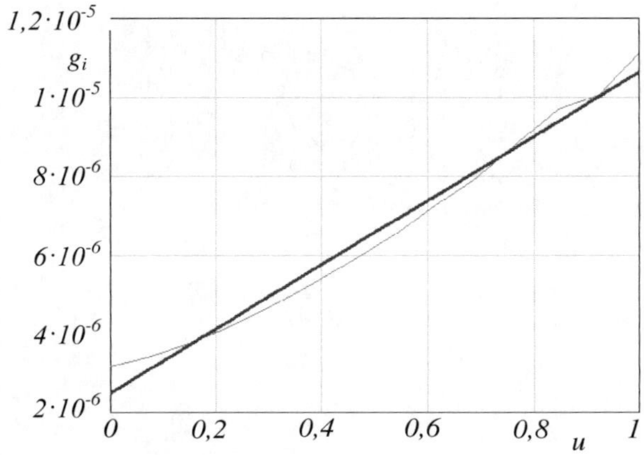

(b)

Figure 8. Conductance dependence on the dimensionless axial gap: (a) first (upper) throttle; (b) second (lower) throttle.

4. Static characteristics and hydrostatic stiffness of the system

First, it is necessary to square both parts (9) $g_1^2(p_1 - p_2 - p_1') = g_2^2(p_2 - p_3 - p_2')$ and find pressure in the chamber *C:*

$$p_2 = \frac{g_1^2 p_1 + g_2^2 p_3}{g_1^2 + g_2^2} = \frac{(a_1 u + b_1)^2 (p_1 - p_1') + (a_2 u + b_2)^2 (p_3 + p_2')}{(a_1 u + b_1)^2 + (a_2 u + b_2)^2} \tag{10}$$

Further, we substitute this expression in (8) and an equation to determine the relative dimensionless gap will look as following:

$$\overline{C}\frac{(a_1 u + b_1)^2 \psi_1 + (a_2 u + b_2)^2 \psi_3}{(a_1 u + b_1)^2 + (a_2 u + b_2)^2} = \overline{T}_0 + (\overline{A} - \overline{B})\psi_1 + (\overline{A}_0 - \overline{D})\psi_3 - \overline{K}(A^2 - B^2 - C^2)\Omega^2 \tag{11}$$

where

$$\overline{A} = \frac{A}{A}, \overline{B} = \frac{B}{A}, \overline{C} = \frac{C}{A}, \overline{D} = \frac{D}{A}, \psi_1 = \frac{p_1}{p_n}, \psi_3 = \frac{p_3}{p_n}, \psi_1' = \frac{p_1'}{p_n}, \psi_2' = \frac{p_2'}{p_n}$$

$$\overline{T}_0 = \frac{T_0}{Ap_n}, \overline{K} = \frac{\rho\omega_n^2\kappa_1^2}{2\pi A p_n} = \frac{K}{Ap_n}, \quad \Omega = \frac{\omega}{\omega_n}.$$

From the equation (10) a dependence of the relative numerical gap on the dimensionless pressure can be derived ψ_1:

$$u = \frac{b_2\sqrt{\dfrac{f(\psi_1) - \psi_3 - \psi_2'}{\psi_1 - \psi_1' - f(\psi_1)}} - b_1}{a_1 - a_2\sqrt{\dfrac{f(\psi_1) - \psi_3 - \psi_2'}{\psi_1 - \psi_1' - f(\psi_1)}}} \tag{12}$$

where

$$f(\psi_1) = \overline{T}_0 + (\overline{A} - \overline{B})\psi_1 + (\overline{A}_0 - \overline{D})\psi_3 - \overline{K}(A^2 - B^2 - C^2)\Omega^2$$

The resulting equation makes it possible to construct a dependence of rotor axial displacement on any of the external influences, and also to evaluate effect on the static characteristics of independent parameters.

Calculation example. Initial data:

$$p_{1n} = p_n = 31{,}7MPa,\ p_3 = 2{,}0MPa,\ \omega = \omega_n = 1963s^{-1}, \rho = 1102kg/m^3,$$
$$r_1 = 131mm,\ r_2 = 119mm,\ r_3 = 77mm,\ r_4 = 46mm,\ r_5 = 50mm.$$

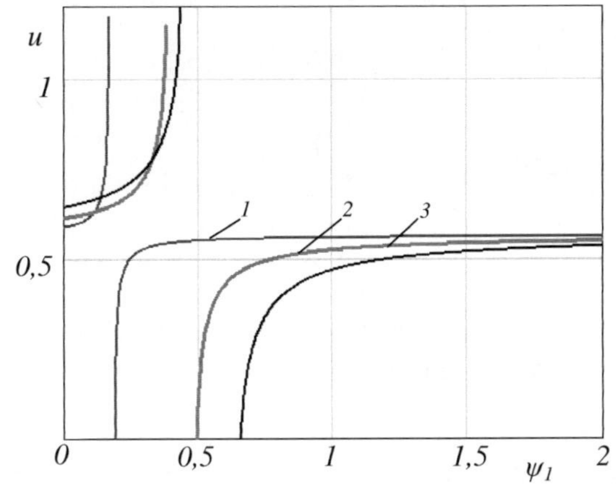

Figure 9. Static characteristics at rotational speeds: 1 -10,500 rpm; 2 -18750 rpm; 3 - 21150 rpm.

It can be seen from Figure 9 that a working area of the discharge pressure when a pump operates at the nominal frequency starts at a value $0{,}5\psi_1$, and at a nominal pressure $p_n = 31{,}7MPa$ it is equal to 16,8 MPa. At $p_1 = p_n = 31{,}7MPa \quad u = 0{,}526$.

Derivative

$$\frac{\partial \psi_2}{\partial u} = \frac{\partial}{\partial u}\left[\frac{(a_1u+b_1)^2(\psi_1-\psi_1')+(a_2u+b_2)^2(\psi_3+\psi_2')}{(a_1u+b_1)^2+(a_2u+b_2)^2}\right]$$

represents hydrostatic stiffness of the support-seal assembly in question. Negative value of stiffness is a sign of stability of the rotor equilibrium position.

$$\frac{\partial \psi_2}{\partial u} = \left[(a_1u+b_1)^2+(a_2u+b_2)^2\right]^{-2}\times$$
$$\times\left\{\begin{aligned}&\left[2a_1(a_1u+b_1)(\psi_1-\psi_1')+2a_2(a_2u+b_2)(\psi_3+\psi_2')\right]\cdot\left[(a_1u+b_1)^2+(a_2u+b_2)^2\right]-\\&-\left[(a_1u+b_1)^2(\psi_1-\psi_1')+(a_2u+b_2)^2(\psi_3+\psi_2')\right]\cdot\left[2a_1(a_1u+b_1)+2a_2(a_2u+b_2)\right]\end{aligned}\right\}$$

$$\frac{\partial \psi_2}{\partial u} = \frac{2(\psi_1-\psi_1'-\psi_3-\psi_2')(a_1b_2-a_2b_1)(a_1a_2u^2+(a_1b_2+a_2b_1)u+b_1b_2)}{\left[(a_1u+b_1)^2+(a_2u+b_2)^2\right]^2}$$

or

$$\kappa_S = \frac{\partial \psi_2}{\partial u} = \frac{2a_1a_2(\psi_1-\psi_1'-\psi_3-\psi_2')(a_1b_2-a_2b_1)\left(u+\frac{b_1}{a_1}\right)\left(u+\frac{b_2}{a_2}\right)}{\left[(a_1u+b_1)^2+(a_2u+b_2)^2\right]^2} \tag{12}$$

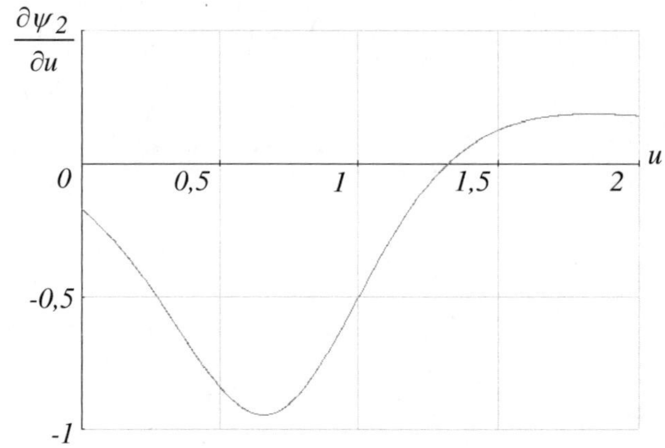

Figure 10. Hydrostatic stiffness.

Figure 10 shows that in the working section where $u < 1{,}32$ hydrostatic stiffness has a negative value and, consequently, equilibrium position of the rotor is stable.

5. Conclusion

The article presents a calculation of rotor axial vibrations for the pump with the automatic machine for unloading axial forces. To calculate static characteristics of a system there a numerical experiment was performed and there mass flow rates and pressure values in the discharge chamber depending on the rotor axial displacement (axial gap of the lower throttle) were obtained. Based on the results of numerical calculations, turbulent conductance of the upper and lower slotted throttles were calculated. This made it possible to obtain a dependence of the axial displacement on the discharge pressure and, as a result, determine axial gaps sizes of the slotted throttles, as well as the mass flow rate at the nominal operating mode of the pump. Therefore, at a nominal frequency of 18750 *rpm* mass flow rate is 25,4 *kg/s*, axial gap of the lower throttle is 0,34 *mm*, the upper one - 0,31 *mm*. At a minimum frequency of 10500 *rpm* mass flow rate is 12,1 *kg/s*, axial gap of the lower throttle is 0,34 *mm*, the upper one - 0,31 *mm*. At a maximum frequency of 21150 *rpm*, mass flow rate is 27,55 *kg/s*, axial gap of the lower throttle is 0,33 *mm*, the upper one - 0,32 *mm*. Thus, it can be stated that a face gap has almost constant value in the entire range of operating frequencies of the rotor. As it can be seen from the static characteristics (Figure 9), if discharge pressure increases, the face gap value increases, and when rotation speed goes up then the face gap value goes down. Analysis of hydrostatic stiffness (Figure 10) makes it possible to conclude that equilibrium position of the rotor is stable.

References

[1] Korczak A, Martsynkovskyy V and Gudkov S 2012 Estimating Influence of Inertial Resistance of Throttle for Hydraulic Balancing Device on Rotor Axial Vibration *Procedia Engineering* **39** pp 261-274
[2] Gurov V 2000 Development of cryogenic turbo pumps (Informconversiya: Moscow) p 132
[3] Martsynkovskyy V A 2005 Gap seals: theory and practice (Sumy: SSU) p 416
[4] Besekersky V and Popov E 1972 Theory of automatic control systems (Moscow: Nauka) p 768
[5] Glikman B 1989 Automatic control of the liquid rocket engines (Moscow: Machinebuilding) p 296
[6] Martsynkovskyy V A 2012 Dynamics of a centrifugal pump rotors (Sumy: SSU) p 562
[7] Baibikov A and Karahanian V 1982 Hydrodynamics of the auxiliary channels of blade machines (Moscow: Machinebuilding) p 112
[8] Shnepp V 1982 Engineering method for calculating disk losses and axial forces in a centrifugal stage, taking into account leakage of seals (Moscow: Machinebuilding)

IOP Conf. Series: Materials Science and Engineering **233** (2017) 012056 doi:10.1088/1757-899X/233/1/012056

[9] Vidishev V, Kanalin U, Kukharev V et al. 2000 Balancing and control of the axial load of the radial-thrust bearings of the feeding units of the GRD RD-170 *Proc. of the GDL-OKB* pp 100-114
[10] Borovskiy B 1989 Energy parameters and characteristics of high-speed rotary pumps (Moscow: Machinebuilding) p 183
[11] Ovsyanikov B and Borovskiy B 1996 Theory and calculation of power units for liquid rocket engines (Moscow: Machinebuilding) p 376
[12] Jędral W 1996 Pompy wirowe odśrodkowe (Warszawa: Oficyna Wydawnicza Politechniki Warszawskiej) p 324

IOP Conf. Series: Materials Science and Engineering **233** (2017) 012057 doi:10.1088/1757-899X/233/1/012057

Evaluation of rotor axial vibrations in a turbo pump unit equipped with an automatic unloading machine

V A Martsynkovskyy[1], A Deineka[1] and V Kovalenko[2]

[1]Sumy State University, Department of General Mechanics and Dynamics of Machines, 2, Rimsky-Korsakov Street, Sumy 40007, Ukraine
[2]Design office "YUZHNOYE", Theoretical department for designing rocket engines, 3, Krivorozhskaya Street, Dnipro 49008, Ukraine

E-mail: a.dejneka@omdm.sumdu.edu.ua

Abstract. The article presents forced axial vibrations of the rotor with an automatic unloading machine in an oxidizer pump. A feature of the design is the use in the autoloading system of slotted throttles with mutually inverse throttling. Their conductivity is determined by a numerical experiment in the ANSYS CFX software package.

1. Introductory remarks

The article presents (Figure 1) a simplified rotor design of the oxidizer pump equipped with an automatic unloading machine (AUM), in which, unlike in traditional devices [1], inverse slotted throttles are used. Static calculation of the system is given in the article "Static calculation of the rotor unloading automatic machine for a high-pressure centrifugal pump" (V A Martsynkovskyy, A Deineka, A Korczak and G Peczkis), published in this issue of the journal.

Unit rotor together with the automatic unloading machine is a complex dynamic system with distributed parameters that is a subject to periodic external perturbations. Each cross section of the rotor performs interrelated radial, angular and axial oscillations [2]. In regards to reliability, primarily axial vibrations can be dangerous, since their relatively large amplitudes limit the operating life of ball bearings wherein rotor is located. In this paper, as a first approximation, a simplified problem is considered: a rotor is studied as a solid body with an automatic loading system which performs one-dimensional axial oscillations along the support axis. Such a simplified model enables to obtain static and dynamic characteristics in an analytical way and accurately describe the main regularities of the system oscillations.

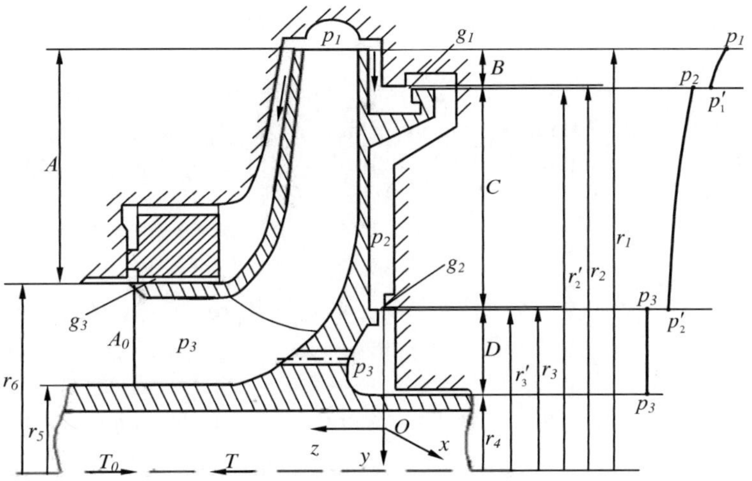

Figure 1. Calculation scheme.

2. Derivation of the axial oscillation equation

Expression of the total axial force acting on the impeller, obtained in the article "Static calculation of the rotor unloading automatic machine for a high-pressure centrifugal pump", has the form:

$$F_z = T_0 - Cp_2 + (A-B)p_1 + (A_0 - D)p_3 - K\left(A^2 - B^2 - C^2\right)\Omega^2 \tag{1}$$

where T_0 - is the residual axial force to be balanced; p_1, p_3 - supply pressure and pressure at the inlet; p_2 - pressure in the chamber C between the upper and lower throttles, depending on the axial position of the rotor

$$K = \frac{\rho\omega_n^2\kappa_1^2}{2\pi} \tag{2}$$

κ_1 - swirl coefficient of the flow in side chambers [3]; ω_n - nominal rotor speed; $\Omega = \omega/\omega_n$ - dimensionless rotation speed; ρ - density of the pumped medium. The last term in expression (1) is the pressure force conditioned by the average angular velocity of liquid in the chambers. Capital letters indicate the flat annular sections shown in Figure 1.

At axial oscillations, the force of viscous resistance $-c\dot{z}$ (c is the damping coefficient) acts on the rotor, therefore, on the basis of the 2nd Newton's law, the equation of axial oscillations will take the form:

$$m\ddot{z} + c\dot{z}_z = -Cp_2 + T_0 + (A-B)p_1 + (A_0 - D)p_3 - K\left(A^2 - B^2 - C^2\right)\Omega^2$$

m -indicated rotor weight. Having devided these equation term-by-term on Ap_n, we obtain the dimensionless form of the equation:

$$T_1^2\ddot{u} + 2\varsigma T_1\dot{u} = \bar{C}\psi_2 - \bar{T}_0 - \left(1 - \bar{B}\right)\psi_1 - \left(\bar{A}_0 - \bar{D}\right)\psi_3 + \bar{K}\left(A^2 - B^2 - C^2\right)\Omega^2 \tag{3}$$

$$T_1^2 = \frac{mz_n}{Ap_n}, 2\varsigma T_1 = \frac{cz_n}{Ap_n}, \bar{K} = \frac{\rho\omega_n^2\kappa_1^2}{2\pi Ap_n}, \bar{A}_0 = \frac{A_0}{A}, \bar{B} = \frac{B}{A}, \bar{C} = \frac{C}{A}$$
$$\bar{D} = \frac{D}{A}, \bar{T}_0 = \frac{T_0}{A}; \psi_1 = \frac{p_1}{p_n}, \psi_2 = \frac{p_2}{p_n}, \psi_3 = \frac{p_3}{p_n}; u = \frac{z}{z_n} \tag{4}$$

z - rotor axial displacement equal to the gap in the lower throttle, z_n - its base value, $u = z/z_n$ - the dimensionless axial displacement of the rotor.

The equation (3) includes an unknown pressure p_2 in the chamber 2, which depends on the gap size and is determined from the balance equation of the flow rate passing through the slotted throttles of the equilibrium system. For automodeling section of the turbulent flow, the flow rate through the upper and lower throttles is

$$Q_1 = g_1(u)\sqrt{p_1' - p_2},\ Q_2 = g_2(u)\sqrt{p_2' - p_3} \tag{5}$$

where $p_1' = p_1 - KB\Omega^2$, $p_2' = p_2 - KC\Omega^2$ i.e. decrease in pressure at the lower radius of the corresponding chamber due to the inertial effect is taken into account. In static calculation $Q_{10} = Q_{20} = Q_0$

According to the results of the numerical experiment, conductance almost linearly depends on the gap:

$$g_1(u) = a_1 u + b_1,\ g_2(u) = a_2 u + b_2 \tag{6}$$

And for the unit under consideration

$$a_1 = -1{,}08 \cdot 10^{-5}\left(m^7/kg\right)^{0,5},\ b_1 = 1{,}43 \cdot 10^{-5}\left(m^7/kg\right)^{0,5}$$

$$a_2 = 0{,}814 \cdot 10^{-5}\left(m^7/kg\right)^{0,5},\ b_2 = 0{,}249 \cdot 10^{-5}\left(m^7/kg\right)^{0,5}$$

Negative value of the angular coefficient a_1 is due to the fact that when the lower gap z increases, the upper gap, and hence the conductance, decreases: inverse throttling takes place.

In the dynamics by rotor axial oscillations, the balance equation includes the displacement flow rate $C\dot{z}$ and compression $\frac{V}{E}\dot{p}_2$ of liquid in the chamber [4].

$$g_1(u)\sqrt{p_1' - p_2} = g_2(u)\sqrt{p_2' - p_3} + C\dot{z} + \frac{V}{E}\dot{p}_2 \tag{7}$$

The obtained first-order differential equation relating to the unknown p_2 is nonlinear. We linearize it in a neighborhood of the equilibrium position, passing to the variational equations. It should be taken into account that inertial forces in the corresponding chambers are:

$$KB\Omega^2 = p_1 - p_1',\ KC\Omega^2 = p_2 - p_2'$$

and their variations

$$2KB\Omega_0\delta\Omega = \delta p_1 - \delta p_1',,\ 2KC\Omega_0\delta\Omega = \delta p_2 - \delta p_2'$$

Further, we go into the quasistatic change in the rotation frequency, therefore $\delta\Omega = 0,\ \delta p_1' = \delta p_1, \delta p_2' = \delta p_2$, and variational equations are (7):

$$\frac{V}{EQ_0}\delta\dot{p}_2 + \frac{1}{2}\left(\frac{1}{p_{10}' - p_{20}} + \frac{1}{p_{20}' - p_{30}}\right)\delta p_2 =$$

$$= -\frac{Cz_n}{Q_0}\delta\dot{u} - \left(\frac{a_2}{g_{20}} - \frac{a_1}{g_{10}}\right)\delta u + \frac{1}{2}\left(\frac{\delta p_1}{p_{10}' - p_{20}} + \frac{\delta p_3}{p_{20}' - p_{30}}\right)$$

Thereafter, we move to the dimensionless pressures, denoting

$$\Delta\psi_{10} = \psi'_{10} - \psi_{20},\ \Delta\psi_{20} = \psi'_{20} - \psi_{30},\ \Psi = \left(\frac{1}{\Delta\psi_{10}} + \frac{1}{\Delta\psi_{20}}\right)^{-1} \tag{8}$$

$$\frac{Vp_n}{EQ_0}\delta\dot{\psi}_2 + \frac{1}{2}\left(\frac{1}{\Delta\psi_{10}} + \frac{1}{\Delta\psi_{20}}\right)\delta\psi_2 = -\frac{Cz_n}{Q_0}\delta\dot{u} - \left(\frac{a_2}{g_{20}} - \frac{a_1}{g_{10}}\right)\delta u + \frac{1}{2\Delta\psi_{10}}\delta\psi_1 + \frac{1}{2\Delta\psi_{20}}\delta\psi_3$$

To obtain the equation in a short form, it is necessary to divide the last equation term-by-term on a coefficient $\frac{1}{\Delta\psi_{10}} + \frac{1}{\Delta\psi_{20}} = \frac{1}{\Psi}$ when desired quantity is $\delta\psi_2$

$$2\frac{Vp_n}{EQ_0}\Psi\delta\dot{\psi}_2 + \delta\psi_2 == -\frac{2Cz_n\Psi}{Q_0}\delta\dot{u} - 2\Psi\left(\frac{a_2}{g_{20}} - \frac{a_1}{g_{10}}\right)\delta u + \frac{\Psi}{\Delta\psi_{10}}\delta\psi_1 + \frac{\Psi}{\Delta\psi_{20}}\delta\psi_3$$

To reduce the recording, variation sign will further be omitted, keeping in mind, however, that this is not about the absolute values of the variables, but about their small deviations from the established values. The latter are marked with an additional zero index and are determined by static calculation. Having denoted the time constants and transmission coefficients as:

$$T_2 = 2\frac{Vp_n}{EQ_0}\Psi, \kappa_s = 2\Psi\left(\frac{a_2}{g_{20}} - \frac{a_1}{g_{10}}\right), \tau_2 = \frac{Cz_n}{\left(\frac{a_2}{g_{20}} - \frac{a_1}{g_{10}}\right)Q_0}, k_1 = \frac{\Psi}{\Delta\psi_{10}}, k_3 = \frac{\Psi}{\Delta\psi_{20}} \tag{9}$$

we obtain the standard form of the linearized flow rate balance:

$$T_2\dot{\psi}_2 + \psi_2 = -\kappa_s(\tau_2\dot{u} + u) + k_1\psi_1 + k_3\psi_3 \tag{10}$$

Pressure in the chamber can be considered as a controlling action, and equation (10) - as the equation of the automatic controller

If we introduce an operator of differentiation in time $s \equiv d/dt$, then equation (10) takes the operator form:

$$(T_2 s + 1)\psi_2 = -\kappa_s(\tau_2 s + 1)u + k_1\psi_1 + k_3\psi_3$$

we can derive the control influence of it:

$$\psi_2 = -\kappa_s\frac{M_2(s)}{D_2(s)}u + \frac{1}{D_2(s)}(k_1\psi_1 + k_3\psi_3) \tag{11}$$

where

$$D_2(s) = T_2 s + 1,\ M_2(s) = \tau_2 s + 1 \tag{12}$$

proper operator of the controller and operator of the actions by error, respectively.

Relation of the controller reaction to the stimulus is the transfer function of the controller. If the stimulus is a harmonic function, then an operator of time differentiation is replaced by an imaginary operator $i\omega$:

$$\frac{\psi_2}{u} = -\kappa_s\frac{\tau_2 i\omega + 1}{T_2 i\omega + 1} = -\kappa_s[U_2(\omega) + i\omega V_2(\omega)] = W_2(i\omega) \tag{13}$$

$$U_2(\omega) = \frac{1 + T_2\tau_2\omega^2}{1 + T_2^2\omega^2},\ V_2(\omega) = \frac{\tau_2 - T_2}{1 + T_2^2\omega^2}$$

and the transfer function $W_2(i\omega)$ becomes the frequency transfer function or dynamic stiffness of the controller. In the steady-state $\omega = 0$ and $W_2(0) = -\kappa_s$, i.e. dynamic rigidity transforms into hydrostatic stiffness of the system. Its negative value $(-\kappa_s < 0,\ \kappa_s > 0)$ is a condition of static stability, therefore the system is functional only in the section where $\kappa_s > 0$. This condition is met if $\frac{a_2}{g_{20}} - \frac{a_1}{g_{10}} > 0$. For the design considered here $a_1 < 0,\ a_2 > 0$, so this condition of static stability is fulfilled at the whole range of changes in external influences.

The real part U_2 characterizes the controller's stiffness itself, and the imaginary part V_2 - its "contribution" to damping [2]. In the absence of external damping c = 0, axial oscillations decay, if $V_2 > 0$ or

$$\tau_2 > T_2 \tag{14}$$

This condition with some margin for the stabilizing effect of external damping can be considered as a condition for rotor axial stability.

Having substituted expression of the regulating effect (11) into the equation of axial oscillations (3), we obtain motion equation of the system "impeller-balancing unit". Preliminary equation (3) must be linearized, i.e. move to variations. Herein the last term on the right-hand part is $2\overline{K}\left(A^2 - B^2 - C^2\right)\Omega_0\delta\Omega = 0$. After some transformations, we obtain:

$$\left[\left(T_1^2 s^2 + 2\varsigma T_1 s\right)\left(T_2 s + 1\right) + \kappa_s \overline{C}\left(\tau_2 s + 1\right)\right]u = -\left(T_2 s + 1\right)\overline{T}_0 -$$
$$-\left[\left(1 - \overline{B}\right)\left(T_2 s + 1\right) - \overline{C}k_1\right]\psi_1 - \left[\left(\overline{A}_0 - \overline{D}\right)\left(T_2 s + 1\right) - \overline{C}k_3\right]\psi_3$$

Having groupped the terms in powers of the differentiation operator in time, we obtain final form of the system equation:

$$D_0(s)u = \left(1 - \overline{B}\right)N_1(s)\psi_1 + \left(\overline{A}_0 - \overline{D}\right)N_3(s)\psi_3 - N_T(s)\overline{T}_0 \tag{15}$$

In this equation, the system's proper operator $D_0(s)$ and the action operators are expressed by the equations:

$$D_0(s) = a_0 s^3 + a_1 s^2 + a_2 s + a_3 \tag{16}$$

$$N_1(s) = \frac{k_1\overline{C}}{\left(1 - \overline{B}\right)} - (T_2 s + 1),\ N_3(s) = \frac{k_3\overline{C}}{\left(\overline{A}_0 - \overline{D}\right)} - (T_2 s + 1),\ N_T(s) = T_2 s + 1 \tag{17}$$

$$a_0 = T_1^2 T_2,\ a_1 = T_1^2 + 2\varsigma T_1 T_2,\ a_2 = 2\varsigma T_1 + \overline{C}\kappa_s\tau_2,\ a_3 = \overline{C}\kappa_s \tag{18}$$

Static rotor displacement relative to the equilibrium position, caused by the deviations of external influences, can be determined using the equation (15), assuming $s = d/dt = 0$. Using coefficients of the proper operator one can determine stability of the axial oscillations. According to the Routh-Hurwitz criterion, a third-order system is stable if the following condition is fulfilled:

$$a_1 a_2 > a_0 a_3 \tag{19}$$

for all positive coefficients. It can be seen from the formulas (18) that all the coefficients are positive if $\kappa_s > 0$, i.e. under condition that the system is statically stable. If we neglect the effect of external damping $(\varsigma = 0)$, then the equation (19) after substitution (18) is reduced to the value which was obtained earlier (14): $\tau_2 > T_2$. Having substituted (9), we obtain a restriction of the chamber volume:

$$V < \frac{ECz_n}{2p_n\left(\dfrac{a_2}{g_{20}} - \dfrac{a_1}{g_{10}}\right)\Psi} \tag{20}$$

3. Analysis of rotor axial vibrations

For the model under consideration, axial vibrations are described by the third-order differential equation. Operator $D_0(s)$ is the system's proper operator, and equation $D_0(s)u = 0$ is the equation of free axial vibrations of the impeller. Here we confine ourselves to analyzing forced oscillations and stability. A similar problem of rotor axial oscillations of a shaftless motor pump with the traditional design of an automatic unloading machine was considered in the article [5].

We will assume that variations of external influences $\psi_1, \psi_3, \overline{T}_0$ change according to a harmonic law $\psi_1 = \psi_{a1}e^{i\omega t}$, $\psi_3 = \psi_{a3}e^{i\omega t}$, $\overline{T}_0 = T_a e^{i\omega t}$. In this case, linear system response to each of these changes (component of the specific solution) has the form:

$$u_1 = u_{1a}e^{i(\omega t + \varphi_1)},\ u_3 = u_{3a}e^{i(\omega t + \varphi_e)},\ u_T = u_{Ta}e^{i(\omega t + \varphi_T)}$$

Amplitudes and phases of harmonic reactions, i.e. relations of the reactions to individual harmonic influences, are determined by the frequency transfer functions $W(i\omega)$ that can be obtained by introducing a substitution $s = i\omega$ into the equation of axial oscillations (15):

$$W_1(i\omega) = \frac{u_1}{\left(1-\overline{B}\right)\psi_1} = \frac{u_{1a}}{\left(1-\overline{B}\right)\psi_{1a}}e^{i\phi_1} = \frac{N_1(i\omega)}{D_0(i\omega)}$$

$$W_3(i\omega) = \frac{u_{3a}}{\left(\overline{A}_0 - \overline{D}\right)\psi_{3a}}e^{i\phi_3} = \frac{N_3(i\omega)}{D_0(i\omega)}.\ W_T(i\omega) = \frac{u_{Ta}}{T_a}e^{i\phi_T} = \frac{N_T(i\omega)}{D_0(i\omega)} \tag{21}$$

We single out real and imaginary parts in the proper operator (16) and the action operators (17), replacing $s = i\omega$:

$$D_0 = U_0 + i\omega V_0,\quad U_0(\omega) = a_3 - \omega^2 a_1,\ V_0(\omega) = a_2 - \omega^2 a_0 \tag{22}$$

$$N_1 = U_1 + i\omega V_1, U_1 = \frac{k_1\overline{C}}{\left(1-\overline{B}\right)} - 1, V_1 = -T_2$$

$$N_3 = U_3 + i\omega V_3,\quad U_3 = \frac{k_3\overline{C}}{\left(\overline{A}_0 - \overline{D}\right)} - 1,\ V_3 = -T_2 \tag{23}$$

$$N_T = U_T + i\omega V_T,\ U_T = 1,\ V_T = T_2$$

Having substituted these expressions into formulas (21), we obtain frequency transfer functions in the form of complex numbers, for example:

$$W_1(i\omega) = \frac{U_1 + i\omega V_1}{U_0 + i\omega V_0} = \frac{u_{1a}}{\left(1-\overline{B}\right)\psi_{1a}}e^{i\varphi_1} = B_1(\omega)e^{i\varphi_1(\omega)}$$

where $B_1(\omega) = u_{1a}/\left(1-\overline{B}\right)\psi_{1a} = \left|W_1(i\omega)\right|$, $\varphi_1(\omega) = \arg W_1(i\omega)$- are the amplitude and phase frequency response characteristics. We divide the real and imaginary parts in the expression W_1:

$$W_1 = \frac{U_1 + i\omega V_1}{U_0 + i\omega V_0} = \frac{U_0 U_1 + \omega^2 V_0 V_1}{U_0^2 + \omega^2 V_0^2} + i\omega \frac{(U_0 V_1 - U_1 V_0)}{U_0^2 + \omega^2 V_0^2}$$

Now the amplitude and phase W_1 have a form:

$$B_1(\omega) = \frac{u_{1a}}{(1-\overline{B})\psi_{1a}} = \sqrt{\frac{U_1^2 + \omega^2 V_1^2}{U_0^2 + \omega^2 V_0^2}}, \varphi_1(\omega) = arctg\omega \frac{U_0 V_1 - U_1 V_0}{U_0 U_1 + \omega^2 V_0 V_1} \quad (24)$$

By analogy, frequency response characteristics according to the influence of ψ_3 and $\overline{T}_0$ can be determined:

$$B_3(\omega) = \frac{u_{3a}}{(\overline{A}_0 - \overline{D})\psi_{3a}} = \sqrt{\frac{U_3^2 + \omega^2 V_3^2}{U_0^2 + \omega^2 V_0^2}}, \ \varphi_3(\omega) = arctg\omega \frac{U_0 V_3 - U_3 V_0}{U_0 U_3 + \omega^2 V_0 V_3}$$
$$B_T(\omega) = \frac{u_{Ta}}{T_a} = \sqrt{\frac{U_T^2 + \omega^2 V_T^2}{U_0^2 + \omega^2 V_0^2}}, \ \varphi_T(\omega) = arctg\omega \frac{U_0 V_T - U_T V_0}{U_0 U_T + \omega^2 V_0 V_T} \quad (25)$$

Calculation formulas can be obtained using the expressions for the real and imaginary parts of the proper operator (16) and the operators of the action (17):

$$B_1(\omega) = \frac{u_{1a}}{(1-\overline{B})\psi_{1a}} = \sqrt{\frac{\left(\frac{k_1\overline{C}}{1-\overline{B}} - 1\right)^2 + \omega^2 T_2^2}{\left(a_3 - \omega^2 a_1\right)^2 + \omega^2\left(a_2 - \omega^2 a_0\right)^2}}$$
$$\phi_1(\omega) = -arctg\omega \frac{\left(a_3 - \omega^2 a_1\right)T_2 + \left(\frac{k_1\overline{C}}{1-\overline{B}} - 1\right)\left(a_2 - \omega^2 a_0\right)}{\left(a_3 - \omega^2 a_1\right)\left(\frac{k_1\overline{C}}{1-\overline{B}} - 1\right) - \omega^2\left(a_2 - \omega^2 a_0\right)T_2} \quad (26)$$

$$B_3(\omega) = \frac{u_{3a}}{(\overline{A}_0 - \overline{D})\psi_{3a}} = \sqrt{\frac{\left(\frac{k_3\overline{C}}{\overline{A}_0 - \overline{D}} - 1\right)^2 + \omega^2 T_2^2}{\left(a_3 - \omega^2 a_1\right)^2 + \omega^2\left(a_2 - \omega^2 a_0\right)^2}}$$
$$\phi_3(\omega) = -arctg\omega \frac{\left(a_3 - \omega^2 a_1\right)T_2 + \left(\frac{k_3\overline{C}}{\overline{A}_0 - \overline{D}} - 1\right)\left(a_2 - \omega^2 a_0\right)}{\left(a_3 - \omega^2 a_1\right)\left(\frac{k_3\overline{C}}{\overline{A}_0 - \overline{D}} - 1\right) - \omega^2\left(a_2 - \omega^2 a_0\right)T_2} \quad (27)$$

$$B_T(\omega) = \frac{u_{Ta}}{T_a} = \sqrt{\frac{1 + \omega^2 T_2^2}{\left(a_3 - \omega^2 a_1\right)^2 + \omega^2\left(a_2 - \omega^2 a_0\right)^2}}$$
$$\phi_T(\omega) = arctg\omega \frac{\left(a_3 - \omega^2 a_1\right)T_2 - \left(a_2 - \omega^2 a_0\right)}{\left(a_3 - \omega^2 a_1\right) + \omega^2\left(a_2 - \omega^2 a_0\right)T_2} \quad (28)$$

IOP Conf. Series: Materials Science and Engineering **233** (2017) 012057 doi:10.1088/1757-899X/233/1/012057

As an example, we calculate the amplitude and phase frequency characteristics of the pump rotor, described in the article "Static calculation of the rotor unloading automatic machine for a high-pressure centrifugal pump". We will also use the results of static calculation presented in this paper. To analyze the rotor dynamics, additional parameters are taken: rotor weight $m = 49{,}6kg$ damping index of the rotor free axial vibrations without AUM $\varsigma = 0{,}05$, elasticity modulus of the pumped medium $E = 2 \cdot 10^9 Pa$. Steady-state values are used from the static calculation results: pressure in the chamber $p_{20} = 21{,}6MPa$, the dimensionless end clearance in the lower throttle $u_0 \approx 0{,}53$ and the volume flow $Q_0 \approx 23 \cdot 10^{-3} m^3/s$.

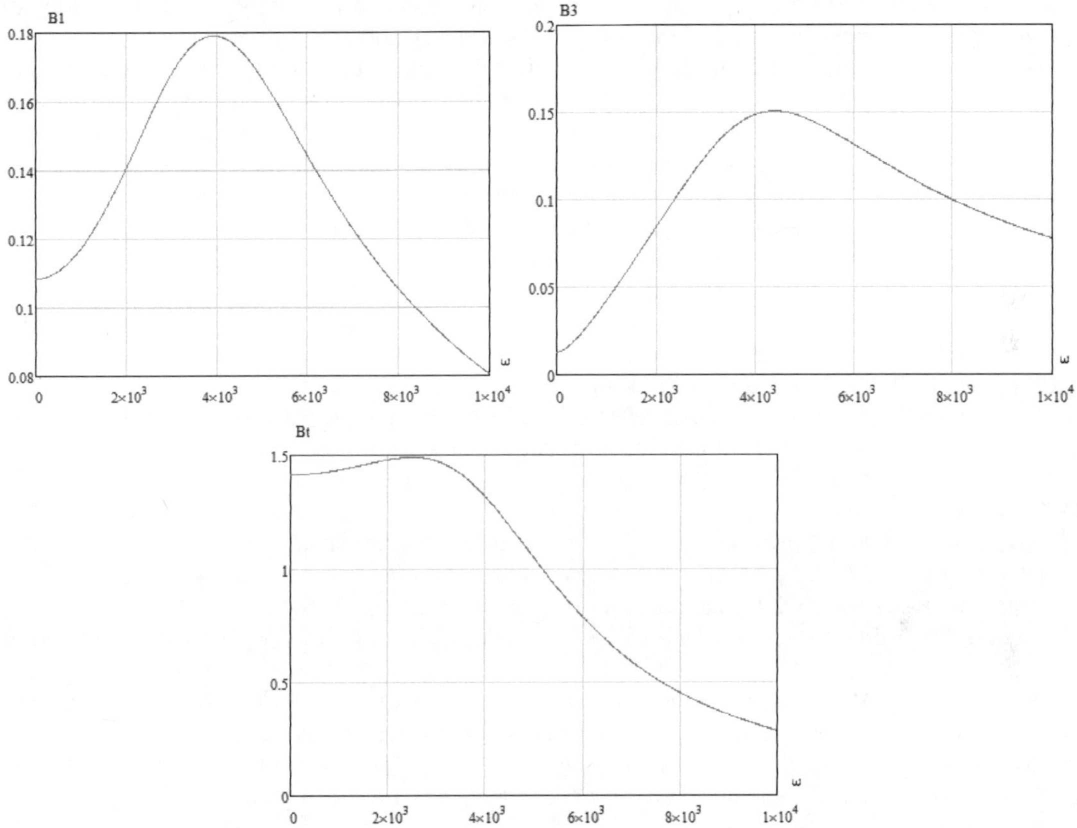

Figure 2. Amplitude frequency characteristics of axial oscillations.

Studying the frequency characteristics (Figure 2) one can see that the first resonance occurs at a frequency of 4000 s^{-1}. The second proper frequency of axial oscillations is far beyond the achievable rotation frequency, so it is of no practical interest. Using formulas (25) - (27), amplitudes of the resonance oscillations can be determined. If the relative amplitudes of the external perturbations are taken as $\psi_{1a} = \psi_{3a} = 0{,}1$, we obtain the following values $z_{1a} = 0{,}014 z_n = 0{,}0091 mm$, $z_{3a} = 0{,}0083 z_n = 0{,}0054 mm$.

4. Conclusions

The article presents a calculation of rotor axial vibrations of the oxidizer pump. To obtain static characteristics of a system there was performed a numerical experiment in the ANSYS CFX software package and there were obtained mass flow rates and pressure values in the discharge chamber depending on the rotor axial displacement (axial gap of the lower throttle). Based on the results of numerical calculations, turbulent conductance of the upper and lower slotted throttles were calculated.

This made it possible to obtain a dependence of the axial displacement on the discharge pressure and, as a result, determine axial gaps sizes of the slotted throttles, as well as the mass flow rate at the nominal operating mode of the pump. Therefore, at a nominal frequency of 18750 rpm mass flow rate is 25.4 kg / s, axial gap of the lower throttle is 0.34 mm, the upper one - 0.31 mm. At a minimum frequency of 10500 rpm mass flow rate is 12.1 kg / s, axial gap of the lower throttle is 0.34 mm, the upper one - 0.31 mm. At a maximum frequency of 21150 rpm, mass flow rate is 27.55 kg / s, axial gap of the lower throttle is 0.33 mm, the upper one - 0.32 mm. Thus, it can be stated that a face gap has almost constant value in the entire range of operating frequencies of the TPU rotor. As it can be seen from the static characteristics (Figure 2), if discharge pressure increases, the face gap value increases, and when rotation speed goes up then the face gap value goes down. Analysis of hydrostatic stiffness (Figure 2) makes it possible to conclude that equilibrium position of the rotor is stable.

To estimate the critical frequencies of the oxidizer pump rotor oscillations, amplitude and phase frequency characteristics were built. The analysis of these characteristics showed that the first proper frequency of axial oscillations is equal to 38200 rpm, the second proper frequency is far beyond the attainable rotational frequencies. The obtained estimates indicate that the critical frequency is more than twice higher than the nominal rotation frequency and 1.8 times higher than the maximum operating frequency. Thus rotor is rigid relating to the axial oscillations. It should be noted that the stability condition of the rotor axial oscillations is also fulfilled, which is proved by the fact that all coefficients of the system characteristic equation are positive and the condition (20) is fulfilled.

Radial, angular and axial oscillations of the TPU oxidizer pump rotor should be a subject of the detailed study, wherein a rotor is studied as a free solid body influenced by a complex system of hydrodynamic forces. It shoud be born in mind that it is necessity to take into account deformed state of the automatic unloading machine affected by the medium pressure on the distribution of this pressure and vice versa. In fact, it is a related task of hydroelasticity.

References

[1] Ovsianikov B and Borovskii B 1986 *Theory and calculation of power units for liquid rocket engines* 3rd ed (Moscow: Machinebuilding) p 376
[2] Martsynkovskyy V 2012 Dynamics of a centrifugal pump rotors (Sumy: SSU)
[3] Jędral W 1996 Vortex centrifugal pumps (Warszawa: Oficyna Wydawnicza Politechniki Warszawskiej) p 324
[4] Martsinkovsky V, Zhulov O and Kundera Cz 2014 Analysis of the Radial-Axial Oscillations of the Rotor of a High-Speed Pump *Applied Mechanics and Materials* **630** pp 341-349
[5] Martsinkovsky V, Korczak A and Kundera Cz 2014 The Centrifugal Pump with the Impeller Supported in Sealing Clearances *Advanced Materials Research* **874** pp 83-88

IOP Conf. Series: Materials Science and Engineering **233** (2017) 012058 doi:10.1088/1757-899X/233/1/012058

Head pulsations in a centrifugal pump

V S Boiko[1], M I Sotnyk[2] and V V Moskalenko[3]

[1] Department of Theoretical Electrical Engineering, National Technical University of Ukraine "Kyiv Politechnic Institute" them. I. Sikorsky, 37, Peremogy pr., Kyiv, 03056, Ukraine
[2] Department of Applied Hydroaeromechanics, Sumy State University, 2, Rymskogo-Korsakova st., Sumy, 40007, Ukraine
[3] Department of Applied Hydroaeromechanics, Sumy State University, 2, Rymskogo-Korsakova st., Sumy, 40007, Ukraine

E-mail: m.sotnik@pgm.sumdu.edu.ua

Abstract. This article investigated the factors, which affect to the character of the head pulsations of a centrifugal pump. We investigated the dependence of the shape and depth of these pulsations from the operation mode of the pump. Was determined, that the head pulsations at the outlet of the impeller (pulsations on the blade passing frequency) cause head pulsations at the outlet of the pump, that have the same frequency, but differ in shape and depth. These pulsations depend on the design features of the flow-through part of the pump (from the ratio of hydraulic losses on the friction and losses on the vortex formation). A feature of the researches that were conducted is also the using of not only hydraulic but also electric modeling methods. It allows determining the values of the components of hydraulic losses.

1. Introduction

The increase in the installed capacity of centrifugal pumps and the increase in the requirements for the efficiency of their operation lead to the reinforcing of the legislation of the European countries, which limits the vibroacoustic characteristics of hydraulic machines and equipment's [1]. More and more attention is paid to researching problems, which related to improving the reliability and operation resource of power machines. In this regard, the analysis of dynamic processes and the development of ways to reduce the pump's vibroactivity is becoming increasingly important, since this is related to the reliability and service life of electromechanical systems of water supply networks.

It is known that the vibrations of blade machines cover a wide frequency range - between 15 Hz to 20 kHz. The reason of this may be different dynamic processes: mechanical, hydrodynamic, electromagnetic and aerodynamic phenomena. Research results show that a significant part of the vibrations is due to hydrodynamic processes, in particular, head pulsations in the flowing part of the pump [2].

The main reason of such pulsations is hydrodynamic oscillations at a blade passing frequency. The pulsations on the blade passing frequency occur due as a result of the passage of the impeller blade through the cutwater, a section where the flow passes from rotational to translational motion.

The frequency and amplitude of the head pulsations are determined by the equations:

$$f_z = z \cdot n/60 \quad (1)$$

where
f_z – pulsations on the blade passing frequency (Hz);
z – the number of blades of the impeller;
n – rotational speed (rev/min).

As you can see, the number of pulsations during the period of the main frequency (the one rotation of the pump impeller) according to the number of impeller blades. The amplitude of these pulsations depends on a number of design factors - the shape and number of impeller blades, the size, and configuration of the retraction of the radial clearance between the impeller and the cutwater, the speed of rotation of the rotor, and etc. An unfavorable relationship of these factors to each other can lead to an increase the amplitude of the head pressure in the volute of a centrifugal pump and in the adjacent discharge pipeline. Proceeding from the foregoing, the calculation of the amplitudes of head pulsations in centrifugal pumps at an early stage of design is an actual task.

Therefore, an important moment in the analysis of head pulsations in electromechanical systems with blade machines is the moment of choice the method of investigation. At the moment, the most common methods for analyzing head pulsations in blade machines is the method that applies the solutions of non-stationary equations of hydrodynamics [3] and the method for determining nonstationary pressure by integrating the Reynolds equations, while this nonstationary velocity field is determined through by method of laser anemometry [4].

A more innovative method is described in [1], which involves the using of acoustic-vortex numerical simulation of head pulsations. As the authors of the publication note, this approach makes it possible to construct an efficient and more accurate calculation method in compared with the usual methods of computational hydrodynamics.

In our article, we propose and apply a different approach that combines analysis of work processes in electromechanical systems (EMC) of water supply networks in the ANSYS CFX 14.0 program package with electrical modeling. During the research, it was established that at the stage of designing blade machines it is also necessary to take into account the possibility of imposing various kinds of pulsations and the probability of occurrence of resonant phenomena. This is can lead to an increase the amplitude of head pulsating at the pump outlet.

2. Investigation object and the geometry of the impeller

To investigate the factors that influence to the character of the head pulsation, we were selected the pump with a two-way suction and with a volute - AD2000-100-2 with the following parameters:
- flow - 2000 m^3/h;
- head – 100 m;
- rotational speed - 980 rev./min.

Simulations carried out concerning the pump unit with the following geometric parameters of the impeller, inlet diameter D_0=334 mm, outlet diameter D_2=850 mm, the number of blades z=6, the angle of the blades $\beta_{2л}$=31° - at the outlet.

The design feature for the pump of type D is impeller of fluid double suction that compensating the axial forces, which act on the rotor. The flow of fluid that out from the impeller is directed to the pressure pipeline and passes through the volute and the diffuser of the pump. In a volute, near the cutwater, is formed unidirectional fluid flow, therefore decreases the velocity of flow and part of the kinetic energy of fluid flow is converted into pressure energy. Based on the kinematics of workflow and design features of the type D pump, flow of fluid in the outlet of the diffuser is pulsating and the number of pulsations has corresponded to the number of impeller blades.

3. Hydrodynamic modeling in ANSYS CFX

To determine the value of head and the amplitude of pulsation, we had created a model of the pump AD2000-100-2 and had conducted a numerical simulation of the flow of fluid in the flowing part of this pump. The numerical simulation had carried out for such pump operation modes:: 0,6Q_{opt} and Q_{opt}. Calculation had carried the non-stationary, was set a step of time, which corresponding to the rotation of the impeller by 5 degrees for the operation modes $0{,}6Q_{opt}$ and Q_{opt}. All results, which given in this article were taken after the fifth rotation of the pump impeller.

To evaluate the change in the values of the amplitude of the head pulsations, the values of the head pulsations for various operating modes were derived. Figure 1 shows the graph of the change in head per rotation of the impeller for the Q_{opt} operation mode. Analyzing the results obtained, it should be noted that for an optimal operating mode, the minimum depth of pulsations is less than 5%.

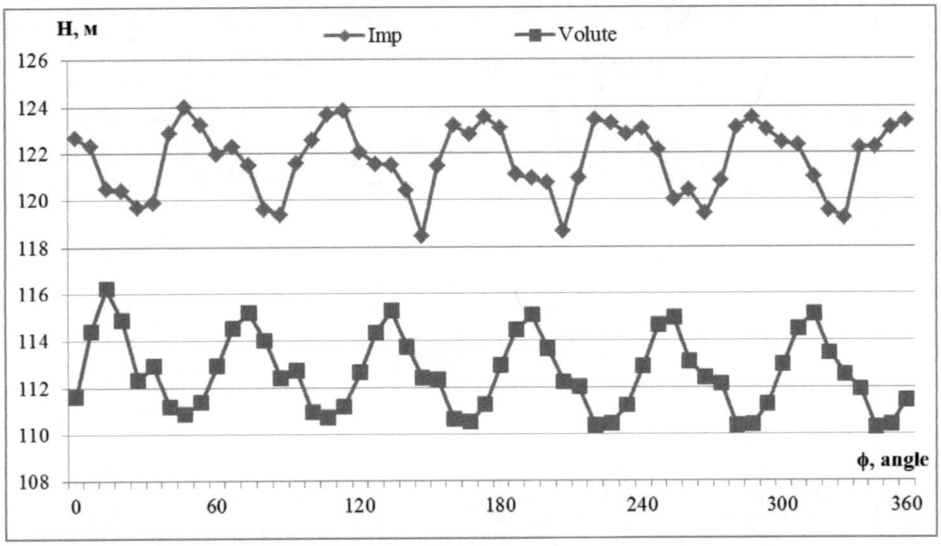

Figure 1. The graph of the head change at the outlet of the impeller and the volute of the pump for the nominal operating mode, Q_{opt}.

At the same time, as the flow of fluid decreases is observed an increase in the amplitude of the head pulsations. For the $0{,}6Q_{opt}$ mode, the maximum value of the head pulsation is about 15%. Figure 2 shows the head pulsations at the outlet from the impeller and the volute of the pump for the operating mode $0{,}6Q_{opt}$.

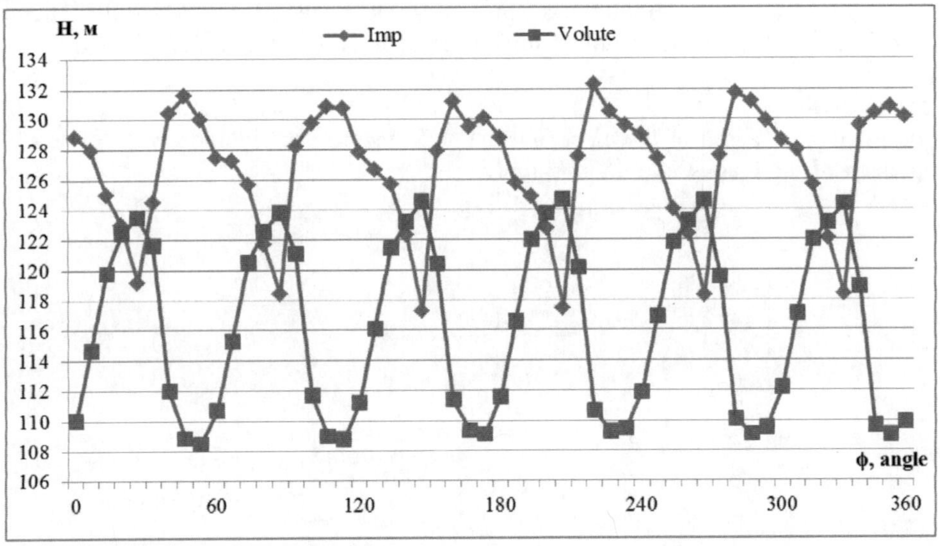

Figure 2. The graph of the head change at the outlet from the impeller and the volute of the pump for the operating mode $0{,}6Q_{opt}$.

Analyzing the results obtained, it should be noted that the value of the amplitude of head pulsations at the outlet from the volute can be greater, than the same value at the source of occurrence, the impeller of the pump. It is likely, that an increase in the amplitude of the head pulsations at the outlet from the volute of the pump is due to the appearance of resonant phenomena [5].

The nature of the change in the amplitude of the head pulsations at the outlet from the impeller and the volute, depending on the pump supply, is shown in Figure 3.

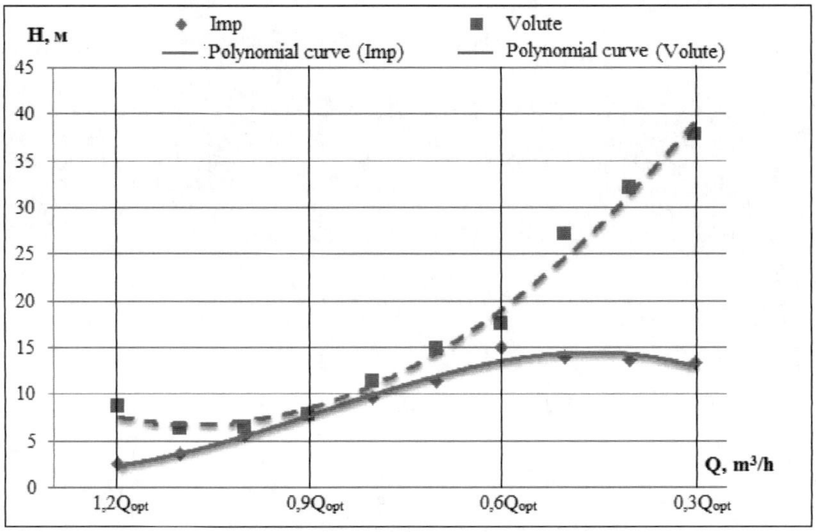

Figure 3. Change in the amplitude of head pulsations at the outlet from the impeller and the volute, depending on pump supply.

Despite the high accuracy and informativeness of the received results of the mathematical modeling of the working processes of a centrifugal pump, it has some disadvantages. They consist in the complexity of creating a computational model, in the use of many computing resources and a considerable time for obtaining results. That is why, for carrying out express researchers, we recommend the use of electrical simulation of the operating modes of the pump unit. The creation of an electric model is much simpler, and the results of modeling require significantly less time.

4. Electrical modeling

The chart of an electric model of the electromechanical system, that includes the volute of the pump and the impeller with six blades, shown in Figure 4.

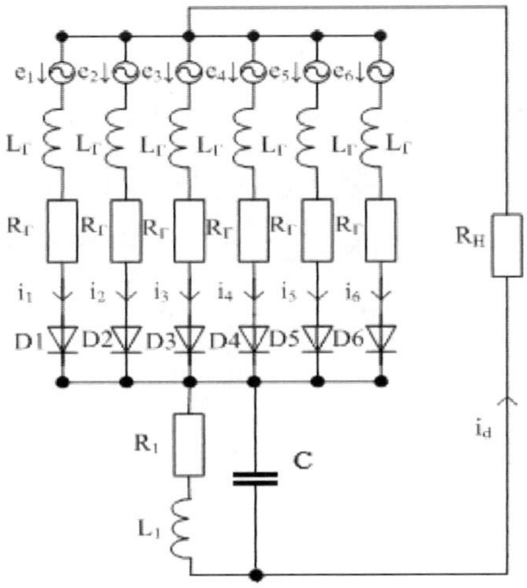

Figure 4. The scheme of the electric model of the electromechanical system with pump unit, that has six blades AD 2000-100-2 (m = 6).

Consider the nature of the application of electrical methods of modeling operate modes in electromechanical systems of water supply systems by the example of the electric circuit in Figure 4 [6, 7]. We will carry out the researches the electromagnetic processes in the electric model EMS in the medium load mode, which is most common for this model, given the range of optimal flow of pump unit AD2000-100-2.

The electromotive forces $e_1 \dots e_6$ model the physical process of energy transfer from the drive motor of the pump into the electromechanical system with simultaneous energy conversion by form and frequency. This EMF is sinusoidal, reflecting a rotating fluid flow in the centrifugal pump. The frequency of this EMF voltage as well as the speed of rotation of the pump impeller is typically different from the frequency of the power supply network. It is $\omega = \omega_M/p$. So instantaneous system sinusoidal EMF sources separate circuit branches EMS electric model (Figure 4), are as follows:

$$\left.\begin{aligned} e_1 &= E_m \sin(\vartheta + \pi/3); \\ e_2 &= E_m \sin\vartheta; \\ e_3 &= E_m \sin(\vartheta - \pi/3); \\ e_4 &= E_m \sin(\vartheta - 2\pi/3); \\ e_5 &= E_m \sin(\vartheta - \pi); \\ e_6 &= E_m \sin(\vartheta - 4\pi/3); \end{aligned}\right\} \quad (2)$$

where $\vartheta = \omega t$ - variable, in which had performed all calculations.

The active hydraulic resistance $R_Г$, which is based on viscous forces between the layers of fluid and between the fluid and the walls of the channel, reflects the fact of dissipation energy as heat. The inductive hydraulic resistance of $X_Г$, which is created by forces of inertia that counteract the change of supply in a centrifugal pump in the between the area of blades channels (in the diagram Figure 4) is fed as hydraulic inductance $L_Г$.

The analysis of electromagnetic processes had conducted with using the conventional approach for devices of power electronics: the classic method of preparation and solution the system of differential equations with calculating the constants of integration. In the result had obtained:

a) system of differential equations for the circuit of the first stage of commutation (Figure 5(a)) taking into account the presence of active and reactive resistance:

$$e_1 - e_6 = x_\gamma \, di_1 / d\vartheta + R_\Gamma i_1 + u_{D1} - u_{D6} - R_\Gamma i_6 - x_\gamma \, di_6 / d\vartheta , \quad (3)$$

$$e_1 = x_\gamma \, di_{1(2)} / d\vartheta + R_\Gamma i_{1(2)} + u_{D1} + i_{1(2)} R_H + i_{1(2)} R_1 , \quad (4)$$

where $x_\gamma = \omega L_\Gamma$ - the inductive resistance of branches,

and the equation for the contour of the second stage of commutation (Figure 5 (b))

$$e_1 = x_\gamma \, di_{1(2)} / d\vartheta + R_\Gamma i_{1(2)} + u_{D1} + i_{1(2)} R_H + i_{1(2)} R_1 . \quad (5)$$

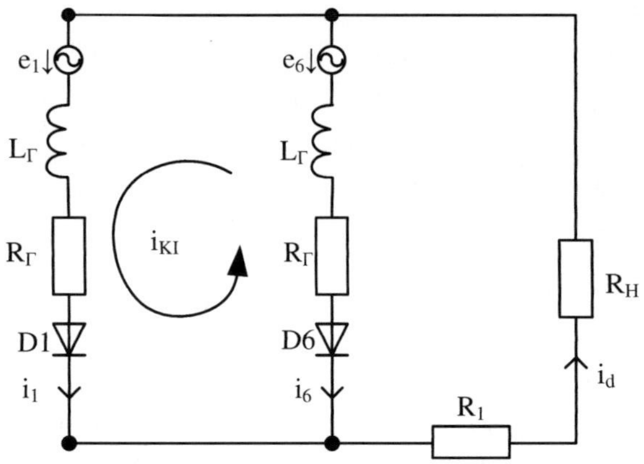

Figure 5 (a). The circuit of the first stage of commutation.

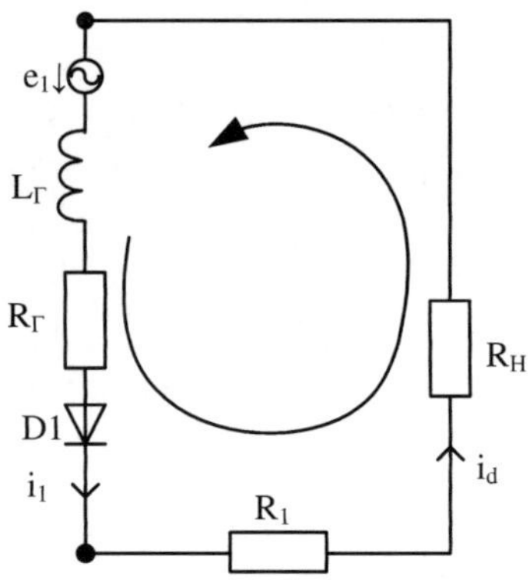

Figure 5 (b). The circuit of the second stage of commutation.

b) regularities of change of currents of diodes and rectified current

$$i_{1(1)} = \frac{E_m}{2x_\gamma}(1 - \cos\vartheta) \quad (6)$$

$$i_{1(2)} = i_{d(2)} = \frac{E_m}{Z}\sin(\vartheta + \frac{\pi}{3} - \varphi) + Ae^{-\frac{R_e}{x_\gamma}(\vartheta - \gamma)} ; \quad (7)$$

$$i_{d(1)} = \frac{\sqrt{3}E_m}{Z'}\sin(\vartheta + \frac{\pi}{2} - \varphi') + Be^{-\frac{(2R_н + 2R_1 + R_\Gamma)}{x_\gamma}\vartheta} , \quad (8)$$

$$\text{де } A = \frac{E_m}{2x_\gamma}(1 - \cos\gamma) - \frac{E_m}{Z}\sin(\gamma + \frac{\pi}{3} - \phi) , \quad (9)$$

$$B = \frac{E_m}{Z}\sin(\frac{2\pi}{3} - \varphi) - \frac{\sqrt{3}E_m}{Z'}\sin(\frac{\pi}{2} - \varphi') + \left[\frac{E_m}{2x_\gamma}(1 - \cos\gamma) - \frac{E_m}{Z}\sin(\gamma + \frac{\pi}{3} - \varphi)\right] e^{-\frac{R_e}{x_\gamma}(\frac{\pi}{3} - \gamma)} \quad (10)$$

$R_e = R_\Gamma + R_1 + R_H$, $Z = \sqrt{R_e^2 + x_\gamma^2}$ - full resistance the circuit of commutation;

$\phi = arctg \dfrac{x_\gamma}{R_e}$ - the angle phase shift on the first stage of commutation;

$Z' = \sqrt{(2R_H + 2R_1 + R_\Gamma)^2 + x_\gamma^2}$, $\varphi' = arctg \dfrac{x_\gamma}{2R_H + 2R_1 + R_\Gamma}$.

c) the equations to calculate the angle of commutation γ:

IOP Conf. Series: Materials Science and Engineering **233** (2017) 012058 doi:10.1088/1757-899X/233/1/012058

$$\cos\gamma + \frac{2(\sin(\gamma+\pi/3-\varphi_k)-\sin\gamma)}{\lambda} + \frac{4\cos(\gamma+\pi/3-\varphi_k)}{\lambda^2} = 1, \tag{11}$$

where $\lambda = R_e / x_\gamma$; $\varphi_k = arctg(1/\lambda)/$

d) the expression of the instantaneous value of rectified voltage [8], which has six identical pulsations during the period, each of which is divided into two intervals with different regularities of changes:

$$u_{d(1)} = \sqrt{3}E_m \sin(\vartheta+\pi/2)/2 - R_\Gamma E_m(1-\cos\vartheta)/2x_\gamma; \tag{12}$$

$$u_{d(2)} = E_m \sin(\vartheta+\pi/3) - x_\gamma E_m \cos(\vartheta+\pi/3-\varphi)/Z + R_н A e^{-\frac{R_e}{x_\gamma}(\vartheta-\gamma)} - (R_\Gamma + R_1)E_m \sin(\vartheta+\pi/3-\varphi)/Z. \tag{13}$$

The limited scope of the article does not allow to give the detailed description of the features theoretical researchers that made in accordance with EMS, which contains the pump unit AD 2000-100-2 and the impeller of the pump, which has six blades.

Was conducted the simulation with using the electric model EMS of pump unit AD2000-100. Below (Figure 6) are shown the curves of the electrical model, where:

1 - the curve of rectified voltage of the electric model, which simulates the head of the impeller;

2 - the curve of rectified voltage at the output from the electric model, which simulates the head of the pump.

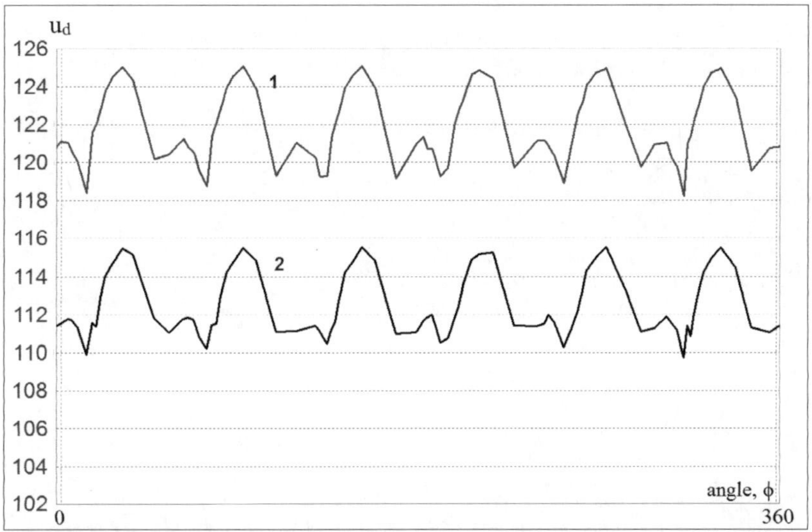

Figure 6. The rectified voltage of electric model of pump unit AD2000-100, operation mode 1,0 Qopt.

5. Conclusions

1. For pump units with single spiral volute, the frequency of head pulsations at the outlet of the impeller and the pump coincide with pulsations on the blade passing frequency.

2. The depth of the head pulsations at the outlet of the impeller is defined by losses on hydraulic friction and on the formation of a vortex, which depends on the supply of pump.

3. The depth of the head pulsation of pump unit at its output depends on the losses of hydraulic friction, on the formation of a vortex and the supply of pump unit. In the range of small feeds (to $0{,}7Q_{opt}$) is observed the greater amplitude of head pulsations than at the outlet of the impeller.

4. According to the results of the mathematical modeling workflow of the pump unit, the electrical modeling method is acceptable for engineering calculations. For curves that show the head at the outlet of the impeller, the difference does not exceed 2%.

6. References

[1] Timushev S F and Klimenko D V 2015 Computation of pressure pulsations in the outlet devise of screw-centrifugal pump with acoustic-vortex method *Vestnik SibGAU* (Moscow, Moscow Aviation Institute) **16** (4) pp 907

[2] Bobkov A V 2012 Damping pressure output centrifugal pump *Modern problems of science and education* **5** pp 81-85

[3] Lugovaya S O and Moskalenko V V 2016 Influence of the blade output section geometry on the non-stationary interaction of the rotating impeller and the spiral volute of the centrifugal pump *Pump and Equipment* **4** pp 64-67

[4] Varaskin A 2003 Laser Doppler anemometry flow "gases-solid particles", problems of achievement, prospects *Thermal physics of high temperatures* **41** pp 746-754

[5] Chen Y N 1961 Pressure Oscillations in the Volute Casing of Storage Pumps *Sulzer Technical Review* pp 21-34

[6] Boiko V S, Boiko V V and Sotnyk M I 2011 Creating method of an electric model of a centrifugal pump: Patent UA № 67781, MPK G06G 7/00. Decl. 01.07.2011. Publ. 12.03.2012, Biul. No. 5

[7] Boiko V S, Sotnyk M I and Khovanskyy S A 2015 Electrical modeling of work processes in electrical systems of water supply networks *Journal of Engineering Sciences* **2** (2) pp 1-12

[8] Boiko V S and Sotnyk M I 2013 Adequacy of electrical modeling of work processes in a centrifugal pump *Technical electrodynamics* **5** pp 90-96